2024 IEEE 7th International Conference on Knowledge Innovation and Invention

2024 IEEE 7th International Conference on Knowledge Innovation and Invention

Volume Editors

Teen-Hang Meen
Chun-Yen Chang
Cheng-Fu Yang

MDPI

Basel • Beijing • Wuhan • Barcelona • Belgrade • Novi Sad • Cluj • Manchester

Volume Editors
Teen-Hang Meen
National Formosa University
Yunlin
Taiwan

Chun-Yen Chang
National Taiwan Normal University
Taipei
Taiwan

Cheng-Fu Yang
National University of Kaohsiung
Kaohsiung
Taiwan

Editorial Office
MDPI AG
Grosspeteranlage 5
4052 Basel, Switzerland

This is a reprint of the Proceedings, published open access by the journal *Engineering Proceedings* (ISSN 2673-4591), freely accessible at: https://www.mdpi.com/2673-4591/89/1.

For citation purposes, cite each article independently as indicated on the article page online and as indicated below:

Lastname, A.A.; Lastname, B.B. Article Title. *Journal Name* **Year**, *Volume Number*, Page Range.

ISBN 978-3-7258-5773-9 (Hbk)
ISBN 978-3-7258-5774-6 (PDF)
https://doi.org/10.3390/books978-3-7258-5774-6

Contents

About the Editors

Teen-Hang Meen

Teen-Hang Meen was born in Tainan, Taiwan on Aug. 1, 1967. He received his BS degree from the Department of Electrical Engineering, National Cheng Kung University (NCKU), Tainan, Taiwan, in 1989, and his MS and PhD degrees from the Institute of Electrical Engineering, National Sun Yat-Sen University (NSYSU), Kaohsiung, Taiwan, in 1991 and 1994, respectively. He was the chairman of the Department of Electronic Engineering from 2005 to 2011 at National Formosa University, Yunlin, Taiwan. He received the excellent research award of National Formosa University in 2008 and 2014. Currently, he is a lifetime distinguished professor with the Department of Electronic Engineering, National Formosa University, Yunlin, Taiwan. In addition, he is the president of the International Institute of Knowledge Innovation and Invention (IIKII), and the chair of the IEEE Tainan Section Sensors Council. He has published more than 200 SCI and SSCI papers in recent years.

Chun-Yen Chang

Chun-Yen Chang is a science education scholar in Taiwan. Currently, he serves as the National Taiwan Normal University (NTNU) Chair Professor, Director of the Science Education Center (NTNU), and Professor of the Graduate Institute of Science Education and the Department of Earth Sciences (NTNU). Since July 2022, he has also served as the Adjunct Professor of the Department of Biology, State University of Malang, Indonesia. Over the past few years, he has been a Visiting Professor at the Taipei Medical University, The Education University of Hong Kong, and the Paris 8 University. Dr. Chang's major research interests include science education, e-Learning, interdisciplinary science learning, and science communication. Dr. Chang has authored and co-authored more than 150 articles, from which more than 130 are indexed in the Science/Social Science Citation Index (SCI/SSCI) database. His series of research on science learning environments, published in *Science Education* in 2005 and 2006, was also reported in the "Taiwanese New Discovery" section of the *Science Development Journal* in 2007. In NSTC's 50th anniversary celebrations, Dr. Chang's research work on problem-solving ability and automatic scoring of science open-ended questions was honored as one of the NSTC's 50 Great Achievements in 2008. Currently, His h-index is 34 in the Scopus database.

Cheng-Fu Yang

Cheng-Fu Yang was born in Taiwan on July 12, 1964. Yang earned his BS, master's degree, and PhD in 1986, 1988, and 1993 from the "Department of Electrical Engineering" of Cheng Kung University. From February 2004 to now he has been a Professor of Chemical and Materials Engineering at the National University of Kaohsiung. He obtained "The Outstanding Contribution Awards of the Chinese Ceramic Society" in 2009. In January, 2010, he was the first one (and the only one) to obtain the distinguished professor of NUK award. From August, 2014, to May, 2015, he was the fellow of the "The Institution of Engineering and Technology" (IET). He was the "The Mingjiang Scholar and Chair Inviting Professor" of Jimei University, Xiamen, Fujian, China, and he has been the "Honorary Chair Professor" of Chaoyang University of Technology since August, 2020. He has been an Editorial Board Member or Guest Editor of more than 15 international journals, for example *Coatings, Materials, Applied Sciences*, etc.; and he is the Chief Editor of the journal "*Applied Functional Materials*". So far, he has published more than 390 SCI and SSCI journal papers and more than 140 EI journal papers and also has published more than 300 domestic and international conference papers.

engineering proceedings

MDPI

Editorial

Preface: 2024 IEEE 7th International Conference on Knowledge Innovation and Invention †

Teen-Hang Meen [1,*], Chun-Yen Chang [2] and Cheng-Fu Yang [3,4,*]

1 Department of Electronic Engineering, National Formosa University, Yunlin 632, Taiwan
2 Graduate Institute of Science Education, National Taiwan Normal University, Taipei 116, Taiwan; changcy@ntnu.edu.tw
3 Department of Chemical and Materials Engineering, National University of Kaohsiung, Kaohsiung 811, Taiwan
4 Department of Aeronautical Engineering, Chaoyang University of Technology, Taichung 413, Taiwan
* Correspondence: thmeen@gs.nfu.edu.tw (T.-H.M.); cfyang@nuk.edu.tw (C.-F.Y.)
† All proceeding papers published in this volume were presented at the 2024 IEEE 7th International Conference on Knowledge Innovation and Invention, Nagoya, Japan, 16–18 August 2024.

This volume presents the proceedings of the 2024 IEEE 7th International Conference on Knowledge Innovation and Invention (IEEE ICKII 2024). This conference was jointly organized by the Science Education Center of National Taiwan Normal University; the Photovoltaic Research and Education Center for Smart Green Energy at National Formosa University; the International Institute of Knowledge Innovation and Invention (IIKII); and the IEEE Tainan Section Sensors Council (IEEE TSSC), in the field of sciences and engineering technologies. It aimed to provide a unified communication platform for researchers across a broad range of topics, including Information Technology, Innovation Design, Bioengineering, Communication Science and Engineering, Industrial Design, Creative Design, Applied Mathematics, Computer Science, Electrical and Electronic Engineering, Mechanical and Automation Engineering, Green Technology and Architecture Engineering, Materials Science, and other related fields. This conference facilitated interdisciplinary and international collaboration among leading technologists in both academia and industry. Ample time was provided for presentations and discussions, and attendants participated in various activities, bringing together a diverse group of engineers and technologists from multiple disciplines to foster new ideas, collaborations, and business opportunities.

ICKII 2024 was held in a hybrid format, offering both on-site and online presentations. Figure 1 shows a group photo taken during the opening ceremony of the conference. Figure 2 captures the first keynote speech, entitled "The Future of Scientific Inquiry: Cutting-Edge Assessments Merging AI and Genomics", delivered by the Chair, Professor Chun-Yen Chang from National Taiwan Normal University, Taiwan. This talk introduced the groundbreaking approaches to transforming scientific inquiry assessments. Professor Chang and his research team have embarked on an ambitious interdisciplinary journey, blending science education, genomics, and neuroscience to create a new generation of AI-assisted online scientific inquiry tests. Their goal is to link assessment results with genomic and neurophysiological characteristics, paving the way for truly integrative research. Some of their most notable achievements over the past two years are as follows: 1. Development of Interactive Test Questions; 2. Innovative Assessment Methods; 3. Latent Profile Analysis; 4. Biological Information Analysis; 5. Learning Environment Planning; 6. Epigenetic Memory Model. This research has led to the creation of innovative assessment tools and has enabled deeper exploration of the biological factors influencing learning and memory. Ultimately, it contributes to the development of more effective educational strategies by leveraging tools such as generative AI to shape the future of scientific inquiry and bridge the gap between science education, genomics, and neuroscience.

Figure 1. Group photo at opening ceremony of ICKII 2024.

Figure 2. Photo of first keynote speech at ICKII 2024.

Figure 3 shows the second keynote speech, entitled "AI-Enhanced Wearable Assistant Device for Biomedical Application", presented by Professor Yi-Chun Du from the Department of Biomedical Engineering, National Cheng Kung University, Taiwan. He discussed four projects on AI-enhanced wearable assistant devices in the biomedical field, specifically designed for hemodialysis patients, pregnant women, individuals with dysarthria, and those requiring hearing loss care. Finally, a project involving integrated mixed reality (MR) wearable technology for surgical assistance was explored. Looking ahead, there is strong anticipation for the continued development of AI-enhanced wearable devices. The commercialization of these technologies has the potential to reduce clinical workloads and improve the quality of care in both healthcare and caregiving industries.

Figure 3. Photo of second keynote speech at ICKII 2024.

Figure 4 displays the third keynote speech, entitled "Synthesis and Development of Bionic Ceramic Scaffolds by 3D Printing Technology", delivered by Associate Professor Wen-Fan Chen from the Institute of Medical Science and Technology, National Sun Yat-Sen University, Taiwan. Each year, millions of patients worldwide undergo bone grafting procedures to treat bone defects. While fractures in young people are often caused by excessive exercise or car accidents, osteoporosis is more common among the elderly. As a result, orthopedic-related products continue to account for a high proportion of the medical device market. Although autografts are currently considered to be the gold standard for grafting, its application is limited by the finite amount of donor bone available and the need for multiple surgeries. Allografts, on the other hand, require various considerations such as the risk of disease transmission and reduced biological activity following sterilization. Therefore, the development of synthetic bone materials has become a promising area of research that requires urgent innovation. In line with this need, Professor Chen shared the main research directions of her team, which included the following: (1) Synthesis of novel ceramic materials with good compatibility with natural bone; (2) integration of 3D printing technology developed by her team to produce bionic scaffolds; and (3) development of magnetic nanoparticles capable of using an inductive heating effect to eliminate diseased cells near the implanted bone area. By combining these innovations, the team aims to create a customizable, degradable bionic scaffold with high strength, high bioactivity, and magneto-thermal properties.

ICKII 2024 provided 15 Regular Sessions and 4 Invited Sessions, showcasing 223 papers. These covered topics such as Information Technology, Innovation Design, Bioengineering, Communication Science and Engineering, Industrial Design, Creative Design, Applied Mathematics, Computer Science, Electrical and Electronic Engineering, Mechanical and Automation Engineering, Green Technology and Architecture Engineering, Materials Science, and other related fields. Figures 5 and 6 show the on-site and online oral presentation sessions.

Figure 4. Photo of third keynote speech at ICKII 2024.

Figure 5. Presentation at an on-site session of ICKII 2024.

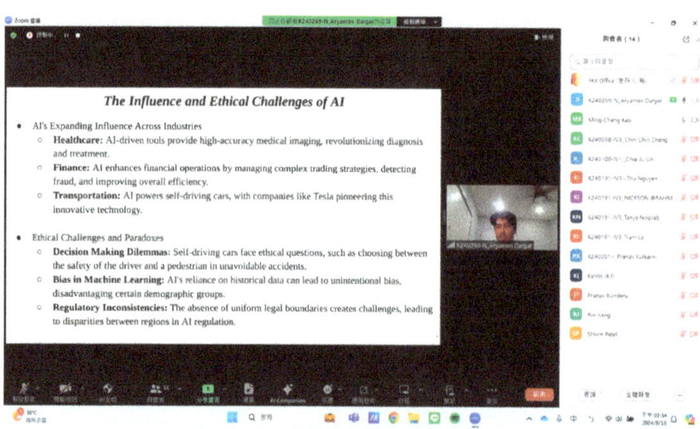

Figure 6. Presentation at an online session of ICKII 2024.

Many significant results were shared at ICKII 2024 by enthusiastic participants. Through a rigorous peer-review process, 55 excellent papers related to engineering fields relevant to the conference were selected for publication in *Engineering Proceedings* (ISSN: 2673-4591, indexed by Scopus). The proceedings of ICKII 2024 are expected to foster interdisciplinary collaboration among science and engineering technologists across academia and industry, while also promoting international networking.

Conflicts of Interest: The authors declare no conflicts of interest.

engineering proceedings

MDPI

Editorial

Statement of Peer Review †

Teen-Hang Meen [1,*], Chun-Yen Chang [2] and Cheng-Fu Yang [3,4,*]

1 Department of Electronic Engineering, National Formosa University, Yunlin 632, Taiwan
2 Graduate Institute of Science Education, National Taiwan Normal University, Taipei 116, Taiwan; changcy@ntnu.edu.tw
3 Department of Chemical and Materials Engineering, National University of Kaohsiung, Kaohsiung 811, Taiwan
4 Department of Aeronautical Engineering, Chaoyang University of Technology, Taichung 413, Taiwan
* Correspondence: thmeen@gs.nfu.edu.tw (T.-H.M.); cfyang@nuk.edu.tw (C.-F.Y.)

† All proceeding papers published in this volume were presented at the 2024 IEEE 7th International Conference on Knowledge Innovation and Invention, Nagoya, Japan, 16–18 August 2024.

In submitting conference proceedings of the 2024 IEEE 7th International Conference on Knowledge Innovation and Invention (IEEE ICKII 2024) to *Engineering Proceedings*, the volume editors of the proceedings certify to the publisher that all papers published in this volume have been subjected to peer review administered by the volume editors. Reviews were conducted by expert referees to the professional and scientific standards expected of a proceedings journal.

- Type of peer review: 2–3 reviewers by double-blind;
- Conference submission management system: online submission system on conference homepage;
- Number of submissions sent for review: 223;
- Number of submissions accepted: 55;
- Number of submissions published: 46;
- Acceptance rate: 0.247;
- Total number of reviewers involved: 50;
- Any additional information on the review process: Please refer to Figure 1.

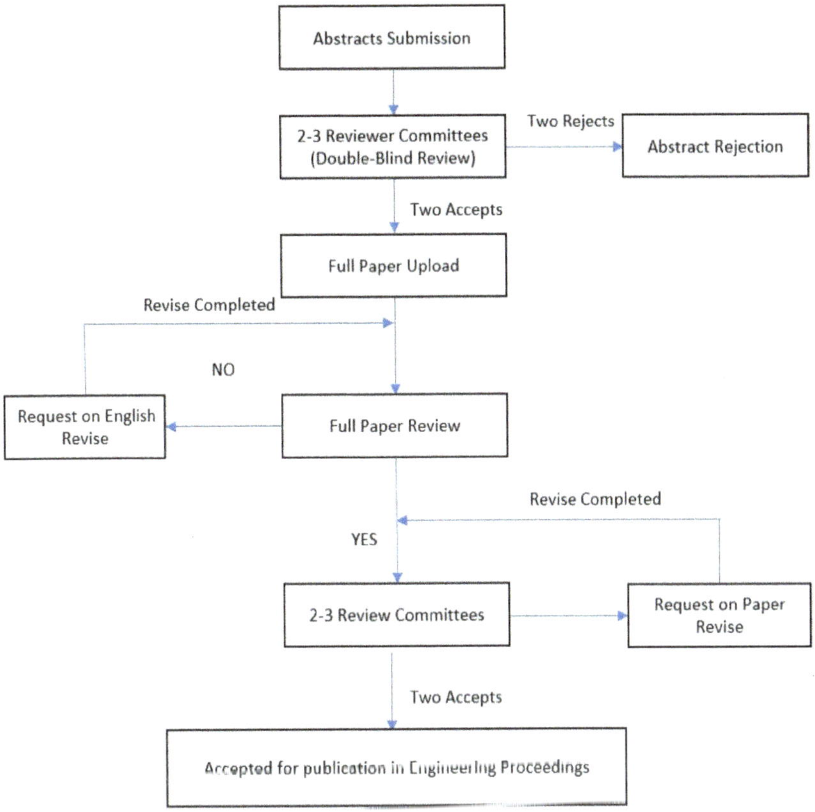

Figure 1. Review process for conference proceedings of 2024 IEEE 7th International Conference on Knowledge Innovation and Invention (IEEE ICKII 2024).

Conflicts of Interest: The authors declare no conflict of interest.

engineering proceedings

Proceeding Paper

A Quantitative Model of Supply Chain Disruption Propagation Dynamics †

Shudong Liu *, Shili Xiang and Lu Wang

Institute for Infocomm Research (I2R), A*STAR, 1 Fusionopolis Way, Singapore 138632, Singapore; sxiang@i2r.a-star.edu.sg (S.X.); wang_lu@i2r.a-star.edu.sg (L.W.)

* Correspondence: liush@i2r.a-star.edu.sg

† Presented at the 2024 IEEE 7th International Conference on Knowledge Innovation and Invention, Nagoya, Japan, 16–18 August 2024.

Abstract: Supply chain disruptions caused by natural disasters and human-made incidents have inflicted substantial losses on numerous companies. The management of supply chain risks, including disruption risk, has garnered significant attention from both academia and industries over the past few decades. Companies must develop effective solutions for disruption risk management. To seek an effective solution with disruption event monitoring and mitigation plans, we investigated the mechanisms and dynamics of disruption propagation along the supply chain. Specifically, we developed a quantitative mathematical model for the entire supply chain using system dynamics to understand the characteristics of disruption propagation, such as time spent for the disruption from the event location to a focal company and its impact. The model provides important management information on the development of a disruption monitoring system to enhance the resilience and robustness of supply chains, develop mitigation strategies, and minimize the adverse effects of disruptions.

Keywords: disruption propagation; supply chain risk management; system dynamics; supply chain resilience; disruption event

1. Introduction

Supply chains are extraordinarily complex, with various companies collaborating closely to improve efficiency and reduce costs. While the close relationship among companies brings significant benefits in developing innovative products and cost reduction, the interdependence among companies increases the potential risk for supply chain disruption. Supply chain disruptions stemming from natural disasters and human-made incidents cause substantial losses. For instance, the devastation caused by Hurricane Harvey in 2017 resulted in an estimated loss exceeding USD 125 billion [1]. Other examples include a fire at the Philips microchip plant in Albuquerque, New Mexico, in 2000; Hurricane Katrina in 2006; the tsunami in Japan and the flood in Thailand in 2011; a fire at the Meridian Magnesium Products of America factory in Eaton Rapids, Michigan, in 2018; and Hurricane Ian in USA in 2022.

Over the past two decades, the management of supply chain risks, including disruption risk, has garnered significant attention. Researchers have explored risk classification and analysis, disruption propagation, disruption recovery, supply chain resilience, supply chain design, and so on. Despite the availability of numerous models and frameworks for supply chain risk management [2–4], the disruption propagation process remains a challenge, needing more quantitative models [5] that provide a deeper understanding of its dynamics.

In this study, we developed a model to capture the essential dynamics of disruption propagation and understand important factors and their interrelationships. It is a core part of developing an effective solution for disruption monitoring, analysis and assessment, and early warning and mitigation plans. To establish the model, we delved into the mechanism and dynamics of disruption propagation. The developed mathematical model was used to analyze the time to propagate disruption from the event location to the focal company and assess the severity of its impact. The model provides important information to develop disruption monitoring systems and enhance the resilience and robustness of supply chains. It also helps to develop mitigation strategies to minimize the adverse effects of disruptions.

The rest of this paper is organized as follows. Section 2 presents the literature review. In Section 3, we describe the problem formulation. Section 4 models the supply chain as a continuous flow system using system dynamics principles. Section 5 describes the characteristics of disruption propagation. A numerical example for calculating the disruption propagation is shown in Section 6. Finally, conclusions and discussions are given in Section 7.

2. Literature Review

Supply chain risk management has attracted intense interest from academics and industries. Various aspects have been researched, including disruption types, disruption propagation and ripple effects, disruption recovery, supply chain resilience, and supply chain design. In this study, we specifically focus on disruption propagation [2,4–8].

Different methods have been used to analyze disruption propagation. Bayesian networks are used to obtain the risks/probabilities of some disruptions [3,4,9]. However, these models often model the disruptions as static events while neglecting the disruption duration and not explicitly considering the inventory levels of companies. Wilson [10] used system dynamics simulation to investigate the effect of a transportation disruption on supply chain performance. Schmitt and Singh [11] developed a discrete-event simulation model for a supply chain to analyze the inventory placement and backup methodologies and assess their impact on reducing supply chain risk. Tan et al. [12] used discrete-event simulation to study the effect of disruption on downstream companies' performance. In these simulations, the disruption duration and inventory levels are incorporated. While simulation results provide valuable information, the results are dependent on the simulation scenarios, parameter settings, and the simulation model details. Therefore, these findings cannot be applied to other situations.

Researchers have also used Petri Nets to model and analyze supply chain disruptions [13–15]. Petri Nets model concurrent events in systems, and they are often combined with discrete-event simulation to analyze disruption. These analyses share similar characteristics with discrete event simulation.

Disruption duration and inventory levels are overlooked in models or considered only in specific scenarios. In this study, we developed a model for disruption propagation using system dynamics principles and considering details in production, inventory, and transportation.

3. Problem Formulation

Company A obtains raw materials/components from its tier 1 supplier. The supplier has its own suppliers (tier 2 supplier of Company A). In this way, a complex, multi-tiered supply chain is formed (Figure 1). Due to the process characteristics of Company A, the limited shelf life of raw materials/components, and the competitive pressure, the reliable and on-time supply of raw materials and components is critical to the company. The company wants to have a good understanding about how the disruption events in the supply chain affect its material supply, more specifically, if a disruption event occurs in an upstream supplier, when will the disruption propagate to the focal company and what is the severity?

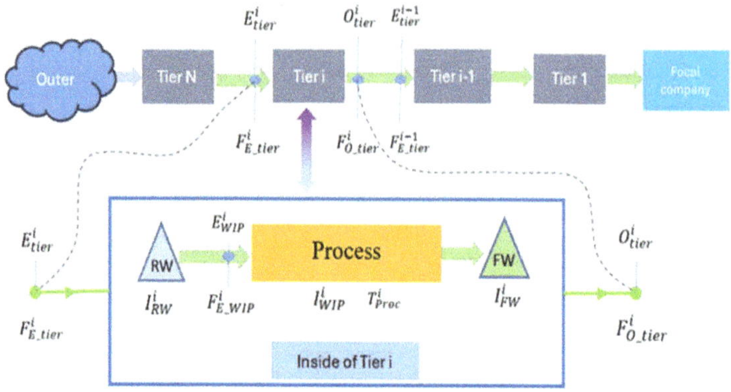

Figure 1. Material flow for one type of raw material of the focal company.

In this analysis, we assume a single disruption event at a specific company for simplicity. This assumption aligns with many cases in the literature. Another assumption is that when a disruption event happens, all companies in the supply chain still continue their normal operational practices without adopting measures to obtain competitive advantages during the supply shortage. For example, if a company offers higher prices to secure more material supply from its supplier, other customers of the same supplier may face more serious shortages. We do not consider these scenarios. In other words, when the disruption event happens, we assume the material flow (among different companies) in the supply chain is still like the normal case until the materials are exhausted. This assumption reflects the current practices, especially in the absence of a disruption monitoring system and active response strategies. It also allows for the fundamental characteristics of disruption propagation.

4. Modeling Process and Material Flow

In supply chains, the purchase orders and material flow are, in general, discrete and intermittent, making many problems difficult to analyze, if not intractable. Continuous models for supply chain management are an important approach to address these challenging problems. Notable examples exist, such as the famous EOQ model and other models in related areas [2,16,17]. These continuous models grasp the essential characteristics of the processes, and they can provide good approximations to discrete cases or serve as a base for developing more complicated or accurate models. Here, we develop a continuous model for the material flow to grasp important factors and obtain essential characteristics of disruption propagation in the supply chain by using system dynamics principles and analogizing material flow to fluid flow in a pipe system.

Unlike flow in a pipe system, the materials in the supply chain are processed and transformed at each stage. In each factory, the raw material/component is processed, converted into a new final product, and then sent to its customers. Different materials have different measurement units, and factories have different consumption/conversion rates. For example, several raw material/component units are needed to produce one unit of the final product [12]. This makes the model more complex.

A model of continuous material flow is shown in Figure 1. The meanings of the symbols are listed in Table 1. The model consists of tiers (factories) and links between tiers. In each factory, the raw materials/components are converted into final products. The final product of tier i serves as the raw material/component for tier $i-1$. Each tier has three major parts: raw material warehouse (RW), process (machines producing final product), and final product warehouse (FW). When the production is in the Make-To-Order (MTO) mode, there is no inventory of the final product. In order to have a unified model for both the MTO and MTS (Make-To-Stock) mode, we assume there is a final product warehouse (FW) for MTO, but its inventory level is always 0. In the process (production line), there is a Work-In-Progress (WIP) inventory.

Table 1. Symbols in the model.

Symbol	Meaning
E^i_{tier}, O^i_{tier}	material flow entry point (location) and output point (location) of tier i, respectively
$F^i_{E\_tier}$, $F^i_{O\_tier}$	material flow rate entering and leaving tier i, respectively
I^i_{RW}, I^i_{FW}, I^i_{WIP}	inventory level for raw material, final product, and work in progress, respectively
T^i_{Proc}	process time converting raw material into products
I^i_{RF}	total inventory of raw material and final product at tier i
Tr_i	transportation time between tiers i and $i-1$
T_{disrp}	duration of disruption

Consider tier i. The raw material and final product warehouse are modeled as two stocks in the system dynamics with unlimited storage capacity. The "process" (converting raw material into the final product) is modeled as a long pipe. The transportation of materials between warehouses and processes are modeled as short pipes, whose length is ignored for simplicity. One critical issue in the material flow is that materials before and after the process are different, and they have different measurement units. To address this, we use the following method to measure the amount of different materials.

We use the normal material flow rate for each material as the benchmark and measurement unit. For example, the final product inventory (in any unit, e.g., kg) is divided by the normal demand per day (e.g., kg/day) to obtain the inventory level in days. Similarly, for both raw material inventory and WIP inventory, we convert the inventory in physical units into inventory in days. In this way, we standardize the unit of measurement across different materials into the same time unit (e.g., days). Moreover, one day's worth of raw material is converted into one day's worth of final product. This simplifies the analysis of disruption propagation to treat the material flow in the supply chain as the flow of a single type of material (e.g., water). In normal situations, the material flow rate is 1 as the inventory is converted into a time unit (e.g., days), and material flow is normalized by the normal situations.

5. Characteristics of Disruption Propagation

5.1. Disruption Signal Generation and Profile

Disruption events (such as earthquakes or floods) cause material flow disruptions, which are classified into production disruption, supply disruption, and transportation disruption. Figure 2 shows an example of the disruption signal for production disruption. In the figure, the material flow disruption start time t_{ds} and end time t_{de} are shown. After production is recovered, there is a period in which the production level is higher than the normal level to refill inventory or satisfy backorders. This extra production capacity is obtained through overwork or other approaches (e.g., purchasing more/new machines or previous surplus capacity). Following this recovery period, production drops back to a normal level from t_{rf}. The above disruption signal represents many typical cases where most companies do not operate at full capacity for maintenance or demand fluctuations (or demand is less than their existing full production capacity).

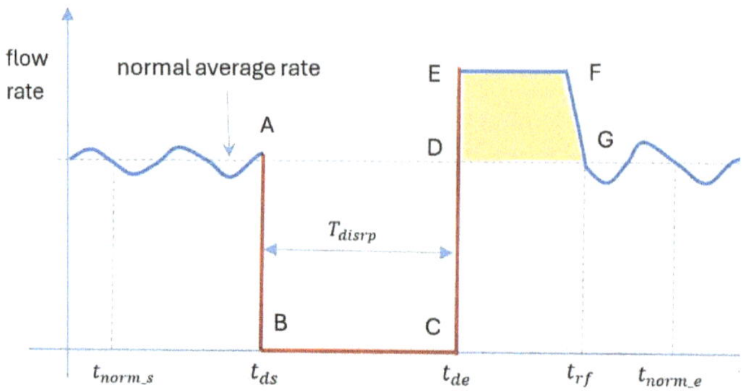

Figure 2. An example of typical disruption signals.

The signals for supply disruption and transportation disruption are similar to the one depicted in Figure 2. For brevity, these signals are not plotted.

5.2. Disruption Propagation Across One Tier

Once we know how disruption propagates across one tier, it is not difficult to extend it to multiple tiers. Now consider tier *i*, which has a supply disruption (shown in Figure 2); we analyze its response signal, i.e., the output material flow. Figure 3 shows an example of internal inventory variation and how the output material flow signal of the tier is generated. The figure illustrates a case where the supply disruption is long enough to cause an output flow disruption. Figure 3a shows the input material flow at the entry point of the tier, with the duration of disruption denoted as T_{disrp}. Since our focus is on the disruption propagation rather than the disruption recovery period, in part (a), we do not show the time (after t_{de}) when the input material flow is dropped back to the normal flow rate. Figure 3b illustrates the inventory variation (a sum of raw material RW and final product FW) with time. Figure 3c depicts the WIP inventory on a production line variation. Finally, Figure 3d shows the output material flow at the tier's exit point.

Before the supply disruption at time t_{ds}, the input material flow into the tier is at the normal level, i.e., the flow rate is 1 by normalization. The inventory levels of the raw material and final product are planned by the inventory management, set as 10 in the example. The WIP inventory is at a normal level, depending on the production process, set as 2 in the example. The output flow from the tier is the same as the input flow, also at the normal level (flow rate 1). Since the disruption occurs at time t_{ds}, the input flow ceases, but the output flow is still ongoing, so the total inventory of raw material and final production decreases with time. We assume the system prioritizes the existing final product inventory to satisfy the demands downstream rather than using newly produced products converted from WIP. At time t_{RF0}, the RF inventory I_{RF} drops to 0.

After t_{RF0}, the final product inventories and raw material inventories are depleted, and the production line converts the remaining WIP into final products. Hence, the inventory of WIP has decreased over time since t_{RF0}. After the T_{Proc} time unit, all WIP is converted into the final product, and the WIP inventory is reduced to 0. The time is denoted as t_{WIP0}. Therefore, $t_{WIP0} - t_{RF0} = T_{Proc}$. During the period $[t_{RF0}, t_{WIP0}]$, the tier still can provide products to the downstream company, maintaining a normal output flow. Since t_{WIP0}, the production has stopped, and there is no output material flow. Hence, the disruption of output flow happens at the exit point of the tier. This time is equal to t_{WIP0}.

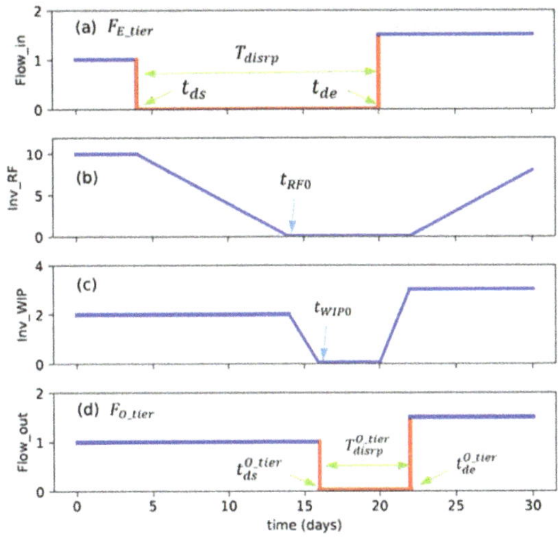

Figure 3. Disruption propagation across one tier.

When the raw material supply is recovered at time t_{de}, the final product flow cannot resume immediately. It takes T_{Proc} to convert raw materials into final products. Hence, the output flow disruption end time is $t_{de} + T_{Proc}$. Since time t_{de}, the raw material supply has been larger than the normal level, and the production level has also become larger than the normal level as a response to the disruption. Therefore, the output flow of the final product has exceeded the normal level since the end time of output flow disruption, as shown in Figure 3d, for the final product flow pattern. The pattern is the same as the raw material flow's pattern in Figure 3a, differing mainly in disruption start time and duration.

The output flow disruption's start time, end time, and duration are as follows:

$$t^{O\_tier}_{ds} = t_{ds} + I_{RF} + T_{Proc}, \tag{1}$$

$$t^{O\_tier}_{de} = t_{de} + T_{Proc}, \tag{2}$$

$$T^{O\_tier}_{disrp} = T_{disrp} - I_{RF}. \tag{3}$$

If the supply disruption duration T_{disrp} is shorter than I_{RF}, there is no output flow disruption. If there is output flow disruption, its duration is the supply disruption duration reduced by I_{RF} (Equation (3)). This highlights the importance of inventory management.

5.3. Disruption Propagation Along Multiple Tiers

The output flow signal of tier i is fed to its next tier as the input, but there is a time delay due to the transportation time Tr_i. As the output disruption signal is similar to the input disruption signal (with differing disruption durations), we repeat the above procedure to calculate its output flow signal until we obtain the signal at the focal company.

When there is a raw material supply disruption at the entry point of tier K, how many tiers the supply disruption can propagate downstream, the time when the disruption arrives at the entry point of each tier, and the severity of disruption (duration) are determined.

Let $I_m = \{K, K-1, \ldots, K-m+1\}$ be the set of indices of tiers starting from tier K. In I_m, there are m tiers, including tier K. Let CI^m_{RF} be the Cumulative Inventory (raw material and final product) for the tiers in I_m. Thus, the number of tiers that the disruption can propagate (including tier K) is

$$N_{prop} = \min\left\{m \middle| CI_{RF}^{m} \geq T_{disrp}^{K}\right\}. \quad (4)$$

For tier $i \in \{K - - - 1, - \ldots, -K - N_{prop} + 1\}$, the input flow disruption start time, end time, and duration are:

$$t_{ds}^{i} = t_{ds}^{i+1} + I_{RF}^{i+1} + T_{Proc}^{i+1} + Tr_{i+1}, \quad (5)$$

$$t_{de}^{i} = t_{de}^{i+1} + T_{Proc}^{i+1} + Tr_{i+1}, \quad (6)$$

$$T_{disrp}^{i} = T_{disrp}^{i+1} - I_{RF}^{i+1}. \quad (7)$$

For tiers after $(K - N_{prop} + 1)$, there is no input/supply disruption.

From the above analysis, we obtain the following insights:

- Inventory is critical in preventing disruption propagation, and it can work as a buffer to reduce the effect of input flow disruption. It is the inventory of raw materials and final products that plays a critical role. When a disruption propagates through a tier, the duration of disruption is reduced by their combined inventory level (in terms of the time unit). The WIP can delay the start time of disruption for the next tiers, but it does not affect the duration of disruption.
- The number of tiers that a material flow disruption propagates depends on the sum of inventories of raw material and final product at different tiers from the source of the disruption. As the disruption propagates, it becomes weaker and eventually ceases at the entry point of a certain tier. Therefore, from a focal company's perspective, it is often unnecessary to look into many tiers upstream.
- For the MTO supplier of a focal company, its disruptions are more likely to propagate to the focal company than those from MTS suppliers. This is because the MTO supplier has no final product inventory and, thus, typically has less total inventory than MTS suppliers. Therefore, when we develop a monitoring system for disruption, if many upstream tiers are of MTO tiers, we may need to monitor more upstream tiers compared to the case in which many upstream tiers are of MTS tiers.
- From the mechanism of disruption propagation, we can estimate the scale of disruption that should be monitored at each tier and establish appropriate thresholds. Based on the formula of propagation, we can derive which scale (threshold) of disruption at each tier will be able to propagate to the focal company. Therefore, if the disruption happening at a given tier is less than its corresponding threshold, the focal company does not need to monitor this disruption. The larger the tier number of suppliers, the larger the threshold. This is useful for developing an effective disruption monitoring system. Nevertheless, in setting such thresholds, other factors, such as the uncertainties, must be considered in upstream parameter estimations and the risk level that the focal company can afford.
- The time is estimated when a disruption at a certain upstream tier is propagated to the focal company. From this time, the focal company knows how much time it has to adopt mitigation plans or implement response measures.

5.4. Stochastic Parameters

Using Equations (1)–(7), we can calculate the time when the initial disruption reaches a focal company and its severity with the values of all parameters at each tier. However, the upper tiers' parameters, such as inventory, are not available to the focal company, and the focal company can only infer the range based on domain knowledge and common practice in the industry. Therefore, the above analysis is applied to the cases where the parameters are modeled as random variables to incorporate the uncertainties.

It is required to estimate the following parameters for each tier: inventory levels for raw material and final product, processing time, and transportation time to the next tier. The transportation time is stable and easier to estimate, and thus, it is set as a constant. The processing time depends on the process characteristics, which may be similar for different companies having the same process. Hence, we may set it as a constant, too.

The inventory level varies significantly. It depends on the inventory management policy and the time when the initial disruption happens. Consider tier i. Let $p^i_{RF}(x)$ be the probability density function of the inventory level (raw material and final product), and the inventory level is in the range [0, $U^i_{I\_RF}$]. We set the lower threshold of the inventory level as 0 to be generic. If the inventory level of the company is unlikely to be lower than a certain threshold L, the inventory probability could be set close to 0 for inventory levels within the range [0, L].

In the initial disruption signal, the start time of the disruption is a fixed known value, and its duration is estimated as a random variable. When the disruption propagates downstream to a certain tier, both the start time and duration of the supply disruption are uncertain. The supply disruption of tier i is defined by the start time t^i_{ds} and the disruption duration T^i_{disrp}. The disruption duration at all tiers is in the range of [0, U_{disrp}], and t^i_{ds} is in the range [L^i_{ds}, U^i_{ds}]. Let $p^i_{ds}(t)$ be the probability density distribution for the start time t^i_{ds} of the supply disruption, $p^i_{disrp}(t)$ be the probability density distribution of its duration T^i_{disrp}, and $F^i_{disrp}(t) = P\left(T^i_{disrp}(t) \leq t\right)$ be the accumulative. The supply disruption at the entry of tier $i+1$ does not pass through the tier when the inventory (including raw material and final product) I^{i+1}_{RF} of tier $i+1$ is more than the disruption period T^{i+1}_{disrp}, i.e.,

$$p^i_{disrp}\left(T^i_{disrp} = 0\right) = \int_0^{U_{disrp}} \left[p^{i+1}_{RF}(t) \cdot F^{i+1}_{disrp}(t)\right] dt. \tag{8}$$

The probability that $T^i_{disrp} = x$ $(x > 0)$ is the probability that the disruption period is x longer than the inventory level, i.e.,

$$p^i_{disrp}\left(T^i_{disrp} = x\right) = \int_0^{U_{disrp}} \left[p^{i+1}_{RF}(t-x) p^{i+1}_{disrp}(t)\right] dt. \tag{9}$$

From Equations (8) and (9), we obtain the probability function of disruption duration for the supply disruption signal at tier i.

Now consider the disruption start time at the entry point of tier i. When there is no disruption, we do not need to consider the start time. When there is a disruption, from Equation (5) and the bounds of t^{i+1}_{ds}, the disruption start time t^i_{ds} is in the range: $R_i = [L^{i+1}_{ds} + T^{i+1}_{Proc} + Tr_{i+1}, U^{i+1}_{ds} + T^{i+1}_{Proc} + Tr_{i+1} + U^{i+1}_{I\_RF}\Big]$. So, when $x \in R_i$, the probability of the disruption start time t^i_{ds} is

$$p^i_{ds}\left(t^i_{ds} = x\right) = \int_{L^{i+1}_{ds}}^{x - T^{i+1}_{Proc} - Tr_{i+1}} \left[p^{i+1}_{ds}(t) p^{i+1}_{RF}(x-t)\right] dt, \ x \in R_i \tag{10}$$

otherwise, the probability is 0.

From Equations (8)–(10), the probability distributions for the disruption start time and duration are obtained when the supply disruption propagates through one tier. Continuing the above process, all distributions at all tiers are obtained. Moreover, from these distributions, it is straightforward to calculate the mean value and standard deviation of these distributions and assess the risks at each tier.

6. Numerical Example

We use an example to illustrate the calculation of disruption propagation. In this example, the materials are produced on an MTO basis. The production period at each tier ranges from 3 to 6 days. All the parameters of each tier, the initial disruption signal, and the propagation details are shown in Table 2. The values italicized are the calculated ones.

Table 2. An example of calculating disruption propagation.

Tier	I_{RF}	T_{Proc}	Tr	t_{ds}	t_{de}	T_{disrp}
K	15	5	4	0	30	30
$K-1$	18	6	2	24	39	15
$K-2$	10	3	2	---	---	0

The columns t_{ds} and t_{de} are for the disruption start time and end time at the entry point of each tier. The initial supply disruption of tier K starts at time 0 and ends at time 30. The initial disruption of tier K propagates to the entry point of tier $K-1$ at time 24 and ends at time 39. The disruption duration at tier $K-1$ is reduced from the initial 30 units (days) to 15 units. The supply disruption at tier $K-1$ does not propagate further. Therefore, at the entry point of tier $K-2$, there is no supply disruption.

7. Conclusions

We study the disruption propagation mechanism in multiple-tier supply chains. A continuous material flow model with production and inventory details based on system dynamics principles is developed for a supply chain. The characteristics of disruption propagation are obtained and discussed using the model. Closed-form expressions are derived to calculate the time when the disruption propagates to the next tier, the disruption duration, and the condition under which the disruption is not propagated further. The total inventory of raw materials and final products is critical to prevent supply disruption from propagating to the subsequent tiers. We also extend the model to the cases with stochastic parameters to incorporate the uncertainty. This study provides valuable managemental insights for monitoring disruption events upstream and managing the disruption effects. The results are also useful in designing a resilient and robust supply chain.

The material flow model serves as a foundation for future research. First, although the continuous model captures the essential properties of material flow, the calculation of disruption propagation time and severity may not accurately reflect the discrete nature of material flow between companies. Therefore, it may be necessary to develop correction mechanisms based on the continuous model to account for practical considerations. Second, it is necessary to develop models for disruption recovery based on similar principles and methods. Finally, the scenarios where the companies actively adopt measures in response to disruption events must be constructed.

Author Contributions: Conceptualization, S.L., S.X. and L.W.; methodology and mathematical model, S.L.; validation and analysis, S.X. and L.W.; writing—original draft preparation, S.L.; writing—review and editing, S.X. and L.W.; funding acquisition, S.X. and S.L. All authors have read and agreed to the published version of the manuscript.

Funding: This research was supported by The Agency for Science, Technology and Research (*A*STAR*) Singapore under A*STAR—MiRXES Agile Diagnostics Manufacturing Joint Lab (Grant No.: I2301E0017).

Institutional Review Board Statement: Not applicable.

Informed Consent Statement: Not applicable.

Data Availability Statement: Example data are provided in this paper.

Acknowledgments: We extend our sincere thanks to Timothy KWANG and D. Puvaneswari Dhanaraj from MiRXES for their invaluable domain expertise and insightful feedback, which helped to shape our research focus. We also express deep appreciation to Ning Liu, Wenda Chen, and Junjie Cai from Advanced Remanufacturing and Technology Centre (ARTC), A*STAR, for their valuable discussions, collaboration, and strong support of this project. Finally, we would like to thank the anonymous reviewers and the editor for their thoughtful efforts in improving this manuscript.

Conflicts of Interest: The authors declare no conflict of interest. The funders had no role in the design of this study; in the collection, analyses, or interpretation of data; in the writing of this manuscript; or in the decision to publish the results.

References

1. Qin, R.; Khakzad, N.; Zhu, J. An overview of the impact of hurricane harvey on chemical and process facilities in Texas. *Int. J. Disaster Risk Reduct.* **2020**, *45*, 101453. [CrossRef]
2. Sokolov, B.; Ivanov, D.; Dolgui, A. The ripple effect in supply chains: Trade of efficiency-flexibility-resilience in disruption management. *Int. J. Prod. Res.* **2014**, *52*, 2154–2172.
3. Hosseini, S.; Ivanov, D. Bayesian networks for supply chain risk, resilience and ripple effect analysis: A literature review. *Expert Syst. Appl.* **2020**, *161*, 113649. [CrossRef] [PubMed]
4. Hosseini, S.; Ivanov, D.; Dolgui, A. Ripple effect modelling of supplier disruption: Integrated Markov chain and dynamic Bayesian network approach. *Int. J. Prod. Res.* **2020**, *58*, 3284–3303. [CrossRef]
5. Ivanov, D.A.; Dolgui, A.; Sokolov, B.V.; Ivanov, M. Literature review on disruption recovery in the supply chain. *Int. J. Prod. Res.* **2017**, *55*, 6158–6174. [CrossRef]
6. Rinaldi, M.; Murino, T.; Gebennini, E.; Morea, D.; Bottani, E. A literature review on quantitative models for supply chain risk management: Can they be applied to pandemic disruptions? *Comput. Ind. Eng.* **2022**, *170*, 108329. [CrossRef] [PubMed]
7. Katsaliaki, K.; Galetsi, P.; Kumar, S. Supply chain disruptions and resilience: A major review and future research agenda. *Ann. Oper. Res.* **2022**, *319*, 965–1002. [CrossRef] [PubMed]
8. Yu, H.; Zeng, A.Z.; Zhao, L. Single or dual sourcing: Decision-making in the presence of supply chain disruption risks. *Omega* **2009**, *37*, 788–800. [CrossRef]
9. Xu, H.-Y.; Fu, X.; Ponnambalam, L.; Namatame, A.; Yin, X.F.; Goh, R.S.M. A model to evaluate risk propagation considering effect of dynamic risk information sharing and multi-sourcing in supply chain networks. In Proceedings of the 2015 IEEE International Conference on Industrial Engineering and Engineering Management (IEEM), Singapore, 6–9 December 2015; pp. 1593–1597.
10. Wilson, M.C. The impact of transportation disruptions on supply chain performance. *Transp. Res. Part E Logist. Transp. Rev.* **2007**, *43*, 295–320. [CrossRef]
11. Schmitt, A.J.; Singh, M. A quantitative analysis of disruption risk in a multi-echelon supply chain. *Int. J. Prod. Econ.* **2012**, *139*, 22–32. [CrossRef]
12. Tan, P.S.; Lee, S.G.; Tan, C.S. Modeling disruption propagation in a complex supply chain. In Proceedings of the 2016 IEEE International Conference on Industrial Engineering and Engineering Management (IEEM), Bali, Indonesia, 4–7 December 2016; pp. 154–158.
13. Liu, R.; Kumar, A.; van der Aalst, W. A formal modeling approach for supply chain event management. *Decis. Support Syst.* **2007**, *43*, 761–778. [CrossRef]
14. Blackhurst, J.; Wu, T.; Ogrady, P. Methodology for supply chain disruption analysis. *Int. J. Prod. Res.* **2007**, *45*, 1665–1682.
15. Zegordi, S.H.; Davarzani, H. Developing a supply chain disruption analysis model: Application of colored petri-nets. *Expert Syst. Appl.* **2012**, *39*, 2102–2111. [CrossRef]
16. Liu, S.; Song, M.; Tan, K.C.; Zhang, C. Multiclass dynamic inventory rationing with stochastic demands and backordering. *Eur. J. Oper. Res.* **2015**, *244*, 153–163. [CrossRef]
17. Liu, S.; Zhang, C. An efficient inventory rationing strategy for systems with two demand classes and backordering. *Int. J. Prod. Res.* **2015**, *53*, 6136–6142. [CrossRef]

engineering proceedings

MDPI

Proceeding Paper

Enhancing Stroke Risk Prediction: Leveraging Machine Learning and Magnetic Resonance Imaging Data for Advanced Assessment †

Pranav Kunderu [1,*], Salik Mian [2] and Shivm Patel [3]

[1] Independent Researcher, Los Angeles, CA 91307, USA
[2] Independent Researcher, Los Angeles, CA 91326, USA; salikmian75@gmail.com
[3] Molecular & Cell Biology, University of California, Berkeley, CA 94720, USA; shivmpatel@berkeley.edu
* Correspondence: pranavkunderu1209@gmail.com
† Presented at the 2024 IEEE 7th International Conference on Knowledge Innovation and Invention, Nagoya, Japan, 16–18 August 2024.

Abstract: As a leading cause of death, strokes have been regarded as a dangerously impactful condition with little to no predictability. Currently, there is no effective method to predict a stroke using warning signs and hereditary factors. We developed a quantitative method to predict strokes before happening. We used MRI scan data obtained from OpenNeuro, specifically images showing the signs of pre-stroke and post-stroke. We trained machine learning models using the data, including support vector machines (SVMs), K-nearest neighbors (KNNs), and random forests. The models predicted the risk of a stroke accurately. The models allow for diagnosing and enable clinicians to care for patients promptly, potentially saving lives and improving outcomes.

Keywords: MRI; stroke; support vector machines (SVMs); k-nearest neighbors (KNNs); random forests

1. Introduction

Hemorrhagic stroke is the deadliest type and occurs when a blood vessel in the brain randomly begins leaking or bursting [1]. Ischemic stroke happens when an artery becomes blocked in the brain [2]. Such strokes harm the nervous system but remain unpredictable. Therefore, families or friends must monitor a patient's speaking or walking, face, arms, eyes, and talking [3]. Such monitoring is troublesome. It is necessary to develop a method utilizing machine learning (ML) and Magnetic Resonance Imaging (MRI) data to assess and predict a stroke.

MRI is used to create an image of the entire structure of the human body [4]. It allows doctors to understand the positions of the organs, bones, and muscles of patients that they are treating. This allows surgeries to diagnose patients. This technology can be used to create prediction models with ML [5]. ML can be used with MRI scans to design algorithms that provide modeling for diseases [6]. We created a model with high accuracy [7] to assist doctors in predicting strokes in patients [8].

To create an ML algorithm, we collected stroke prediction confidence scores using MRI scans. We ran practical tests on the model to diagnose strokes quickly and accurately.

2. Literature Review

Cerebrovascular accidents can cause strokes, which are either ischemic or hemorrhagic. Hemorrhagic strokes account for 10−20% of strokes and are caused by intracerebral hemorrhage and subarachnoid hemorrhage. Often caused by hypertension or another underlying condition, they are characterized by an initial compression from hematoma raising ICP, followed by secondary injury due to inflammation and oxidative stress. Diagnosis is made by using computed tomography (CT) and MRI. The prognosis depends upon hematoma volume and age. If managed and rehabilitated well, the recovery from strokes can be satisfactory [1]. Ischemic stroke occurs when there is a sudden onset of focal neurological deficits due to the impaired blood flow to the brain caused by an embolic or thrombotic event. Prevention through lifestyle modification and management of risk factors such as hypertension and diabetes is vital to prevent it [2]. Reference [3] explores how stroke is the second leading cause of death and disability worldwide. Substantial advancements in the understanding of stroke pathophysiology and treatment have concentrated on the restoration of blood flow and management of neurological damage. Recently, since clinical trials reported limited success, researchers have developed refined animal models of stroke and targeted study designs, along with advanced technologies.

MRI provides relevant information for diagnosis. Though tPA is the only proven therapy for acute ischemic stroke, there are hemorrhagic risks. MRI is used to identify patients with thrombolysis by differentiating stroke from stroke mimics, particularly in diffusion-weighted imaging and perfusion-weighted imaging. The assessment of salvageable tissue and the prediction of clot characteristics and hemorrhagic risk can also be allowed. The dependence on MRI-based selection of patients for thrombolysis necessitates incorporating findings using MRI for effective treatment [4]. ML can be used to improve stroke treatment, as it enables efficient analysis of large data volumes and supports strategies in personalized medicine. Recent advancements in ML research allow for high accuracy using the analyses of imaging, diagnosis of stroke subtypes, risk stratification, guidance of treatments, and prediction of patient outcomes. ML assists stroke medicine with potential enhancement of diagnosis, treatment selection, and prognostication to improve outcomes and quality of life [5].

ML applications in stroke rehabilitation include random forests (RFs), logistic regression, deep neural networks, and convolutional neural networks. They are used to predict motor function recovery using clinical data and Magnetic Resonance Imaging effectively. These models have the potential to predict functional outcomes, interpret fluoroscopic swallowing study results, and prognose depression, language, cognitive, and sensory recovery from stroke [6]. Reference [7] used ten different classifiers to train logistic regression, decision tree, AdaBoost, and XGBoost and maximize accuracy using weighted voting. The accuracy of the proposed model was 97%, while areas under the curve (AUC) were high, and the rates of false positive and negative rates were low. Thus, the weighted voting classifier is used for early stroke prediction, efficiently and effectively benefiting physicians and patients [7].

Reference [8] explored how strokes need immediate diagnosis and intervention to prevent permanent damage to the brain. An advanced stroke detection algorithm was proposed for predicting strokes using ML techniques. The dataset used contained parameters such as age, body mass ratio (BMI), gender, heart disease, and smoking status. The data were preprocessed for missing values, categorical features, and balance. Comparisons were made with different classifiers, including Naïve Bayes, logistic regression, XGBoost, decision trees, AdaBoost, K-Nearest Neighbor, RFs, voting classifier, and support vector machines. In the research, the performance of the models was evaluated using accuracy, F1-score, recall, precision, and an extra metric of specificity. The SVM algorithm showed an accuracy of 99.5%, precision of 99.9%, recall of 99.1%, F1-score of 99.5%, and specificity of 99%.

There is an absence of research on using MRI scans for diagnosing strokes with ML methods, which necessitates in-depth exploration.

3. Methods

The data were obtained from OpenNeuro which is a publicly available database. The database included information on post-stroke patients and healthy pre-stroke patients. The post-stroke images were collected from MRI scans [9]. The pre-stroke images were acquired from Ref. [10].

3.1. Data Description

Twenty-nine scans were collected from the post-stroke database and twenty-nine scans were obtained from the healthy pre-stroke database. Three scans from both databases were used for training. The stroke risk confidence score ranged from 0 to 1, 0 being healthy and 1 being stroke.

3.2. Data Preprocessing

Each image was transformed for ML. Images were transformed into a grayscale image and reshaped in a resolution of 128 $\times$ 128 to ensure uniformity. Each pixel of each image was represented in 2D. The grayscale numbers ranged from 0 to 255. The 128 $\times$ 128 array was reduced into a 1-dimensional array.

3.3. Variables

The dependent variable was the similarity score. The similarity score ranged between 0 and 1, with 0 for a healthy patient and 1 for a stroke patient. By determining how similar a pre-stroke patient's MRI scan was to a post-stroke patient, the model predicted the risk of stroke.

3.4. Methods

3.4.1. Transfer Learning

After training a neural network using a large dataset of images, the model was fine-tuned using a smaller dataset. Such a strategy makes use of the features and patterns learned from the original task to optimize performance on the new task with less data. We utilized the VGG16 transfer learning model for feature extraction. We inputted the 1D array of pixel values, and the VGG16 base model returned features such as edges and textures that were fed into the classifier models. The classifier models included SVMs, RFs, KNNs, and Adam optimizers.

3.4.2. Cross Validation

The benefits of cross-validation include a reduction in overfitting and reduced variance in performance. The cross-validation was conducted by splitting the data into several groups. We used 5-fold cross-validation. The dataset was split into 5 groups; 4 were used to train while the 5th was used to test. The model was retrained 5 times. The performance of the model was evaluated in 5 iterations. By training with different subsets of the data, the model improved its ability to classify new data showing improved generalization.

3.5. Algorithms

3.5.1. SVM

SVMs adopt supervised learning for classification and regression. They ensure a maximum distance between an optimized hyperplane (that separates the different classes of data) and the closest data points to the hyperplane. SVMs are effective when the dimensionality of the data is high and the number of samples is limited. They allow for linear and nonlinear classification using kernel functions. We utilized the Gaussian kernel, also known as a radial basis function (RBF) kernel. This kernel is suited for classification problems where the relationship between the independent and dependent data is highly nonlinear.

3.5.2. K-Nearest Neighbors

KNN is a simple instance-based learning algorithm for classification and regression. When implementing KNNs, all the input data are mapped into a feature space. A new data point is classified based on the majority class of its 'k' nearest neighbors in the feature space. For regression, this output is usually an average of values obtained from those neighbors. The algorithm uses metrics of distances, among which is the Euclidean distance, to determine how similar data points are to each other. The Euclidean distance was used in this study (1).

$$distance\ (x_1,\ x_2) = sqrt(\sum_{i=1}^{n} (\mathrm{x}_{1i} - \mathrm{x}_{2i})^2) \tag{1}$$

where x_1 and x_2 are data points in the feature space and x_{1i} and x_{2i} are the ith features of x_1 and x_2.

Although KNNs are easy to implement, they require large datasets since computing numerous distances among data points is involved.

3.5.3. RF

RF is an ensemble learning method where many decision trees combine to increase the accuracy and power of prediction. A decision tree is another type of ML model consisting of three main components: nodes, branches and leaf nodes. The full set of data starts at the first node (also called the root node) and the dataset is split based on a decision represented by the node. For example, a set of human individuals is split based on the age at a particular node. Then, each subset travels along a branch until it encounters the next node, which further splits the data into smaller subsets. This cycle repeats until the data reach a leaf node, which represents the classification of the data. In RFs, each tree in the forest is trained on a random subset of the full training data. This randomness reduces overfitting and increases generalization ability. In prediction, all the predictions of individual trees are aggregated by RF for a final decision. In other words, the most popular classification by the trees in RF is what the model decides as the final classification.

3.5.4. Adaptive Moment Estimation (Adam) Optimizer

The Adam optimizer is a popular ML algorithm. The algorithm guides the gradient of the model. When training a model, the difference between the predicted and actual result is used, called the error. The gradient is used to change the parameters (or weights) and reduce the error as much as possible. The Adam optimizer uses past gradients instead of only current gradients. By evaluating the average direction and size of the previous gradients, the Adam optimizer chooses when to make large or small changes to the weights intelligently. This allows for smooth changes that do not disrupt the accuracy of the model or the learning process. Due to this feature, it requires less manual tuning than many other models. Along with this optimizer, we used binary cross entropy for our loss type to calculate the error of the model in the learning process. Binary cross entropy was well suited for classification models with two classes, which in our case are stroke and healthy.

3.6. Statistical Measures

To evaluate the performance of the models, we used the following measures: precision, recall, accuracy, and F-1 score.

$$Precision = \left(\frac{TP}{TP + FP}\right) \tag{2}$$

$$Recall = \left(\frac{TP}{TP + FN}\right) \tag{3}$$

$$Accuracy = \left(\frac{TP + TN}{TP + FP + FN + TN}\right) \tag{4}$$

$$F1\ Score = \left(2 \times \frac{Precision \times Recall}{Precision + Recall}\right) \tag{5}$$

where TP refers to a true positive, TN refers to a true negative, FP means false positive, and FN means false negative.

In addition to the statistical values recorded in the classification report, the model outputs the R^2 value and Mean-Squared Error (MSE). MSE is calculated using (6).

$$MSE = \frac{1}{n}\sum_{i=1}^{n}(y_i - \hat{y}_i)^2 \tag{6}$$

where n is the number of data points, y_i is the actual value corresponding to a data point, and $\hat{y}_i$ is the predicted value from the model. Of course, a smaller MSE value suggests the predictions made by the model are accurate to the actual classifications.

The R^2 value of a model is an indicator of how well the model understands the patterns and relationships between our independent and dependent variables. This value ranges between 0 and 1, with 1 representing a perfect model of the data and 0 being an unreliable model. The R^2 value is calculated using (7).

$$R^2 = 1 - \frac{\sum_{i=1}^{n}(y_i - \hat{y}_i)^2}{\sum_{i=1}^{n}(y_i - \bar{y})^2} \tag{7}$$

where y_i corresponds to the actual value associated with a datapoint, $\hat{y}_i$ is the predicted value the model generates, $\bar{y}$ is the mean of the actual values across all data points, and n is the number of data points.

4. Results

We used Adam optimizer, SVM, RF, and KNN in this study. The confidence levels for the different scans using the Adam optimizer model showed that the confidence values ranged from 0.0039 to 0.0112 in Scan 1 with an average of 0.0064, indicating a high probability that the patient was healthy. Scan 2 had higher confidence values from 0.0196 to 0.0293, averaging 0.0243, still indicating a relatively healthy patient but with rising certainty of a possible condition. Scan 3 also showed a confidence of 0.0222, identical to that of Scan 2. Scan 4 demonstrated high confidence levels of 0.9741–0.9854 with an average of 0.9806, a strong indicator for a stroke condition. Scan 5 showed confidence levels ranging from 0.9968 to 0.9993, with an average of 0.9982, indicating a very high likelihood that the patient was a stroke patient. Scan 6 showed high confidence, averaging 0.9966, thus strongly indicating the likelihood of a stroke. Scan 7 presented considerably lower confidence, averaging 0.1068, indicating a much healthier patient compared to the previous scans. Scan 8 showed a moderately high level of confidence, averaging 0.7591, indicating a higher likelihood of a stroke than the initial scans but less certain than Scans 4, 5, and 6.

In the SVM model, Scan 1 showed confidence ranging from 0.0489 to 0.0503, with an average of 0.0497. This indicated a healthy patient. In Scan 2, the confidence level ranged from 0.1091 to 0.1113 with an average of 0.1110. Thus, this was a relatively healthy patient. Scan 3 showed a confidence level with an average of 0.1208. Scan 4 showed a high confidence level between 0.9059 and 0.9098 with an average of 0.9082, hence strongly indicating a stroke patient. Scan 5's confidence level was from 0.9686 to 0.9702, averaging

0.9695, indicative of a stroke. Scan 6 showed a high confidence level with an average of 0.9560. Scan 7 showed a lower confidence level, averaging 0.3136, indicating a healthier patient but approaching a worrying level. Scan 8 presented medium confidence levels, with an average of 0.7717, indicating a higher probability of stroke but less confidence compared to Scan 4 and Scan 5.

The RF model presented the same confidence levels for all scans. Scan 1 showed a confidence level of 0.06. This means it was a healthy patient. Scan 2 maintained a confidence level of 0.24. Scan 3 showed a confidence level of 0.25. Scan 4 showed a high confidence level of 0.86, indicating a stroke. Scan 5 showed an even superior confidence level of 0.92, signifying a likelihood of stroke. Scan 6 maintained a confidence level of 0.87, also indicating a high likelihood of stroke. Scan 7 showed a confidence value of 0.43, indicative of a healthier patient compared to Scans 4 and 5 but less healthy than earlier scans. Scan 8 presented a confidence value of 0.55, showing a moderate risk of stroke.

In the KNN model, Scans 1, 2, 3, and 7 showed a confidence level of 0.0; thus, these patients were healthy. Confidence levels of 1.0 were observed for Scans 4, 5, 6, and 8 indicating a stroke.

The accuracy was 1.0. In cross-validation, a recall and F-1 score of 1.0 were observed for health and stroke patients. The R^2 value was 1.0, and MSE was 0.0. Therefore, the four models predicted stroke perfectly.

5. Conclusions/Discussions

The RF model showed consistent levels of confidence, being robust, reliable, and insensitive to intermediate risk levels. The Adam optimizer model was better at predicting stroke than the SVM and RF models. The result obtained by the Adam optimizer model presented various confidence levels. This model was good at sorting low- from high-risk cases, but the mixed confidence levels suggested room for improvement in identifying varying degrees of stroke risks. The SVM model presented consistent levels of confidence and perfectly distinguished a stroke case. Even though this model was stable, it was less discriminatory on intermediate risk levels compared to the Adam optimizer. Further development and testing of the model's robustness and generalizability are necessary.

Many strokes occur without obvious symptoms, making strokes an unpredictable disease. Each person's stroke risk is influenced by a combination of genetic, environmental, and lifestyle factors, which make it difficult to create a one-size-fits-all predictive model. While risk factors such as high blood pressure, diabetes, and smoking are known to increase stroke risk, the prediction of a stroke remains complex. Biomarkers include specific proteins or genetic markers in blood or cerebrospinal fluid. By combining and creating models that consider biometrics and patient medical history, we can create more accurate predictive models.

Risk factors can be managed but are not controlled completely. Further research is needed to integrate biomarkers. Such integration can increase predictive capabilities. Using ML, a model to predict strokes can be developed. The model's ability to predict stroke can be enhanced if trained with more data. More MRI scans could make the model prediction more accurate in predicting the risk of having a stroke.

Author Contributions: Conceptualization, P.K. and S.M.; methodology, P.K. and S.M.; software, P.K., S.M. and S.P.; validation, P.K., S.M. and S.P.; formal analysis, P.K.; investigation, P.K. and S.M.; resources, S.P.; data curation, P.K. and S.M.; writing—original draft preparation, P.K., S.M. and S.P.; writing—review and editing, P.K., S.M. and S.P.; visualization, S.M.; supervision, P.K. and S.M.; project administration, P.K. and S.M. All authors have read and agreed to the published version of the manuscript.

Funding: This research received no external funding.

Institutional Review Board Statement: Not applicable.

Informed Consent Statement: Not applicable.

Data Availability Statement: Data are contained within the article.

Conflicts of Interest: The authors declare no conflict of interest.

References

1. Unnithan, A.K.A.; Mehta, P.; Das, J.M. Hemorrhagic Stroke. National Library of Medicine. 8 May 2023. Available online: https://www.ncbi.nlm.nih.gov/books/NBK559173/ (accessed on 1 July 2024).
2. Hui, C.; Tadi, P.; Patti, L. Ischemic Stroke. National Library of Medicine. 2022. Available online: https://www.ncbi.nlm.nih.gov/books/NBK499997/ (accessed on 1 July 2024).
3. Kuriakose, D.; Xiao, Z. Pathophysiology and Treatment of stroke: Present Status and Future Perspectives. *Int. J. Mol. Sci.* **2020**, *21*, 7609. [CrossRef] [PubMed]
4. Kim, B.J.; Kang, H.G.; Kim, H.-J.; Ahn, S.-H.; Kim, N.Y.; Warach, S.; Kang, D.-W. Magnetic Resonance Imaging in Acute Ischemic Stroke Treatment. *J. Stroke* **2014**, *16*, 131–145. [CrossRef]
5. Daidone, M.; Ferrantelli, S.; Tuttolomondo, A. ML applications in stroke medicine: Advancements, challenges, and future prospectives. *Neural Regen. Res.* **2024**, *19*, 769–773. [CrossRef] [PubMed]
6. Choo, Y.J.; Chang, M.C. Use of ML in Stroke Rehabilitation: A Narrative Review. *Brain Neurorehabilit.* **2022**, *15*, e26. [CrossRef] [PubMed]
7. Emon, M.U.; Keya, M.S.; Meghla, T.I.; Rahman, M.M.; Al Mamun, M.S.; Kaiser, M.S. Performance Analysis of ML Approaches in Stroke Prediction. In Proceedings of the 2020 4th International Conference on Electronics, Communication and Aerospace Technology (ICECA), Coimbatore, India, 5–7 November 2020. Available online: https://ieeexplore.ieee.org/document/9297525 (accessed on 1 July 2024).
8. Mushtaq, S.; Saini, K.S.; Bashir, S. ML for Brain Stroke Prediction. In Proceedings of the 2023 International Conference on Disruptive Technologies (ICDT), Greater Noida, India, 11–12 May 2023. Available online: https://ieeexplore.ieee.org/document/10151148 (accessed on 1 July 2024).
9. OpenNeuro. openneuro.org. Available online: https://openneuro.org/datasets/ds003999/versions/1.0.2 (accessed on 20 July 2024).
10. OpenNeuro. openneuro.org. Available online: https://openneuro.org/datasets/ds003469/versions/1.0.0 (accessed on 20 July 2024).

engineering proceedings

MDPI

Proceeding Paper

Application of XGBoost Algorithm to Develop Mutual Fund Marketing Prediction Model for Banks' Wealth Management †

Jen-Ying Shih

Graduate Institute of Global Business and Strategy, National Taiwan Normal University, Taipei 106, Taiwan; jyshih@ntnu.edu.tw

† Presented at the 2024 IEEE 7th International Conference on Knowledge Innovation and Invention, Nagoya, Japan, 16–18 August 2024.

Abstract: Competition in Taiwan's banking industry is becoming fierce. Banks' traditional income based on interest rates is insufficient to support their growth. Therefore, banks are eager to expand their wealth management business to increase profits. The fee income from the sale of mutual funds is one of the major sources of banks' wealth management business. The problem is how to look for the right customers and contact them effectively. Therefore, it is necessary to develop classification prediction models for these banks to evaluate their customers' potential to buy mutual fund products sold by commercial banks and then deploy marketing resources on these customers to increase banks' profits. Recently, the XGBoost algorithm has been widely used in conducting classification tasks. Therefore, using the eXtreme Gradient Boosting algorithm, a mutual fund marketing prediction model is developed based on a commercial bank's data in this study. The results show that whether a customer has an unsecured loan, a customer's amount of assets in the bank, the number of months for transactions, a place of residence, and whether the bank is the main bank for the total amount of credit card bills in the past six months are the top five factors for the models, providing valuable information for effective wealth management and marketing.

Keywords: XGBoost; mutual fund; marketing prediction model; wealth management; commercial bank

1. Introduction

Competition in Taiwan's banking industry is becoming intense because 70 banks are serving 23 million people in Taiwan [1]. Conventional income from interest rate spreads between deposits and loans is no longer sufficient to sustain growth for Taiwan's banks. Therefore, banks are eager to expand their wealth management to increase profits. The fee income from the sale of mutual funds and after-sale asset management services is one of the major sources of banks' wealth management business. However, the fee income can be obtained after banks have successfully sold the mutual funds to customers. Therefore, it is required for them to look for the right customers and contact them effectively to realize the fee income.

Direct marketing models are used to choose the right customers [2]. The factors that explain whether a bank's customers buy mutual fund products recommended by banks are the model's input variables. The classification algorithm is important to study the relationship between the target variable (whether to buy or not to buy the mutual fund) and the explained factors. Then, marketing resources can be precisely allocated to customers with a high potential to buy mutual funds. As a result, banks can earn fee income from the sale of mutual funds and then increase their profits.

In this research, a classification prediction model is developed for banks' marketing decisions. The model evaluates their customers' potential to buy mutual fund products and then deploys marketing resources to these customers.

Recently, the extreme gradient boost (XGBoost) algorithm has been widely used in credit scoring, bankruptcy prediction, and fraud detection in the banking sector owing to its exceptional performance in conducting classification [3,4]. By recognizing the potential of the XGBoost algorithm and applying it to develop mutual fund marketing prediction models based on a commercial bank's data, the application and effectiveness of the algorithm in the banking industry are validated.

In this research, the following are explored:

- Factors to develop a direct marketing model.
- Effectiveness of the XGBoost algorithm in building a mutual fund marketing model.
- Comparison of the XGBoost algorithm with the other algorithms.

The organization of this paper is as follows: After introducing the research background and purpose in Section 1, Section 2 reviews the related literature. Section 3 describes the data set and research process. Section 4 presents the results and discussion. Finally, concluding remarks are presented in Section 5.

2. Literature Review

2.1. Direct Marketing of Financial Commodities

Past research highlighted the importance of direct marketing and proposed that direct marketing is critical for businesses to engage customers and drive sales [2,3]. Reference [2] adopted a system perspective to understand how different models interact within the broader marketing system and proposed that quantitative models help optimize these marketing efforts. The quantitative models of direct marketing include statistical methods, such as logistic regression, and machine learning techniques, including support vector machines (SVMs), artificial neural networks (ANNs), decision trees, and related enhanced algorithms (e.g., random forest (RF), a method that operates by constructing a multitude of decision trees at training time and outputting the mode of the classes). These algorithms are used to predict the likelihood of a customer buying a product or service.

Commercial banks are eager to apply financial technology (Fintech) and marketing technology (Martech) to boost their revenues and profits [3]. Reference [4] found that marketing activities have positive impacts on commercial banks' profit.

2.2. XGboost Algorithm

The XGBoost model, a tree-based Gradient Boosting ensemble, is renowned for its high efficiency and predictive accuracy. Its application in banking includes credit card fraud detection, bankruptcy prediction, and personal credit risk assessment, underscoring its significance and potential. The model performance of various algorithms is evaluated using two data sets (European data set and German data sets) in Ref. [5], including SVMs, logistic regression, RFs, K-Nearest Neighbors, J48 decision trees, ANNs, and boosting frameworks for decision trees (such as XGBoost), for credit card fraud detection. The results show that XGBoost cannot outperform ANNs and RFs based on the F1 measure. The XGBoost algorithm with a feature importance selection mechanism is applied to construct a bankruptcy prediction model [6]. The model performance is compared with other algorithms and feature selection approaches. The feature importance selection mechanism in XGBoost calculates how each feature contributes to the model's predictions. When evaluating forecast accuracy using the area under curve (AUC), XGBoost demonstrated superior discrimination power compared to other feature selection techniques, including stepwise discriminant analysis, stepwise logistic regression, and partial least square-discriminant analysis. XGBoost is used to explain bank default [7]. They suggested that lower values of retained earnings to average equity, pretax return on assets, and total risk-based capital ratio are linked to a higher risk of bank failure. Additionally, an excessively high yield on earning assets increases banks' likelihood of financial distress.

Logistic regression is applied to select features, and the XGBoost algorithm is used to construct a classification model for assessing personal credit risk [8]. Decision tree and K-nearest neighbor algorithms were used as benchmarks for comparison. Their results demonstrated that XGBoost performed well in terms of AUC, accuracy, and Type II error.

Past research on the application of the XGBoost algorithm focused on credit evaluation, whether for individual customers or corporate customers. Few studies reported the algorithm's performance on marketing decisions in the banking industry. The performance and feasibility of banks' marketing decisions were evaluated using a real data set from a commercial bank in this study.

3. Data Set and Research Process

In this research, XGboost, a python-based package, was used to develop a classification model.

3.1. Data Set

The data set was obtained from a commercial bank. It contained 156,666 records. 14,075 records presented the data of customers who bought mutual funds within three months, and 142,591 records presented the data of customers who did not buy mutual funds within three months. A 1:1 undersampling approach was used to avoid the problem of rare events being ignored by the training model [9]. As a result, 28,150 records remained and were split into a training set (21,112) and a testing set (7038).

The summary of the data set is shown in Table 1. The target variable is a binary variable that defines whether a customer buys a mutual fund within three months. Four facets include the customer's transaction with the bank, the customer's democratic profile, the channel usage profile, and the holding situation of financial commodities. They were used to determine whether a customer would buy mutual funds. For example, Reference [10] identified that e-banking channels could enhance banks' market shares. Thus, features reflecting channel preference were determined in this study.

Table 1. Summary of data set.

Description	Number of Records
Customers who do not buy mutual funds within three months (Y = 0)	142,591
Customers who buy mutual funds within three months (Y = 1)	14,075
1:1 undersampling (14,075 × 2)	28,150
Training set (75% of data points)	21,112
Test set (25% of data points)	7038

A total of 36 features were determined to describe every data point (i.e., every potential customer and the customer's outcome), as shown in Table 2. After screening out features that contained missing values over 20%, 32 features remained. Then, categorical variables were transformed into dummy variables. Finally, 56 variables were used to develop the XGBoost model.

Table 2. Definition of features (input variables).

Feature Name	Definition	Data Type
add_now	Place of residence	NOMINAL
age	Customer's age	INTERVAL
allprice_6mon	Total amount of credit card bills in the past six months	INTERVAL
allprice_6mon_bank	Whether the total amount of credit card bills in the past six months was incurred in this bank	BINARY
asset	Customer's assets in the bank	INTERVAL
card	Whether the customer has a debit card in the bank	BINARY
chi_num	Number of customers' children	INTERVAL
Credit	Whether the customer is a new credit card user within three months in the bank	BINARY
credit_old	Whether the customer has a credit card in the bank	BINARY
cus_mon	Number of months of transactions with this bank	INTERVAL
degree	Customer's education level	ORDINAL
delete_insurance_creditcard	Whether the customer used a credit card to pay insurance premiums in the past year	BINARY
delete_insurance_deposit	Whether premiums are withheld from the customer's deposits	BINARY
ebank_mon	The customer's online banking usage period	INTERVAL
f_rate_atm	ATM usage ratio by customers in the past year—foreign currency deposits	INTERVAL
f_rate_counter	Customer usage ratio of over-the-counter counters in the past year—foreign currency deposits	INTERVAL
f_rate_ebank	Internet banking usage ratio in the past year—foreign currency deposits	INTERVAL
f_rate_phone	Rate of voice calls used by customers in the past year—foreign currency deposits	INTERVAL
foreign_currency_loan	Whether the customer has a foreign currency loan	BINARY
gender	Gender	BINARY
income	Income	INTERVAL
insurance	Whether the customer has insurance products	BINARY
job_g	Customer's occupation	NOMINAL
liability	Customer's liabilities within the bank	INTERVAL
marry	Customer's marital status	BINARY
prefer_f	Preferred channel in the past year—foreign currency deposits	NOMINAL
prefer_tw	Preferred channel in the past year—Taiwan dollar deposits	NOMINAL
pro_num	Total number of items held by customers	INTERVAL
sec_1monprice_1yr	Average monthly security transaction amount in the past year	INTERVAL
sec_freq_1yr	Number of security transactions in the past year	INTERVAL
secured_loan	Whether the customer has a secured loan	BINARY
tw_rate_atm	ATM usage ratio in the past year—New Taiwan dollar deposits	INTERVAL
tw_rate_counter	Ratio of customers using over-the-counter counters in the past year—New Taiwan dollar deposits	INTERVAL
tw_rate_ebank	Internet banking usage ratio in the past year—New Taiwan dollar deposits	INTERVAL
tw_rate_phone	Rate of voice calls used by customers in the past year–New Taiwan dollar deposits	INTERVAL
unsecured_loan	Whether the customer has an unsecured loan	BINARY

3.2. Research Process

The XGBoost algorithm was used to develop the classification model. Using the trained XBGoost model, the feature importance was determined to explain the impact of each feature on the target variable (whether a customer buys the mutual fund within three months). The testing data set was used for model evaluation. Finally, to understand XGBoost model performance, logistic regression, decision trees, support vector machines, and neural network algorithms were applied for model comparison. Figure 1 shows the research process of this research.

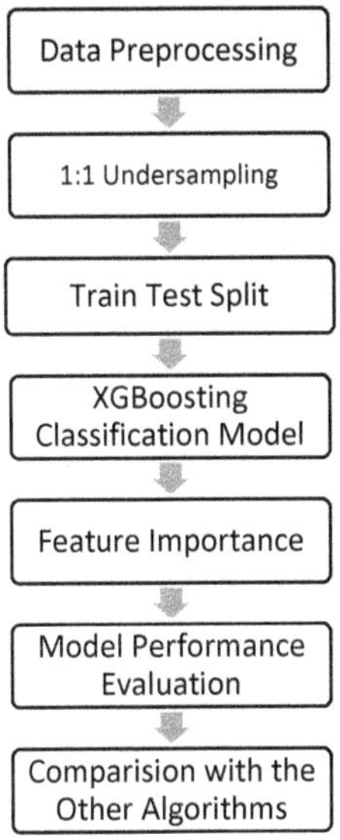

Figure 1. Research process.

3.3. XGBoost

XGBoost operates in a series of steps to enhance predictive accuracy and efficiency. The input data and parameters for the algorithm are as follows:

Input data: Explanatory feature matrix (X) and target variable (y).

3.3.1. Parameters

XGBoost has many parameters to optimize the model's performance. These parameters are categorized into three groups: general parameters, booster parameters, and learning task parameters. These parameters offer flexibility to fine-tune the XGBoost model for various types of data and tasks to achieve optimal performance and generalization. The parameter tuning strategy was referred to the web resource [11].

The parameters in this research included the following:

- Booster specifies options including tree-based models and linear models. Tree-based models were used as the booster in this research.
- Eta (learning_rate) is the step size shrinkage to prevent overfitting. The value range is between 0 and 1. It controls the weight of the new tree's contribution.
- Lambda (Reg_lambda) is the regularization parameter of the algorithm. L2 regularization term on weights is specified to avoid overfitting.
- The "objective" parameter is used for the learning task and the corresponding loss function. It was defined as "binary:logistic", for logistic regression to the binary classification task in this research.
- Eval_metric: Evaluation metrics for validation data. In this research, eval_metric were specified as "error", which used the binary classification error rate as evaluation metrics.
- n_estimators presented the number of boosting rounds (trees to grow).
- In early_stopping_rounds, training stops when a metric is not improved for a certain number of rounds.

3.3.2. Algorithm

The algorithm works in the following steps:

1. Initialization: The process starts with an initial prediction for each data point, often using the base class probabilities. For binary classification, the probability belongs to class 1, typically set to 0.5.
2. Iterative Boosting:

Residual Calculation is used to compute the residuals for each data point. In this research's classification task, the residuals were defined as the difference between the actual class labels (0 or 1) and the predicted probabilities.

Gradient Calculation: The gradient of the loss function is calculated concerning the current predictions. Logistic regression loss involves computing the derivative of the logistic loss function.

3. Model Training on Gradients

A new decision tree is trained on these gradients. The goal of the new tree is to predict the gradient of the loss function, effectively focusing on the errors made by the current iteration model.

4. Weighted Summation: Predictions are updated by adding the weighted output of the new tree to the current predictions. The learning rate controls the weight of the new tree's contribution. For logistic regression, the model adds the predicted log odds (logits) from the new tree to the existing log odds and then applies the logistic function to obtain updated probabilities.
5. Regularization: XGBoost applies regularization to penalize the complexity of the trees (e.g., the number of leaves and the leaf weights), prevent overfitting, and ensure that the model generalizes well to new data.
6. Tree Pruning: The algorithm grows each tree to a specified maximum depth and then prunes it by removing branches that do not significantly improve the loss function. This further helps prevent overfitting.
7. Iteration: Steps 2–6 are repeated iteratively to add new trees until the specified number of trees is reached or until further improvements in the loss function become minimal.
8. Final Prediction: For classification, the final prediction is made as the probability of class 1. This is obtained by applying the logistic function to the sum of the log odds predicted by all trees. For binary classification, a threshold (commonly 0.5) is applied to the probability of assigning class labels.

By iteratively adding trees that focus on the residuals (errors) of the current model, XGBoost enhances the model's ability to correctly classify data points, leading to high accuracy and robustness.

3.3.3. Feature Importance

In XGBoost, feature importance is measured using gain, cover, frequency, SHapley Additive exPlanations (SHAP) values, and permutation importance. These methods are used to understand which features contribute the most to the model's predictions.

In this research, the gain was measured as the improvement in accuracy brought by a feature to the branches. It measures how much each feature contributes to reducing the loss function (error) in the model. The higher the gain, the more important the feature is considered.

4. Results and Discussion

In the data exploration phase, interesting relationships between explanatory features and the target variables were observed. For features with a numerical data type, after conducting a statistical F-test on the two groups (y = 0 and y = 1) of data points, the top one significant feature was the "asset" (F-value = 6923.65, $p < 0.01$). The y = 1 group had a higher asset level than the y = 0 group, which revealed that customers with more assets were more likely to buy mutual funds. The second significant feature was the "unsecured_loan" (F-value = 3598.33, $p < 0.01$). The y = 0 group had a higher unsecured loan level than the y = 1 group, which showed that customers with more unsecured loans were less likely to buy mutual funds. The third significant feature is the "secured_loan" (F-value = 893.70, $p < 0.01$). The y = 1 group had a higher secured loan level than the y = 0 group, which revealed that customers with more secured loans were more likely to buy mutual funds. As the customers have assets as collateral, they have the financial resources to purchase funds for financial purposes.

Table 3 shows the top 15 important features that indicate whether a customer has an unsecured loan, a customer's amount of assets in the bank, the number of months a customer has dealt with the bank, the place of residence, and whether the bank is the main bank for the total amount of credit card bills in the past six months are top five factors for the models, providing crucial insights for effective wealth management marketing. The first two important features calculated by the XGBoost model also echo the analysis results of the F-test, as mentioned earlier.

Table 3. Feature importance.

Features	Feature Importance
unsecured_loan	0.5033
Asset	0.0728
cus_mon	0.0189
add_now_NANTOU	0.0170
add_now_MIAOLI	0.0142
add_now_PINGTUNG	0.0136
add_now_KAOHSIUNG CITY	0.0135
foreign_currency_loan	0.0123
gender_F	0.0122
prefer_f_3	0.0122
Marry	0.0117
prefer_f_2	0.0116
allprice_6mon_bank	0.0114
Liability	0.0108
delete_insurance_deposit	0.0108

Figure 2 shows the confusion matrix of the XGBoost model. The accuracy was 0.6371, the precision was 0.5956, the recall rate was 0.8756, and the F1 score (the harmonic mean of precision and recall) was 0.7090. The classification performance was slightly better than the other statistical methods and machine learning models (Table 4).

Table 4. Model performance on test set.

Model	Accuracy Rate	F1	Lift (Top 10%)
Logistic Regression	0.6279	0.70	1.5
Decision Tree	0.6367	0.71	1.6
XGBoost	0.6371	0.71	1.6
Neural Networks	0.6294	0.70	1.5
Support Vector Machines	0.6265	0.70	1.4

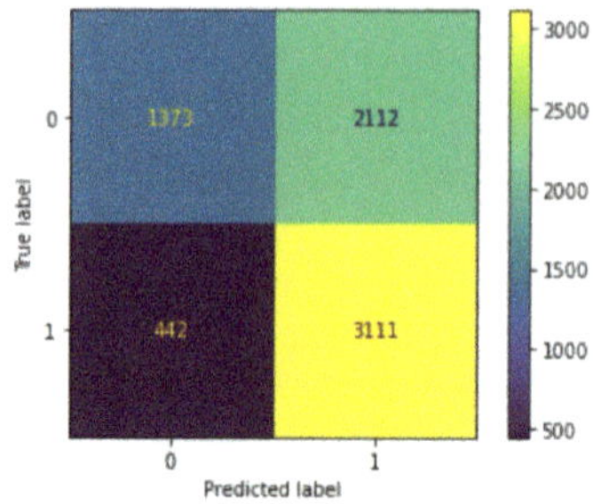

Figure 2. Confusion matrix of XGBoost model.

Lift indicates how well a model predicts positive outcomes better than random guessing. It is the ratio of the cumulative gain (true positives captured) to the expected gain if no model is used. Based on the lift performance metric, the top 10% lift of the XGBoost model was 1.6, which was also slightly better than the other models. Figure 3 shows the curve of the lift of the top five factors of the XGBoost model.

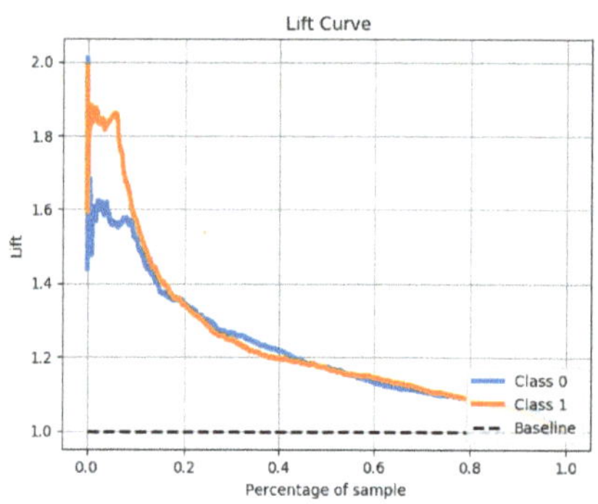

Figure 3. Lift of XGBoost Model.

5. Conclusions

Past research on direct marketing has mainly focused on the retailing sector and seldom discussed financial commodities in the banking industry. This research's findings provide information to help direct marketing decisions in the banking industry. Five factors influencing the model include a customer with an unsecured loan, the amount of assets that a customer holds in the bank, the duration of the customer's relationship with the bank, the customer's place of residence, and whether the bank is the primary institution for their total credit card bill payments over the past six months. These factors are important to formulate effective wealth management marketing strategies. The XGBoost algorithm's classification performance in judging whether a potential customer buys mutual funds is slightly better than that of the other models.

Funding: This research received no external funding.

Institutional Review Board Statement: Not applicable.

Informed Consent Statement: Not applicable.

Data Availability Statement: Restrictions apply to the availability of these data. Data were obtained from the case bank and are available from the author with the permission of the case bank.

Acknowledgments: The author appreciate the case bank for providing data and domain expertise in this research.

Conflicts of Interest: The author declares no conflicts of interest.

References

1. WLIST.pdf. Available online: https://www.cbc.gov.tw/public/data/EBOOKXLS/WLIST.pdf (accessed on 29 July 2024).
2. Olson, D.L.; Chae, B. Direct marketing decision support through predictive customer response modeling. *Decis. Support Syst.* **2012**, *54*, 443–451. [CrossRef]
3. Shih, J.-Y.; Chen, W.-H.; Chang, Y.-J. Developing target marketing models for personal loans. In Proceedings of the 2014 IEEE International Conference on Industrial Engineering and Engineering Management, Selangor, Malaysia, 9–12 December 2014; pp. 1347–1351. [CrossRef]
4. Chen, K. The effects of marketing on commercial banks' operating businesses and profitability: Evidence from US bank holding companies. *Int. J. Bank Mark.* **2020**, *38*, 1059–1079. [CrossRef]
5. Sontakke, A.; Yewale, M.; Zambare, S.; Tendulkar, S.; Chaudhari, A. Credit Card Fraud Detection Using Machine Learning and Predictive Models: A Comparative Study. In *Hybrid Intelligent Systems*; Abraham, A., Siarry, P., Piuri, V., Gandhi, N., Casalino, G., Castillo, O., Hung, P., Eds.; Springer International Publishing: Cham, Switzerland, 2022; pp. 171–180. [CrossRef]
6. Jabeur, S.B.; Stef, N.; Carmona, P. Bankruptcy Prediction using the XGBoost Algorithm and Variable Importance Feature Engineering. *Comput. Econ.* **2023**, *61*, 715–741. [CrossRef]
7. Carmona, P.; Climent, F.; Momparler, A. Predicting failure in the U.S. banking industry: An extreme gradient boosting approach. *Int. Rev. Econ. Financ.* **2019**, *61*, 304–323. [CrossRef]
8. Wang, K.; Li, M.; Cheng, J.; Zhou, X.; Li, G. Research on personal credit risk evaluation based on XGBoost. *Procedia Comput. Sci.* **2022**, *199*, 1128–1135. [CrossRef]
9. Ling, C.X.; Li, C. Data Mining for Direct Marketing: Problems and Solutions. In Proceedings of the Fourth International Conference on Knowledge Discovery and Data Mining, New York, NY, USA, 27–31 August 1998.
10. Nazaritehrani, A.; Mashali, B. Development of E-banking channels and market share in developing countries. *Financ. Innov.* **2020**, *6*, 12. [CrossRef]
11. Notes on Parameter Tuning—XGBoost 2.1.1 Documentation. Available online: https://xgboost.readthedocs.io/en/stable/tutorials/param_tuning.html (accessed on 11 August 2024).

engineering proceedings

MDPI

Proceeding Paper

Deepening Mathematical Understanding Using Visualization and Interactive Learning for Deaf Students †

Stefanie Amiruzzaman [1], Md Amiruzzaman [2,*], Heena Begum [2], Deepshikha Bhati [3] and Tsung Heng Wu [3]

1 Department of Languages and Cultures, West Chester University, West Chester, PA 19383, USA; samiruzzaman@wcupa.edu
2 Department of Computer Science, West Chester University, West Chester, PA 19383, USA; hb965928@wcupa.edu
3 Department of Computer Science, Kent State University, Kent, OH 44242, USA; dbhati@e-kent.edu (D.B.); twu10@e-kent.edu (T.H.W.)
* Correspondence: mamiruzzaman@wcupa.edu
† Presented at the 2024 IEEE 7th International Conference on Knowledge Innovation and Invention, Nagoya, Japan, 16–18 August 2024.

Abstract: Learning mathematical concepts is challenging for several students. Paying attention to class lectures and following instructions for different steps to solve a problem are the keys to success in learning mathematical concepts. However, deaf and hard-of-hearing (DHH) students have challenges in focusing on their teacher's mouths to lipread or depend on an interpreter's sign language. Visualization is important in learning as it enhances attention and keeps students focused on a subject. Interactive learning or a hands-on approach can help learners engage in a topic and provide an opportunity to better understand a concept. Combining these two techniques (i.e., visualization and interactive learning), we present an interactive number line (INL) tool to help DHH students understand mathematical concepts such as mean, median, mode, and range. This tool has proven to be useful for visual learning and/or for new learners as it helps provide an activity-based learning environment and feedback.

Keywords: measures of central tendency; statistical concepts; manipulative; visualization; interactive

1. Introduction

Mathematical concepts are interesting; however, many students struggle to learn mathematical concepts [1]. For example, deaf and hard-of-hearing (DHH) students struggle to learn mathematical concepts as they cannot focus on class lectures and interpreters at the same time. Teachers use manipulatives to help students learn mathematical concepts, such as number lines and Cuisenaire rods. Manipulatives help to construct different concepts, such as representations and distributions. A number line is a representation of a mathematical concept used to help students visualize a mathematical concept that represents the concept, and it plays the role of a mathematical model. Mathematical models and manipulatives are helpful for assisting learners to grasp a concept and develop a deeper understanding [2–4]. A number line is a mathematical manipulative. Similar to other manipulatives, a student can use a number line to develop their understanding of different mathematical concepts such as range, mean, mode, and median.

In this study, we propose a web-based interactive number line incorporating visualization techniques. This will allow DHH students to visualize data and interact with a system to discover information about the data such as how the data are distributed and what the ranges of the data are.

Visualization is important in learning as it enhances attention and keeps a learner focused on a subject [5,6]. It is used to create imaginary scenarios within one's mind, and when used in the process of learning, it helps achieve results faster than traditional methods. Interactive learning or a hands-on approach can help learners engage in a topic and better understand a concept rather than depend on their memory [7,8]. In addition to that, interactive learning turns everything into a game and keeps it interesting for learners. Combining these two techniques, we developed an interactive number line tool to help DHH students understand mathematical concepts including mean, median, mode, and range. We explored how beneficial this tool is in helping to provide an activity-based learning environment. Existing studies have indicated that the use of number lines in statistics helps students understand different mathematical concepts [3,9]. Past research has indicated that number lines help students understand basic math concepts, such as range, mean, mode, and median [3,4]. Using the success of a number line in learning mathematics, Weng and Bouck [4] recommended using it as a tool to visualize and understand the magnitudes and directions of numbers (i.e., negative and positive). These studies have suggested that visualization and understanding the magnitudes of numbers help students understand different mathematical concepts. Past studies have also found that number lines are helpful for DHH students [2,10].

Previous studies have recommended the use of manipulatives as mathematical models or as tools to help students visualize mathematical operations [10,11]. For example, a number line manipulative has been used as a mathematical model [12]. Mathematical models and visualization tools help students understand the relationship between new and previously learned concepts and make learning meaningful [13]. Many educators use paper-based number lines in their classrooms to help students understand mathematical concepts, as they graphically represent numbers and their distributions [14]. Paper-based number lines are limited in size and availability as static manipulatives. On the other hand, a virtual number line manipulative can be used in versatile ways. Also, it is easy to manipulate based on contextual needs. Therefore, a web-based interactive number line (INL) is required to be developed so that students can use it to explore their data and understand different mathematical concepts.

Therefore, in this study, a visual analytic-based number line was developed for DHH students. A pilot study was conducted to understand how DHH students customize number lines based on their needs. By adjusting the start and end point and fraction sizes, levels of feedback and color choices could be made to collect performance data. The data were analyzed to help the students develop confidence and performance.

The contributions of this study are as follows (see Figure 1):

- A visual analytic-based number-line, where DHH students can obtain real-time feedback on their work, was developed.
- We developed the requirements of the system from a pilot study and refined them from a thorough study of the participants' needs.
- A customized system that allowed the DHH users to customize a number line based on their needs was created. The customizations included adjusting the start and end points and fraction sizes. Also, levels of feedback and color choices were shown.
- We collected the DHH students' performance data, clustered them into groups, and provided exercises that helped them to develop their confidence and performance. A prototype was developed (see Figure 2).

2. Related Work

Visualization and understanding the magnitudes of numbers can help students to understand the measures of central tendency (MOCT). Manipulatives are a tool to help students visualize mathematical operations [9,15]. Mathematical models and visualization help students understand the relationship between new and learned concepts [9]. Paper-based number lines are limited in size and availability. Virtual number line manipulatives are used in versatile ways. According to Santos and Cordes and Schindler et al. [16,17], DHH children are not exposed to sign language at an early age. Therefore, they fall behind in mathematics compared to their classmates. Due to limited/reduced language access, early number concepts and the ability to solve math problems are lacking.

Less exposure to learning opportunities for learning mathematical concepts in everyday activities and reduced quality of education for DHH students have been observed [17]. DHH students have idiosyncratic difficulties in basic mathematics processing, too. However, DHH students can express concepts visually, which is critical for learning mathematics [18]. The number line is used to visualize and understand the magnitude of the number and the directions of numbers (i.e., negative and positive) [19]. DHH students are visual learners because they depend on their visual senses to absorb information, which means the learning resources must be simple and equipped with visual representations [20]. Nunes [21] suggested that if teachers can find visual and spatial ways to teach mathematics concepts to DHH students, positive changes in average achievements in mathematics can be observed.

3. Method

We conducted a requirements analysis and assessed the design, development, and usability of the proposed visual analytic system (see Figure 1). Participants were invited to participate in the study. A prototype was developed to conduct a pilot study to refine the questionnaire and update the requirements for the visual analytic system (see Figure 2). The participants were presented with a series of basic mathematical problems, and they were asked to solve the problems. Those who were able to solve the problems were then asked to explain how they solved them and what approach they used. If participants were unable to solve the problems, then research staff helped them understand the related concepts. After that, the participants were asked to solve the problems again based on the knowledge they acquired.

In the first phase of this study, we conducted a literature review and studied existing applications related to number lines in developing mathematical knowledge. Also, we prepared an initial requirements analysis to develop a new interactive computer application. In the second phase, we refined the requirements and outlined the design of the proposed visual analytic system. The prototype was created to develop a visual analytics system and collect participants' data and analyze it using the performance metrics of finding solutions for the given problems. In the third phase, we incorporated a machine-learning-based method to cluster the participants based on their approaches to solving mathematical problems, as well as their performance. Based on these clusters, we designed appropriate instructions to improve their performance in understanding the concepts (see Figure 1).

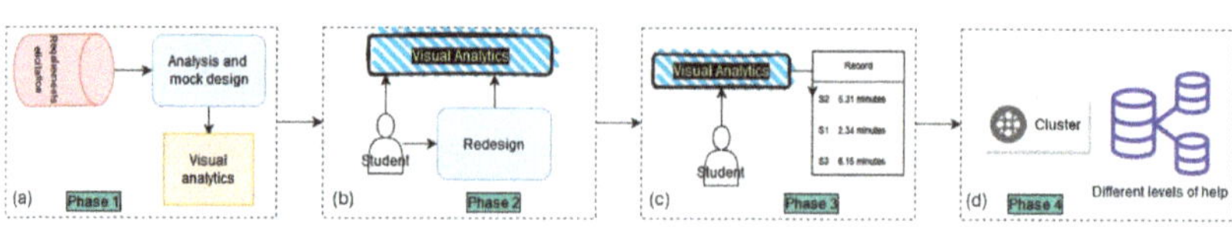

Figure 1. Framework of the study. The different phases of the study: (**a**) phase 1, analyzing the requirements and developing the analytics system; (**b**) phase 2, redesigning the analytics after receiving feedback from the pilot study; (**c**) phase 3, collecting the students' data; and (**d**) phase 4, dividing the students based on their performances.

4. Interface Development and Overview

Hyper-text markup language (HTML) and JavaScript were used to develop the interface. The HTML canvas element was employed to draw graphics dynamically using JavaScript. The interface consisted of the key components designed to enhance user interaction and facilitate the data analysis, as illustrated in Figure 2. The number line interface featured a horizontal number line ranging from 0 to 20, with circular markers indicating the selected data points. The pointer on the line, set at 8 in Figure 2 below, represented the selected value. The selected value displayed the numerical value at the pointer's position, which is 11.0 in the current example shown below.

Feedback messages assisted the users in adjusting the pointer, with five types of guidance employed as follows: (1) "You are going in the right direction"; (2) "You are almost there"; (3) "You have found the right answer"; (4) "Try going back in the opposite direction"; and (5) "You are going in the wrong direction". The data selection and sorting displayed the selected numbers and their sorted values. Finally, the statistical calculation and explanation offered options for calculating the key statistical measures such as mean, median, mode, and range, along with an explanation and the calculated value of the mean, which is 8.2 in Figure 2.

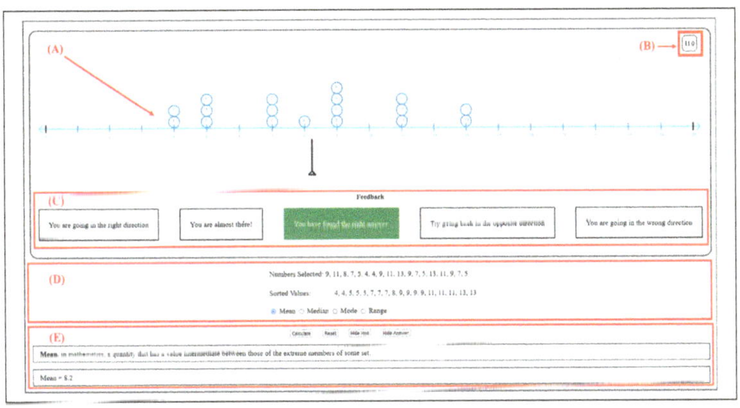

Figure 2. Interface of the application where students could practice their knowledge of measures of central tendency: (**A**) drawing the number line; (**B**) interval times; (**C**) feedback mechanisms; (**D**) statistical calculations; and (**E**) validation controls.

5. Drawing the Number Line

5.1. Creating a Blank Canvas

A blank canvas was created based on the heights and widths of the windows using the offsetHeight and offsetWidth properties of the canvas. This ensured that the canvas size was dynamically adjusted to fit the screen dimensions, providing a responsive user experience.

5.2. Drawing the Number Line on a Canvas

The number line was drawn on the canvas using a function, which set the context of the canvas instance. The beginPath function was then used to signal the start of a new path. To define the x and y coordinates for the line's start and end points, the moveTo and lineTo functions were employed. Finally, the stroke function was used to render the line on the canvas.

5.3. Intervals

To draw intervals on the number line, the interval size was calculated based on the start and end points of the number line. The horizontal line was traversed, and vertical lines were drawn at each calculated interval point using a combination of the moveTo, lineTo, and stroke functions.

6. User Interaction

6.1. Entering Values

The users entered values into the dataset by clicking on the numbers denoting each vertical line on the number line. When a number was clicked, a circle was drawn at the corresponding point using the beginPath, arc, and stroke functions with the canvas context reference.

6.2. Movable Pointer

The canvas contained a movable pointer that moved horizontally along the number line in both forward and backward directions. To create the movable pointer, the with function was called again, passing the canvas context. The beginPath function was used to start the drawing, followed by defining the x and y coordinates with moveTo and lineTo. The stroke function was then used to render the pointer. The canvas was redrawn every 10 ms using the setInterval function, which called the draw function at each interval. This ensured smooth pointer movement.

7. Feedback Mechanism

Below the canvas, there was a feedback section that provided messages to users when they attempted to predict the correct answer, as shown in Figure 2. As the users moved the movable pointer along the number line, the feedback section would display the following messages:

- **'You are going in the right direction'**. This message indicated that the user was moving the pointer correctly towards the answer, encouraging them to continue.
- **'You are almost there'**. This message was displayed when the user was very close to the correct answer, motivating them to make small adjustments to reach the exact value.
- **'You have found the right answer'**. This feedback provided the following options: (a) mean, (b) median, (c) mode, and (d) range. The definitions of these concepts are presented below.
- **'Try going back in the opposite direction'**. This message suggested that the user had given a wrong answer and needed to reverse direction to find the correct answer.
- **'You are going in the wrong direction'**. This feedback alerted the user that they were moving away from the correct answer, prompting them to reconsider their direction and adjust accordingly.

8. Dataset Management

A user would select a number on the number line. These fields maintained the same dataset, as follows: one list maintained the order in which the user entered the numbers, and it was editable for entering or removing numbers manually, while the other was a sorted list of all entries. The number of entries in both lists was the same.

Selecting Calculation Type

Once a user had entered all the numbers, they were required to select the type of calculation to perform. The application provided the following options: (a) mean, (b) median, (c) mode, and (d) range. The definitions of these concepts are presented as follows:

- A mean in mathematical quantity that has a value that is intermediate between the extreme members of a set of numbers, as follows:

$$\overline{x} = \frac{\sum_{i=1}^{n} x_i}{n} \tag{1}$$

where, $x_1, x_2, \ldots, x_n \in X$ are the numbers and n is the number of items.

- A median is the middle number in a sorted list of numbers. To determine the median value in a sequence of numbers, the numbers must first be sorted, or arranged, in value order from lowest to highest or highest to lowest, which is calculated as follows:

$$\tilde{x} = \begin{cases} X\left[\frac{n}{2}\right] \\ \frac{X\left[\frac{n-1}{2}\right]+X\left[\frac{n+1}{2}\right]}{2} \end{cases} \begin{pmatrix} \text{if } n \text{ is even} \\ \text{if } n \text{ is odd} \end{pmatrix}, \quad (2)$$

where X = a is the ordered list of values in dataset (i.e., $x_{(1)}, x_{(1)}, \ldots, x_{(n)}$) and n is the number of values in the dataset ($x_{(1)}$ = min(X), and $x_{(n)}$ = max(X)).

- A mode is the value that appears most frequently in a dataset. A set of data may have one mode, more than one mode, or no mode at all, and it is calculated as follows:

$$\hat{x} = \left(\frac{f_m - f_1}{(f_m - f_1) + (f_m - f_2)}\right), \quad (3)$$

where L is the lower limit of the modal class, h is the size of the class interval, fm is the frequency of the modal class, f_1 is the frequency of the class preceding the modal class, and f_2 is the frequency of the class succeeding the modal class.

- A range is the difference between the highest and lowest values in a set of numbers. To find it, the lowest number in the distribution is subtracted from the highest, as follows:

$$r = x_{(n)} - x_{(1)} \quad (4)$$

where r is the range and $x_{(1)}, x_{(2)}, \ldots, x_{(n)} \in X$ are the ordered values.

9. Application

Figure 3 shows the predictive analysis of the time taken to solve the problems by applying K-means clustering, with $k = 2$. The x-axis denotes the number of problems, while the y-axis represents the time taken to solve them in seconds. Each data point on the plot signifies an observed scenario. Two distinct clusters of data points are visualized, and they are distinguished by different colors (blue and red) and letter labels ("X"). These clusters represent the different prediction scenarios or outcomes derived from the K-means algorithm (see Figure 3). The modifications applied to the y-coordinates of the data points indicate the varying predictions or adjustments made during the analysis. This scatter plot serves as a visual aid to understand the relationship between the number of problems and the corresponding time taken for their resolution, providing insights into predictive modeling outcomes in problem-solving scenarios (see Figure 3).

When the user selected one of these calculation types and clicked on 'Calculate Answer', the answer was calculated and stored in a variable called 'finalAnswer'. Each time the user selected a new calculation type, they were required to click the 'Calculate Answer' button to indicate that the correct calculation type had been selected. After clicking Calculate Answer and dragging the movable pointer on the canvas, the feedback section became active and started providing feedback based on the pointer's position on the number line. The pointer's position on the number line was calculated by the myMove function. When the user clicked the mouse button, the onmousedown event was triggered within the canvas, calling the myDown custom function. This function first checked if the mouse click occurred on the movable pointer or the numbers present on the number line. If the click was on the movable pointer, a flag dragok was set to true, and the pointer position was set to the user's current position. Once these values were set, the myMove custom function was called. The myMove function checked if the mouse was still being dragged based on the dragok flag. Each time the mouse was dragged while holding the click button, the current position of the pointer was calculated using the pageX property of the mousemove event and the offsetLeft property of the canvas. Using the pointer's current position, the relative pointer value (RPV) on the number line was calculated and rounded to one decimal place. The formula for calculating the relative pointer value used in this application is as follows:

$$\text{RPV} = \frac{(\text{Current point ervalue} - 140)}{65} + 0.15 \tag{5}$$

The above equation provided us with a value that had up to 15 decimal places. To round this value to one decimal place, a fixed function argument 1 of JavaScript was used. The relative pointer value was compared against the final answer each time the mouse was clicked and dragged. Using this comparison, the application provided feedback to the user. The different categories of the feedback section were decided based on the normal distribution. The normal distribution equation was based on mean and standard deviation to quantify the characteristics of the given dataset.

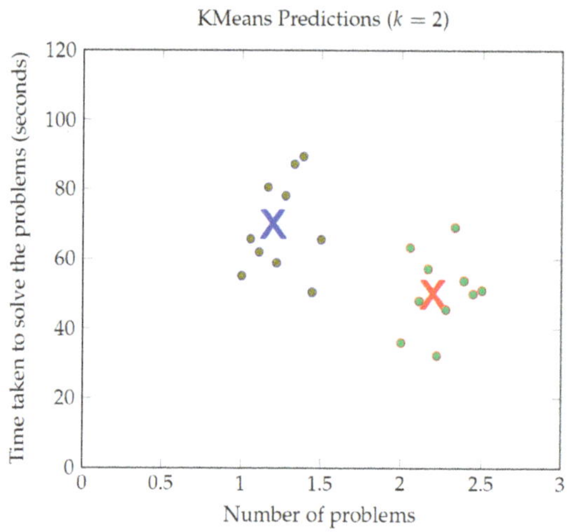

Figure 3. The K-means prediction results, with $k = 2$.

The scatter plot shows in Figure 3 the predictive analysis of the time taken to solve problems by applying K-Means clustering, with $k = 2$. The x-axis denotes the number of problems, while the y-axis represents the time taken to solve them in seconds. Each data point on the plot signifies an observed scenario. Two distinct clusters of data points are visualized, and they are distinguished by different colors (blue and red) and letter labels ("X"). These clusters represent the different prediction scenarios or outcomes derived from the K-means algorithm. The modifications applied to the *y*-coordinates of the data points indicate the varying predictions or adjustments made during the analysis. This scatter plot serves as a visual aid to understand the relationship between the number of problems and the corresponding time taken for their resolution, providing insights into predictive modeling outcomes in problem-solving scenarios.

When a user selected one of these calculation types and clicked on Calculate Answer, the answer was calculated and stored in a variable called *finalAnswer*. Each time a user selected a new calculation type, they were required to click the Calculate Answer button to indicate that the correct calculation type had been selected.

After clicking Calculate Answer and dragging the movable pointer on the canvas, the feedback section became active and started providing feedback based on the pointer's position on the number line. The pointer's position on the number line was calculated by the *myMove* function. The standard deviation indicated the "spread" or variations in the data points around that mean value.

As can be seen from the graph in Figure 4, the standard deviation represented the following:

- 68.3% of the data values were within one standard deviation of the mean (−1 to +1)
- 95.4% of the data values were within two standard deviations of the mean (−2 to +2)
- 99.7% of the data values were within three standard deviations of the mean (−3 to +3)

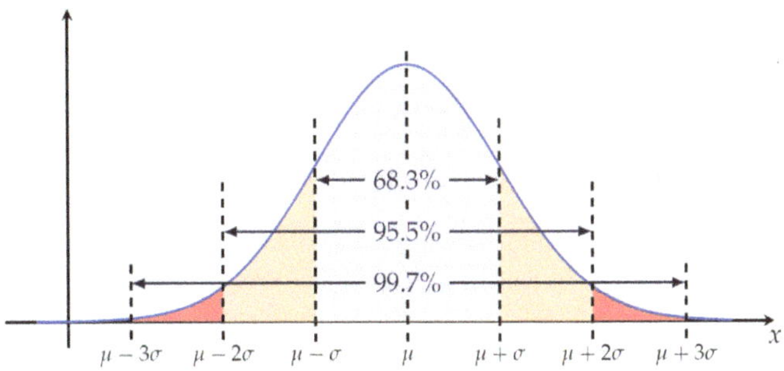

Figure 4. Normal curve of the distribution.

Using the above information, visual feedback was generated. If the RPV was within one standard deviation on the left, then we used "light green color", and if it was within two standard deviations, then we used "light light green color". Therefore, the idea was to show more of the lighter green color as the RPV moved further away from the mean value. If the RPV was exactly in the place where the value should have been, then the "green" color was used. The opposite side was used, with "light red" and "light light red" colors used to show that the RPV was moving away from the desired value.

10. Validations and Controls

There were validations in place to verify whether the numbers were entered and a calculation type had been selected. When the user clicked the Calculate Answer button, these checks ensured that the values are selected and a calculation type had been set, facilitating a smooth learning process. Along with the Calculate Answer button, a Reset button was created to clear all fields and the canvas. All entries made in the canvas were removed, and a fresh number line was drawn again on the screen. All respective fields for calculation type and the sorted list of numbers would be cleared and set to null.

When the entered value matched the final answer precisely, the message "You have found the right answer" was highlighted. If the pointer was within one standard deviation to the left of the final answer, "You are almost there!" was highlighted, and similarly, if it was within one standard deviation to the right, "Try going back in the opposite direction" was highlighted. For pointer values within two standard deviations to the left and right of the final answer, the messages "You are going in the right direction" and "You are going in the wrong direction" were highlighted, respectively.

For new learners, Show Hint and Reveal Answer buttons were provided. When the Show Hint button was clicked, a hint box appeared containing the hint or meaning of the calculation type to help the user. The box was closed by clicking the Hide Hint button in place of the Show Hint button. Similarly, the Reveal Answer button showed the final answer in the answer box if the user was unable to guess the correct answer. Once clicked, the answer box would show up, and it could be hidden again by clicking the Hide Answer button.

11. Conclusions

We incorporated modern technological ideas to develop an interactive visual number line. The development of this number-line visualization tool integrated HTML, JavaScript, and the HTML canvas element to create an interactive and educational interface. The validation mechanisms, user controls, and feedback systems ensured a robust and user-friendly experience, aiding learners in understanding and performing statistical calculations. This tool is useful for visual and/or new learners as it helps to provide an activity-based learning environment and feedback. The tool is available for learners to track their learning process based on accuracy and the time taken to predict the correct answer. New features need to be added to enable users to learn other mathematical concepts.

Author Contributions: Conceptualization, S.A. and M.A.; methodology, S.A. and M.A.; software, T.H.W.; validation, S.A., M.A. and D.B.; formal analysis, S.A., M.A. and D.B.; investigation, S.A. and M.A.; resources, S.A., M.A. and D.B.; data curation, S.A., M.A. and D.B.; writing—original draft preparation, S.A., M.A. and D.B.; writing—review and editing, S.A., M.A. and D.B.; visualization, S.A., M.A., H.B. and T.H.W.; supervision, S.A. and M.A.; project administration, S.A. and M.A.; funding acquisition, S.A., M.A. and D.B. All authors have read and agreed to the published version of the manuscript.

Funding: This work was partially supported by funds from the Professional Activities Advisory Committee (PAAC) at Kent State University.

Data Availability Statement: Data are contained within the article.

Conflicts of Interest: The authors declare no conflict of interest.

References

1. Xin, Y.P.; Kim, S.J.; Lei, Q.; Liu, B.Y.; Wei, S.; Kastberg, S.E.; Chen, Y.V. The effect of model-based problem solving on the performance of students who are struggling in mathematics. *J. Spec. Educ.* **2023**, *57*, 181–192. [CrossRef]
2. Gonsalves, N.; Krawec, J. Using number lines to solve math word problems: A strategy for students with learning disabilities. *Learn. Disabil. Res. Pract.* **2014**, *29*, 160–170. [CrossRef]
3. Sidney, P.G.; Thompson, C.A.; Rivera, F.D. Number lines, but not area models, support children's accuracy and conceptual models of fraction division. *Contemp. Educ. Psychol.* **2019**, *58*, 288–298. [CrossRef]
4. Weng, P.L.; Bouck, E.C. An evaluation of app-based and paper-based number lines for teaching number comparison. *Educ. Train. Autism Dev. Disabil.* **2016**, *51*, 27–40.
5. Contero, M.; Naya, F.; Company, P.; Saorin, J.L.; Conesa, J. Improving visualization skills in engineering education. *IEEE Comput. Graph. Appl.* **2005**, *25*, 24–31. [CrossRef] [PubMed]
6. Gilbert, J.K. Visualization: A metacognitive skill in science and science education. In *Visualization in Science Education*; Springer: Berlin/Heidelberg, Germany, 2005; pp. 9–27.
7. Regan, M.; Sheppard, S. Interactive multimedia courseware and the hands-on learning experience: An assessment study. *J. Eng. Educ.* **1996**, *85*, 123–132. [CrossRef]
8. Lin, K.Y.; Williams, P.J. Two-stage hands-on technology activity to develop preservice teachers' competency in applying science and mathematics concepts. *Int. J. Technol. Des. Educ.* **2017**, *27*, 89–105. [CrossRef]
9. Amiruzzaman, M. Exploring Preservice Teachers' Understanding of Measures of Central Tendency. Ph.D. Thesis, Kent State University, Kent, OH, USA, 2016.
10. Long, H.M.; Bouck, E.C.; Kelly, H. An evidence-based practice synthesis of virtual manipulatives for students with asd and idd. *Focus Autism Other Dev. Disabil.* **2022**, *38*, 10883576221121654. [CrossRef]
11. Bartolini, M.G.; Martignone, F. Manipulatives in mathematics education. In *Encyclopedia of Mathematics Education*; Springer: Cham, Switzerland, 2020; pp. 487–494.
12. Ahrendt, S.; Monson, D.; Cramer, K. Promoting discourse: Fractions on number lines. *Math. Teach. Learn. Teach. PK-12* **2021**, *114*, 284–289.
13. Burkhardt, H. Modelling in Mathematics Classrooms: Reflections on past developments and the future. *ZDM* **2006**, *38*, 178–195. [CrossRef]
14. Bouck, E.C.; Long, H.; Park, J. Using a virtual number line and corrective feedback to teach addition of integers to middle school students with developmental disabilities. *J. Dev. Phys. Disabil.* **2021**, *33*, 99–116. [CrossRef]

15. Carraher, D.W.; Schliemann, A.D.; Brizuela, B.M.; Earnest, D. Arithmetic and algebra in early mathematics education. *J. Res. Math. Educ.* **2006**, *37*, 87–115.
16. Santos, S.; Cordes, S. Math abilities in deaf and hard of hearing children: The role of language in developing number concepts. *Psychol. Rev.* **2022**, *129*, 199. [CrossRef] [PubMed]
17. Schindler, M.; Doderer, J.H.; Simon, A.L.; Schaffernicht, E.; Lilienthal, A.J.; Schafer, K. Small number enumeration processes of deaf or hard-of-hearing students: A study using eye tracking and artificial intelligence. *Front. Psychol.* **2022**, *13*, 909775. [CrossRef] [PubMed]
18. Adler, H.; Jacob, B.; Kurz, K.; Kushalnagar, R. Undergraduate research in mathematics with deaf and hard-of-hearing students: Four perspectives. *Involv. J. Math.* **2014**, *7*, 247–264. [CrossRef]
19. Bakker, A.; Gravemeijer, K.P. Learning to reason about distribution. In *The Challenge of Developing Statistical Literacy, Reasoning and Thinking*; Springer: Berlin/Heidelberg, Germany, 2004; pp. 147–168.
20. Suarsana, I.M.; Mahayukti, G.A.; Sudarma, I.K.; Yoga, I.N.B.A. Development of interactive mathematics learning media on statistics topic for hearing-impaired student. *Int. Res. J. Eng. IT Sci. Res.* **2018**, *4*, 55–66. [CrossRef]
21. Nunes, T. Deaf children, special needs, and mathematics learning. In *Encyclopedia of Mathematics Education*; Springer: Cham, Switzerland, 2020; pp. 181–183.

engineering proceedings

MDPI

Proceeding Paper

Algorithm Design for Multicellular Molecular Communication Simulations †

Shohei Imanaka and Tadashi Nakano *

Graduate School of Informatics, Osaka Metropolitan University, Osaka 558-8585, Japan
* Correspondence: tnakano@omu.ac.jp

† Presented at the 2024 IEEE 7th International Conference on Knowledge Innovation and Invention, Nagoya, Japan, 16–18 August 2024.

Abstract: Computing interactions among cells is computationally complex in multicellular molecular communication simulations, making the design of efficient algorithms crucial. We review simulation algorithms for computing interactions among cells in multicellular molecular communication simulations. The naïve algorithm, the Cell List algorithm, the Barnes–Hut algorithm, and a hybrid of the Cell List and Barnes–Hut algorithms are explored.

Keywords: molecular communication; bio-nanomachine; simulation algorithm; Barnes–Hut algorithm; cell list algorithm

1. Introduction

Molecular communication is an emerging communication paradigm for bio-nanomachines (e.g., naturally occurring and artificially synthesized cells) [1–3]. In molecular communication, bio-nanomachines act as communication nodes and exchange information through biochemical and biophysical interactions. This form of communication is highly parallelizable, biocompatible, and energy-efficient. Applications of molecular communication and bio-nanomachines encompass in-body molecular sensing, targeted drug delivery, biological robotics, and 6G wireless molecular communication [4,5].

Computer simulation is promising in advancing research in molecular communication [6–8]. While wet laboratory experiments are essential, they can still be costly. Thus, simulation experiments, which can be conducted on personal computers, serve as an alternative approach. In previous work, we designed and implemented a multicellular molecular communication simulator [9,10]. The simulator was designed to be extensible, allowing users to design, model, and simulate their systems, such as cell-based drug delivery systems. Within the simulator, bio-nanomachines can be engineered from biological cells and exhibit cellular functions such as sensing the environment, sending and receiving molecules, migrating, dividing, and dying.

This article provides a comprehensive overview of algorithms for multicellular molecular communication simulations. In multicellular molecular communication simulations, computing interactions among cells is expensive, making the design of efficient algorithms crucial. Therefore, we review four algorithms, the naïve algorithm, the Cell List algorithm, the Barnes–Hut algorithm, and a hybrid of the Cell List and Barnes–Hut algorithms, as they vary in cost and accuracy depending on user needs. The Cell List algorithm is proposed in Ref. [9], the Barnes–Hut algorithm is proposed in Ref. [10], and a hybrid algorithm is designed in this study to simulate based on the Barnes–Hut algorithm.

The remaining sections of this article are organized as follows. Section 2 provides an overview of simulation algorithms, including the procedure to compute cell–cell interactions. We then provide detailed descriptions of the four algorithms: the naïve algorithm in Section 3, the Cell List algorithm in Section 4, the Barnes–Hut algorithm in Section 5, and a hybrid of the Cell List and Barnes–Hut algorithms in Section 6. Finally, in Section 7, we discuss future challenges and conclude this study.

2. Simulation Algorithms

In simulations, a two-dimensional simulation space is considered where the set N of cells (or bio-nanomachines) interact through molecular communication and move collectively. Time is denoted as t, and it discretely advances by Δt. For simplicity, cells do not divide, and the set N is time-invariant.

In simulations, we assume that cells receive a force as a result of interactions with other cells. The magnitude of the force depends on cell–cell distance. Cells move based on the force in the simulation space; the velocity of a cell increases linearly as the force acting on it increases, with α being the proportionality constant (i.e., we neglect inertia).

Let $\boldsymbol{x}_i(t)$ be the position of cell i at time t, $\boldsymbol{v}_i(t)$ be the velocity of cell i at time t, $\boldsymbol{f}_i(t)$ be the force acting on cell i at time t, $\boldsymbol{f}_{ij}(t)$ be the force acting on cell i from cell j at time t, $f(d)$ be the magnitude of the force between cells separated by a distance d, and d_{ij} be the distance between cell i and cell j at time t. Note that we use boldface to express a vector.

Under such assumptions, the position $\boldsymbol{x}_i(t+\Delta t)$ of cell i at time $t+\Delta t$ is given by the following equations:

$$
\begin{aligned}
&\boldsymbol{x}_i(t+\Delta t) = \boldsymbol{x}_i(t) + \boldsymbol{v}_i(t)\Delta t, \\
&\boldsymbol{v}_i(t) = \frac{1}{\alpha}\boldsymbol{f}_i(t), \\
&\boldsymbol{f}_i(t) = \textstyle\sum_{j\in\mathcal{N},\, j\neq i}\boldsymbol{f}_{ij}(t), \\
&\boldsymbol{f}_{ij}(t) = \frac{\boldsymbol{x}_j(t) - \boldsymbol{x}_i(t)}{d_{ij}(t)} f(d_{ij}(t)), \\
&d_{ij}(t) = \left|\boldsymbol{x}_j(t) - \boldsymbol{x}_i(t)\right|.
\end{aligned}
$$

Algorithm 1 outlines the simulation step. After initialization, it contains two for-loops: one for calculating interactions between cells (lines 2–4) and the other for updating cell positions (lines 5–7). In the first for-loop, we iteratively compute the forces acting on the cells. The 'calcForce()' function depends on the algorithm being used. In the second for-loop, we iteratively update the positions of all cells.

In the following, we describe four algorithms to compute 'calcForce()' in Algorithm 1: the naïve algorithm, the Cell List algorithm, the Barnes–Hut algorithm, and a hybrid of the Cell List and Barnes–Hut algorithms.

Algorithm 1 Simulation step

1: Initialization
2: **for** cell $i \in N$ **do**
3: $\boldsymbol{f}_i$:= calcForce(cell i)
4: **end for**
5: **for** cell $i \in N$ **do**
6: Update i's position using $\boldsymbol{f}_i$
7: **end for**

3. Naïve Algorithm

The naïve algorithm is the simplest algorithm for calculating the force acting on a cell by computing the interactions with all other cells. Figure 1 illustrates the computational process of the algorithm. In the figure, cells are represented as circles, and the cells exerting force on the red cell are depicted with arrows connecting them. While the naïve algorithm is the most accurate, its computational cost increases sharply as the number of cells increases.

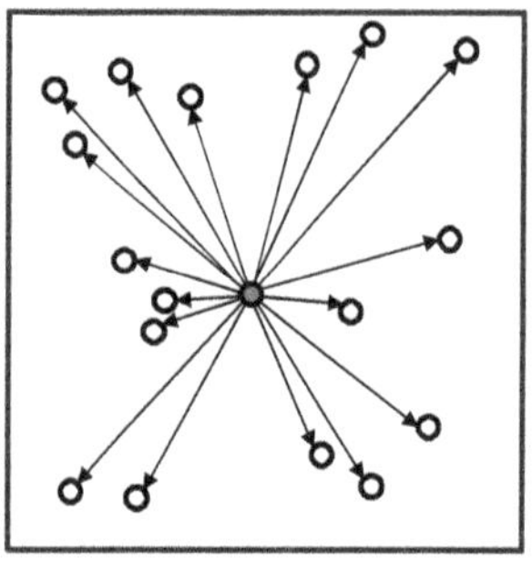

Figure 1. Naïve algorithm.

In the naïve algorithm, the initialization process (line 1) in Algorithm 1 is unnecessary. Algorithm 2 presents the pseudocode for the 'calcForce()' function in the naïve algorithm. Since we calculate the force acting on cell i from cell j $|N|$ times for both cells i and j, the time complexity of this algorithm is $O(|N|^2)$.

Algorithm 2 calcForce(cell i) in the naïve algorithm

for cell $i \in N$ **do**
    $f_i += f_{ij}$
end for

4. Cell List Algorithm

The Cell List algorithm accelerates simulations by limiting the number of cells that exert force on a cell. The Cell List algorithm is an accelerated algorithm that reduces computation time by limiting the calculation to nearby cells when determining the force acting on a cell. However, there is a trade-off between computation speed and accuracy, depending on the range considered. Figure 2 illustrates the computational process of the algorithm. In the Cell List algorithm, the simulation space is divided into grid units, with each grid maintaining the addresses of all cells within it. When calculating the force acting on the red cell, the algorithm searches for cells within a grid search range (yellow area) and then further limits the calculation to those cells within the red cell's search radius (blue circle). The grid search range determines how many grid cells from the grid containing the red cell are included in the search area. For example, when the grid search range is w (a non-negative integer), the search area forms a square of $(2w+1) \times (2w+1)$ grid cells. As shown in Figure 2, the grid search range is set to 1. The grid search range should be set to accommodate the search radius.

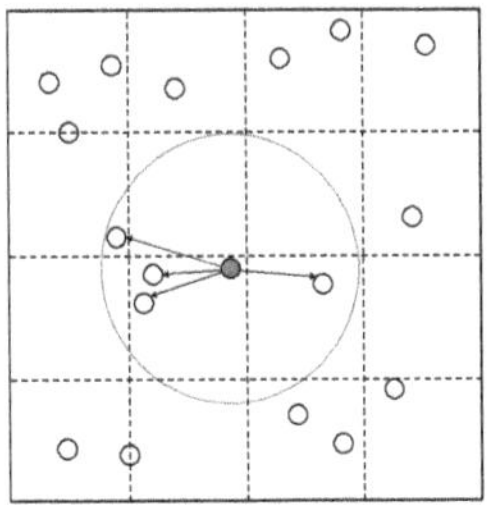

Figure 2. Cell List algorithm.

Algorithm 1 shows the Cell List algorithm, including the initialization of the cells within the grids (line 1). Algorithm 3 presents the pseudocode for the 'calcForce()' function of the Cell List algorithm. We obtain the list of surrounding cells to be calculated using the 'getCellList()' function and stored in the 'CellList' list (line 1). We then calculate the interaction only for the cells in the 'CellList' list (lines 2–4). Algorithm 4 presents the pseudocode for the 'getCellList()' function of the Cell List algorithm. Based on the address of the input cell, the corresponding grid is identified (line 5), and a list of cells belonging to the surrounding grids is created based on the grid search range (lines 6–8). From this list, only the cells within the search radius are selected for inclusion in a new list (lines 9–11) and returned to the caller (line 16). Each grid is assigned a grid number for identification. Here, columns are denoted as 'gridX' and rows are denoted as 'gridY', defining the grid as '(gridX, gridY)'.

Algorithm 3 calcForce(cell *i*) in the Cell List algorithm

1: cellList = getCellList(*i*)
2: **for** cell $j \in$ cellList **do**
3:     $f_i += f_{ij}$
4: **end for**

Algorithm 4 getCellList(cell *i*) in the Cell List algorithm

1: INPUT cell *i*
2: SR = search radius
3: SW = search width
4: cellList = empty list
5: gridX, gridY = the grid containing cell *i*
6: **for** cell $j \in$ cellList **do**
7:   **for** y in [gridY − SW, gridY + SW] **do**
8:     **for** x in [gridX − SW, gridX + SW] **do**
9:     cellsInGrid = list of cells in grid (x, y)
10:       **for** j in cellsInGrid **do**
11:         **if** distance from cells i to $j \leq$ SR **then**
          add cell j to cellList
12:         **end if**
13:       **end for**
14:     **end for**
15:   **end for**
16: **return** cellList

Since the Cell List algorithm's processing speed and accuracy are influenced by parameters like search range and search radius, the time complexity can vary from $O(|N|)$ to $O(|N|^2)$.

5. Barnes–Hut Algorithm

The Barnes–Hut algorithm is commonly used for n-body problems in physics [11]. In our previous work, we applied it to multicellular molecular communication simulations [10]. Unlike the Cell List algorithm, the Barnes–Hut algorithm allows a cell to receive forces from all other cells in multicellular molecular communication simulations, therefore providing improved accuracy. In the Barnes–Hut algorithm, distant cells were grouped and treated as a unified cell located at the center of mass of the cell group. The Barnes–Hut algorithm consists of the following three steps: Constructing the Barnes–Hut tree, computing centers of mass, and calculating cell–cell interaction forces.

5.1. Construction of the Barnes–Hut Tree

In the Barnes–Hut algorithm, the simulation space is divided into grids based on the distribution of cells and managed as a tree structure. This approach shares similarities with the Cell List algorithm, but the number and size of the grids in the Barnes–Hut algorithm are dynamic and updated at each step. Figure 3a illustrates how the simulation space is divided into grids using the Barnes–Hut algorithm. The grid division process continues until the number of cells in a grid is less than one, as depicted in Figure 3a,b. The figures also present the tree structure referred to as the Barnes–Hut tree, with nodes corresponding to these grids.

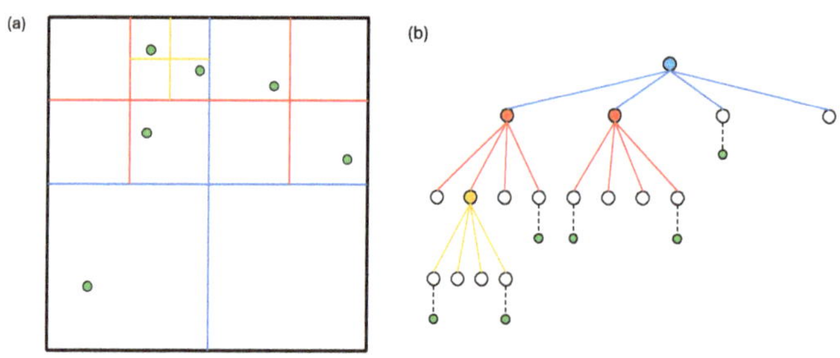

Figure 3. (**a**) Example simulation space divided into grids; (**b**) quad-tree representation where nodes correspond to the grids in (**a**).

In general, for an n-dimensional space, the partitioning is 2^n. Since we focus on a two-dimensional space, it is divided into four (2^2) partitions, resulting in a quad-tree. Hence, we consider a quad-tree in the rest of this section.

5.2. Computation of Centers of Mass

After constructing the quad-tree, the center of mass for each node is calculated. Each node corresponds to a grid in the simulation space, and the center of mass is determined based on the positions of all cells within that grid. As shown in Figure 3b, leaf nodes either have no cells or contain only one cell. In the former case, nodes without cells are not assigned a center of mass (or the mass of the center of mass was set to 0). In the latter case, nodes containing a single cell had their cell positions as their center of mass.

5.3. Computation of Cell–Cell Interaction Forces

In the Barnes–Hut algorithm, distant groups of cells are calculated using centers of mass, while calculations between cells that are not part of such groups are performed directly between individual cells. A group of cells in this context refers to all the cells contained within a specific grid. Additionally, the center-of-mass points associated with each grid are used to determine distances. In the algorithm, nodes in the quad-tree are traversed in a depth-first manner starting from the root node, and calculations are performed using the center of mass associated with the corresponding distant grid node. Since the cells within the grid are collectively accounted for by the center of mass, there is no need to search for subsequent descendant nodes during this process. For leaf nodes containing cells, the center of mass is also utilized in the calculations. As mentioned earlier, when a node has only one cell, the cell position is treated as a center of mass.

In the algorithm, we use the indicator θ given below to determine whether a node is sufficiently distant from the cell:

$$\theta > \frac{d}{r}, \tag{1}$$

where θ is a threshold, d is the grid size, and r is the distance from the cell to the node (i.e., the center of mass). When the given expression evaluates to true, the force is computed by utilizing the center of mass. Algorithm 5 presents the pseudocode for the 'getCellList()' function of the Barnes Hut algorithm.

Algorithm 5 calcForce(cell i, node n) in the Barnes Hut algorithm

```
Force f = (0, 0)
if node n is a leaf node containing one cell then
    f = force that node n's center of mass exerts on cell i
else
    r = distance from cell i to node n's center of mass
    d = n's grid size
    if d/r < θ then
    f = force that node n's center of mass exerts on cell i
    else
        for child node m ∈ n do
            f+ = calcForce(i, m)
        end for
    end if
end if
return f
```

6. Hybrid Algorithm

We designed a hybrid algorithm combining the Barnes–Hut algorithm and the Cell List algorithm to further accelerate the original Barnes–Hut algorithm. The original Barnes–Hut algorithm computes interactions with all cells distributed in the simulation space. In cases where the impact of interactions decreases sharply with increasing distance between cells, the original Barnes–Hut algorithm becomes inefficient. In the hybrid algorithm, we introduce the search radius used in the Cell List algorithm and exclude cells beyond a certain distance from the computation of cell–cell interactions.

The hybrid algorithm consists of the following four steps.

- Construction of the tree structure
- Calculation of centers of mass
- Coloring of the tree
- Calculation of interactions using the colored tree

The first two steps are the same as those in the original Barnes–Hut algorithm, and the last two steps are different, whereby the hybrid algorithm creates and uses a colored tree. In the following, we describe the last two steps in detail.

6.1. Coloring Tree

After constructing a tree and calculating the centers of mass, the Hybrid algorithm colors the tree. Figure 4a shows cell–cell interactions considered in the original Barnes–Hut algorithm. As an example, Figure 4b shows those in the hybrid algorithm; cells within the grids and the search radius are considered. Figure 5 shows the colored tree to compute the cell–cell interactions shown in Figure 4b.

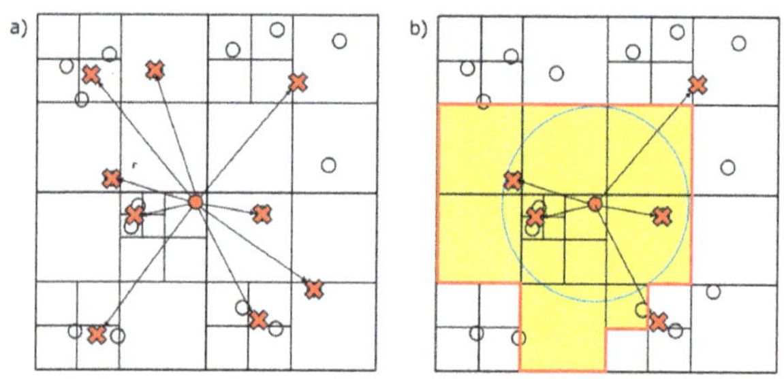

Figure 4. (**a**) Barnes–Hut algorithm and (**b**) hybrid algorithm.

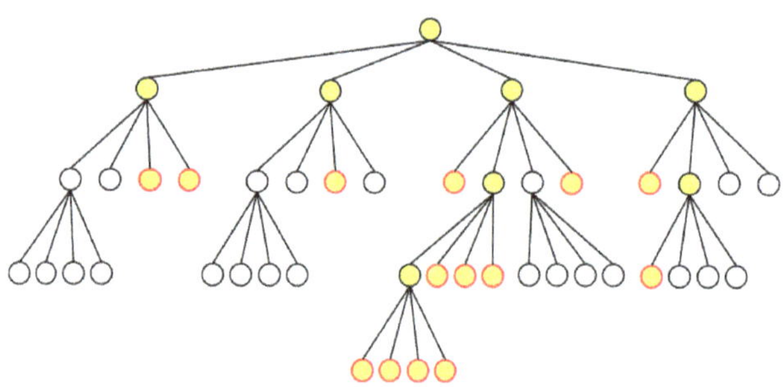

Figure 5. Colored quad-tree.

To color a tree, we use the functions 'colorTree()' and 'colorNode()' shown in Algorithms 6 and 7, respectively. By executing 'colorTree()' with cell i and the root node as arguments, grids within the search radius centered on cell i are searched. When a grid is found, 'colorNode()' is called to color the node. 'colorNode()' colors the node passed as an argument, and then recursively calls itself with the parent node as an argument, thereby coloring all nodes containing the leaf nodes.

Algorithm 6 colorTree(cell i, node n)

```
if node n is not empty then
    for child node m ∈ n do
        colorTree(i, m)
    end for
    if node n is a leaf node then
        if the grid corresponding to node n is within the search radius then
            colorNode(n)
        end if
    end if
end if
```

Algorithm 7 colorNode(node n) in Algorithm 6

```
if node n is not colored then
    color node n
    if node n has a parent then
        colorNode(parent of node n)
    end if
end if
```

6.2. Calculating Interactions Using Colored Tree

The basic calculation of interactions follows the Barnes–Hut algorithm in Algorithm 2. In exploring child nodes at line 11 of Algorithm 2, we check whether the node is colored, and continue to explore only if the node is colored. Specifically, we insert a conditional statement right before line 11 of Algorithm 2 to check if the child node to be explored is colored.

7. Conclusions

Computing interactions among cells is computationally expensive in multicellular molecular communication simulations, making the design of efficient algorithms crucial. In this article, we provide an overview of simulation algorithms for computing interactions among cells. We explain the naïve algorithm, the Cell List algorithm, and the Barnes–Hut algorithm. We also designed a hybrid of the Cell List and Barnes–Hut algorithms. Currently, we are developing a multicellular molecular communication simulator [9,10] in which the four algorithms are incorporated, enabling users to select the algorithm best suited to their simulations. In addition, we are implementing parallel computations using multicore computers to further accelerate molecular communication simulations.

Author Contributions: Conceptualization, S.I. and T.N.; methodology, S.I.; writing—original draft preparation, S.I. and T.N. All authors have read and agreed to the published version of the manuscript.

Funding: This research was supported by JSPS KAKENHI Grant Number JP21H04876.

Institutional Review Board Statement: Not applicable.

Informed Consent Statement: Not applicable.

Data Availability Statement: Data are contained within the article.

Conflicts of Interest: The authors declare no conflict of interest.

References

1. Nakano, T.; Eckford, A.; Haraguchi, T. *Molecular Communication*; Cambridge University Press: Cambridge, UK, 2024.
2. Farsad, N.; Yilmaz, H.B.; Eckford, A.; Chae, C.-B.; Guo, W. A comprehensive survey of recent advancements in molecular communication. *IEEE Commun. Surv. Tutor.* **2016**, *18*, 1887–1919. [CrossRef]
3. Bi, D.; Almpanis, A.; Noel, A.; Deng, Y.; Schober, R. A survey of molecular communication in cell biology: Establishing a new hierarchy for interdisciplinary applications. *IEEE Commun. Surv. Tutor.* **2021**, *23*, 1494–1545. [CrossRef]
4. Akyildiz, I.F.; Kak, A.; Nie, S. 6G and beyond: The future of wireless communications systems. *IEEE Access* **2020**, *8*, 133995–134030. [CrossRef]
5. Guo, W.; Abbaszadeh, M.; Lin, L.; Charmet, J.; Thomas, P.; Wei, Z.; Li, B.; Zhao, C. Molecular physical layer for 6G in wave-denied environments. *IEEE Commun. Mag.* **2021**, *59*, 33–39. [CrossRef]
6. Noel, A.; Cheung, K.C.; Schober, R.; Makrakis, D.; Hafid, A. Simulating with accord: Actor-based communication via reaction-diffusion. *Nano Commun. Netw.* **2017**, *11*, 44–75. [CrossRef]
7. Llatser, I.; Demiray, D.; Cabellos-Aparicio, A.; Altilar, D.T.; Alarcón, E. N3sim: Simulation framework for diffusion-based molecular communication nanonetworks. *Simul. Model. Pract. Theory* **2014**, *42*, 210–222. [CrossRef]
8. Yilmaz, H.B.; Chae, C.-B. Simulation study of molecular communication systems with an absorbing receiver: Modulation and isi mitigation techniques. *Simul. Model. Pract. Theory* **2014**, *49*, 136–150. [CrossRef]
9. Saiki, T.; Nakano, T. Design and implementation of a multicellular molecular communication simulator. In Proceedings of the 2022 Joint 12th International Conference on Soft Computing and Intelligent Systems and 23rd International Symposium on Advanced Intelligent Systems (SCISISIS), Ise, Japan, 29 November–2 December 2022.
10. Imanaka, S.; Saiki, T.; Nakano, T. Accelerating multicellular molecular communication simulations using the Barnes-Hut algorithm. In Proceedings of the 2024 IEEE 4th International Conference on Electronic Communications, Internet of Things and Big Data (ICIEB 2024), Taipei, Taiwan, 19–21 April 2024.
11. Barnes, J.; Hut, P. A hierarchical o(n log n) force-calculation algorithm. *Nature* **1986**, *324*, 446–449. [CrossRef]

engineering proceedings

MDPI

Proceeding Paper

Disaster Response System Dynamic Assessment Model Based on Questionnaire Investigation: A Case Study of Nuclear Accident Preparedness Education †

Ching-Yi Wu [1,2] and Yi-Lung Yeh [1,2,*]

1 Department of Civil Engineering, National Pingtung University of Science and Technology, Pingtung County 91201, Taiwan; fishpop571@gmail.com
2 Disaster Prevention and Mitigation Technology Research Center, General Research Service Center, National Pingtung University of Science and Technology, Pingtung County 91201, Taiwan
* Correspondence: yalung@mail.npust.edu.tw; Tel.: +886-8-7740425; Fax: +886-8-7740122
† Presented at the 2024 IEEE 7th International Conference on Knowledge Innovation and Invention, Nagoya, Japan, 16–18 August 2024.

Abstract: Based on the results of the nuclear accident response education questionnaire, we use system dynamics to develop a nuclear accident response education effectiveness evaluation model. A questionnaire survey was conducted for the pre- and post-tests of the course to understand course effectiveness and explore the interactive relationship between factors. The system dynamics is used to develop a nuclear accident response education effectiveness evaluation model. Age and education are significantly related to education and training experience (number of times), and the age, occupation, and education level are significantly related to individuals' experience (number of times) of participating in a nuclear accident evacuation exercise. In the workshop courses, the public's awareness of self-protection against nuclear accidents was significantly improved. Experience in educational training was significantly related to evacuation exercises. Individuals' age and experience of participating in education and training can be used to predict their willingness to participate in evacuation exercises. Using systems thinking and analysis, an evaluation model for nuclear incident response education effectiveness is constructed as a reference for evaluating effectiveness. The designed education and training courses can increase public participation in nuclear accident response education and strengthen the cognitive benefits of response protection.

Keywords: disaster risk management; emergency planning zone; nuclear accident; radiation protection; nuclear safety education; systems thinking and analysis

1. Introduction

According to the International Atomic Energy Agency (IAEA) "2022 Annual Report", there are 438 nuclear power plants in operation in 32 regions around the world [1]. The only operating nuclear power plant in Taiwan is in the decommissioning stage and will cease operation in 2025. However, the decommissioning process in nuclear power plants takes a long time, and it is necessary to strengthen people's concept of and their response to nuclear accidents continuously. The concept of disaster prevention in education is based on reducing the risk of disaster, improving preparedness, and strengthening recovery rapidly after the disaster.

Education is important as it can change people's conceptions and attitudes about the risks of nuclear power. People who are opposed to nuclear power misunderstand nuclear

 https://doi.org/10.3390/engproc2025089006

power [2]. Tsai et al. [3] pointed out that the application of gamified courses in higher education has received widespread attention. Compared with regular school courses, experiential learning helps participants reduce their feelings of rejection and boredom. The use of the game mode improves the learners' understanding of nuclear power [4]. Han et al. [5] also proposed that the attitudes of students towards nuclear power can be changed through education.

When a nuclear accident requires the evacuation of people, their degree of cooperation affects the smoothness of the evacuation operation. Seddighi et al. [6] mentioned that the capabilities developed by education can reduce the risk of disaster. Shaw et al. [7] believe that disaster prevention education includes the education of students in schools and the education of adults in the community. Ismail-Zadeh et al. [8] believe that one of the ways to reduce the risk of disaster is to convey knowledge about disaster response and its impact, such as disaster reduction and preparedness, to decision-makers and those directly exposed to risks.

Using interactive teaching materials, we develop a method to instruct public courses and survey the government's public protection plans for nuclear accidents before and after the course to evaluate the effectiveness of interactive nuclear safety education.

2. Materials and Methods

2.1. Questionnaire Design

A nuclear safety education and training questionnaire was created for the pre- and post-test surveys based on the course content and the public's understanding of it (Table 1).

Table 1. Nuclear safety education and training course content and teaching.

Item	Course Content	Comprehension Level of Nuclear Safety Education and Training Courses
1	Knowing the nuclear accident public warning system	Do you know the nuclear accident siren?
2	Planning the nuclear accident assemble location and functions	Do you know the public assembly location of nuclear accidents?
3	Nuclear accident evacuation bus and route planning	Do you know how to take the nuclear accident evacuation buses?
4	The plan for nuclear accident's public evacuation shelters	Do you know the plan of nuclear accident evacuation shelters for people?
5	Who and when to take iodine tablets in a nuclear accident	Do you know the standard for taking iodine tablets during a nuclear accident?

Those who participated in nuclear safety education and training in the emergency planning zone of the third nuclear power plant participated in this study. In total, 296 questionnaires were recovered, with 291 being valid questionnaires, resulting in a recovery rate of 98.3%.

2.2. Data Processing and Analysis

We used a *t*-test to compare the pre-and post-test results of the education and training courses. A paired samples *t*-test was conducted to compare the mean difference between two groups of dependent samples.

We collected the questionnaire results and conducted data processing and analysis using SPSS 22. The analysis methods included descriptive statistics, the chi-square test, correlation analysis, a paired samples *t*-test, and regression analysis. The significance level of each statistical test was set at $\alpha = 0.05$.

2.3. System Dynamics Assessment Model

System dynamics is based on system thinking. It is mainly used to understand and analyze dynamic systems composed of multiple interactions. It is a process-oriented research method. The system dynamics assessment model simulates and predicts the system behavior by building models. The construction of the model relies on Causal Loop Diagrams (CLDs) and Stock Flow Diagrams (SFDs). Causal loop diagrams are used to describe the interactions between variables in a system and their feedback loops. Feedback loops can be positive or negative, enhancing system behavior or weakening system behavior.

3. Results and Discussion

The demographic variables and past participation experience were analyzed using a chi-square test, bivariate correlation analysis, a nuclear safety education and training course effectiveness *t*-test, and stepwise multiple regression analysis.

3.1. Demographic Variables and Participation Experience Analysis

Of the 291 participants, 66.3% were females, 86.2% were middle-aged and elderly, 43.6% were homeowners, 20.6% worked as freelancers, and 57.7% had the education level of junior high school or below.

In total, 200 participants (68.6%) had experienced nuclear safety education and training; 46.4% had participated in nuclear safety evacuation exercises; and 53.6% had partaken in neither.

We conducted a chi-square test on past participation experience to verify whether different demographic variables are related to their participation behavior. Age ($x^2(12) = 28.432$, $p = 0.005 < 0.05$) and education level ($x^2(6) = 12.965$, $p = 0.044 < 0.05$) were correlated with "nuclear safety education and training".

Age ($x^2(12) = 28.336$, $p = 0.004 < 0.05$), occupation ($x^2(15) = 25.784$, $p = 0.040$), and education level ($x^2(6) = 14.794$, $p = 0.0022$) were correlated with the participants' experience participating in "Nuclear Safety Evacuation Exercise" (Table 2).

Table 2. Cross-contingency analysis table of past participation experience among nuclear safety education and training participants.

Variables	Nuclear Safety Education and Training Experience (Number of Times)	Variables	Nuclear Safety Evacuation Exercise Experience (Number of Times)
	x^2		x^2
Gender	1.555	Gender	7.120
Age	28.432 **	Age	28.336 **
occupation	13.853	occupation	25.784 *
Education level	14.794 *	Education level	14.794 *
Gender	1.555	Gender	7.120

* as $p < 0.05$, ** as $p < 0.01$.

Based on the participants' experience of undertaking nuclear safety education and training courses, the correlation between variables showed a significant correlation ($r = 0.683$, $p = 0.000 < 0.001$) between the participant's experience of undertaking nuclear safety education and training and nuclear safety evacuation exercises.

3.2. Bivariate Analysis of Training Benefits

The scores of the participants' questions before and after the nuclear safety education training were evaluated using the *t*-test. The averages of the two samples were 7.964 and 9.233, and the correlation coefficient between the samples was 0.406 ($p = 0.000 < 0.001$, reaching significance). In the paired analysis of the pre-test and post-test, $t(290) = -18.760$, $p = 0.000$, the test result was found to be significant. The results of the paired sample analysis before and after the nuclear safety education training are shown in Table 3.

Table 3. Paired sample analysis before and after nuclear safety education training.

Mean	Standard Deviation	Standard Error	95% CI	t-Value	Degree of Freedom	Significance
−1.540	1.400	0.082	[−1.701, −1.378]	−18.760	290	0.000 ***

*** as $p < 0.001$.

3.3. Prediction and Analysis of Nuclear Safety Evacuation Exercise

Stepwise regression analysis was used to load the predictor variables, check the changes in the explained regression coefficient R^2 of the predictor variables, and select the most appropriate model. Multi-collinearity was tested, and the tolerance and variation inflation factor (VIF) were used for observation. The tolerance was between 0.721 and 0.964, and the variation inflation factor was between 1.037 and 1.388. The tolerances of this model were higher than 0.2, and the VIF was less than 4, showing multivariate collinearity. The Durbin–Watson test value was 1.733, indicating that the residuals were independent with no autocorrelation.

Model 1 showed the prediction and analysis results of the demographic variables with regard to participation in nuclear safety evacuation exercises. The prediction variables were gender, age, occupation, and education level. Before performing regression analysis, the population background was normalized to meet the requirements of the regression analysis. The occupation category was divided into two categories: "Housekeeper or Retired" and "Employed". Age ($\beta = 0.225$, $t = 3.722$, $p = 0.000 < 0.001$) explained the variation in the participants' participation in nuclear safety evacuation exercises ($R^2 = 0.059$, $F(4,286) = 4.502$, $p = 0.000 < 0.001$).

Model 2 used past nuclear safety education and training experience as predictor variables. Age ($\beta = 0.099$, $t = 2.179$, $p = 0.030 < 0.05$) and past nuclear safety education experience ($\beta = 0.667$, $t = 15.436$, $p = 0.000 < 0.001$) explained the variation in the participants' participation in nuclear safety evacuation exercises ($R^2 = 0.488$, $f(5,285) = 54.242$, $p = 0.000 < 0.001$). The explanatory power of Model 2 was 48.8%. The analysis results are shown in Table 4.

Table 4. Regression analysis of demographic variables, nuclear safety education and training on nuclear safety evacuation exercise.

	Model 1			Model 2		
	B	SE B	β	B	SE B	β
Constant	1.045	0.274		0.189	0.210	
Gender	−0.091	0.122	−0.048	0.006	0.091	0.003
Age	0.185	0.050	0.225 ***	0.081	0.037	0.099 *
Occupation	0.068	0.122	0.037	−0.037	0.091	−0.020
Education level	−0.072	0.076	−0.057	−0.097	0.056	−0.077
Training experience				0.645	0.042	0.667 ***
R^2	0.059	0.488	0.059	0.488	0.059	0.488
Adj R^2	0.046	0.479	0.046	0.479	0.046	0.479
F for change in R^2	4.502 **	54.242 ***	4.502 **	54.242 ***	4.502 **	54.242 ***

* as $p < 0.05$, ** as $p < 0.01$, *** as $p < 0.001$.

3.4. System Dynamics Evaluation

A system dynamics model was used to develop an evaluation model of nuclear safety education effectiveness based on the five major themes of the public nuclear safety education content: public warning system, assemble locations, evacuation methods, evacuation planning, and the time at which iodine tablets should be taken. The factors related to public nuclear safety education were used to construct an institutional structure model for evaluating the effectiveness of nuclear safety education (Figure 1).

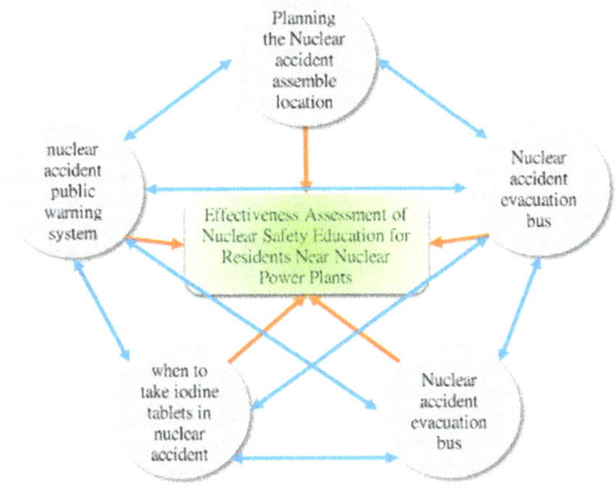

Figure 1. Model for effectiveness assessment of nuclear safety education using systems thinking

3.5. Discussion

The chi-square test was used to verify the correlation between the demographic variables and past participation in nuclear safety education and nuclear safety evacuation exercises. Age and education level were significantly related to the number of times the participants had participated in nuclear safety education and training. Age, occupation, and education level were significantly related to the number of times the participants' had participated in nuclear safety evacuation exercises.

Most of the participants in nuclear safety education and training were middle-aged/elderly people (over 51 years old), with education levels below junior high school (inclusive). Those who had participated in nuclear safety evacuation exercises were middle-aged/elderly people (over 51 years old) who were professionals or who had family members who were professionals. Most of them were managers, retired, or self-employed, and their education level was junior high school or below. Those who participated in nuclear safety education courses and nuclear safety evacuation exercises had less need to consider their employment status, so they were more willing to participate in the courses or exercises.

The post-test scores were higher than the pre-test scores, indicating that nuclear safety education and training courses led to a significant difference in people's understanding of self-protection against nuclear accidents. There was a significant correlation between the participants' experience participating in nuclear safety education and training and their experience participating in nuclear safety evacuation exercises ($r = 0.683$, $p = 0.000 < 0.001$, moderate correlation).

Multiple stepwise regression analysis was used to explore whether the demographic variables and the participants' past experience of nuclear safety education had an impact on their participation in nuclear safety evacuation exercises. Age ($\beta = 0.099$), past participation ($\beta = 0.667$), and the standardized regression coefficients were positive, showing a positive relationship. The older they were and the more experience they had, the more experience (willingness) they had in participating in nuclear safety evacuation exercises.

To understand the factors that affect the effectiveness of nuclear safety education and their correlations, an analysis framework was constructed to evaluate its effectiveness in improving disaster prevention for the public in the future.

4. Conclusions

Nuclear accidents are unfamiliar to the public in disaster prevention promotion education courses. Since the launch of the public protection course for nuclear accidents, the public's understanding of nuclear accident preparedness and response measures has gradually improved. However, when promoting courses about nuclear accident public protection in the community, demographic statistics must be considered. Therefore, the course design and production of teaching materials for disaster prevention education must be "life-oriented" and "interesting". A "content localization" curriculum needs to be offered to improve the goal of nuclear accident prevention education.

Author Contributions: Y.-L.Y.: conceptualization and formal analysis, Writing—review and editing, supervision, project administration, and funding acquisition. C.-Y.W.: methodology, software, investigation, data curation and writing—original draft preparation. All authors have read and agreed to the published version of the manuscript.

Funding: This research was funded by a research project of the Pingtung County Government in Taiwan. The project is titled 'Community-Based Public Awareness and Evacuation Drills within the Emergency Planning Zone for Nuclear Accidents'.

Institutional Review Board Statement: Not applicable.

Informed Consent Statement: Not applicable.

Data Availability Statement: The data presented in this study are available upon request from the corresponding author.

Conflicts of Interest: The authors declare no conflicts of interest. The founders assist in rice cultivation and field management.

References

1. International Atomic Energy Agency. *Nuclear Power Status Worldwide-2022*; IAEA Annual Report; IAEA: Vienna, Austria, 2022.
2. Yim, M.S.; Vaganov, P.A. Effects of education on nuclear risk perception and attitude: Theory. *Prog. Nucl. Energy* **2023**, *42*, 221–235. [CrossRef]
3. Tsai, M.H.; Chang, Y.L.; Shiau, J.S.; Wang, S.M. Exploring the effects of a serious game-based learning package for disaster prevention education: The case of battle of flooding protection. *Int. J. Disaster Risk Reduct.* **2020**, *43*, 101393. [CrossRef]
4. Huang, Y.P.; Wei, H.W.; Chen, T.Y.; Jhang, N.Y.; Chen, C.; Cheng, Y.C.; Yang, C.W.; Shih, W.K. The effectiveness of integrating somatosensory technology into nuclear energy education learning. *Procedia Soc. Behav. Sci.* **2015**, *176*, 476–482. [CrossRef]
5. Han, E.O.; Kim, J.R.; Choi, Y.S. Korean student' behavioral change toward nuclear power generation through education. *Nucl. Eng. Technol.* **2014**, *46*, 707–718. [CrossRef]
6. Seddighi, H.; Lopez, M.L.; Zwitter, A.; Muldoon, M.L.; Sajjadi, H.; Yousefzadeh, S. Non-formal disaster education programs for school students in Iran: A qualitative study of the challenges experienced by stakeholders. *Int. J. Disaster Risk Reduct.* **2023**, *86*, 103531. [CrossRef]
7. Shaw, R.; Mallick, F.; Takeuchi, Y. Essentials of higher education in disaster risk reduction: Prospects and challenges. *Community Environ. Disaster Risk Manag.* **2011**, *7*, 95–113.
8. Ismail-Zadeh, A.T.; Cutter, S.L.; Takeuchi, K.; Paton, D. Forging a paradigm shift in disaster science. *Nat. Hazards* **2017**, *86*, 969–988. [CrossRef]

engineering proceedings

MDPI

Proceeding Paper

Advanced Machine Learning for Preschooler Magnetic Resonance Imaging Analysis in Classification of Anxiety Disorders †

Salik Mian [1,*], Pranav Kunderu [2] and Shivm Patel [3]

1 Independent Researcher, Los Angeles, CA 91326, USA
2 Independent Researcher, Los Angeles, CA 91307, USA; pranavkunderu1209@gmail.com
3 Molecular & Cell Biology, University of California, Berkeley, CA 94720, USA; shivmpatel@berkeley.edu
* Correspondence: salikmian75@gmail.com

† Presented at the 2024 IEEE 7th International Conference on Knowledge Innovation and Invention, Nagoya, Japan, 16–18 August 2024.

Abstract: Numerous individuals suffer from anxiety disorder. Treatments for anxiety usually involve psychologists and counselors based on qualitative data from interviews and conversations to make an educated guess to classify their anxiety. We built a quantitative method for the diagnosis of anxiety, which can be used by psychologists and doctors to obtain accurate data to treat it. The data were obtained from OpenNeuro and Magnetic Resonance Imaging (MRI) scan images of preschoolers with different types of anxiety: generalized anxiety and separation anxiety. These data were used to train machine learning models: Support Vector Machines (SVMs), decision trees, and Logistic Regression. The dataset consisted of MRI images. The method was refined until the desired accuracy was obtained. The model can be used to diagnose anxiety disorders for patients to be treated with a personalized approach.

Keywords: MRI; anxiety disorder; support vector machines (SVMs); decision trees; logistic regression; transfer learning

1. Introduction

Magnetic Resonance Imaging (MRI) is a noninvasive imaging technique [1]. By utilizing nuclear magnetic resonance, it creates images of details of the inner structures of the body, including tissues, organs, and the nervous system. This technology uses the magnetic properties of hydrogen atoms in the body by aligning them in a strong magnetic field and then disrupting this alignment with radiofrequency pulses. The resulting signals are captured and retransformed into high-resolution images by sophisticated computer algorithms.

Generalized anxiety disorder (GAD) [2] is a psychiatric disorder characterized by excessive worry about various daily routine activities. Psychologically, patients may feel deep anxiety, periods of intense fear, and rumination. Behaviorally, hypervigilance and irritability are common and thus pose an inability to relax. Cognitive symptoms in the form of obsessive thoughts and concentration deficits are commonly described by patients. Other common complaints include sleeplessness, emotional distress, excessive worry, headaches, nausea, palpitations, and trembling. Effective treatment options include cognitive behavioral therapy, medication like SSRIs, and certain modifications to lifestyle, such as regular exercise and ample sleep. Separation anxiety [3] refers to a feeling of excessive fear or worry related to being separated from one or more attachment figures, usually parents or other primary caregivers. Separation anxiety is characterized by headaches, stomachaches, nausea, and palpitations physically and depression as well as intense fear, particularly

https://doi.org/10.3390/engproc2025089007

while contemplating or experiencing separation from attachment figures psychologically. Crying, nightmares, palpitations, and rapid breathing are the common physical and behavioral symptoms, adding to the general feeling of distress and poor coping related to separation.

Anxiety disorders [4] affect millions of people across the world. Anxiety disorders [5], from generalized anxiety to separation anxiety, each differently affect a person's life and brain.

Traditionally, in the diagnosis and characterization of an anxiety disorder, discussions with psychologists and counselors are used [6]. This method needs to be improved by using machine learning techniques and neuroimaging data in mental health diagnostics [7].

To create a machine learning model that analyzes the MRI scans of preschool students for the classification of anxiety disorder, we compared the performance of several models including Support Vector Machines (SVMs), decision trees, and Logistic Regression to improve diagnostic accuracy. Using the developed machine learning-based model, anxiety disorders can be diagnosed. The model was trained and evaluated using MRI scan data from OpenNeuro.

The model allows psychiatrists, mental health professionals, and pediatricians to diagnose faster and more personally and allows advanced neuroimaging and machine learning technologies to be used in mental health research.

2. Literature Review

MRI is a noninvasive diagnostic procedure that involves the use of strong magnetic fields and radio waves in the production of images without radiation [1]. However, it is hazardous for people who have medical implants or metallic objects in them. These risks include projectile effects, tissue heating, and the malfunction of devices. The absolute contraindications include cardiac implants and metallic intraocular foreign bodies, whereas relative contraindications such as coronary stents and joint replacements need to be treated cautiously. This entails special caution in patients with kidney problems, a high body mass index (BMI), claustrophobia, or pregnancy. This means that ensuring safety in using MRI involves careful patient screening and interprofessional collaboration.

Anxiety is associated with fear, hosting cognition and affective processes and physiological and behavioral responses. Pathological anxiety results from exaggerated perceived threats and excessive responses. It is a psychiatric disorder, resulting from gene–environment interactions and psychological factors induced by medication, substance abuse, and trauma. Manifestations include cognitive fears, physiological changes in the form of a racing heartbeat, and behavioral avoidance. The diagnosis of anxiety requires an assessment of symptoms ruling out other medical conditions. Management includes pharmacotherapy with Selective Serotonin Reuptake Inhibitors (SSRIs), Serotonin–Norepinephrine Reuptake Inhibitors (SNRIs), benzodiazepines, beta-blockers, and psychotherapy, particularly cognitive behavioral therapy, which manages anxiety through structured interventions and exposure therapy. Interprofessional coordination is important for better treatment.

The global and regional prevalence of anxiety disorders was observed from 1990 to 2019 [5]. In total, 4.05% of the world's population, or 301 million people, have anxiety disorders. Since 1990, the number has increased to 55%, with high-income regions reporting the highest rates, in which more women (1.66 times) show it than men. It is correlated with socioeconomic development, urbanization, and an increased elderly population. When the rates are age-standardized, they are stable, but the total number of cases has increased, requiring targeted prevention and management strategies. Cross-cultural metrics are used to provide further insight into the trends in anxiety disorder, including through population-based studies.

Anxiety is a normal human emotion. When the emotion is disproportionate, distress and impairment are caused, which becomes a disorder. In total, 4 to 20% of children and adolescents are affected by anxiety disorders including specific phobias, social phobia, generalized anxiety disorder, and separation anxiety [8]. Girls exhibit more acquired anxiety than boys. Middle- and upper-class children in metropolises report higher rates of anxiety, which is affected by academic pressure, family dynamics, or peer relationships. Early diagnosis and treatment, including the severe forms of cognitive behavioral therapy (CBT), present a concern. Untreated anxiety causes further mental problem generation if they remain isolated without the appropriate detection and intervention.

A cross-sectional study is used to assess the inter-relationship of self-management with anxiety and quality of life in patients with anxiety disorder [6]. In the study, 245 patients from West China Hospital were assessed, which showed that the better the level of self-management, the lower the anxiety level and the higher the quality of life. Self-management mediates the relationship between anxiety and quality of life; therefore, enhancing self-management is associated with better patient treatment. Healthcare providers need to provide self-management interventions to improve the quality of life of patients living with anxiety disorders. Digital health tools such as smartphones and wearable sensors are used to improve existing psychiatric diagnosis and treatment. Technologies are used to retrieve real-time data of self-reported symptoms, behavior, and physiological parameters to create new diagnostic methods. While this represents promising results, obstacles to the translation of these tools in clinical practice remain, such as the validation and standardization of the data and assurance of practical use. This calls for further research to explore the full exploitation of these technologies in mental healthcare.

GAD is a common mental illness characterized by excessive, persistent, and unrealistic worry regarding ordinary things. It affects 20% of adults accompanied by psychological and physical complaints. Diagnosis is based on criteria, including the anxiety prevailed for six months combined with other symptoms such as continual restlessness, easily becoming tired, and sleep disturbances. The causes include stress, genetics, and environmental factors. Cognitive behavioral therapy and medications such as SSRIs and SNRIs are used for treatment. An interprofessional team approach is also used including patient education, medication adherence, and the modification of lifestyle for better outcomes and improved quality of life. Separation anxiety disorder (SAD), one of the more common childhood anxiety disorders, presents excessive fear or anxiety about separation from attachment figures [3]. Although developmentally appropriate in early childhood, SAD becomes problematic when the intensity of the behavior is inappropriate for the child's age or situation. The disorder impacts daily life as it produces bad mental and physical health outcomes. Diagnosis is given using DSM-5-TR criteria, and treatment involves cognitive behavioral therapy. In severe cases, SSRIs are prescribed. It requires interprofessional coordination to enhance clinical outcomes and provide educational and therapeutic support to the patient and his or her family.

Yet, machine learning and MRI scan data have not been integrated. Therefore, we developed a prediction model using machine learning models.

3. Methods

3.1. Data Collection

The MRI scans available in an open-source neural imaging database called OpenNeuro were used in this study [9]. We used the scans for feature extraction. We classified the scans into an individual folder. In total, 45 subjects of preschool age took part in the study.

We utilized the scans of 8 separation anxiety patients, 9 generalized anxiety patients, and 11 randomly selected healthy patients. We used 11 of the 23 healthy scans available to balance the potency of the data from each class.

3.2. Data Preprocessing

We resized each image into a 128 × 128 image. Then, we converted each image into grayscale. The brightness of each pixel was represented with numbers from 0 to 255. We flattened the array of numbers into a 1-dimensional array for further processing. Due to the lack of data to train the model, we used data augmentation. This technique increases the size of the dataset by modifying data using rotations, translations, and dilations. To augment images, we reshaped the image back into a 128 × 128 resolution. To maintain accuracy, we utilized rotation and random scaling. We selected scan images randomly to augment them.

We represented pixels of images in a 2D array. Each pixel in each image is assigned a coordinate (x, y). Then, we used a rotation matrix Equation (1).

$$R = \begin{bmatrix} \cos(\theta) & -\sin(\theta) \\ \sin(\theta) & \cos(\theta) \end{bmatrix} \quad (1)$$

The matrix is applied to each pixel's coordinate (represented as a vector).

$$\begin{bmatrix} x' \\ y' \end{bmatrix} = \mathbf{R} \cdot \begin{bmatrix} x \\ y \end{bmatrix} \quad (2)$$

Each image was rescaled to create a scaling matrix Equation (3).

$$S = \begin{bmatrix} s_x & 0 \\ 0 & s_y \end{bmatrix} \quad (3)$$

where s_x and s_y correspond to the factors by which we wish to scale. The matrix is then applied to the coordinate vector Equation (4).

$$\begin{bmatrix} x' \\ y' \end{bmatrix} = \mathbf{S} \cdot \begin{bmatrix} x \\ y \end{bmatrix} \quad (4)$$

In addition to data augmentation, we applied a normalization function. Normalization is useful in machine learning since it allows all features of the data to be analyzed with equal weight. Then, the machine learning model is not affected by extremely small or large values. The impact of outliers is also omitted. We used Z-score normalization in which the Z-score of each element is replaced with an element with the corresponding Z-score Equation (5).

$$\mu = \frac{1}{N}\sum_{i=1}^{N} x_i \quad (5)$$

where μ is the mean and N is the number of data points.

The standard deviation of the data is calculated as in Equation (6).

$$\sigma = \sqrt{\frac{1}{N}\sum_{i=1}^{N} (x_i - \mu)^2} \quad (6)$$

where σ is the standard deviation.

The Z-score is calculated using

$$z_i = \frac{x_i - \mu}{\sigma} \quad (7)$$

The new list of data points consists of calculated Z-scores. The benefit of this is that the mean of the data points is 0 and the standard deviation is 1 to improve convergence rates and model accuracy.

3.3. *Algorithms*

The three algorithms used in this study include SVM, decision trees, and Logistic Regression.

3.3.1. SVM

A SVM maximizes the margin of a hyperplane to separate the classes of data. The SVM maps the input data in higher dimensional space to create a separating hyperplane. The input point is encoded as a kernel. The Gaussian kernel, also known as the radial basis function (RBF) kernel, is used in the model of this study. The RBF kernel is optimal for data points that are not linearly separable and would be easier to separate in higher dimensional spaces.

3.3.2. Decision Tree

A decision tree resembles a flowchart. There are three main features of trees, nodes, branches, and leaf nodes. Each node represents a point in the decision tree where a decision is made. Every decision is made concerning the features that are extracted from all data. The set of the input data is split at every node and follows a branch to the next node, where another decision is made, and the dataset splits again. This cycle continues until the data reach leaf nodes, which represent the classification that the decision tree makes for given input data. Each decision tree begins with a root node corresponding to the first decision made by the model to split the dataset.

3.3.3. Logistic Regression

Logistic Regression is used to compute the probability that an input belongs to a certain class. As there were 3 classes in this study, we used multinomial Logistic Regression. Each class has its set of weights and biases, which are linear transformations applied to an input. Equation (8) was used to calculate the probability that input x belongs in class A

$$P(A \mid x) = \frac{e^{w_A \cdot x + b_A}}{e^{w_A \cdot x + b_A} + e^{w_B \cdot x + b_B} + e^{w_C \cdot x + b_C}} \quad (8)$$

The model was trained by optimizing the weights and biases for each class so that the prediction made was accurate. The measure of the prediction accuracy matched the actual classification called cross-entropy loss.

3.3.4. Transfer Learning and Feature Extraction

To improve model performance and reduce computational load, we employed transfer learning. Transfer learning uses a previously trained model on large datasets as a starting point. We used VGG16 as this is useful for images and extracting features such as edges and textures with high accuracy. This model feeds the features extracted from all data into the model (known as the classifier layer) to output a more accurate prediction.

3.3.5. Cross-Validation

To prevent overfitting and decrease the variance in performance from small datasets and random initialization, we utilized cross-validation to split the input dataset into multiple sections called "folds". We used 5-fold validation. Four of these sections were used to train the dataset and the fifth was used to test the performance. This whole process was repeated five times with a different fold used for each time. The performances of the five models were then averaged. The model with the best performance was selected to predict the results.

3.4. Measures

In evaluating the suitability of the model, we employed four different statistical measures, namely accuracy, precision, recall, and F-1 score. The four statistical measures are the results of estimating true positives (TPs), true negatives (TNs), false positives (FPs), and false negatives (FNs) (Equations (9)–(12)).

$$\text{Precision} = \frac{\text{True Positive}}{\text{True Positive} + \text{False Positive}} = \frac{\text{True Positive}}{\text{Total Predicted Positive}} \tag{9}$$

$$\text{Recall} = \frac{\text{True Positive}}{\text{True Positive} + \text{False Negative}} = \frac{\text{True Positive}}{\text{Total Actual Positive}} \tag{10}$$

$$F_1 = 2 \times \frac{\text{Precision} \times \text{Recall}}{\text{Precision} + \text{Recall}} \tag{11}$$

$$\text{Accuracy} = \frac{TP + TN}{TP + FP + FN + TN} \tag{12}$$

Two other measures used were R^2 and Mean Squared Error (MSE). R^2 highlights how well the model fits the data's relationship. R^2 ranges from 0 to 1, and a higher R^2 value means that the model predicts better. The MSE is calculated by taking the difference between the predicted value and the actual value of each data point. A lower MSE implies a better prediction with fewer mistakes.

4. Results

4.1. Logistic Regression

The Logistic Regression model showed an accuracy of 66.67%. For generalized anxiety, the model presented a precision of 0.67, a recall of 1.00, and an F1-score of 0.80. For the healthy class, the precision and recall were 0.50, and the F1 score was 0.50. For separation anxiety, the precision was 1.00, the recall was 0.50, and the F1-score was 0.67. The weighted averages were 0.72, 0.67, and 0.66 for the precision, recall, and F1-score. The R^2 score was 0.50 and the MSE was 0.33.

4.2. SVM

The tuned hyperparameter of the SVM model was C, which had a best value of 0.1 and resulted in a test accuracy of 66.67%. The precision and recall were 1.00, and the F1 score was 1.00. For separation anxiety, the precision was 1.00, the recall was 0.00, and the F1 score was 0.00. For generalized anxiety, the precision was 0.50, the recall was 1.00, and the F1 score was 0.67. The weighted averages were 0.83, 0.67, and 0.56 for the precision, recall, and F1-score. The R^2 score of the model was 0.50 and the MSE was 0.33.

4.3. Decision Tree

For generalized anxiety, the precision was 1.00, the recall was 0.50, and the F1-score was 0.67. For the healthy class, the precision was 0.33 and recall was 0.50. The F1-score was 0.40. For separation anxiety, the precision and recall were 0.50, with an F1-score of 0.50. The weighted averages were 0.61, 0.50, and 0.52 for the precision, recall, and F1-score. The R^2 score for the model was 0.25, and the MSE equaled 0.50.

5. Conclusions/Discussion

The Logistic Regression model showed an accuracy of 66.67% with an R^2 of 0.5 and MSE of 0.33. The SVM model showed an accuracy of 66.67%, an R^2 score of 0.50, and an MSE of 0.33. The decision tree showed an accuracy of 50%, R^2 of 0.25, and MSE of 0.50. Logistic Regression and the SVM showed better performances than decision trees. The SVM performed better than Logistic Regression.

The model can be used for the classification of anxiety disorders and for addressing class imbalance issues. Using the model, a precise diagnosis was enabled to treat patients better. We classified individuals with multiple anxieties based on their predominant symptoms, either separation or generalized anxiety. Using data from patients with well-defined diagnoses, a better model can be established. The model can be used for the classification of chronic stress, traumatic events, genetic predispositions, chronic illnesses, and significant life changes.

Comorbid conditions such as depression and PTSD often overlap with anxiety, complicating diagnosis and treatment. Pediatricians and psychiatrists can apply machine learning techniques to see how the brain responds to anxiety using MRI scans. Future research is necessary to consider health issues or the root causes of anxiety and create a more accurate model. More data are needed to encompass individuals with both single and multiple diagnoses to improve diagnostic precision and personalized treatment. Using multimodal data, MRI data can be combined for cognitive assessments. Behavioral data can be used to capture a broader spectrum of issues relating to anxiety. Through the integration of patient-reported outcomes and feedback, perceptions of symptoms, the effects of treatment, and quality of life can be improved. Such a personalized approach ensures better treatment. The most important data are patient information. Ethical consideration and stringent measures for data privacy need to be undertaken. Privacy measures and informed consent are required to protect the rights of participants and preserve the integrity of the research process. Developing user-friendly tools, resources, and strategies for practitioners ensures the models are accessible and actionable in clinical practice. The developed model allows for more personalized treatment plans and early intervention strategies, ultimately improving patient outcomes.

Author Contributions: Conceptualization, S.M. and P.K.; methodology, S.M. and P.K.; software, S.M., P.K. and S.P.; validation, S.M., P.K. and S.P.; formal analysis, S.M.; investigation, S.M. and P.K.; resources, S.P.; data curation, S.M. and P.K.; writing—original draft preparation, S.M., P.K. and S.P.; writing—review and editing, S.M., P.K. and S.P.; visualization, S.M.; supervision, S.M. and P.K.; project administration, S.M. and P.K.; All authors have read and agreed to the published version of the manuscript.

Funding: This research received no external funding.

Institutional Review Board Statement: Not applicable.

Informed Consent Statement: Not applicable.

Data Availability Statement: Data are contained within the article.

Conflicts of Interest: The authors declare no conflicts of interest.

References

1. Ghadimi, M.; Sapra, A. Magnetic resonance imaging contraindications. In *StatPearls*; NCBI Bookshelf. 1 May 2023. Available online: https://www.ncbi.nlm.nih.gov/books/NBK551669/ (accessed on 1 July 2024).
2. Munir, S.; Takov, V. Generalized Anxiety Disorder. In *StatPearls*; NCBI Bookshelf. 17 October 2022. Available online: https://www.ncbi.nlm.nih.gov/books/NBK441870/ (accessed on 1 July 2024).

3. Feriante, J.; Torrico, T.J.; Bernstein, B. Separation anxiety disorder. In *StatPearls*; NCBI Bookshelf. 26 February 2023. Available online: https://www.ncbi.nlm.nih.gov/books/NBK560793/#:~:text=Separation%20anxiety%20disorder%20(SAD)%20is,separation%20from%20an%20attachment%20figure (accessed on 1 July 2024).
4. Chand, S.P.; Marwaha, R. Anxiety. In *StatPearls*; NCBI Bookshelf. 24 April 2023. Available online: https://www.ncbi.nlm.nih.gov/books/NBK470361/ (accessed on 1 July 2024).
5. Javaid, S.F.; Hashim, I.J.; Hashim, M.J.; Stip, E.; Samad, M.A.; Ahbabi, A.A. Epidemiology of anxiety disorders: Global burden and sociodemographic associations. *Middle East Curr. Psychiatry* **2023**, *30*, 44. [CrossRef]
6. Cao, X.; Feng, M.; Ge, R.; Wen, Y.; Yang, J.; Li, X. Relationship between self-management of patients with anxiety disorders and their anxiety level and quality of life: A cross-sectional study. *PLoS ONE* **2023**, *18*, e0284121. [CrossRef] [PubMed]
7. Roberts, L.W.; Chan, S.; Torous, J. New tests, new tools: Mobile and connected technologies in advancing psychiatric diagnosis. *NPJ Digit. Med.* **2018**, *1*, 20176. [CrossRef]
8. Bhatia; Goyal, A. Anxiety disorders in children and adolescents. *J. Postgrad. Med.* **2018**, *64*, 75–76. [CrossRef] [PubMed]
9. Preschool Anxiety Disorders. Available online: https://openfmri.org/dataset/ds000144/ (accessed on 1 July 2024).

engineering proceedings

MDPI

Proceeding Paper

Developing Weka-Based Image Classification Learning Model: Enhancing Novice Designers' Recognition of Brand Typicality †

Hung-Hsiang Wang and Ching-Yi Chen *

Department of Industrial Design, National Taipei University of Technology, Taipei 10617, Taiwan; wanghh@mail.ntut.edu.tw
* Correspondence: kelly512727@gmail.com

† Presented at the 2024 IEEE 7th International Conference on Knowledge Innovation and Invention, Nagoya, Japan, 16–18 August 2024.

Abstract: Brand typicality is crucial in shaping consumer perceptions of brands and poses challenges for novice designers to capture due to their limited tacit knowledge. Using Weka's image classification, we developed a brand product classification model. A dataset with 600 images was obtained from Asus and MSI, the leading eSports brands, covering various products such as controllers, mouse devices, headsets, and PC gaming components. The random forest classifier achieved an accuracy of 81 to 85%, slightly higher in the PC gaming category. The design features from Asus ROG and MSI game series products were extracted to generate 36 test images. We used keywords as prompts in Midjurney and Stable Diffusion to generate 36 test images. The developed brand product classification model in this study correctly classified 30 images. However, in the OP category, two graphics card images and one casing image were misclassified. In the PC category, two mouse images and a laptop picture were misclassified. Discrepancies between AI-generated images and personal expertise were improved in terms of the efficiency of the model for new designers. The developed model deepens the understanding of brand characteristics, maintains brand coherence, and strengthens product design innovation and market competitiveness. The model effectively assesses brand characteristics in product appearances using AI, highlighting its role in improving early design processes and new product development strategies.

Keywords: eSports product; brand typicality; AI-generated images; image classification; Weka

1. Introduction

In the competitive eSports market, product design must blend technological innovation with brand identity to stay competitive. Brand image significantly influences consumer decisions, with product design playing a crucial role in shaping this image [1]. Moreover, deep learning and artificial intelligence (AI) technologies are increasingly used in product design, particularly for replicating brand styles [2]. We examined the application of deep learning and image recognition in eSports product design of Asus and MSI to create an aided design system aimed at enhancing design efficiency and fostering innovation.

With the continuous advancement of deep learning and AI technology, these tools are applied to product design, especially in understanding and replicating brand style. Furthermore, many product companies are leveraging AI-generated features for initial proposals or as a source of inspiration for ideas. The application of such technology in design helps novice designers who have a limited understanding of the typical characteristics of the brand better understand and apply these tools and promote product design innovation and efficiency which is the key to driving market competitiveness [3].

We established an image classification learning model based on Weka which is used as an auxiliary tool for novice designers in the initial design. This system helps them evaluate whether their designs match the typical characteristics of the brand, thereby improving the quality and innovation of their design proposals. The system uses AI-generated images as test cases, which are generated based on big data containing uncertain data sources and features. By using products available on the market as training data, the objectivity and accuracy of the assessment can be ensured.

Experts were invited to classify these images and verify the effectiveness of the system in this study. They also evaluated whether the AI-generated image classification results matched Weka's classification results. The developed model was authenticated from the perspective of a professional designer. It improves the productivity of novice designers and enhances their accuracy in judging AI-generated images. The model reduces the need for novice and experienced designers in the early stages of design.

2. Literature Review

In the literature review, three key areas were identified. We investigated the link between brand identity and product design, emphasizing how design enhances brand image. We also explored the use of deep learning and image recognition in product design, showing how these technologies help designers replicate brand styles. AI-generated images were classified to illustrate the efficacy of these technologies in automatically assessing design schemes.

2.1. *Brand Identity and Product Design*

Brand image in eSports product design transcends logos, infusing design aspects with visual style and user experience. Schmitt posited that design fundamentally shapes brand strategy [4], a concept exemplified by Apple's minimalist esthetics and focus on user-centric features. Kapferer emphasized visual representation's role in establishing a durable brand image [5]. Balancing performance and esthetics is crucial in gaming hardware, targeting professional gamers and a wider audience. We examined how deep learning enhances product design by aligning brand identity with innovative, consistent design approaches.

2.2. *Deep Learning and Image Recognition*

Advances in deep learning, particularly using convolutional neural networks (CNNs), transform product design by enhancing image classification. These networks excel in handling complex image datasets, facilitating designers in capturing and replicating brand-specific styles for gaming hardware designs [6]. CNNs support the consistent application of design languages and foster innovation by generating new design possibilities [7,8]. The potential of deep learning was explored to elevate innovation and strengthen market competitiveness in eSports product design [9].

2.3. *AI-Assisted Product Design*

AI has been a crucial aid in product development, performing precise visual analysis and generating brand-specific design sketches. Recent advances in generative AI technology have allowed for the realism of created artifacts, making AI a valuable source of inspiration and a creative partner [10,11]. These advancements are new tools for designers to enhance creativity and efficiency. AI-generated imagery and design concepts significantly boost designers' productivity and creativity in creating customized, brand-related designs [8]. AI-assisted designs help brands discover trends and stay market relevant by accurately incorporating brand elements [12]. In this study, three AI image generation platforms, ChatGPT 4.0 (DALL-E 2), Midjourney, and Stable Diffusion were compared to evaluate their potential and technical features in design education.

- ChatGPT 4.0 (DALL-E 2) excels at generating high-resolution, detailed images from textual descriptions, supporting educational processes by illustrating concepts and styles [13].
- Midjourney focuses on artistically inspired images with powerful stylized expressive capabilities, ideal for creative visual expression [14].
- Stable Diffusion uses diffusion model technology to fine-tune styles while maintaining image quality, making it suitable for generating diverse high-quality images, crucial for visual concepts and projects [15].

Each platform's strengths and limitations guide selecting the appropriate AI technology in eSports hardware design. Such scientific assessment enhances design innovation and improves market competitiveness [16].

2.4. Case Selection: Asus and MSI

In the competitive global gaming hardware market, Asus and MSI impact the gaming community with their innovative designs and strong brand images. Asus targets high-end gamers with advanced technological products, while MSI emphasizes high performance and reliability for a stable gaming experience. By encroaching on Asus's market share, MSI showcases the intense rivalry with other brands [17]. Figures 1 and 2 highlight the distinct product designs and strategies of Asus and MSI, providing insights into their marketing approaches and future development strategies.

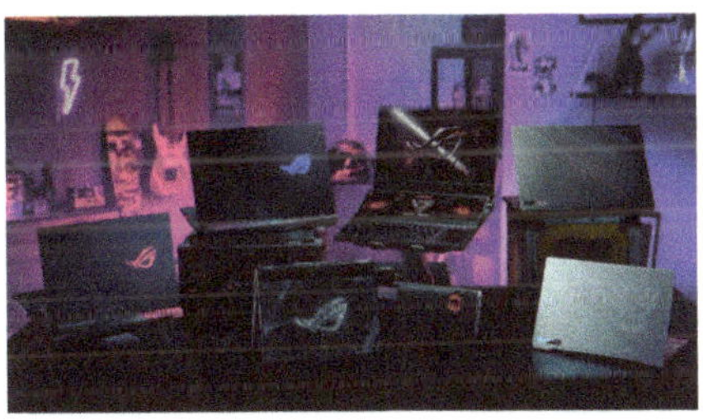

Figure 1. Asus gaming hardware products.

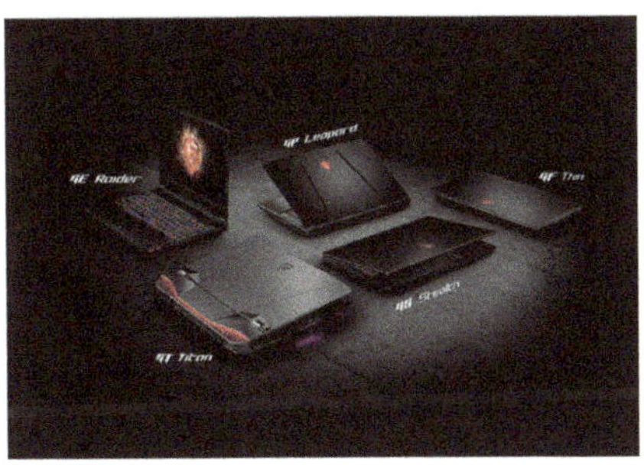

Figure 2. MSI gaming hardware products.

3. Research Procedure

In this study, we designed a Weka-based image classification system to assist novice designers in accurately grasping brand characteristics at the proposal stage. The research methodology consists of a series of steps, as displayed in Figure 3.

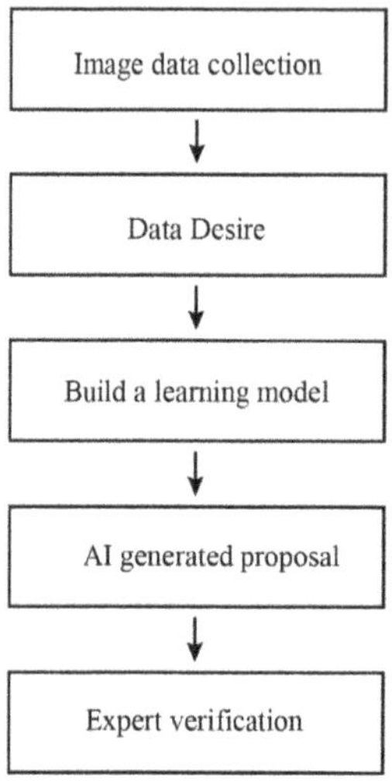

Figure 3. Research procedure.

3.1. Data Collection and Preprocessing

eSports products are divided into two categories: OP and PC gaming, each with unique design features. The OP gaming category includes routers, cases, monitors, and graphics cards, with 300 images collected for each (Figure 4). The PC gaming category includes handlebars, headphones, laptops, and mice, each with 300 images (Figure 5). Figures 4 and 5 display a selection, with the remaining images stored in the cloud. The dataset used in this study comprised 1200 images from Asus and MSI and analyzed for the visual characteristics of gaming hardware to enhance design efficiency and innovation. By adjusting the image resolution to 70–100 DPI and size to 700 × 700 pixels, the model effectively identified key visual features without excessive computational load. The model refined eSports product design details and optimized the design process.

Figure 4. Asus and MSI gaming OP.

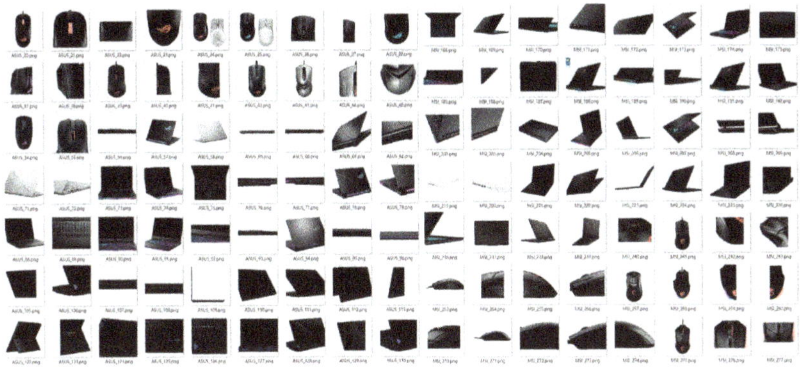

Figure 5. Asus and MSI gaming PC.

3.2. Model Training and Testing

3.2.1. Training Process

We evaluated classifiers and the combination of BinaryPatternsPyramidFilter and RandomForest classifier. Accuracies ranged from 81 to 85% on pre-processed eSports product images. The classifiers outperformed other classifiers, such as the SimpleColorHistogramFilter paired with SMO, which yielded the lowest accuracy. The superior performance of BinaryPatternsPyramidFilter and RandomForest is attributed to their ability to handle complex visual data with texture features and ensemble learning, unlike SMO which relies solely on color features. We used 60% of the images for training. In testing, the model's robustness and consistency were verified with an accuracy of 81% in the OP gaming category and 82% in the PC gaming category. Figures 6 and 7 show the details of the training and testing. These findings demonstrate the feasibility of deep learning for eSports hardware design in model optimization and enhanced classification efficiency and accuracy.

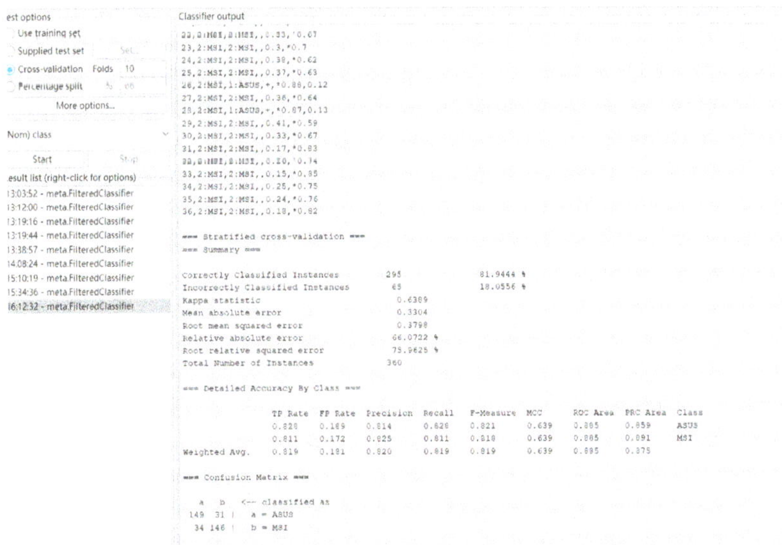

Figure 6. Training for OP gaming.

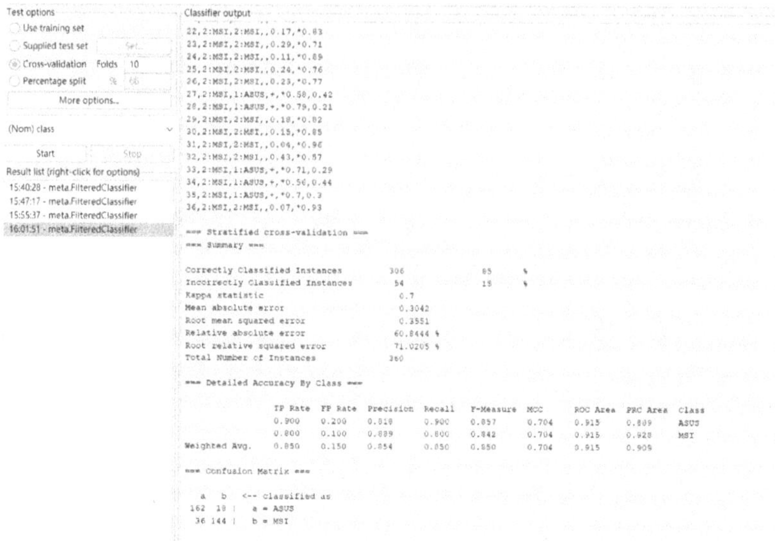

Figure 7. Training for PC gaming.

3.2.2. Keyword Preparation

We collected product data from Asus ROG and MSI on websites, news, and reviews to refine keywords for an AI image generation platform. Key terms including "thin and stylish" and "high-performance GPU" were used to ensure that they accurately reflected product features (Table 1). We conducted a preliminary test on the AI platform to assess how well it interpreted and generated images, making necessary adjustments to enhance accuracy and relevance. This method boosts the experiment's validity and refines image evaluation and classification.

Table 1. Keywords used in This Study.

Product Category	Brand	Keyword Descriptions
Mouse	Asus	Precise sensors, customizable RGB lighting, ergonomic design, responsive buttons.
Laptop	Asus	Light and stylish, gray color scheme, subtle texture, elegant feel.
Headphones	Asus	Immersive sound, comfortable design, noise cancelation, customizable RGB lighting.
Monitor	Asus	Curved ultra-wide display, immersive gaming design, elegant silhouette.
Display Card	Asus	High performance, advanced cooling system, sleek design.
Chassis	Asus	Minimalist and stylish design, eye-catching housing, RGB lighting, modern esthetics.
Laptop	MSI	High-performance GPU, advanced cooling, dynamic design.
Mouse	MSI	High precision, fast reaction time, ergonomic design.
Head-phones	MSI	Restrained and stylish, shielded, durable construction.
Monitor	MSI	QHD resolution, ultra-narrow bezels, suitable for competitive gaming.
Chassis	MSI	Sharp facets, advanced components, modular housing design.

3.2.3. Image Generation

In image generation, we used such keywords as "ergonomic design", "customizable RGB lighting", and "high-performance GPU" to create images that reflect the core design and functional features of Asus and MSI gaming products. To ensure authenticity, we edited brand logos using image editing software to calculate the accuracy of the AI-generated images using 36 images. Table 2 illustrates the generated images of each product category.

Table 2. Generated Images.

Product Category	Brand	Chat GPT	Midjourney	Stable Diffusion
Mouse	Asus	 Mouse.1	 Mouse.2	 Mouse.3
Laptop	Asus	 Laptop.1	 Laptop.2	 Laptop.3
Headphones	Asus	 Headphone.1	 Headphone.2	 Headphone.3
Monitor	Asus	 Monitor.1	 Monitor.2	 Monitor.3
Display Card	Asus	 Display Card.1	 Display Card.2	 Display Card.4
Chassis	Asus	 Chassis.1	 Chassis.2	 Chassis.3
Mouse	MSI	 Laptop.4	 Laptop.5	 Laptop.6
Laptop	MSI	 Mouse.4	 Mouse.5	 Mouse.6

Table 2. *Cont.*

Product Category	Brand	Chat GPT	Midjourney	Stable Diffusion
Headphones	MSI	 Headphone.4	 Headphone.5	 Headphone.6
Monitor	MSI	 Monitor.4	 Monitor.5	 Monitor.6
Chassis	MSI	 Chassis.4	 Chassis.5	 Chassis.6
Display Card	MSI	 Display Card.4	 Display Card.5	 Display Card.6

3.2.4. Machine Learning Classification Results

The results of image classification using Weka showed that two graphics cards and one case were misclassified in the OP category, while two mice and one laptop were misclassified in the PC category (Table 3).

Table 3. Weka Classification Error Table.

 Display Card.4	 Display Card.2	 Chassis.6
 Laptop.6	 Mouse.5	 Mouse.6

3.2.5. Analysis of WEKA Experimental Results

Based on the Weka image classification results, we speculated that the classification errors were caused by the similarities between Asus and MSI gaming series products. The AI-generated database were used to mix the features of both brands. Additionally, the sample size in the Weka database was insufficient, lacking enough training data for products with similar features. To improve classification accuracy the data volume needs to increase for these similar products to enhance the model's ability to accurately distinguish between different brand features.

3.2.6. Expert Qualitative Analysis

The AI-generated images and the classification results of the Weka system were submitted to the design expert for evaluation, and the manual classification results of the experts were compared with the automatic classification results of the system. Table 4 presents a brief introduction to the experts' data.

Table 4. Designer Profile.

Designer	Age	Category	Job Title	Working Years
A	29	PC	Senior Engineer	2
B	35	PC	Senior Engineer	5
C	30	OP	Senior Engineer	2
D	32	OP	Senior Engineer	4

Tables 5–8 depict the contents of the summary of interviews.

Table 5. Designer A's Design Style Description for PC parts.

Image Number	Brand Speculation	Design Style Description
Headphone 1, 2, 4	Asus	Simple, modern, and tech-savvy with smooth lines and colorful LED lighting for visual appeal.
Headphone 3, 5, 6	MSI	Distinct gaming style with red and black color combinations, and red LED lighting for strong personality.
Laptop 1, 3, 4	Asus	Minimalist and industrial esthetics with low-key colors and simple shapes, reflecting high-end technology.
Laptop 2, 5, 6	MSI	Designed for gaming performance with bold red stripes and intricate engravings, offering an aggressive and dynamic look.
Mouse 1, 2, 5	Asus	Modern and clean with smooth design lines and a functional yet esthetic appearance.
Mouse 3, 4, 6	MSI	Emphasizes playstyle with red and black color combinations and an aggressive design.

Table 6. Designer B's Design Style Description for PC parts.

Image Number	Brand Speculation	Design Style Description
Headphone 1, 2, 4	Asus	Simple, modern, and tech-savvy with smooth lines and colorful LED lighting for visual appeal.
Headphone 3, 5, 6	MSI	Distinct gaming style with red and black color combinations, and red LED lighting for strong personality.
Laptop 1, 3, 4	Asus	Minimalist and industrial esthetics with low-key colors and simple shapes, reflecting high-end technology.
Laptop 2, 5, 6	MSI	Designed for gaming performance with bold red stripes and intricate engravings, offering an aggressive and dynamic look.
Mouse 1, 2, 5	Asus	Modern and clean with smooth design lines and a functional yet esthetic appearance.
Mouse 3, 4, 6	MSI	Emphasizes playstyle with red and black color combinations and an aggressive design.

Table 7. Designer C's Design Style Description for OP parts.

Image Number	Brand Speculation	Design Style Description
Monitor 1, 2, 3	Asus	Modern technology style with cool colors, clear geometric lines, and lighting effects that emphasize a tech-savvy appearance.
Monitor 4, 5, 6	MSI	Distinct gaming style with a red and black color palette, lighting, and angle cuts to enhance visual impact.
Display Card 1, 3, 6	Asus	Asus's signature streamlined design and round logo for a clean, high-performance look.
Display Card 2, 4, 5	MSI	Highly functional and playful with MSI's characteristic red design and symmetrical fan layout.
Chassis 1, 3, 6	Asus	Minimalist design with a high-tech feel, streamlined shape, and LED strips that accentuate the modern look of Asus products.
Chassis 2, 4, 5	MSI	Strong playstyle with red lighting effects, intricate panel engravings, and sharp corners that highlight MSI's gaming esthetics and mechanics.

Table 8. Designer D's Design Style Description for OP parts.

Image Number	Brand Speculation	Design Style Description
Monitor 1, 2, 3, 4	Asus	Designed with Asus's signature beveled elements, tripod design, and symmetrical screen for a harmonious visual effect.
Monitor 5, 6	MSI	Features MSI's characteristic sharp edges and dragon-like shapes, contrasting with Asus style.
Display Card 1, 2, 3	Asus	Emphasizes technology and modernity with varied lighting effects aligned with Asus's current design style.
Display Card 4, 5, 6	MSI	Features MSI's signature black and red color scheme, presenting a strong and visually impactful style.
Chassis 1, 2, 3, 6	Asus	Showcases Asus's slash design and colorful ROG lighting effects, incorporating Hyperion GR701 elements, with Chassis 6 referencing ROG Z11 styling.
Chassis 4, 5	MSI	Exhibits MSI's black and red color tones with a triangular facet design, offering a distinctive look rarely seen in rounded shapes.

3.3. Expert Qualitative Results Analysis

We used Weka tools and expert evaluations to perform product classification comparisons and explored the effectiveness of machine learning applications in eSports product design. Specifically, we conducted detailed classification assessments for graphics cards, chassis, laptops, mice, and headsets.

- Graphics Cards and Chassis: The inconsistency between expert and Weka classifications in these categories, especially the misclassification of Graphics Card 2 and Chassis 6, highlights the cognitive challenges posed by visual style similarities. This indicates that even experts are confused by design elements with subtle differences.
- Laptops: Weka made a classification error with Laptop 6, and designers A and B also misjudged Laptops 2 and 4, reflecting the risk of misjudgment even with significant brand designs, particularly when product designs are close to or mimic competitors.
- Mice and Headsets: The classification errors for mice and headsets demonstrate the limitations of using visual identification to distinguish brand characteristics in

highly similar product categories. Particularly in the headset category, Designer B's misclassification of Headsets 3 and 4 highlights the subjective differences in brand style interpretation even among experts.

Although the Weka classification model was evaluated as effective in helping novice designers, the qualitative analysis by experts showed that professional experience is indispensable in maintaining brand visual consistency and handling subtle style differences. This emphasizes the importance of combining machine learning tools and expert esthetics in the product development process and the need for strengthening machine learning models in detail recognition and style analysis.

4. Conclusions

We classified the images of eSports hardware products and compared the results with professional designers' evaluations using the Weka tool. The Weka tool accurately classified products with brand-specific features but struggled with products having similar styles or subtle design details. The potential of machine learning in visual product classification was validated by highlighting the need for real-time feedback in design evaluation. The developed model helps designers better understand and utilize brand-specific design languages and enhance product market competitiveness. This machine learning model can be applied to the automotive, consumer electronics, and fashion industries, supporting innovation and market trend analysis. Further research is necessary for optimization and applicability in different market conditions and more efficient data-driven decision-making and product development.

Author Contributions: Conceptualization, C.-Y.C.; methodology, C.-Y.C.; software, C.-Y.C.; validation, C.-Y.C. and H.-H.W.; formal analysis, C.-Y.C.; investigation, C.-Y.C.; resources, C.-Y.C.; data curation, C.-Y.C.; writing—original draft preparation, C.-Y.C.; writing—review and editing, H.-H.W.; visualization, C.-Y.C.; supervision, H.-H.W.; project administration, H.-H.W.; funding acquisition, H.-H.W. All authors have read and agreed to the published version of the manuscript.

Funding: This research was funded by the National Science and Technology Council (NSTC), Taiwan, R.O.C., under grant number 111-2410-H-027-019-MY2.

Institutional Review Board Statement: Not applicable.

Informed Consent Statement: Not applicable.

Data Availability Statement: Data are contained within the article.

Conflicts of Interest: The authors declare no conflict of interest.

References

1. Hsu, C.-L.; Chen, Y.-C.; Yang, T.-N.; Lin, W.-K.; Liu, Y.-H. Does product design matter? Exploring its influences in consumers' psychological responses and brand loyalty. *Inf. Technol. People* **2018**, *31*, 886–907. [CrossRef]
2. Zhang, Y. Research on key technologies of remote design of mechanical products based on AI. *J. Vis. Commun. Image Represent.* **2019**, *60*, 250–257. [CrossRef]
3. Calò, T. Rethinking Design Workflows with Artificial Intelligence. In Proceedings of the CHItaly 2023: Crossing HCI and AI, Doctoral Consortium, Turin, Italy, 20–22 September 2023; pp. 8–14.
4. Schmitt, B. Experiential marketing. *J. Mark. Manag.* **1999**, *15*, 53–67. [CrossRef]
5. Kapferer, J.-N. *The New Strategic Brand Management: Advanced Insights and Strategic Thinking*; Kogan Page Publishers: London, UK, 2012.
6. Chen, W.; Xie, D.; Zhang, Y.; Pu, S. All you need is a few shifts: Designing efficient convolutional neural networks for image classification. In Proceedings of the IEEE/CVF Conference on Computer Vision and Pattern Recognition, Long Beach, CA, USA, 15–20 June 2019; pp. 7241–7250.
7. LeCun, Y.; Bengio, Y.; Hinton, G. Deep learning. *Nature* **2015**, *521*, 436–444. [CrossRef] [PubMed]

8. Goodfellow, I.; Pouget-Abadie, J.; Mirza, M.; Xu, B.; Warde-Farley, D.; Ozair, S.; Courville, A.; Bengio, Y. Generative adversarial nets. *Commun. ACM* **2020**, *63*, 139–144. [CrossRef]
9. Wang, H.-H.; Chen, C.-P. A case study on evolution of car styling and brand consistency using deep learning. *Symmetry* **2020**, *12*, 2074. [CrossRef]
10. Maher, M.L.; Weisz, J.D.; Chilton, L.B.; Geyer, W.; Strobelt, H. HAI-GEN 2023: 4th Workshop on Human-AI Co-Creation with Generative Models. In *Companion Proceedings of the 28th International Conference on Intelligent User Interfaces;* Association for Computing Machinery: New York, NY, USA, 2023; pp. 190–192.
11. Weisz, J.D.; Maher, M.L.; Strobelt, H.; Chilton, L.B.; Bau, D.; Geyer, W. Hai-gen 2022: 3rd workshop on human-ai co-creation with generative models. In Proceedings of the 27th International Conference on Intelligent User Interfaces, Helsinki, Finland, 22–25 March 2022; pp. 4–6.
12. Agersborg, C.; Månsson, I.; Roth, E. Brand Management and Artificial Intelligence—A World of Man Plus Machine: A Qualitative Study Exploring How Artificial Intelligence Can Contribute to Brand Management in the B2C Sector. Master's Thesis, University of Gothenburg, Gothenburg, Sweden, 2020. Available online: https://gupea.ub.gu.se/handle/2077/64924 (accessed on 1 July 2024).
13. Derevyanko, N.; Zalevska, O. Comparative analysis of neural networks Midjourney, Stable Diffusion, and DALL-E and ways of their implementation in the educational process of students of design specialities. *Sci. Bull. Mukachevo State Univ. Ser. Pedagog. Psychol.* **2023**, *9*, 36–44. [CrossRef]
14. Wasielewski, A. "Midjourney can't count": Questions of representation and meaning for text-to-image generators. *Interdiscip. J. Image Sci.* **2023**, *37*, 71–82.
15. Borji, A. Generated faces in the wild: Quantitative comparison of stable diffusion, midjourney and dall-e 2. *arXiv* **2022**, arXiv:2210.00586.
16. Caggiani, L.; Camporeale, R. Toward sustainability: Bike-sharing systems design, simulation and management. *Sustainability* **2021**, *13*, 7519. [CrossRef]
17. Wang, C.-H.; Yun, Y. Demand planning and sales forecasting for motherboard manufacturers considering dynamic interactions of computer products. *Comput. Ind. Eng.* **2020**, *149*, 106788. [CrossRef]

engineering proceedings

MDPI

Proceeding Paper

CAVE Automatic Virtual Environment Technology: A Patent Analysis †

Fatma Beji [1], William de Paula Ferreira [1,*], Isabelle Pivotto Dabat [1] and Vitor Matias [2]

1 Department of Systems Engineering, École de Technologie Supérieure, Montreal, QC H3C 1K3, Canada; fatma.beji.1@ens.etsmtl.ca (F.B.); isabelle.pivotto-dabat@etsmtl.ca (I.P.D.)
2 Department of Specialized Education and Training, University of Quebec in Montreal (UQAM), Montreal, QC H2X 3R9, Canada; matias.vitor@uqam.ca
* Correspondence: william.ferreira@etsmtl.ca
† Presented at the 2024 IEEE 7th International Conference on Knowledge Innovation and Invention, Nagoya, Japan, 16–18 August 2024.

Abstract: Cave automatic virtual environment (CAVE) technology provides a highly immersive experience in virtual reality (VR) environments, transcending traditional boundaries of VR head-mounted devices. CAVE is applied to many fields, including education, construction, healthcare, and manufacturing. Despite its relevance, studies examining CAVE technology evolution and research directions are still lacking. To address this research gap, we analyzed patents using CAVE to understand the technology's development and identify opportunities for future research, development, and innovation. Patent data were collected from the Lens database and analyzed using data mining techniques. An increasing number of CAVE patents were granted, reflecting significant growth and investments in this field. The results highlight emerging trends in the development of CAVE systems, emphasizing various technical configurations and innovative applications across a wide range of fields.

Keywords: CAVE; patent analysis; data mining; virtual reality; immersive technologies

1. Introduction

A cave automatic virtual environment (CAVE) system consists of a cube-shaped virtual reality (VR) room with floor, ceiling, and walls as projection surfaces on which multiple projectors display images, creating an immersive 3D VR environment [1]. It represents an advance in VR, offering 3D immersion for users that transcends the traditional limits of the virtual experience provided by VR head-mounted devices [2,3]. The Electronic Visualization Laboratory first developed a VR CAVE system at the University of Illinois, in Chicago, in the early 1990s [4]. It has evolved significantly since then and has been used for different applications in various sectors such as education [5], medicine and healthcare [6–8], entertainment, leisure and culture [9], and manufacturing [10].

Despite CAVE's relevance, studies examining its technological evolution and research directions are still lacking. Accordingly, a patent analysis provides information on the competitive dynamics and innovations that characterize its evolution. It encompasses essential steps, such as technology exploration to identify new advances, technology landscape assessment to guide research and development, competitive analysis to position patents against competitors, and patent rankings to quantify their strength and identify new patent applications [11]. Even though patent analyses on augmented reality (AR) and VR have been conducted [12,13], there is a notable gap in the analyses of patents related to CAVE technology.

This study aims to fill this gap by applying an innovative data mining-based approach to quantitative analyses to understand patents associated with CAVE systems. The results provide information on the main CAVE technological developments, existing CAVE application domains, companies investing in and developing CAVE technology, and the classification of CAVE patents.

The remainder of this article is organized as follows: Section 2 provides a background review of CAVE systems. Section 3 introduces the research methodology. Section 4 presents the results, and Section 5 contains a discussion of the results. Lastly, Section 6 concludes this study.

2. Background

CAVE systems are classified into fully immersive versions, which plunge the user into a 360-degree virtual universe, and partially immersive versions, offering an immersive experience on a restricted selection of surfaces [1]. An example of a CAVE system is illustrated in Figure 1. The implementation of CAVE systems relies on direct or indirect projection techniques, allowing a virtual environment to be projected onto the walls of the dedicated space [8]. As described by Manjrekar et al. [4], a CAVE system includes six main components: (1) audio system; (2) video system; (3) cave calibration; (4) stereographic rendering; (5) tracking systems; and (6) graphic engine. The audio system enriches immersion with spatial sound environments, and the video system—which uses high-quality projectors—creates an illusion of depth by displaying stereoscopic images on walls and floors. Precise calibration of these systems is essential in aligning projections and tracking user movements perfectly. Stereoscopy offers two slightly different perspectives for each eye, and tracking systems adjust visual perspectives according to user movements, contributing to the immersive effect. Graphic engines—optimized for the high performance required by CAVEs—manage the rendering of virtual environments.

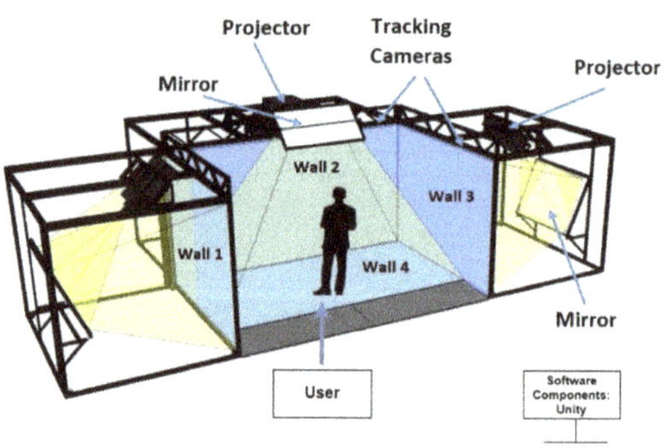

Figure 1. Illustration of a CAVE system [14].

For complete immersion, CAVE systems incorporate 360-degree cameras capable of capturing the entire surrounding space for a complete virtual reconstruction. For partially immersive configurations, 180-degree cameras are preferred, focusing immersion on specific areas of the environment [8,15,16]. Typically, four to eight cameras are used to ensure complete coverage and precise tracking of user movements. Projectors also play a key role in creating the illusion of a virtual world. Their number varies according to the configuration of the CAVE. A fully immersive system may require between four and six projectors to cover all the walls, floor, and sometimes the ceiling. Projectors must have high resolution to maintain superior image quality and can be distributed to avoid overlapping and unwanted shadows. Adopted projection technologies vary from digital light processing projectors to liquid-crystal display projectors, each with advantages in terms of brightness, contrast, and color fidelity [1]. The tracking system in a CAVE is essential for a successful immersive experience.

Various sensors are used in a CAVE, including optical, magnetic, and inertial motion sensors. These sensors detect user movements and adjust the virtual environment accordingly. The number of sensors required depends on the complexity of the tracking system and the desired precision. A standard configuration includes a dozen sensors strategically distributed throughout the space to capture all aspects of user movements. Necessary hardware includes high-resolution projectors, motion sensors, 3D glasses, and multidirectional audio systems, while the software includes rendering engines such as Unity or Unreal Engine, calibration software, and VR middleware such as VR Juggler or CAVElib, facilitating interaction between hardware and applications. The combination of these hardware and software components makes CAVE systems an exceptional platform for data mining and environmental simulations for different purposes.

3. Methodology

We adopt a quantitative-based patent analysis methodology for the identification of emerging technological trends, assessment of competitor activities, and detection of development and collaboration opportunities [11,17]. It also aids in mitigating risks of infringement and litigation by clarifying a technology's intellectual property [18]. Moreover, patent analyses are conducted to uncover unexplored research areas and guide research and development investments, ensuring a strategic alignment with market developments [19]. Furthermore, this methodology helps formulate strategies for protecting and capitalizing on inventions, thereby strengthening the innovative position of both academia and industry in the long term [17].

The Lens patent database was adopted as the main data source of the patents associated with CAVE technology, for specific reasons. First, Lens offers free access to a vast worldwide patent database, facilitating the exploration of a wide range of documents relevant to our analysis. Lens compiles data from various sources, including government patent offices in several jurisdictions and other international institutions. In addition, Lens offers a user-friendly and intuitive search interface and advanced data visualization features that facilitate patent analysis.

Table 1 presents the search protocol in this study. We identified 264 patents, from which 215 were selected for analysis, excluding redundant patents filed in different jurisdictions. After reviewing each patent, abstract, and claim, we excluded 73 patents that were beyond the scope of this research. These included patents involving water caves, wine cellars, plant containers called caves, or medical devices used in CAVE environments. The review results included 142 patents for further analysis using descriptive statistics and data mining (e.g., clustering techniques) based on the Python programming language and its specialized libraries (e.g., Pandas, Matplotlib, Seaborn, Scikit-learn).

Table 1. Search protocol.

Data source:	Lens
Search string:	(CAVE AND immersive) OR (CAVE AND virtual)
Period:	From 1992 (emergence of the 1st CAVE) to 7 June 2024
Search fields:	Title, abstract, and claims
Language:	English
Document:	Patents

4. Results

Of the patents included in the sample, 39.4% are active patents; 32.4% are patents pending evaluation, which means that they are still under examination by the patent office and have yet to be granted; 19.7% are patents interrupted or abandoned by their applicants before being granted or rejected; 5.6% are inactive patents, indicating that they are not commercially launched or awaiting renewal of their rights. Finally, 2.6% are expired patents, meaning that their legal validity period has ended and they are no longer protected. All these non-active patents were included in the analysis since they provide information on the development of CAVE technology.

Figure 2 shows the evolution of CAVE patents over time. From the first filings observed in the early 1990s corresponding to the creation of the first CAVE, a gradual increase in patent publications is observed until the early 2000s. From then on, a marked acceleration is visible, with a notable peak in 2022 showing 21 patents, which suggests an increased interest in CAVE technology due to key technological advances, the emergence of new application markets, or wider recognition of the immersive technology's commercial potential, and applications. Moreover, patent citations indicate a technology not recognized as essential by the community but that continues to impact current innovation.

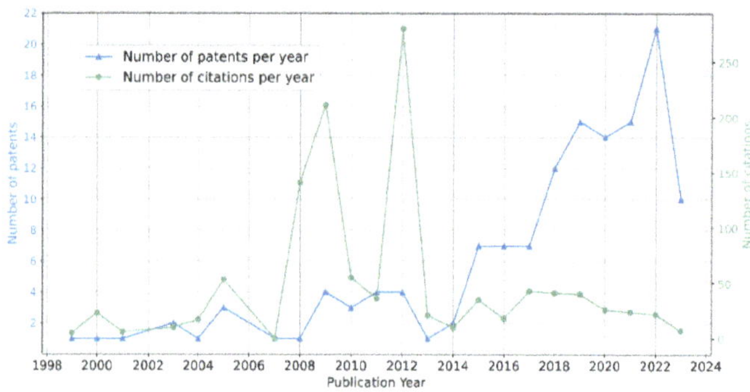

Figure 2. Number of patents and citations for CAVE.

The geographical distribution of patents relating to CAVE technology highlights the global scope of innovation and the variety of players involved. Figure 3 shows that China and the USA largely dominate the field with 54% and 24% of patents, respectively. Other countries including the UK and Japan have 1–3 patents, indicating a global interest and investment in this technology.

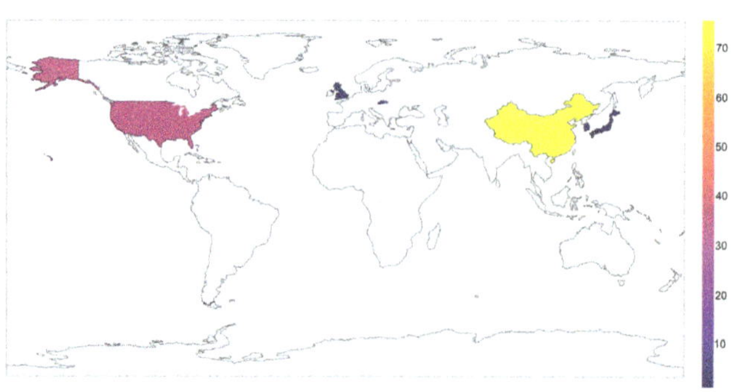

Figure 3. Geographical distribution by affiliation of authors.

Figure 4 shows the main patent applicants on CAVE systems. Well-known companies in the aeronautics and manufacturing sectors—such as Lockheed Corporation based in Bethesda, Maryland, United States with four patents, and Geely Holding Group, headquartered in Hangzhou, Zhejiang, China, with five patents—are among the leading applicants. Other entities—such as Hangzhou Yiyuqianxiang Technology, also located in Hangzhou, Zhejiang, China, and 3Di LLC, based in Unites States, —have three patents each. Industries and universities are contributing to the development of CAVE technology. For example, Jinan University, located in Guangzhou, Guangdong, China, has filed three patents, demonstrating academia's involvement in advancing this field.

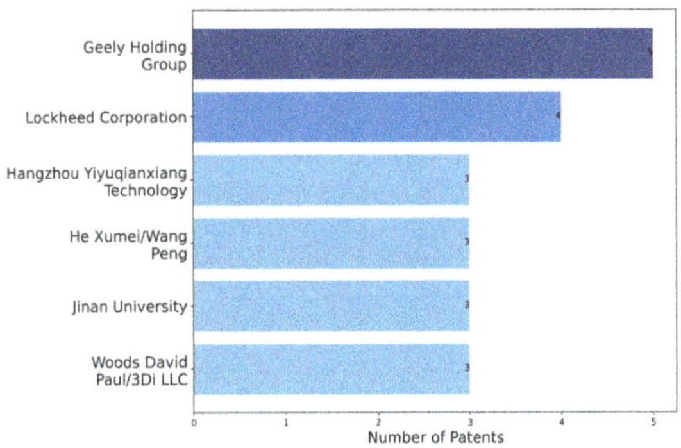

Figure 4. CAVE patent applicants.

Figure 5 summarizes the CAVE patent application domains. It indicates that the two leading areas are VR technology (audio and video equipment manufacturing), with 88 patents, followed by the manufacturing industry, with 16 patents. The gaming industry is also making considerable strides in innovation, with eight patents. This is closely followed by research and development in biotechnology and the medical and healthcare industries, as well as the automotive industry, each with seven patents. Other notable domains include education, human–machine interface (HMI), mining engineering, theater, video processing, photography, marketing consulting, and geosynchronization. The diverse range of domains illustrates the widespread applicability of CAVE systems.

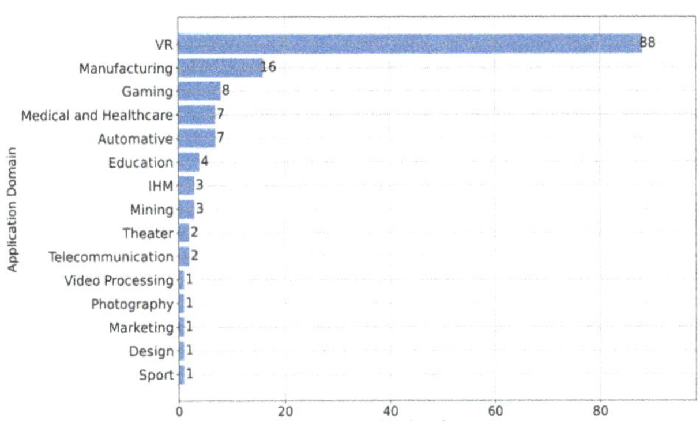

Figure 5. Patent application domains.

Two systems are used to classify patents according to their technical fields: cooperative patent classification (CPC) and international patent classification (IPC). CPC is a relatively recent system developed jointly by the European Patent Office (EPO) and the United States

Patent and Trademark Office (USPTO) and is designed to be more detailed and progressive than the reformed version of IPC. CPC is updated more frequently to reflect the rapid evolution of technology and innovation. Figure 6 shows the number of patents classified under different subcategories of CPC to identify the main innovations in CAVE systems. For example, physics-related sub-classes such as image-processing systems or head-tracking devices are predominant. This corresponds to the key aspects of CAVE technology, which relies on user interaction and complex image processing to create immersive environments.

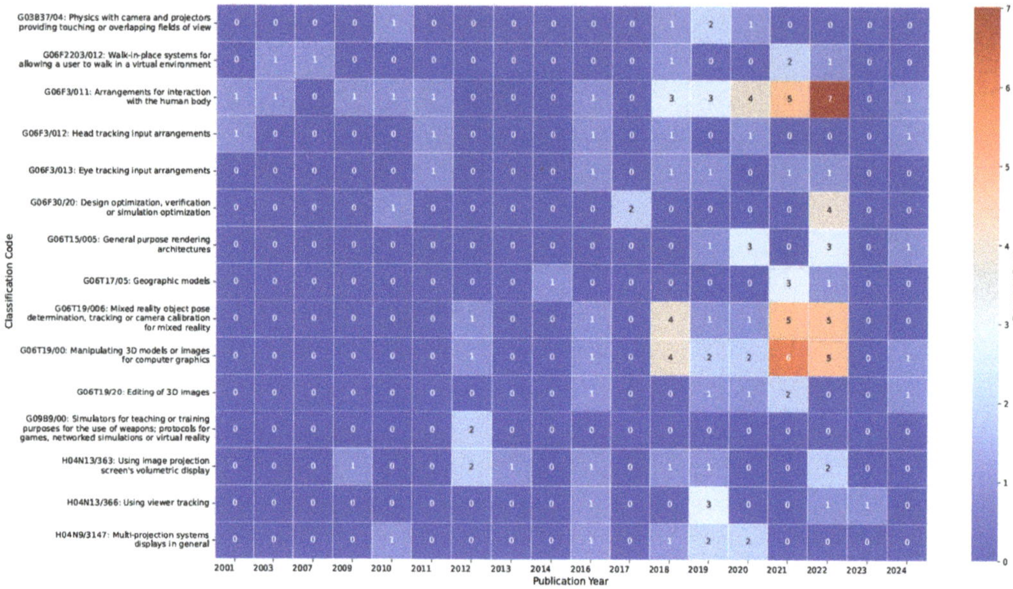

Figure 6. Evolution of patents on CAVE based on CPC.

Figure 7 presents a cluster analysis based on the classification of Manjrekar et al. [4], which describes a CAVE system in terms of its different components. The audio and video systems cluster (36 patents) includes technologies related to both audio and visual components essential for creating an immersive CAVE experience with advancements in sound systems, speakers, display screens, projectors, and integration methods. The CAVE calibration cluster (22 patents) focuses on the technical aspects necessary to ensure the accuracy and reliability of immersive environments, encompassing hardware and software crucial to precision calibration. The stereographic rendering cluster (21 patents) is dedicated to technologies that enable the creation of three-dimensional visuals, providing depth perception and enhancing virtual environment realism through stereoscopic display methods. The tracking systems cluster (34 patents) includes technologies tracking eye, hand, and body movements essential for natural and intuitive interactions within virtual environments to enhance the user experience through precise tracking capabilities. Finally, the graphic engine cluster (35 patents) includes software and hardware components responsible for rendering and managing graphics in a CAVE system, covering innovations in GPUs, rendering algorithms, and software platforms that support high-performance graphics rendering. Furthermore, we identified the CAVE clusters (four patents) to determine an entire CAVE variant, including portable or low-cost systems. It is important to note that some patents may fall into multiple clusters.

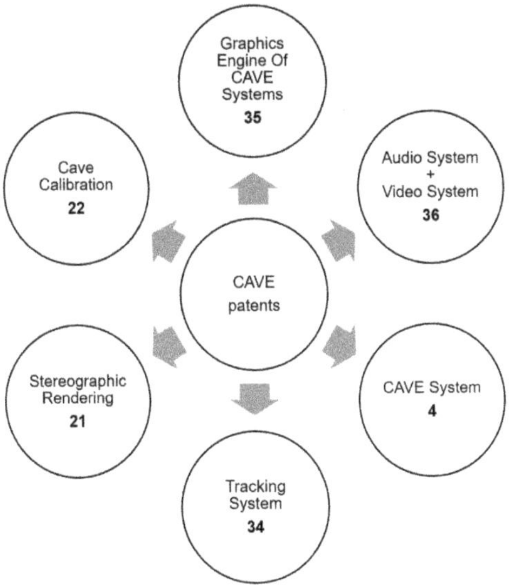

Figure 7. Cluster analysis.

5. Discussion

5.1. CAVE Technological Developments

The cluster analysis results show three major technological development areas: audio/video systems, graphic engines, and motion-tracking systems. Audio and video systems are crucial for creating an immersive experience in a CAVE environment. This includes projection technologies [20], high-resolution displays [21], spatial audio systems [20], and audio/video synchronization technologies [22]. These systems enable the simulation of realistic and immersive virtual environments, essential for a wide range of applications such as immersive training. Graphic engines are at the core of CAVE systems, enabling real-time generation of 3D images. They include advanced rendering algorithms, illumination techniques, and optimizations for handling massive data volumes. These engines ensure a smooth and realistic experience, essential for complex simulations and interactive environments. Motion-tracking systems capture user movements within the CAVE environment. They include camera-based motion capture, inertial sensors, and optical localization systems [23]. Precise motion tracking is essential for natural and intuitive interaction with virtual environments, thus enhancing user engagement and efficiency.

CPC also highlights the CAVE system's technological development aspects, suggesting two major development areas: human-body interaction devices and determining the pose of mixed-reality objects, tracking, or cameras. These devices include human–machine interfaces such as haptic gloves, force-feedback vests, and gesture recognition systems. These technologies enable more natural and immersive interaction with CAVE environments, increasing their utility in various applications such as medical training and manufacturing. Pose determination technologies for mixed-reality objects allow tracking of the spatial position and orientation of objects. They include AR and motion-tracking cameras, merging virtual and real worlds, and enhancing CAVE system capabilities for mixed- and augmented-reality applications.

In summary, the main technological developments in CAVE technology focus on improving immersion and interactivity in virtual environments. Advanced audio/video systems and graphic engines play a crucial role in creating immersive experiences. Motion-tracking systems and human-body interaction devices allow natural and intuitive interaction, which is essential for user engagement. Moreover, mixed-reality technologies and 3D manipulations are critical for advanced applications in various fields, ranging from education to industrial simulations. Finally, different CAVE system configurations are being developed to make it more affordable and portable [24]. These advancements indicate a growing interest in creating functional CAVE systems with immersive rendering capabilities.

5.2. CAVE Application Domain

Six major sectors were analyzed including VR, manufacturing, medical and healthcare, automotive, and education. VR (audio and video equipment manufacturing) and the manufacturing industry are the two leading application areas, highlighting their significant role in enhancing immersive experiences. The gaming industry, particularly electronic video game manufacturing, also leverages CAVE technology, emphasizing its role in creating immersive and interactive gaming experiences. Other domains include HMI, mining, theater, video processing, photography, marketing, resynchronization, design, and sports. CAVE technology is widely used in research, development, and training in the medical and healthcare sectors. Such broad applications underscore the versatility of CAVE technology and its potential for innovation across numerous sectors.

5.3. Development of CAVE Technology

The companies developing CAVE technology are located in the United States and China. In China, the leading company is Geely Holding Group, an automotive group that has developed VR simulation tools for vehicle models since 2018, eventually proposing a new CAVE system in 2020 to meet their technological needs. This product development approach based on specific user needs known as user-centeredness is a key element of the increasingly popular concept of design thinking. The approach of this company has contributed to developing a CAVE system addressing its challenges in automotive design.

In the USA, the leading company is Lockheed Corporation, a manufacturer in the aerospace and defense sector. It focused on developing immersive environments for CAVE technology with user tracking and motion capture. Although the company has since abandoned the development of CAVE technology, its patents have been frequently cited, ranging from 15 to 187 citations. Companies with highly cited patents contribute significantly to technological progress in their specific fields [25]. These highly cited patents signal key innovations or major technological advances, emphasizing their importance for the companies themselves and the overall evolution of technology.

However, even though private companies file a large number of patents, 23 patents are held by academics for various applications, including the development of a portable CAVE. Most of these academics are also found in the United States and China.

5.4. Trends and Dynamics

The CPC classification and the CAVE components/system taxonomy can be used to better understand CAVE technology (Figure 7). Consequently, 22 variants of the initial system developed by Cruz et al. in 1992 [26] and 1993 [27] and many application domains for CAVE technology were found. Additionally, numerous patents target CAVE technological components, such as combined audio/video systems, graphic engines, motion-tracking systems, visual rendering, and calibration systems.

For technological innovation, we focused on under-explored clusters, such as systems that include portable or low-cost CAVE systems. CAVE technology is expensive in terms of initial costs and maintenance fees [8]. Examining patents in the less common technical domains based on CPC classifications is beneficial. This approach provides manufacturers and companies with valuable information on the less-developed components of CAVE technology. Companies need to strategize their research and development efforts more effectively by identifying gaps and emerging areas within the patent portfolios. For instance, patents related to portable and cost-effective CAVE systems could address the current limitations and market needs [8].

Patent analysis helps in identifying technological gaps and understanding the competitive dynamics. Highly cited patents, such as those of Lockheed Corp., often indicate significant technological advancements and influence [25]. Companies can use this information to benchmark their innovations and identify key players and potential collaborators or competitors.

6. Conclusions

We analyzed the evolution and current state of CAVE technology through a patent analysis. The findings indicate a notable increase in granted CAVE patents since its inception in 1992, reflecting growing interest and advancements in the field. Moreover, a cluster analysis based on CPC provides information on the technological trends and competitive dynamics of CAVE technology. By focusing on under-explored areas, companies can drive innovation, reduce costs, and enhance the efficacy and accessibility of CAVE systems. The broad range of applications, from healthcare and education to industrial manufacturing and entertainment, demonstrates the versatility and impact of CAVE technology. The geographical distribution of patents and patent applicants reveals that the USA and China lead the patent filings on CAVE. While current innovations focus on enhancing immersive experiences and user interaction, future research is necessary to explore the development of more affordable and portable CAVE systems. As CAVE technology continues to evolve, it can revolutionize multiple sectors by providing immersive virtual environments to redefine user experience.

Author Contributions: Conceptualization, F.B., W.d.P.F. and I.P.D.; methodology, F.B., W.d.P.F. and I.P.D.; validation, F.B., W.d.P.F., I.P.D. and V.M.; formal analysis, F.B.; investigation, F.B.; resources, W.d.P.F., I.P.D. and V.M.; data curation, F.B.; writing—original draft preparation, F.B.; writing—review and editing, F.B., W.d.P.F., I.P.D., V.M.; visualization, F.B. and W.d.P.F.; supervision, W.d.P.F., I.P.D. and V.M.; funding acquisition, W.d.P.F., I.P.D. and V.M. All authors have read and agreed to the published version of the manuscript.

Funding: This research was supported by the Ministère de l'Economie, de l'Innovation et de l'Energie (MEIE) of the Government of Quebec [N/Réf.: 2022-2026-PSOv2a-ETS-IS-63675].

Institutional Review Board Statement: Not applicable.

Informed Consent Statement: Not applicable.

Data Availability Statement: Data will be made available on request.

Conflicts of Interest: The authors declare that they have no known competing financial interests or personal relationships that could have appeared to influence the work reported in this paper.

References

1. Muhanna, M.A. Virtual reality and the cave: Taxonomy, interaction challenges and research directions. *J. King Saud Univ.-Comput. Inf. Sci.* **2015**, *27*, 344–361. [CrossRef]
2. Kageyama, A.; Tomiyama, A. Visualization framework for cave virtual reality systems. *Int. J. Model. Simul. Sci. Comput.* **2016**, *7*, 1643001. [CrossRef]
3. Flynn, C. An Open-Source Framework for Cave Automatic Virtual Environments. Ph.D. Thesis, Dublin City University, Dublin, Ireland, 2014.
4. Manjrekar, S.; Sandilya, S.; Bhosale, D.; Kanchi, S.; Pitkar, A.; Gondhalekar, M. Cave: An emerging immersive technology—A review. In Proceedings of the 2014 UKSim-AMSS 16th International Conference on Computer Modelling and Simulation, Cambridge, UK, 26–28 March 2014; pp. 131–136.

5. Freina, L.; Ott, M. A literature review on immersive virtual reality in education: State of the art and perspectives. In Proceedings of the International Scientific Conference Elearning and Software for Education, Bucharest, Romania, 23–24 April 2015; Volume 1, pp. 10–1007.
6. Wiebe, A.; Kannen, K.; Selaskowski, B.; Mehren, A.; Thöne, A.K.; Pramme, L.; Blumenthal, N.; Li, M.; Asché, L.; Jonas, S.; et al. Virtual reality in the diagnostic and therapy for mental disorders: A systematic review. *Clin. Psychol. Rev.* **2022**, *98*, 102213. [CrossRef] [PubMed]
7. Gromer, D.; Madeira, O.; Gast, P.; Nehfischer, M.; Jost, M.; Müller, M.; Mühlberger, A.; Pauli, P. Height simulation in a virtual reality cave system: Validity of fear responses and effects of an immersion manipulation. *Front. Hum. Neurosci.* **2018**, *12*, 372. [CrossRef] [PubMed]
8. Pivotto, I.D.; Matias, V.; de Paula Ferreira, W. Cave automatic virtual environment technology to enhance social participation of autistic people: A classification and literature review. *Res. Autism Spectr. Disord.* **2024**, *117*, 102453. [CrossRef]
9. Ishii, A.; Tsuruta, M.; Suzuki, I.; Nakamae, S.; Suzuki, J.; Ochiai, Y. Let your world open: Cave-based visualization methods of public virtual reality towards a shareable vr experience. In Proceedings of the 10th Augmented Human International Conference 2019, Ser. AH2019, Reims, France, 11–12 March 2019; Association for Computing Machinery: New York, NY, USA, 2019; pp. 1–8.
10. Yang, X.; Malak, R.C.; Lauer, C.; Weidig, C.; Hagen, H.; Hamann, B.; Aurich, J.C.; Kreylos, O. Manufacturing system design with virtual factory tools. *Int. J. Comput. Integr. Manuf.* **2015**, *28*, 25–40. [CrossRef]
11. Shalaby, W.; Zadrozny, W. Patent retrieval: A literature review. *Knowl. Inf. Syst.* **2019**, *61*, 631–660. [CrossRef]
12. Taotao, S.; Yun, L. Development of virtual reality technology research via patents data mining. In *Proceedings of the 2009 3rd International Conference on Teaching and Computational Science (WTCS 2009) Volume 2: Education, Psychology and Computer Science; Advanced Technology in Teaching*; Springer: Berlin/Heidelberg, Germany, 2012; pp. 111–116.
13. Evangelista, A.; Ardito, L.; Boccaccio, A.; Fiorentino, M.; Petruzzelli, A.M.; Uva, A.E. Unveiling the technological trends of augmented reality: A patent analysis. *Comput. Ind.* **2020**, *118*, 103221. [CrossRef]
14. Visbox, Inc. V isCubeT M C4-4K (Legacy). 2020. Available online: https://www.visbox.com/products/cave/viscube-c4-4k/ (accessed on 7 June 2024).
15. de Paula Ferreira, W.; Armellini, F.; de Santa-Eulalia, L.A.; Rebolledo, C. Modelling and simulation in industry 4.0. In *Artificial Intelligence in Industry 4.0*; Springer: Cham, Switzerland, 2021; pp. 57–72.
16. de Paula Ferreira, W.; Armellini, F.; de Santa-Eulalia, L.A.; Thomasset-Laperrière, V. A framework for identifying and analysing industry 4.0 scenarios. *J. Manuf. Syst.* **2022**, *65*, 192–207. [CrossRef]
17. Míguez, J.L.; Porteiro, J.; Pérez-Orozco, R.; Patiño, D.; Gómez, M.Á. Biological systems for ccs: Patent review as a criterion for technological development. *Appl. Energy* **2020**, *257*, 114032. [CrossRef]
18. Abbas, A.; Zhang, L.; Khan, S.U. A literature review on the state-of-the-art in patent analysis. *World Pat. Inf.* **2014**, *37*, 3–13. [CrossRef]
19. Feng, L.; Li, Y.; Liu, Z.; Wang, J. Idea generation and new direction for exploitation technologies of coal-seam gas through recombinative innovation and patent analysis. *Int. J. Environ. Res. Public Health* **2020**, *17*, 2928. [CrossRef] [PubMed]
20. Zhou, X.; Li, S. Immersive Virtual Display System and Display Method of Cave (Cave Automatic Virtual Environment). 10 December 2014. Available online: https://lens.org/199-805-588-984-464 (accessed on 7 June 2024).
21. Mayer, E.; Odaker, T.; Kolb, D.; Müller, S.; Kranzlmüller, D. Led cave-new dimensions for large-scale immersive installations. In Proceedings of the 2024 IEEE Conference on Virtual Reality and 3D User Interfaces Abstracts and Workshops (VRW), Orlando, FL, USA, 16–21 March 2024; pp. 515–519.
22. Zhang, C.; Hu, Y.; Lyu, Y.; Chen, D.; Liang, Y.; He, L.; Liu, S. Multi-Channel Cave Type Projection Method. 6 November 2020. Available online: https://lens.org/084-003-292-225-085 (accessed on 7 June 2024).
23. Tseng, C.H. System and Method for Immersive Cave Application. 1 November 2018. Available online: https://lens.org/089-823-476-843-553 (accessed on 7 June 2024).
24. Neira, C.C.; Reiners, D. Portable Cave Automatic Virtual Environment System. 10 September 2019. Available online: https://lens.org/085-457-751-703-377 (accessed on 7 June 2024).
25. Breitzman, A.F.; Mogee, M.E. The many applications of patent analysis. *J. Inf. Sci.* **2002**, *28*, 187–205. [CrossRef]
26. Cruz-Neira, C.; Sandin, D.J.; DeFanti, T.A.; Kenyon, R.V.; Hart, J.C. The CAVE: Audio visual experience automatic virtual environment. *Commun. ACM* **1992**, *35*, 64–72. [CrossRef]
27. Cruz-Neira, C.; Leigh, J.; Papka, M.; Barnes, C.; Cohen, S.M.; Das, S.; Engelmann, R.; Hudson, R.; Roy, T.; Siegel, L.; et al. Scientists in wonderland: A report on visualization applications in the CAVE virtual reality environment. In Proceedings of the 1993 IEEE Research Properties in Virtual Reality Symposium, San Jose, CA, USA, 25–26 October 1993; pp. 59–66.

engineering proceedings

MDPI

Proceeding Paper

Design of Maximally Permissive Controllers for Solving Deadlock Problems in Flexible Manufacturing Systems †

Yen-Liang Pan [1,*], Wen-Yi Chuang [2], Kuang-Hsiung Tan [2] and Ching-Yun Tseng [3]

1 Department of Aeronautical Engineering, Chaoyang University of Technology, Taichung 413310, Taiwan
2 Department of Electrical and Electronic Engineering, Chung Cheng Institute of Technology, National Defense University, Taoyuan 335009, Taiwan; king731107@gmail.com (W.-Y.C.); s913115@gmail.com (K.-H.T.)
3 Mathematics and Physics Division, General Education Center, R.O.C. Air Force Academy, Kaohsiung City 820009, Taiwan; xxzz5801@gmail.com
* Correspondence: peterpan960326@gmail.com
† Presented at the 2024 IEEE 7th International Conference on Knowledge Innovation and Invention, Nagoya, Japan, 16–18 August 2024.

Abstract: Industry 5.0 aims to integrate humans and machines to achieve greater productivity, personalization, and sustainable development in the production process. Built on the foundation of Industry 4.0 which emphasizes automation, digitalization, and intelligent production processes, Industry 5.0 highlights the importance of human resources in modern manufacturing. Robotic arms have replaced traditional manpower, particularly in flexible manufacturing systems. However, integrating advanced machinery into workflows has increased competition in terms of securing resources, resulting in frequent deadlocks. Various deadlock prevention policies have been proposed to address this issue. Despite these efforts, resolving system deadlocks while achieving the optimal number of reachable states remains challenging. Based on existing research, we have developed a novel deadlock recovery method applicable to various flexible manufacturing systems. We designed an adaptable system and a controller that can restore the system to its fully operational state.

Keywords: Petri Net; flexible manufacturing systems; controller synthesis

1. Introduction

System deadlocks are caused repeatedly in flexible manufacturing systems (FMSs) [1,2], resource allocation systems (RASs) [3,4], and automated manufacturing systems (AMSs) [5–7], so solving problems related to these deadlocks is a significant and urgent task. The Petri Net (PN) theory is the one most widely utilized for dealing with deadlock problems. In this study, we adopted reachability graph analysis as our main technique, which with structural analysis are the two main analyzing techniques in PN theory. Our deadlock recovery strategy has demonstrated permissiveness and an optimal performance. Unlike the traditional deadlock prevention strategy, our deadlock recovery strategy improves the system liveness of deadlock states rather than forbidding them. Lu et al. [8] developed a resource flow graph (RFG)-based deadlock recovery method to mitigate resource competition. This method features a low computation cost and an intuitive application, with no need to calculate the whole reachability graph. Tseng et al. [9] improved the RFG-based method to reduce the number of controllers and maintain the same performance.

We have developed an optimal deadlock recovery method to solve system deadlock problems. The existing RFG-based methods [8,9] built transition-based controllers that release the resources from all operations of the RFG and return them to idle and resource places. However, this method requires a lot of reallocating of whole resources. We rearranged the partial resources in the RFG to maintain permissiveness and reachability. An example of a system of simple sequential processes with resources(S^3PR) net, modeled on the PN theory, was introduced for demonstration.

The rest of this article is laid out as follows: Section 2 provides an overview of the basic PN theory. Section 3 clarifies the central concept and the definition of an HRC. Section 4 introduces a classic PN model of S^3PR net to demonstrate the proposed method and compare it with existing works. The final section gives a summary and our conclusions.

2. Preliminaries

According to basic PN theory [10], the PN model of S^3PR net is a four-tuple $N = (P, T, F, W)$. A PN is illustrated as a directed bipartite graph, which is also called a Place/Transition net. P and T are two nonempty disjoint finite sets of places and transitions, respectively, where $P \cap T = \emptyset$ and $P \cup T \neq \emptyset$. F denotes the set of arcs connecting places and transitions, such that $F \subseteq (P \times T) \cup (T \times P)$. $W : F \to \mathbb{N}^+$ is the weighted value of each arc in F. Each place of N is marked with one or more tokens at the current time to represent its property. A marking (or state) is a one-dimensional array mapping $M : P \to \mathbb{Z}$ and M_0 is the initial marking of N. $M(p)$ denotes the marking of any place $p \in P$ within M.

Identical to graph theory, ${}^\bullet x$ and $x^\bullet$ indicate the pre- and post-set of the node x of a PN model, no matter whether x is a place or transition. Given a transition $t \in T$ for firing at marking M, $\forall p \in {}^\bullet t : M(p) \geq W(p,t)$, which is also denoted as $M[t\rangle$. $\neg M[t\rangle$ means t is disabled for the firing at marking M. An S^3PR net contains idle, operation, and resource places, which are denoted as P^0, P_A, and P_R. Each sequential process of an S^3PR consists of one idle place and certain operation places, and different processes are connected by various shared resources.

Definition 1. *Let* $N = \left(P^0 \cup P_A \cup P_R,\ T,\ F,\ W,\ M_0\right)$ *be an* S^3PR *net which holds the following:*

(1) P_A *is the set of operation places,* $\forall p_a \in P_A$, $M_0(p_a) = 0$.

(2) P_R *is the set of resource places,* $\forall p_r \in P_R$, ${}^\bullet p_r \cap p_r^\bullet = \varnothing$; $\forall p_a \in P_A$, $t_i \in {}^\bullet p_a$, $t_j \in p_a{}^\bullet$, $\exists p_r \in P_R$, ${}^\bullet t_i \cap P_R = t_j^\bullet \cap P_R = \{p_r\}$; $\forall p_r \in P_R$, ${}^{\bullet\bullet} p_r \cap P_A = p_r^{\bullet\bullet} \cap P_A \neq \varnothing$.

(3) P^0 *is the set of idle places and each sequential process contains one idle place* p^0.

Using reachability graph analysis, all markings that were reachable from the initial one were found for further study. A reachability graph consists of nodes, which are labeled as markings, and arrows, labeled with transitions. $R(N, M_0)$ denotes the reachability graph of the PN model N with the initial marking M_0. $\mathfrak{M}$ indicates the set of all markings, including M_0. A reachable marking M can be reached by firing a sequence of transitions $\sigma = t_1t_2t_3 \ldots t_{n-1}t_n$, where $M_0[\sigma\rangle M$. A deadlock marking M_D does not reach any other ones due to all transitions being disabled. To develop an optimal deadlock recovery solution, we used a control transition instead of a control place to deal with the deadlock problem. A control transition is a generated transition that is enabled at certain deadlock markings.

3. Hold and Request Circuit-Based Deadlock Recovery Policy

In this study, we built an optimal deadlock recovery method to provide maximum permissiveness and system liveness. A structure named a hold and request circuit (HRC) was used to enhance efficiency.

3.1. HRC

In S^3PR systems, multiple production processes usually request shared resources to follow the steps of production. When any resource is held by one of the relevant processes while it waits for other resources, the others cannot obtain this resource again. System deadlocks caused by a lack of shared resources are undesirable and require full recovery. Via the specific subnet HRC, the flow between the resources and operation is checked, and then a set of well-designed control transitions is added to the system to reallocate those held resources.

An HRC contains operation and resource places, with connecting arrows between them used to present the flow relationships of the competing resources. As aforementioned, a deadlock occurs when resources are held by certain processes and these processes are waiting for other resources to carry out their next operation. When partial operation places and resource places are at a standstill and keep waiting, circular holding and waiting occur, which is one of the conditions of a deadlock. It is necessary to check all HRCs and build corresponding control transitions.

Definition 2. *Given an S^3PR net $N = (P^0 \cup P_A \cup P_R,\ T,\ F,\ W,\ M_0)$, a valid HRC $\mathbb{G}_{HRC}(\mathcal{V},\mathcal{U})$ holds:*

(1) *$\mathcal{V}$ is the set of nodes in the HRC, including operation places $\mathcal{V}_A \subseteq P_A$ and resource places $\mathcal{V}_R \subseteq P_R$, such that $\mathcal{V} = \mathcal{V}_A \cup \mathcal{V}_R$.*

(2) *$\mathcal{U}$ is the set of edges in the HRC, implying a flow of circular holding and waiting. There are two types of resource flow: requesting resources and releasing them from operation, where $\mathcal{U} = \{v_a \to v_r \mid \exists t \in T,\ (v_r,\ t) \cup (t,\ v_a) \subseteq F\} \cup \{v_a \to v_r \mid \exists t \in T,\ (v_r,\ t) \cup (v_a,\ t) \subseteq F\}$*

(3) *Given an HRC with $\mathcal{V}_A = \bigcup_{i \in \mathbb{N}^+}^{n} v_{a_i}$ and $\mathcal{V}_R = \bigcup_{i \in \mathbb{N}^+}^{n} v_{r_i}$, $\forall v_a \in \mathcal{V}_A$, $v_r \in \mathcal{V}_R$, ${}^\bullet v_a \cup v_a^\bullet \subseteq \mathcal{V}_R$, ${}^\bullet v_r \cup v_r^\bullet \subseteq \mathcal{V}_A$.*

(4) *It is noteworthy that there is no repetition of any node in each HRC.*

3.2. Control Transition

Each valid HRC can form one corresponding control transition by reallocating resources. According to the concept of circular holding and waiting, the products and resources from idle and resource places are held in operation places. Therefore, control transitions are designed to solve these standstill conditions, release the products and the resources from operation places, and send them to idle places and resource places.

Definition 3. *Given an HRC $\mathbb{G}_{HRC}(\mathcal{V},\mathcal{U})$, its corresponding control transition t_c has input arcs from all $v_a \in \mathcal{V}_A$, output arcs into all of $v_r \in \mathcal{V}_R$, and an idle place $p^0 \in P^0$ belonging to the same process as v_a.*

3.3. Efficiency Enhancement

In the HRC-based deadlock recovery method, all valid HRCs of the system are found, and related control transitions are generated. All the operation places and resource places of the HRCs are considered within obtaining solutions, but they also bring the requirement of higher consumption for the rearrangement of resources. We assume that not all nodes of the HRCs are necessary for developing solutions. An S^3PR net contains two or more sequential processes with shared resources, with each process being a circuit of one idle place and one operation place. Each pair of operation places from the same HRC belongs to the same process, i.e., they share the same idle place p^0. In a deadlock marking, all operation places in the HRC cannot progress any further. Given two operation places v_a^1 and v_a^2 belong to the same HRC and share the same idle place p^0, the resources in v_a^1 and v_a^2

are not sent to the next operation place in this production circuit. As v_a^1 is closer to the end of this process than v_a^2, it releases its resources to partially recover the processing sequence. That is to say, in any sequential process, only one operation place needs to be considered when generating a control transition.

4. Experimental Results

A classic example of S^3PR net was used for the demonstration of our proposed HRC-based method.

Example 1. *Considering the PN model in Figure 1, there are 19 places and 14 transitions in total, where* $P^0 = \{p_1, p_8\}$, $P_A = \{p_2, p_3, p_4, p_5, p_6, p_7, p_9, p_{10}, p_{11}, p_{12}, p_{13}\}$, *and* $P_R = \{p_{14}, p_{15}, p_{16}, p_{17}, p_{18}, p_{19}\}$. *The initial marking* $M_0 = 6p_1 + 6p_8 + p_{14} + p_{15} + p_{16} + p_{17} + p_{18} + p_{19}$.

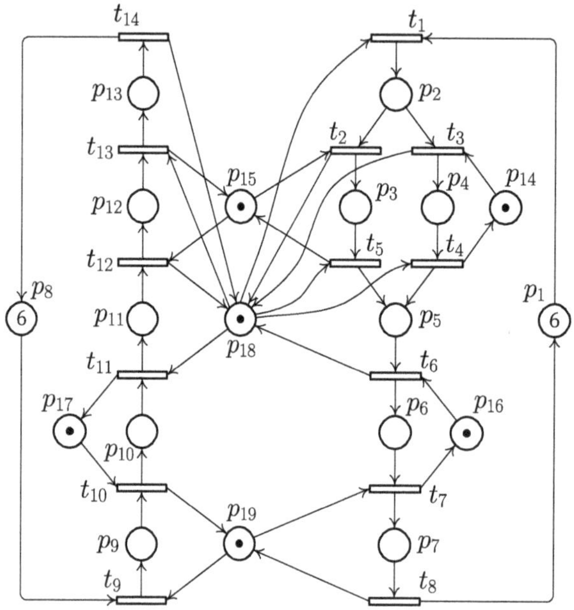

Figure 1. Example of S^3PR net.

According to HRC-based deadlock recovery policies [8,9], there are four HRCs in Example 1. Their contents are shown in Table 1.

Table 1. Contents of all HRCs.

HRC	$\mathcal{V}_A$	$\mathcal{V}_R$	$\mathcal{U}$
$\mathbb{G}_{HRC_1}$	p_2, p_4	p_{14}, p_{18}	$p_2 \to p_{14}, p_{14} \to p_4, p_4 \to p_{18}, p_{18} \to p_2$
$\mathbb{G}_{HRC_2}$	p_3, p_{11}	p_{15}, p_{18}	$p_{11} \to p_{15}, p_{15} \to p_3, p_3 \to p_{18}, p_{18} \to p_{11}$
$\mathbb{G}_{HRC_3}$	p_{11}, p_{12}	p_{15}, p_{18}	$p_{11} \to p_{15}, p_{15} \to p_{12}, p_{12} \to p_{18}, p_{18} \to p_{11}$
$\mathbb{G}_{HRC_4}$	p_5, p_6, p_9, p_{10}	$p_{16}, p_{17}, p_{18}, p_{19}$	$p_{18} \to p_5, p_5 \to p_{16}, p_{16} \to p_6, p_6 \to p_{19}$, $p_{19} \to p_9, p_9 \to p_{17}, p_{17} \to p_{10}, p_{10} \to p_{18}$

There are two sequential processes in the net in Figure 1, including $p_1t_1p_2t_2p_3t_5(t_3p_4t_4)$ $p_5t_6p_6t_7p_7t_8$ and $p_8t_9p_9t_{10}p_{10}t_{11}p_{11}t_{12}p_{12}t_{13}p_{13}t_{14}$. Only $\mathbb{G}_{HRC_4}$ has more than one operation place belonging to each sequential process; the other HRCs have one operation place each. According to Section III-C, $\mathbb{G}_{HRC_4}$ can be replaced by $\mathbb{G}^{\star}_{HRC_4} = (\mathcal{V}^{\star}_4, \mathcal{U}^{\star}_4)$ when $\mathcal{V}^{\star}_4 = \{p_6, p_{10}, p_{16}, p_{17}\}$. The newly generated control transitions are shown in Table 2.

Table 2. Generated control transitions.

t_{c_i}	$^{\bullet}t_{c_i}$	$t^{\bullet}_{c_i} \cap P_R$	$t^{\bullet}_{c_i} \cap P^0$
t_{c_1}	p_2, p_4	p_{14}, p_{18}	$2p_1$
t_{c_2}	p_3, p_{11}	p_{15}, p_{18}	p_1, p_8
t_{c_3}	p_{11}, p_{12}	p_{15}, p_{18}	$2p_8$
t_{c_4}	p_6, p_{10}	p_{16}, p_{17}	p_1, p_8

After four control transitions $t_{c_1}, t_{c_2}, t_{c_3}, t_{c_4}$ are added to the original net in Figure 1, the controlled net is checked with no any reachable deadlock marking by simulation software. The comparative results of the HRC-based methods are shown in Table 3, where the proposed method provides a better performance.

Table 3. Comparison of all RFG-based methods.

Methods	$\lvert \mathfrak{M} \rvert$	$\lvert T_c \rvert$	$\lvert ^{\bullet}T_c \cup T^{\bullet}_c \rvert$	Liveness
[8]	282	6	42	Deadlock-free
[9]	282	4	30	Deadlock-free
Proposed	282	4	24	Deadlock-free

5. Conclusions

We developed an HRC-based deadlock recovery policy that led to an improvement in performance and efficiency. The experimental and comparative results showed that the proposed method is better than other HRC-based ones. In future research, it will be necessary to consider a more complicated system and enhance that system's performance.

Author Contributions: Conceptualization, Y.-L.P. and C.-Y.T.; methodology, W.-Y.C.; software, W.-Y.C. and K.-H.T.; validation, W.-Y.C. and C.-Y.T.; formal analysis, Y.-L.P. and W.-Y.C.; investigation, Y.-L.P.; resources, K.-H.T.; data curation, W.-Y.C.; writing—original draft preparation, W.-Y.C. and C.-Y.T.; writing—review and editing, Y.-L.P. and K.-H.T.; visualization, C.-Y.T.; supervision, Y.-L.P.; project administration, Y.-L.P. All authors have read and agreed to the published version of the manuscript.

Funding: This research received no external funding.

Institutional Review Board Statement: Not applicable.

Informed Consent Statement: Not applicable.

Data Availability Statement: The original contributions presented in this study are included in the article. Further inquiries can be directed to the corresponding author.

Conflicts of Interest: The authors declare no conflicts of interest.

References

1. Košťál, P.; Sobrino, D.R.D. Flexible Manufacturing System for Drawingless Manufacturing. *Key Eng. Mater.* **2013**, *581*, 527–532. [CrossRef]
2. Liu, D.; Li, Z.; Zhou, M. Hybrid Liveness-Enforcing Policy for Generalized Petri Net Models of Flexible Manufacturing Systems. *IEEE Trans. Syst. Man Cybern. Syst.* **2013**, *43*, 85–97. [CrossRef]
3. Lee, S.; Tilbury, D.M. Deadlock-Free Resource Allocation Control for a Reconfigurable Manufacturing System With Serial and Parallel Configuration. *IEEE Trans. Syst. Man Cybern. C* **2007**, *37*, 1373–1381. [CrossRef]

4. Reveliotis, S.; Fei, Z. Robust Deadlock Avoidance for Sequential Resource Allocation Systems With Resource Outages. *IEEE Trans. Automat. Sci. Eng.* **2017**, *14*, 1695–1711. [CrossRef]
5. Chen, C.; Hu, H. Extended Place-Invariant Control in Automated Manufacturing Systems Using Petri Nets. *IEEE Trans. Syst. Man Cybern. Syst.* **2022**, *52*, 1807–1822. [CrossRef]
6. Alzalab, E.A.; El-Sherbeeny, A.M.; El-Meligy, M.A.; Rauf, H.T. Trust-Based Petri Net Model for Fault Detection and Treatment in Automated Manufacturing Systems. *IEEE Access* **2021**, *9*, 157997–158009. [CrossRef]
7. Feng, Y.; Xing, K.; Zhou, M.; Wang, X.; Liu, H. Robust Deadlock Prevention for Automated Manufacturing Systems With Unreliable Resources by Using General Petri Nets. *IEEE Trans. Syst. Man Cybern. Syst.* **2020**, *50*, 3515–3527. [CrossRef]
8. Lu, Y.; Chen, Y.; Li, Z.; Wu, N. An Efficient Method of Deadlock Detection and Recovery for Flexible Manufacturing Systems by Resource Flow Graphs. *IEEE Trans. Automat. Sci. Eng.* **2022**, *19*, 1707–1718. [CrossRef]
9. Tseng, C.-Y.; Chen, J.-C.; Pan, Y.-L. An Improved Deadlock Recovery Policy of Flexible Manufacturing Systems Based on Resource Flow Graphs. *IEEE Access* **2024**, *12*, 65202–65212. [CrossRef]
10. Murata, T. Petri Nets: Properties, Analysis and Applications. *Proc. IEEE* **1989**, *77*, 541–580. [CrossRef]

engineering proceedings

MDPI

Proceeding Paper

Improved Control Mechanism of Bottleneck Bandwidth and Round-Trip Propagation Time v3 Congestion with Enhanced Fairness and Efficiency †

Hung-Chi Chu * and Hao-Chu Chiang

Department of Information and Communication Engineering, Chaoyang University of Technology, Taichung 413310, Taiwan; acbhui4061@gmail.com
* Correspondence: hcchu@cyut.edu.tw

† Presented at the 2024 IEEE 7th International Conference on Knowledge Innovation and Invention, Nagoya, Japan, 16–18 August 2024.

Abstract: The widespread adoption and popularity of various applications have led to large and frequent data transmissions, resulting in network congestion, high packet delays, and packet loss. In 2016, Google proposed the Bottleneck Bandwidth and Round-trip propagation time (BBR) algorithm to mitigate network congestion. However, its network fairness is poor. Consequently, BBRv2 and BBRv3 were introduced in 2018 and 2023 as improved versions. Although BBRv2 exhibited enhanced fairness, its bandwidth utilization rate was lower than that of other existing methods. Meanwhile, BBRv3 still lacked bandwidth fairness in its initial transmission. Therefore, we have improved the fairness based on BBRv3 by considering the maximum sending rate and utilizing connections at different times. Good fairness and bandwidth utilization are maintained on the bottleneck bandwidth with the improved method. The method outperforms Cubic Binary Increase Congestion Control (CUBIC) and BBRv3 in terms of bandwidth utilization and network usage fairness.

Keywords: BBR; congestion control

1. Introduction

With the widespread adoption of the Internet, people's demand for and reliance on it has increased significantly. The proliferation of real-time and multimedia transmissions has exposed the limitations of traditional network structures. Although Transmission Control Protocol/Internet Protocol (TCP/IP) [1,2] has successfully addressed the challenge of connecting data and messages between different computers, the growing network demand and increased data transfer from connected devices have caused network overload. This results in congestion or even paralysis, causing high latency, packet traffic congestion, and packet loss.

The TCP congestion control mechanism has evolved to address these congestion issues. Early TCP implementations used a timeout-based mechanism, assuming that packet loss occurred when a packet took too long to traverse the network, reducing the transmission rate. However, this approach proved too conservative, causing the TCP transfer rates to drop too rapidly when real congestion occurred, thus undermining the overall network performance. As time progressed, congestion control mechanisms such as Tahoe, Reno, CUBIC, and their related extended versions were developed to better address the growing demands of modern networks. However, these congestion control mechanisms continue to compete with one another in terms of bandwidth utilization and fairness.

In this context, Google proposed BBR in 2016, a new congestion control algorithm. BBR aims to maximize network transmission rate utilization while keeping latency low to provide a better user experience. One of BBR's key features is its ability to proactively probe the network environment. It monitors the network's bandwidth and round-trip propagation time in real time, adjusting the transmission rate based on this information. This active probing capability allows BBR to adapt quickly to network changes and effectively avoid congestion. BBR incorporates advanced congestion control algorithms, including ProbeBW and ProbeRTT, to better handle various network environments in steady-state situations. The successful application of BBR in Google's services, such as YouTube and Google Cloud, demonstrates its excellent performance in modern network environments. Due to its effectiveness, BBR is gradually gaining attention from other ISPs and has become an important trend in network congestion control.

However, BBR has shortcomings in terms of fairness during transmission. In BBRv1, connections with different congestion control mechanisms such as CUBIC resulted in almost 10 s of near-maximum bandwidth for each side, leading to unfair behavior. BBRv2 improved the ProbeBW stage and introduced Explicit Congestion Notification (ECN) support and 3% packet loss tolerance, resulting in better fairness but with much lower bandwidth utilization compared to that of CUBIC. BBRv3 reduces the sending rate and adjusts parameters to better optimize fairness, but unfair situations still occur in the early and mid-term stages of connections.

This study aims to address the drawbacks of BBRv3, which struggles to achieve fair bandwidth allocation under different congestion control scenarios. The proposed method involves restricting various parameters during the transmission process. Specifically, the StartUP stage is used to probe the bandwidth status parameters, which are then adjusted to optimize the network bandwidth fairness. This method mitigates the deficiencies observed in BBRv3.

2. Related Work

Congestion controls based on packet loss and active probing are developed in this study.

2.1. Congestion Control Based on Packet Loss

CUBIC is a congestion control algorithm stemming from BIC [3]. It differs from the previously mainstream Reno version as its primary feature is the use of a cubic function to adjust the congestion window. This adjustment is conducted based on the time elapsed since the last decrease in the congestion window. The cubic function equation is defined as shown in (1):

$$W(t) = C \times (t - K)^3 + W_{max} \tag{1}$$

where C is the CUBIC parameter, t is the time elapsed since the last window reduction, and K is the time required to increase W to W_{max} when no further packet loss occurs.

When congestion occurs, CUBIC records the current congestion window and decreases it in a multiplicative manner to alleviate network congestion. Then, it rapidly increases the window to W_{max} using a concave segment of a cubic function to restore the congestion window to the maximum traffic level that existed before packet loss occurred. Once the congestion window reaches W_{max}, CUBIC maintains this level for a period, carefully probing for packet loss to determine if the limits of the available bandwidth have been reached. If no packet loss occurs, CUBIC uses convex functions to start increasing the congestion window until another packet loss event occurs. Notably, CUBIC's design is not related to the Round-Trip Time (RTT) but to the time interval between two packet loss events, as shown in Figure 1. CUBIC has three phases: to find current bandwidth, to avoid congestion, and to probe more bandwidth. When a timeout occurs, packets will be lost, and the process ends the "to find the current bandwidth" phase and enters the "to avoid congestion" phase. At this time, W is W_{max}. In the "to avoid congestion" phase, when $W = W_{max}/2$, it enters the "to probe more bandwidth" phase. If a Timeout occurs during this phase, the process returns to the "to avoid congestion" phase. The threshold is defined as the difference between the congestion window values at the end of the most recent "to avoid congestion" phase.

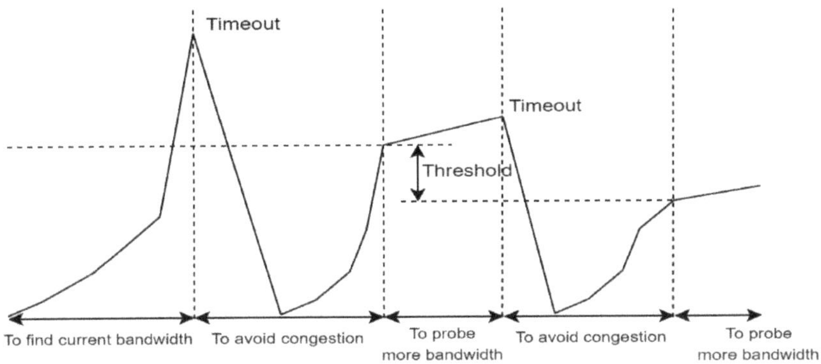

Figure 1. Schematic of CUBIC congestion control.

2.2. Congestion Control Based on Active Probing

BBRv1 probes the maximum available bandwidth and delay state to constantly detect the maximum bandwidth (maxBw) and the minimum RTT (minRTT) of packet transmission [4–6]. It controls the pacing_gain and cwnd_gain parameters to increase the packet transmission rate and expand the congestion window, allowing more packets to be transmitted. BBRv1 permits a 3% packet loss rate to probe for the maximum bandwidth. The transmission bandwidth size is determined based on the ratio of the total transmission to delay to reduce buffer congestion and increase the transmission speed. This approach enables better adaptation to various network environments.

BBRv2 optimizes the bandwidth-hogging in multiple connections to achieve higher fairness and reduce transmission delay [7]. Key additions include the calculation of the packet loss rate and the incorporation of ECN in congestion determination. It allows for a packet loss of 3% and an ECN of 50% to minimize the accumulation in the buffer queue. BBRv2 introduces four stages to ProbeBw and adds limits on the minimum (inflight_lo) and maximum (inflight_hi) number of packets transmitted. These changes reduce bandwidth-hogging issues when interacting with other congestion-controlled connections, thus improving bandwidth sharing. The modifications address the problem of slow convergence to optimal fairness in larger buffer zones. However, while BBRv2 makes significant improvements, it does not fully realize optimized fair sharing in all scenarios.

Based on the result of Ref. [8], BBRv3 makes the following adjustments. The packet sending rate gain in the StartUP stage is changed from 2.89 to 2.77, the packet loss threshold is modified, and the ECN threshold parameter is adjusted. Additionally, the congestion window gain is altered from 2.89 to 2.0, along with parameter settings for bandwidth stabilization. Upon exiting the StartUP stage, the inflight_hi parameter is set based on the greater value between the estimated Bandwidth-Delay Product (*BDP*) and the maximum number of packets successfully transmitted in the last RTT. The criteria for exiting the StartUP stage due to packet loss have also been revised.

3. Improved BBR

3.1. Congestion Control Based on BBR

Congestion control relies on actively detecting the bottleneck bandwidth and estimating the round-trip propagation delay. The first version of BBR uses four stages to adjust congestion control. There are two important gain parameters, pacing_gain and cwnd_gain. Pacing_gain is used to control the delivery rate, while cwnd_gain manages the number of packets being transmitted in flight. When the StartUP stage begins detecting the bottleneck bandwidth and estimating the round-trip propagation delay, BBR sets both pacing_gain and cwnd_gain to $2/ln\ 2$, which is approximately 2.89. BBR considers the maximum transmission rate in the last 10 RTTs as the bottleneck bandwidth (*BtlBw*) and the minimum latency measured within the last 10 s as the round-trip propagation time (*RTprop*). BBR controls its transmission behavior using pacing_gain and cwnd_gain, exponentially increasing the sending rate to fully utilize the bottleneck link. If the sending rate is not increased significantly (less than 25% change within three RTTs) after three sending attempts, the actual sending rate reaches *BtlBw*. The bottleneck bandwidth is defined as shown in (2):

$$BtlBw = \max\{dt\}\ \forall t \in [T - WB\ ,\ T] \tag{2}$$

where d_t is the delivery rate, W_B is 6 to 10 RTTs, and T is the transmission time during probing. The final result obtained is the BDP as the baseline for determining the congestion control. It is assessed whether congestion is present. The congestion control operation is expressed as shown in (3):

$$BDP = BtlBw\ \times RTprop \tag{3}$$

where $RTprop$ is a physical characteristic of the connection.

The approximate value is obtained indirectly by monitoring the RTT of the packets, taking the maximum transmission among 10 RTTs as the criterion. After confirming the congestion control point, in the Drain stage, pacing_gain is set to the reciprocal of $ln\ 2/2$ from the StartUP stage since buffering occurs during detection and there is also the problem of packet loss. This clears the queue that appears during reprobing. When the number of inflight packets matches the BDP, it is determined if the queue is emptied, allowing BBR to leave the Drain stage and enter the ProbeBW stage.

In the ProbeBW stage, BBR remains in this stage for an extended period. The value of cycle_gain = pacing_gain is used to detect the bottleneck bandwidth and adjust the sending rate with values of 1.25, 0.75, and 1 to achieve continuous detection. To detect more bandwidth, the sending rate is increased by 1/4. If $RTprop$ is not updated for more than 10 s, BBR enters the ProbeRTT stage. In this stage, cwnd is reduced to four packets and maintained for 200 ms or one RTT. If the time exceeds this value, the network channel becomes full.

The key feature of BBRv2 is the enhancement of ProbeBW and the addition of four stages to adjust congestion control under steady-state conditions, as shown in Figure 2. Different pacing_gain values are used in various sub-stages of ProbeBW, and the bandwidth detection time is adaptive. This improves the coexistence of BBR and CUBIC. Additionally, an ECN threshold greater than 50% is used to alert the receiving end that congestion is imminent. A 3% packet loss rate is used to determine whether to exit the StartUP stage early and inflight_hi.

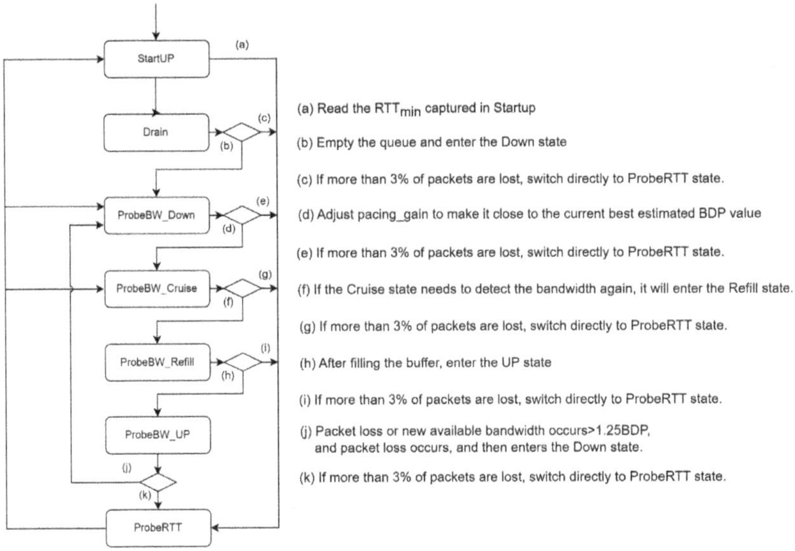

Figure 2. BBRv2 ProbeBW and ProbeRTT detailed status diagram.

The down stage is illustrated in Figure 2(b,c). In the Cruise stage, several bandwidths are reserved for other connections, with inflight kept between inflight_lo and inflight_hi (Figure 2(f,g)). In the Refill stage, the sending rate gradually increases to approach the upper limit of the estimated bandwidth (Figure 2(h,i))

In the ProbeBW_UP stage, the sending rate is accelerated until it reaches the upper limit of the estimated bandwidth to maintain steady-stage behavior. If packet loss occurs, the status transitions to the ProbeRTT stage (Figure 2(j,k)). This allows BBRv2 to maximize the use of the available bandwidth. If an ECN mark is detected or a packet is lost, the sending rate is reduced by 30%. BBRv2 uses an exponential growth method, starting slowly and then accelerating, to detect the available bandwidth. The inflight_hi value is not updated until packet loss occurs or the newly available bandwidth is greater than 1.25 times the estimated bandwidth, coinciding with packet loss.

In the adjustments mentioned, although BBRv2 sets inflight_hi and inflight_lo, the detection bandwidth cannot be effectively increased due to the influence of inflight_hi when striving for bandwidth. With a larger buffer, CUBIC detects the bandwidth based on packet loss, but BBRv2 sets inflight_hi to limit the maximum detection bandwidth. As a result, BBRv2 deviates further from fairness and cannot converge back.

BBRv3 [8,9] adjusts cwnd_gain to 2.0 during the StartUP stage and also modifies pacing_gain from 2.89 to 2.77. When exiting the StartUP stage, BBRv3 sets the inflight_hi parameter based on the maximum value between the estimated BDP and the highest number of successfully transmitted packets during the last RTT. Additionally, the criteria for exiting the StartUP stage due to packet loss are updated to reduce queuing delays during both the StartUP stage and subsequent stages, thereby decreasing the packet loss rate. Bandwidth detection continues until the packet loss rate or ECN exceeds the predefined 1% threshold. Detection stops if no packets are transmitted and the available bandwidth reaches the maximum inflight_hi.

While CUBIC congestion control offers excellent fairness, it falls short compared to the various versions of BBR in terms of bandwidth utilization. Each version of BBR actively detects and increases the bandwidth, but this aggressive approach leads to significantly poorer fairness compared to CUBIC. Although BBRv2 improves fairness compared to

BBRv1 by setting inflight_hi to limit the maximum transmission bandwidth, it does not achieve optimal fairness in larger buffers. In BBRv3, pacing_gain and cwnd_gain are reduced in the StartUP stage, and bandwidth detection is stopped if inflight_hi is not reached. This approach balances fairness and bandwidth utilization better, as illustrated in Figure 3.

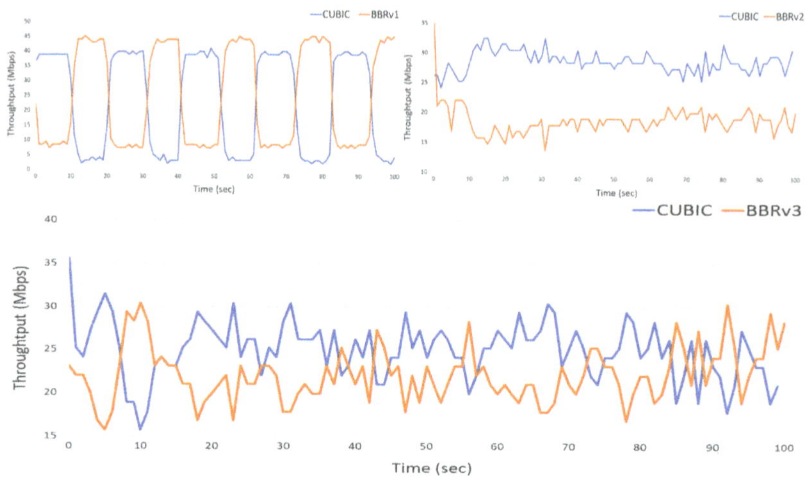

Figure 3. Comparison of throughput between CUBIC and BBR congestion control.

3.2. Fair Congestion Control Mechanism

BBRv3 and CUBIC are used in this study because CUBIC is based on packet loss and offers superior bandwidth detection and fairness compared to other packet loss-based congestion controls. To obtain better bandwidth fairness, we adjust pacing_gain to observe its impact on fairness with the generally low bandwidth of BBRv3. In the StartUP stage, BBRv3 sets pacing_gain to $4ln(2)$ for bandwidth detection. Increasing pacing_gain to $2/ln(2)$ significantly reduces fairness. The proposed pacing_gain setting in the StartUP stage is shown in Algorithm 1.

Algorithm 1: Pacing_gain setting in StartUP stage

Set $\alpha = 4$, $\beta = 2$, $\gamma = 1$ //Initial Setup for BBRv3
Flag = False
$Pacing\_gain = \alpha \times (\ln \beta)^{\gamma}$
Do {
Increasing/decreasing Pacing_gain with different α, β, γ
If Fairness of network increase and it closed to 1 then
Store α, β and γ
Accepted the Paceing_gain with α, β and γFlag = True
End if
} While (Flag == True)
Return Paceing_gain

---

4. Results

The network setup includes four virtual machines (two sending nodes and two receiving nodes) and two switches (SW1 and SW2), connected through bottleneck links for congestion control experiments (Figure 4). The GNS3 network simulator is used for the experiment. The bottleneck bandwidth is 50 Mbps, and the transmitter sending rate is 150 Mbps. Additionally, iperf3 is used for transmitting packets.

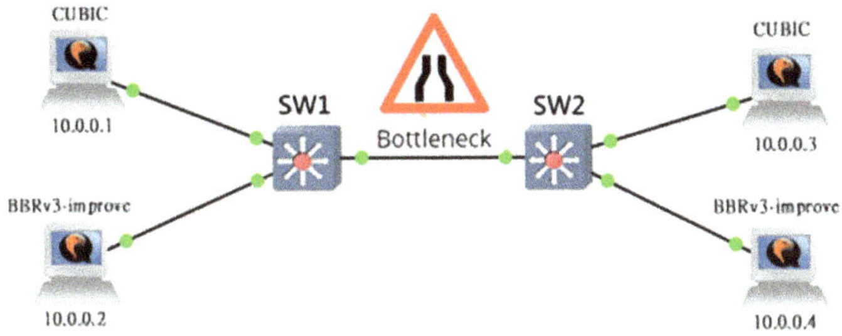

Figure 4. Network topology diagram.

In the first stage of the experiment, Flow1 and Flow2 start transmitting at the same time. The pacing_gain is set to $3ln2$, approximately 2.07, in the StartUP stage to obtain better bandwidth fairness. As shown in Figure 5, when there are two flows, Flow1 uses CUBIC and Flow2 uses BBRv3+ simultaneously in stage 1, and the bandwidth of the two flows is close to 25 Mbps with similar performance (Figure 4).

In the second stage of the experiment, Flow1 begins transmission at the 0 s mark, while Flow2 commences transmission 2 s later. The pacing_gain is set to $4ln2$, using BBRv3 in the StartUP stage of Flow2. The throughput of the two flows is shown in Figure 6. At the 0 s mark, only Flow1 uses the connection, fully utilizing the bandwidth. When Flow2 begins with the 2 s delay, bandwidth competition starts. An obvious fairness issue is visible at the 40 s mark. By the 100 s mark, Flow2 utilizes the entire bandwidth (Figure 6).

Similarly, the proposed BBRv3+ is implemented in Flow2, which began transmission two seconds after Flow1 started. At the 0 s mark, Flow1 exclusively utilizes the connection, maximizing bandwidth usage. The competition for bandwidth begins at the 2 s mark when Flow2 initiates. The proposed congestion control mechanism enables improved bandwidth fairness between the flows. As Flow1 concludes its transmission at the 100 s mark, Flow2 subsequently occupies the entire bandwidth (Figure 7).

The experimental results are shown in Table 1. With a bandwidth of 50 Mbps, each flow is allocated 25 Mbps if the bandwidth is fairly distributed between the two flows. BBRv3+ produces an output of nearly 25 Mbps, regardless of whether the transmissions start simultaneously or with a 2 s delay. This demonstrates that BBRv3+ shows a better output and fairness performance compared to the other methods.

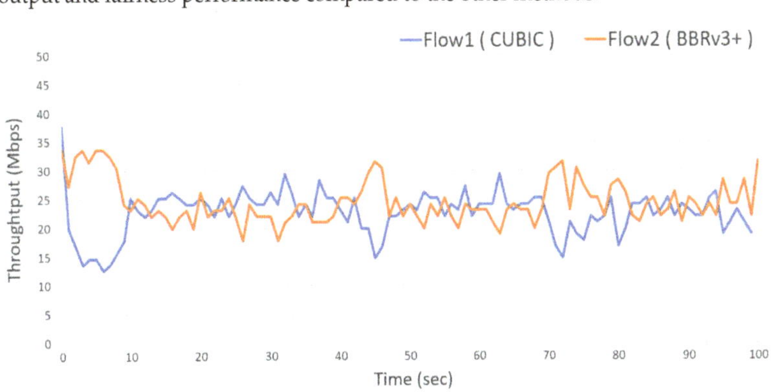

Figure 5. Comparison of throughput between CUBIC and BBRv3+ (1st stage).

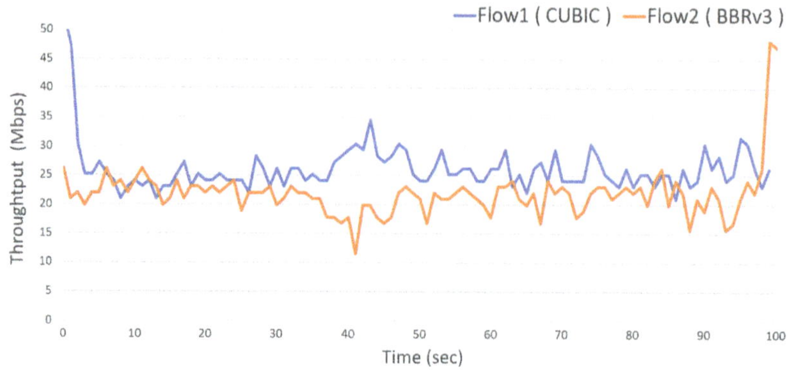

Figure 6. Comparison of throughput between CUBIC and BBRv3 (2nd stage).

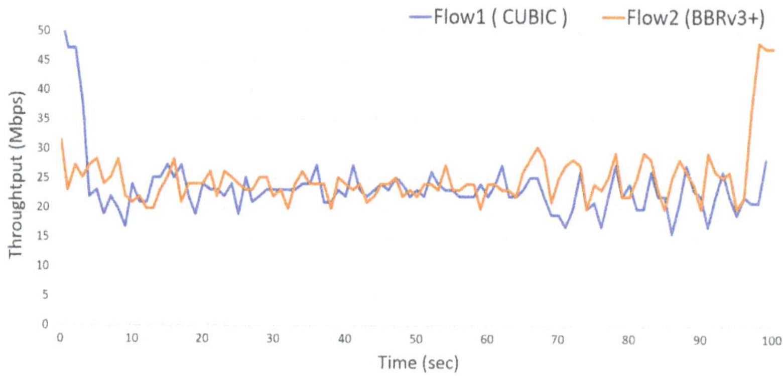

Figure 7. Comparison of throughput between CUBIC and BBRv3+ (2nd stage).

Table 1. Throughput with different congestion controls.

Flow2	Throughput	Flow1: CUBIC Max.	Flow1: CUBIC Min.	Flow1: CUBIC Avg.	Flow2 Max.	Flow2 Min.	Flow2 Avg.
1st Stage	BBRv1	40.9	2.1	23.0	45.1	7.34	24.3
	BBRv2	32.5	24.1	28.5	35.6	13.6	18.6
	BBRv3	35.6	15.7	24.9	30.4	15.7	22.1
	BBRv3+	37.7	12.6	22.7	33.6	17.8	24.5
2nd Stage	BBRv3	52.4	21	26.1	48.2	11.5	21.9
	BBRv3+	52.4	15.7	23.6	48.2	19.9	25.1

5. Conclusions

BBRv3+, the improved congestion control algorithm addresses the unfair bandwidth distribution of BBRv3. When connections are established at different times, fairness in the early and mid-term stages is significantly improved, surpassing the performance of previous BBRs. However, unfair situations persist compared to CUBIC. Therefore, it is required to improve fairness in the early and middle stages of connections by incorporating adaptive adjustments to sending rates and congestion windows across various congestion control scenarios.

Author Contributions: Conceptualization, H.-C.C. (Hung-Chi Chu) and H.-C.C. (Hao-Chu Chiang); methodology, H.-C.C. (Hao-Chu Chiang); software, H.-C.C. (Hao-Chu Chiang); validation, H.-C.C.

(Hung-Chi Chu) and H.-C.C. (Hao-Chu Chiang); formal analysis, H.-C.C. (Hung-Chi Chu) and H.-C.C. (Hao-Chu Chiang); investigation, H.-C.C. (Hung-Chi Chu) and H.-C.C. (Hao-Chu Chiang); resources, H.-C.C. (Hung-Chi Chu) and H.-C.C. (Hao-Chu Chiang); data curation, H.-C.C. (Hao-Chu Chiang); writing—original draft, H.-C.C. (Hung-Chi Chu) and H.-C.C. (Hao-Chu Chiang); writing—review and editing, H.-C.C. (Hao-Chu Chiang); visualization, H.-C.C. (Hung-Chi Chu) and H.-C.C. (Hao-Chu Chiang); supervision, H.-C.C. (Hung-Chi Chu). All authors have read and agreed to the published version of the manuscript.

Funding: This research was funded by the Ministry of Science and Technology, Taiwan, under grant NSTC 113-2221-E-324-015.

Institutional Review Board Statement: Not applicable.

Informed Consent Statement: Not applicable.

Data Availability Statement: The datasets used in the current study are available from the corresponding author upon request.

Conflicts of Interest: The authors declare no conflicts of interest. The funders had no role in the design of the study; in the collection, analyses, or interpretation of the data; in the writing of the manuscript; or in the decision to publish the results.

References

1. Cerf, V.; Kahn, R. A protocol for packet network intercommunication. *IEEE Trans. Commun.* **1974**, *22*, 637–648. [CrossRef]
2. Rose, K. *Computer Network a Top-Down Approach*, 7th ed.; Pearson: London, UK, 2018.
3. Ha, S.; Rhee, I.; Xu, L. Cubic: A new tcp-friendly high-speed tcp variant. *ACM SIGOPS Oper. Syst. Rev.* **2008**, *42*, 64–74. [CrossRef]
4. Cardwell, N.; Cheng, Y.; Gunn, C.S.; Yeganeh, S.H.; Jacobson, V. BBR: Congestion-based congestion control. *Commun. ACM* **2017**, *60*, 58–66. [CrossRef]
5. Cardwell, N.; Cheng, Y.; Gunn, C.S.; Yeganeh, S.H.; Jacobson, V. BBR Congestion Control. IETF Draft draft-cardwell-iccrg-bbr-congestion-control-00. 2017. Available online: https://datatracker.ietf.org/meeting/101/materials/slides-101-iccrg-an-update-on-bbr-work-at-google-00 (accessed on 15 May 2024).
6. BBR-Congestion Control (draft-cardwell-iccrg-bbr-congestion-control-02). 8 September 2022. Available online: https://datatracker.ietf.org/doc/html/draft-cardwell-iccrg-bbr-congestion-control (accessed on 25 June 2024).
7. Cardwell, N.; Cheng, Y.; Yeganeh, S.H.; Jha, P.; Seung, Y.; Yang, K.; Jacobson, V. BBRv2: A model-based congestion control performance optimization. In Proceedings of the IETF 106th Meeting, Singapore, 16–22 November 2019; pp. 1–32.
8. Cardwell, N.; Cheng, Y.; Yang, K.; David, M.; Yeganeh, S.H.; Jha, P.; Seung, Y.; Jacobson, V.; Swett, I.; Wu, B.; et al. BBRv3: Algorithm Bug Fixes and Public Internet Deployment. In Proceedings of the IETF 117th Meeting, San Francisco, CA, USA, 22–28 July 2023; pp. 1–36.
9. Gomez, J.; Kfoury, E.F.; Crichigno, J.; Srivastava, G. Evaluating TCP BBRv3 performance in wired broadband networks. *Comput. Commun.* **2024**, *222*, 198–208. [CrossRef]

engineering proceedings

MDPI

Proceeding Paper

Synergies Between Virtual Commissioning and Digital Twins †

Hermann Boris Djeulezeck Tanegue [1], William de Paula Ferreira [1,*], Rodrigo Furlan de Assis [1] and David Brodeur [2]

1 Department of Systems Engineering, École de Technologie Supérieure (ÉTS), 1100 Notre Dame Street, West, Montreal, QC H3C 1K3, Canada; hermann-boris.djeulezeck-tanegue.1@ens.etsmtl.ca (H.B.D.T.); rodrigo.furlan-de-assis@etsmtl.ca (R.F.d.A.)
2 Productique Québec, Sherbrooke, QC J1G 4L3, Canada; david.brodeur@productique.quebec
* Correspondence: william.ferreira@etsmtl.ca
† Presented at the 2024 IEEE 7th International Conference on Knowledge Innovation and Invention, Nagoya, Japan, 16–18 August 2024.

Abstract: Virtual commissioning (VC) and digital twins (DTs) are two key enabling technologies of Industry 4.0 (I4.0), which can enhance operational efficiency, optimize resource utilization, streamline the development and deployment of automation systems, and accelerate innovation in manufacturing and industrial processes. However, studies examining the synergy between these two technologies are still lacking. To address this research gap, this study aims to investigate the technological relationships and synergies between VC and DTs in the context of I4.0. The results suggest a strong relationship between VC and DTs since they share similar technological components, such as the digital model. Moreover, the result indicates that the concurrent use of VC and DTs can help leverage both technologies and their added value by increasing model reusability. Different strategies are proposed for combining VC and DT functionalities, highlighting how VC can support DTs, how DTs can support VC, and how both technologies can be integrated. It also provides future research directions.

Keywords: industry 4.0; virtual commissioning; digital twins; digital model; simulation

1. Introduction

Virtual commissioning (VC) and digital twins (DTs) are two primary simulation-based technologies adopted in the context of Industry 4.0 (I4.0) [1–4]. VC facilitates virtual testing and systems validation before physical commissioning, reducing integration problems and saving time for optimization or early delivery [5]. DTs involve creating digital counterparts of a physical system, which can simulate, monitor, and optimize system performance [6]. The synergistic application of VC and DTs facilitates the application of both technologies and increases the accuracy and reusability of digital models [7]. This integration is increasingly recognized as transformative, attracting increased interest from academia and industry alike [8].

VC is mainly used in designing and developing automation control systems to facilitate the testing, verification, and validation of programmable logic controller (PLC) codes and systems integration. This approach enhances system reliability from the earliest stages of development [9]. VC methods for production systems include virtual commissioning Software-in-the-Loop (SIL) and Hardware-in-the-Loop (HIL) [10]. SIL establishes a fully virtualized test environment by integrating a virtual plant with a virtual controller to enhance design accuracy and operational readiness. On the other hand, HIL involves a real controller, enabling real-time testing of the controller's physical components within a simulated operational context. Integrated into the DT framework, these VC techniques can complement and enhance the accuracy of design simulations and the operational efficiency of systems.

DTs are mainly used for product and asset life cycle management. It is based on three essential components: physical models in the real world, virtual models in a digital environment, and data connections that facilitate interactions between the two [11]. Furthermore, in exploring DT-related concepts, Kritzinger et al. [12] emphasize the importance of understanding the distinction between three key terms: "digital model", "digital shadow", and "digital twins". The digital model refers to a simplified representation of a physical object, often static and without continuous interaction with its physical counterpart. The digital shadow is more dynamic, so it reflects changes in the physical object in real time but with limited interaction. Finally, DTs represent the most advanced integration, with a digital replica constantly interacting (in real or near real-time) with its physical counterpart.

Although considerable research has been carried out on the individual applications of VC and DTs in the context of I4.0, more research is needed on their use in an integrated way. The combined architecture of VC and DTs represents an advanced integrated approach in I4.0 [10–12]. Studies generally focus on optimizing the stand-alone capabilities of VC and DTs, such as improving system testing through VC or enhancing life cycle management with DTs [13,14]. However, these works do not explore the possible synergies that arise from a cohesive application of both technologies. The intersection of VC and DTs where virtual testing and real-time system monitoring could theoretically converge thus remains to be explored.

This study aims to fill this gap by investigating how VC can complement DTs and vice versa, through a systematic review of studies implementing VC and DTs. The review identifies key statistics and trends that illustrate the current status and impact of such integrations using bibliometric and content analysis. The analysis also focuses on the methodologies and spin-offs to find out (1) how VC supports the development of DTs, (2) how DTs support the development and effectiveness of VC, and (3) the consequences of their mutual option or integration.

The remainder of this article is organized as follows. Section 2 introduces the methodology. Section 3 presents the results, followed by a discussion in Section 4. Section 5 presents avenues for future research. Lastly, Section 6 concludes this study.

2. Methodology

We adopted the preferred reporting items for systematic review and meta-analyses (PRISMA) methodology [15,16]. Its flowchart details the process during the various stages of the review (identification, eligibility assessment, and inclusion) and its checklist provides guidelines on the elements. In total, 136 articles were initially retrieved using the research protocol in Table 1. After screening, 75 records were discarded since they were outside the scope of this research—not exploring both VC and DTs narrowed down the scope to 61 studies. Six articles were excluded because DTs and VC were only used as keywords or quoted expressions. The remaining 55 articles were then prepared for in-depth review.

Table 1. Search Protocol.

Data source:	Scopus
Search string:	"Virtual commissioning" AND "Digital twin"
Period:	From 2011 (emergence of the term Industry 4.0) to 31 January 2024
Search fields:	Title, abstract, and keywords
Language:	English
Document:	Journal and Conference articles

3. Results

3.1. Bibliometric Analysis

Figure 1 shows the distribution of publications over time. This result was used to infer a growing trend in the scientific literature on the adoption or integration of VC and DTs.

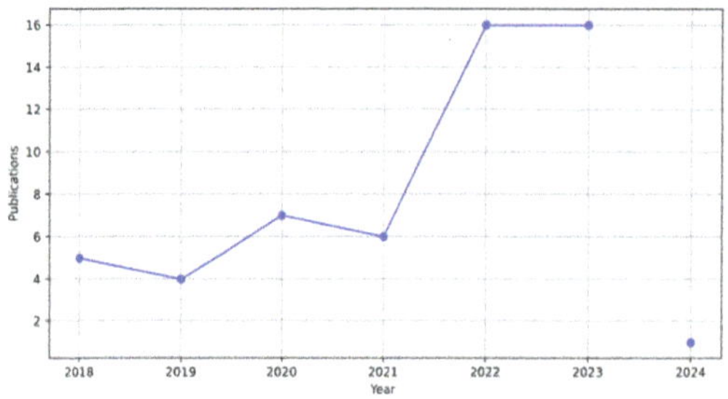

Figure 1. Number of publications per year.

We analyzed 55 articles, 38 of which were published in 26 conferences, highlighting various forums. Among these, publications in the Procedia Conference series—including Procedia CIRP and Procedia Manufacturing—totaled 11 articles, corresponding to 29%. IEEE conferences, such as the "IEEE International Conference on Emerging Technologies" and "Factory Automation", and the "IEEE International Symposium on Industrial Electronics", contributed 9 articles (23%). The remaining 18 articles (48%) were distributed among various other conferences. The review included 17 articles (31%) published in peer-reviewed journals. The journal "Machines" stands out with three articles (5.45%). Other journals, such as "Sensors", "IEEE Access" and the "International Journal of Computer Integrated Manufacturing" contributed one article, reflecting the diversity of academic forums dealing with VC and DTs.

Figure 2 illustrates the distribution of the most frequent terms taken from the publications' titles, abstracts, and keywords, using VOSviewer to visualize word density. This representation reveals three main clusters, focusing on key themes: VC, DTs, and production system.

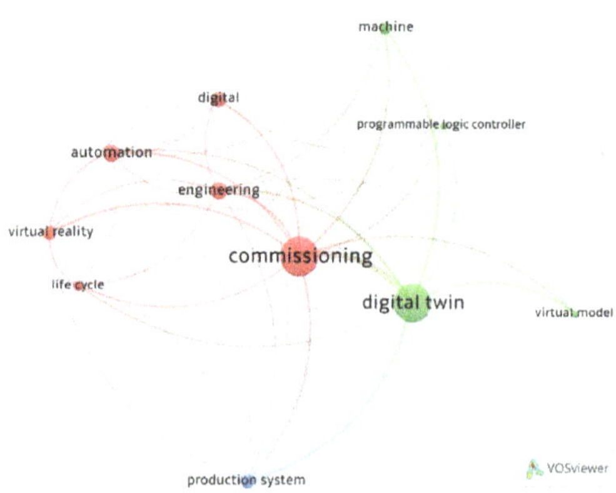

Figure 2. Keywords network visualization. The green nodes represent the *Digital Twins* cluster, the red nodes represent the *Virtual Commissioning* cluster, and the blue nodes represent the *Production Systems* cluster.

Figure 3 shows the geographical distribution by institutional affiliation of authors. Germany leads with 70 affiliations, followed by China and Spain with 29 and 19 affiliations, respectively. Belgium and Sweden also stand out with 16 and 14 affiliations, while the United Kingdom and Russia have 13 authors each. The United States and Italy have eight authors each, and Finland and Mexico have six authors. Japan and France have one and five authors.

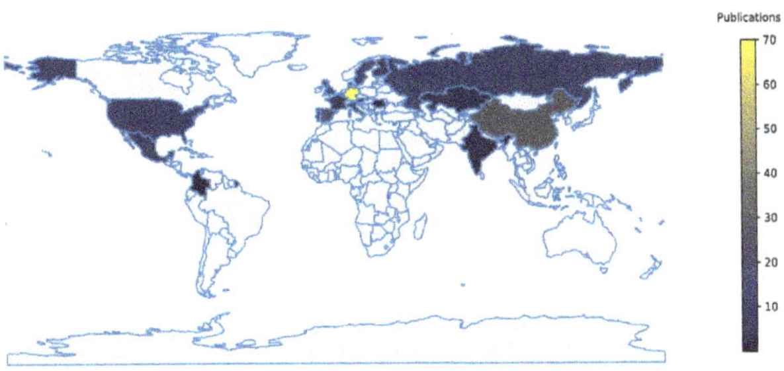

Figure 3. Geographical distribution by affiliation of authors.

Most of the research on VC combined with DTs is carried out in Europe, with Germany having the highest author contributions (Figure 3). The number of authors from China and Spain indicates a growing interest in the research area. The diversity of affiliations/countries suggests a global recognition of the importance of these technologies. Figure 4 represents a collaboration matrix of the authors' affiliations to understand frequent interactions and prominent scientific collaborations. The publications by Germany, China, and Belgium highlight national research networks. The heatmap suggests that future initiatives foster diverse collaborations, particularly between countries with significant individual contributions but less involved in collaborative efforts.

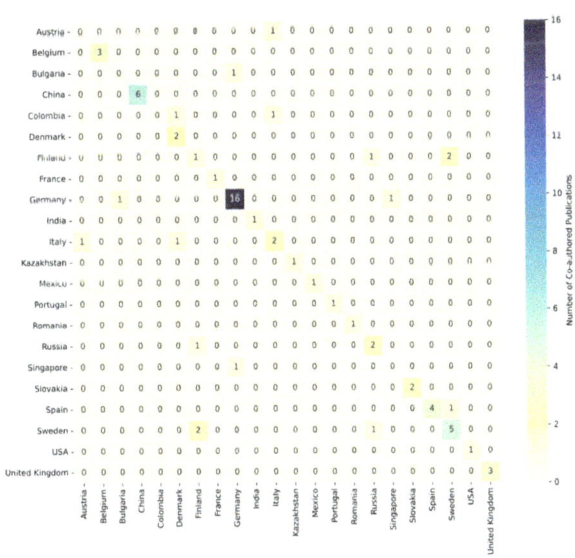

Figure 4. Collaboration. The color intensity represents the number of co-authored publications between countries, with darker shades indicating stronger collaboration.

VC allows a manufacturing system to be verified by performing an emulation involving a virtual factory and a controller or a virtual factory with a simulated controller [17]. Figure 5 shows the predominance of the SIL method with 51% of the publications. This is followed by HIL and the combination of SIL and HIL with 18% and 16%, respectively.

Figure 5 shows a preference for SIL due to its flexibility for detailed simulation without specific hardware. Despite offering a more comprehensive and integrated simulation, the combined use of SIL and HIL appears in lower proportion, suggesting limitations such as complexity and higher costs.

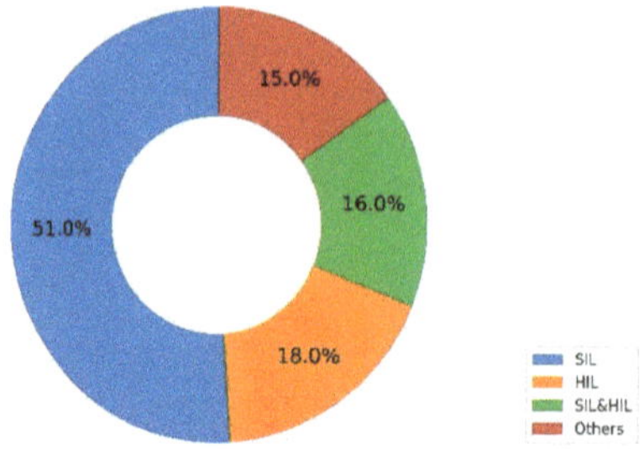

Figure 5. Virtual commissioning methods.

In total, 42% adopt the open platform communications unified architecture (OPC UA) protocol which has interoperability and security and is essential for efficient systems integration in industrial environments. The TCP/IP and PROFINET protocols are less frequent with 5 and 2%, respectively. Interestingly, 49% of the publications do not specify the communication protocol. Furthermore, 18 dealt with "Hypothetical Problems", while 17 addressed "Real Problems", indicating an almost equal split between theoretical investigations and applications. Finally, 17 articles were categorized as "Conceptual Studies". Three publications are "Literature Reviews", with different scopes from this research.

3.2. *Content Analysis*

To better grasp the concepts of VC and DTs, an in-depth comparative analysis of these technologies was carried out, highlighting their main similarities and differences (Table 2).

Table 2. Differences and Similarities of VC and DTs.

Features	VC	DTs
Purpose	Simulate and test systems before commissioning for factory acceptance tests.	Create and maintain an up-to-date digital replica of the real system.
Life cycle	Used at specific points in the plant life cycles.	It can be used throughout the product and plant life cycles.
System	Does not require an existing system.	Requires an existing system with connected devices.
Data types	Use hypothetical or simulated data.	Use real and historical data for simulation and analysis.
Optimization	Offline	Both offline and online
Technologies	Simulation models, PLCs, and communication protocols.	Mainly simulation models, sensors, actuators, smart devices, communication protocols and data service.

The link between VC and DTs reveals its potential and attracts increasing attention in research from academia and industry alike [18]. It is essential to create an environment where these applications are optimally explored [19]. This integrated approach has growing interest, encouraging further research. As illustrated in Figure 6, three strategies combining VC and DT technologies were identified.

- VC supports DTs: VC refines DTs using high-precision simulations to model realistic behavior, thus strengthening life cycle management and facilitating continuous improvement. The simulation models developed for VC are reused to facilitate DT development.
- DTs support VC: This strategy uses DTs to improve VC by verifying designs and validating control configurations before deployment, simplifying testing, and enhancing reliability.
- Integration of VC and DTs: This strategy harnesses the symbiotic relationship between VC and DTs to maximize the benefits of each technology. It ensures real-time synchronization with operational data, increasing their effectiveness from the initial design phases. One of the main approaches is to make VC a subcomponent of DTs, whose scope of application is vast and extends across the entire product and plant life cycles [20].

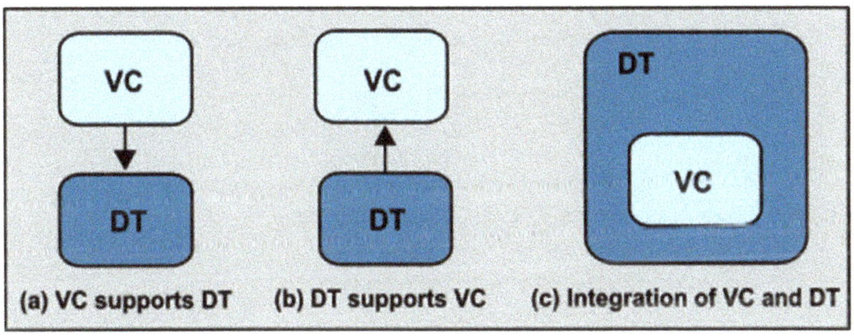

Figure 6. Taxonomy of approaches of combining VC and DTs.

Table 3 shows the examples of applications of these three approaches to combining VC with DTs. DTs reinforce VC by verifying initial designs, testing, and validating new PLC and HMI controls. At the same time, VC contributes to DT development by enabling accurate simulations and immersive experiences. The techniques used for this integration vary, from combining mechanical and electrical systems to create DTs to using advanced platforms such as Matlab Simulink and CoDeSys to simulate the behavior of complex systems. In addition, VC models are used to support and improve the efficiency of DTs, since designing DTs involves creating a virtual prototype connected to a system simulator and an industrial controller [21]. As an example, Khan et al. [22] introduce a model analysis-based platform that offers manufacturing companies an initial means of integrating a VC model that functions as a DT. These approaches demonstrate how VC and DT models are used strategically to improve the design and functionality of industrial systems.

Table 3. Examples of Approaches Combining VC and DTs.

Class	Approach	Ref.
VC supports DT	Use CoDeSys Win Control V3 and OPC with Experior to test and validate the DT designed in Matlab, facilitating optimization, simulation, and debugging in a virtual environment integrated into manufacturing systems.	[23]
	Use Automation Studio for its precision in closed-loop simulation and testing (HIL, SIL), to develop a DT with Unity, benefiting from an immersive experience and improving the interaction between the DT and its real-life counterpart.	[24]
	Integrate CoDeSys for PLC logic and OPC UA for signal exchange between Simulink and Webots, ensuring precise information processing and instant visual feedback, while developing a modular VC architecture to prepare a DT.	[25]
	Proposes a three-phase methodology for the DT: digital simulation of real-world behavior, communication via PLC code between the real and virtual worlds (VC), and integration of the simulation with real-world automation, followed by the complete implementation of the DT of a production line.	[26]
DT supports VC	Use a DT developed with Simumatik3D to facilitate reverse engineering via VC to verify initial designs and test new PLC and HMI controls, reducing cost and time to actual commissioning.	[27]
	Combine mechanical and electrical systems to create a DT, to which control logic is applied via a virtual debugging interface, solving the problems of long development cycles, high costs, and incompatibility between research and development.	[28]
	Combines semantically DT, industrial controller, and SCADA via OPC UA to describe an efficient VC method for the paint industry.	[29]
Integration of VC and DT	Proposes a progression methodology from VC simulation to DT and vice versa, establishing a seamless bidirectional process flow between software and virtual/real interaction for machine development in a dynamic industrial context.	[30]
	Integrates discrete-event simulation and agent-based modeling to connect DT Emulation (DTE) with a PLC, boosting the efficiency and reliability of modular production systems.	[31]
	Examines recent methods for developing VC-enhanced DT and the use of DT-optimized VC models, enabling accurate simulation during both design and operation.	[32]

4. Discussion

The results of the study highlight the predominant use of SIL VC methodologies and their flexibility and capacity to conduct detailed simulations in the absence of specific hardware requirements. Nonetheless, the sparse application of hybrid SIL and HIL approaches indicates challenges such as increased complexity and elevated costs. The results also emphasize the distinct but complementary roles of VC and DTs. While VC is predominantly used in pre-operational phases to simulate and test systems for validation purposes, DTs extend their usefulness throughout the system's operational life cycle, continuously interacting with and adapting to the real system. This delimitation encompasses the scope and application of VC and DTs in industrial practice, as it reflects the need for technologies that can accurately initialize systems and dynamically adapt over time.

VC can support DT development by offering high-fidelity simulations that accurately reflect real-world behavior. These simulations based mainly on SIL methods allow detailed modeling without specific hardware to reduce costs and increase the flexibility of the development process. As highlighted by Lechler et al. [20], VC simulation models serve as the basis for the development of DT models. Also, DTs can support VC by improving the accuracy of post-commissioning simulations and responding adaptively in real time to production challenges. The combination of DTs with VC allows the accurate verification of designs and the validation of control configurations before deployment and continuous improvement after commissioning [33,34].

The integration of VC and DTs allows for optimizing the development and management of industrial systems [20]. This integration is manifested through a progression methodology that enables a seamless two-way process flow, connecting simulation paradigms to bring the virtual and physical spheres together. In addition, this collaboration facilitates better anticipation and management of potential problems, both before and during production, as indicated by Shen et al. [35] and Wang et al. [36], highlighting the critical relevance of this integration for I4.0. By enabling real-time synchronization with operational data, the strengths of both technologies are used to optimize system design and functionality, from the initial phases to maintenance. Moreover, VC is used at particular points in the plant life cycle and depends heavily on simulation model availability, which requires significant effort to develop, preventing VC from being widely adopted in the industry due to its cost [37,38]. However, by integrating the adoption of VC with DTs which relies on simulation models, the use of VC is prolonged to prorate the development cost with DTs, making its adoption appealing to the industry. As highlighted in reference [37], simulation models developed for VC cannot be reused without major effort for DTs due to initial design choices. However, by considering the concurrent adoption of both technologies, engineers can develop simulation models in a way that they can be reused more easily.

5. Future Research

This study also reveals some directions for future research efforts (Figure 7). The use of VC and DT simulation models throughout the life cycle reveals a significant research opportunity in developing methodologies that transition between pre-operational VC simulations and continuous DT operational management [5]. These methodologies allow for the reuse of simulation models and real-time feedback from DT operations to refine VC simulations, optimize configuration, and make constant adjustments. Another gap exists in data management, where there is a need for hybrid models that integrate the hypothetical or design data of VC with the historical data of DTs. This integration increases the accuracy of simulations and the predictive capabilities of DTs. For this, real-time interaction/synchronization mechanisms also represent a challenge.

Other areas for future research are the automatic generation and reuse of simulation models, which represent an opportunity for sustainability and efficiency in VC and DT applications. Strategies for effective model reuse can significantly reduce resource consumption and accelerate the development process. This addresses the gap in current practices, where models are built from scratch for each new project, preventing the widespread adoption of VC and DTs. Lastly, education and training on VC and DTs must be conducted to prepare engineers to develop more complex automated systems in I4.0 and broaden the use of these technologies in industry.

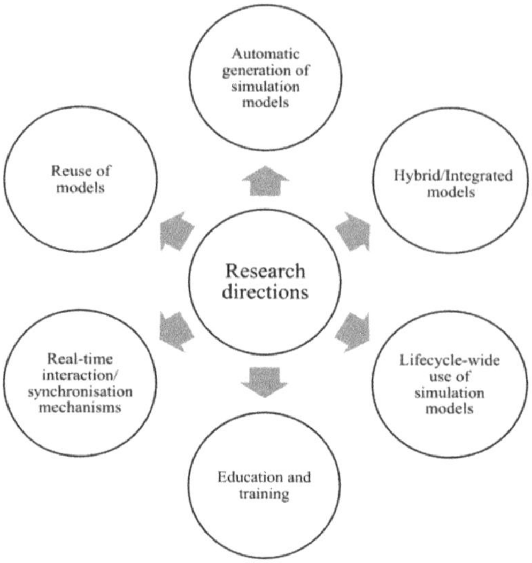

Figure 7. Main research directions.

6. Conclusions

We analyzed the synergies between VC and DT technologies in the context of I4.0 through a systematic literature review, enriching academic discourse on the subject. VC can support the development of DTs and vice versa, and the two technologies can be integrated with VC becoming a component of DTs. It highlights how these technologies significantly improve industrial system design, development, and improvement. By leveraging both VC's ability to perform detailed pre-operational testing and DTs' continuous monitoring and updating of system performance, substantial improvements in systems management can be achieved.

Combined VC and DTs improve simulation accuracy, facilitate its reuse, and enhance the life cycle management of industrial assets. This integrated approach promotes a more dynamic interaction between simulated models and real-world data, which contributes to the development of resilient and adaptable industrial systems. Specifically, the integration of VC and DTs facilitates real-time data flows that significantly improve decision-making processes and allow timely corrections and updates. These capabilities are key in today's fast-changing industrial environments, where the ability to respond quickly and accurately to operational fluctuations significantly affects overall efficiency and productivity.

Author Contributions: Conceptualization, H.B.D.T. and W.d.P.F.; methodology, H.B.D.T. and W.d.P.F.; validation, H.B.D.T., W.d.P.F., R.F.d.A. and D.B.; formal analysis, H.B.D.T.; investigation, H.B.D.T.; resources, W.d.P.F. and D.B.; data curation, H.B.D.T.; writing—original draft preparation, H.B.D.T.; writing—review and editing, H.B.D.T., W.d.P.F., R.F.d.A. and D.B.; visualization, H.B.D.T., W.d.P.F. and R.F.d.A.; supervision, W.d.P.F.; project administration, W.d.P.F.; funding acquisition, W.d.P.F. All authors have read and agreed to the published version of the manuscript.

Funding: This research was supported by the Natural Sciences and Engineering Research Council of Canada (NSERC) through the Alliance Grants program [grant number ALLRP 592520-23] and the Discovery Grant program [grant number RGPIN-2024-03880], as well as by Mitacs through the Mitacs Accelerate program in Canada.

Institutional Review Board Statement: Not applicable.

Informed Consent Statement: Not applicable.

Data Availability Statement: Data will be made available on request.

Conflicts of Interest: The authors declare that they have no known competing financial interests or personal relationships that could have appeared to influence the work reported in this paper.

References

1. de Paula Ferreira, W.; Armellini, F.; de Santa-Eulalia, L.A.; Rebolledo, C. Modelling and simulation in industry 4.0. In *Artificial Intelligence in Industry 4.0: A Collection of Innovative Research Case-Studies That Are Reworking the Way We Look at Industry 4.0 Thanks to Artificial Intelligence*; Springer: Cham, Switzerland, 2021; pp. 57–72.
2. de Paula Ferreira, W.; Armellini, F.; de Santa-Eulalia, L.A.; Thomasset-Laperrière, V. A framework for identifying and analysing industry 4.0 scenarios. *J. Manuf. Syst.* **2022**, *65*, 192–207. [CrossRef]
3. de Assis, R.F.; Guerrini, F.M.; Santa-Eulalia, L.A.; de Paula Ferreira, W. An agent-based model for regional market penetration of electric vehicles in brazil. *J. Clean. Prod.* **2023**, *421*, 138477. [CrossRef]
4. de Paula Ferreira, W.; Armellini, F.; de Santa-Eulalia, L.A.; Thomasset-Laperrière, V. Extending the lean value stream mapping to the context of industry 4.0: An agent-based technology approach. *J. Manuf. Syst.* **2022**, *63*, 1–14. [CrossRef]
5. Striffler, N.; Voigt, T. Concepts and trends of virtual commissioning–a comprehensive review. *J. Manuf. Syst.* **2023**, *71*, 664–680. [CrossRef]
6. Jones, D.; Snider, C.; Nassehi, A.; Yon, J.; Hicks, B. Characterising the digital twin: A systematic literature review. *CIRP J. Manuf. Sci. Technol.* **2020**, *29*, 36–52. [CrossRef]
7. Marques, R.; de Paula Ferreira, W.; Nassif, G.; Armellini, F.; Dungen, J.; de Santa-Eulalia, L.A. Exploring the application of IoT in the service station business. *IFAC-PapersOnLine* **2021**, *54*, 402–407. [CrossRef]
8. Leng, J.; Wang, D.; Shen, W.; Li, X.; Liu, Q.; Chen, X. Digital twins-based smart manufacturing system design in industry 4.0: A review. *J. Manuf. Syst.* **2021**, *60*, 119–137. [CrossRef]
9. Mueller, R.K.; Großmann, D.; Schmied, S.; Jumar, U. Co-Simulation im Planungsprozess für die Informationsmodell-basierte virtuelle Inbetriebnahme. *at-Automatisierungstechnik* **2022**, *70*, 612–622. [CrossRef]
10. Lee, C.G.; Park, S.C. Survey on the virtual commissioning of manufacturing systems. *J. Comput. Des. Eng.* **2014**, *1*, 213–222. [CrossRef]
11. Grieves, M. Digital twin: Manufacturing excellence through virtual factory replication. *White Pap.* **2014**, *1*, 1–7.
12. Kritzinger, W.; Karner, M.; Traar, G.; Henjes, J.; Sihn, W. Digital twin in manufacturing: A categorical literature review and classification. *IFAC-PapersOnLine* **2018**, *51*, 1016–1022. [CrossRef]
13. Gasiyarov, V.R.; Bovshik, P.A.; Loginov, B.M.; Karandaev, A.S.; Khramshin, V.R.; Radionov, A.A. Substantiating and implementing concept of digital twins for virtual commissioning of industrial mechatronic complexes exemplified by rolling mill coilers. *Machines* **2023**, *11*, 276. [CrossRef]
14. Serik, A.; Zhetpissov, Y.; Mosadeghzad, M.; Alizadeh, T. Digital twins development of automatic storage and retrieval station in a production line and an integrated robotic manipulator. In *International Conference on Robot Intelligence Technology and Applications*; Springer: Berlin/Heidelberg, Germany, 2021; pp. 210–223.
15. Page, M.J.; Moher, D.; Bossuyt, P.M.; Boutron, I.; Hoffmann, T.C.; Mulrow, C.D.; Shamseer, L.; Tetzlaff, J.M.; Akl, E.A.; Brennan, S.E.; et al. Prisma 2020 explanation and elaboration: Updated guidance and exemplars for reporting systematic reviews. *BMJ* **2021**, *372*, n160. [CrossRef] [PubMed]
16. Moher, D.; Liberati, A.; Tetzlaff, J.; Altman, D.G.; The PRISMA Group. Preferred reporting items for systematic reviews and meta-analyses: The prisma statement. *Ann. Intern. Med.* **2009**, *151*, 264–269. [CrossRef] [PubMed]
17. Schamp, M.; Hoedt, S.; Claeys, A.; Aghezzaf, E.-H.; Cottyn, J. Impact of a virtual twin on commissioning time and quality. *IFAC-PapersOnLine* **2018**, *51*, 1047–1052. [CrossRef]
18. Yin, Y.; Zheng, P.; Li, C.; Wang, L. A state-of-the-art survey on augmented reality-assisted digital twin for futuristic human-centric industry transformation. *Robot. Comput.-Integr. Manuf.* **2023**, *81*, 102515. [CrossRef]
19. Christ, L.; Milloch, E.; Boshoff, M.; Hypki, A.; Kuhlenkötter, B. Implementation of digital twin and real production system to address actual and future challenges in assembly technology. *Automation* **2023**, *4*, 345–358. [CrossRef]
20. Lechler, T.; Fischer, E.; Metzner, M.; Mayr, A.; Franke, J. Virtual commissioning–scientific review and exploratory use cases in advanced production systems. *Procedia CIRP* **2019**, *81*, 1125–1130. [CrossRef]
21. Toivonen, V.; Lanz, M.; Nylund, H.; Nieminen, H. The fms training center-a versatile learning environment for engineering education. *Procedia Manuf.* **2018**, *23*, 135–140. [CrossRef]
22. Khan, A.; Dahl, M.; Falkman, P.; Fabian, M. Digital twin for legacy systems: Simulation model testing and validation. In Proceedings of the 2018 IEEE 14th International Conference on Automation Science and Engineering (CASE), Munich, Germany, 20–24 August 2018; IEEE: New York, NY, USA, 2018; pp. 421–426.

23. Barbieri, G.; Bertuzzi, A.; Capriotti, A.; Ragazzini, L.; Gutierrez, D.; Negri, E.; Fumagalli, L. A virtual commissioning-based methodology to integrate digital twins into manufacturing systems. *Prod. Eng.* **2021**, *15*, 397–412. [CrossRef]
24. Dosoftei, C.-C. Simulation power vs. immersive capabilities: Enhanced understanding and interaction with digital twin of a mecha- tronic system. *Appl. Sci.* **2023**, *13*, 6463. [CrossRef]
25. Albo, A.; Svedlund, L.; Falkman, P. Modular virtual preparation method of production systems using a digital twin architecture. In Proceedings of the 2021 26th IEEE International Conference on Emerging Technologies and Factory Automation (ETFA), Vasteras, Sweden, 7–10 September 2021; IEEE: New York, NY, USA, 2021; pp. 1–8.
26. Jeon, S.; Schuesslbauer, S. Digital twin application for production optimization. In Proceedings of the 2020 IEEE International Conference on Industrial Engineering and Engineering Management (IEEM), Singapore, 14–17 December 2020; IEEE: New York, NY, USA, 2020; pp. 542–545.
27. Ayani, M.; Ganeba, M.; Ng, A.H. Digital twin: Applying emulation for machine reconditioning. *Procedia Cirp* **2018**, *72*, 243–248. [CrossRef]
28. Wu, B.; Zhang, Z.; Liu, K.; Wu, Z. Collaborative design and virtual commissioning of automated pharmacy based on digital twin. In Proceedings of the 2023 IEEE 16th International Conference on Electronic Measurement & Instruments (ICEMI), Harbin, China, 9–11 August 2023; IEEE: New York, NY, USA, 2023; pp. 335–340.
29. Abassi, A.; Lugo-Calles, J.H.; Balderas-Hill, R.; Lassabe, N. Virtual commissioning implementation of industrial painting process. In Proceedings of the 2022 2nd International Conference on Computers and Automation (CompAuto), Paris, France, 18–20 August 2022; IEEE: New York, NY, USA, 2022; pp. 65–68.
30. Hansen, J.P.D.O.; da Silva, E.R.; Bilberg, A.; Bro, C. Agile machine development from virtual to real. In Proceedings of the Changeable, Agile, Reconfigurable and Virtual Production Conference and the World Mass Customization & Personalization Conference, Aalborg, Denmark, 1–2 November 2021; Springer: Berlin/Heidelberg, Germany, 2021; pp. 389–395.
31. Mykoniatis, K.; Harris, G.A. A digital twin emulator of a modular production system using a data-driven hybrid modeling and simulation approach. *J. Intell. Manuf.* **2021**, *32*, 1899–1911. [CrossRef]
32. Phanden, R.K.; Sharma, P.; Dubey, A. A review on simulation in digital twin for aerospace, manufacturing and robotics. *Mater. Today Proc.* **2021**, *38*, 174–178. [CrossRef]
33. Hu, Y.; Zhang, Y.; Ma, X.; Du, X.; Wang, W.; Zhang, H. Virtual commissioning and process parameter optimization of rolling mill based on digital twin. *Int. J. Adv. Manuf. Technol.* **2024**, *130*, 705–716. [CrossRef]
34. Fernandez, I.A.; Eguía, M.A.; Echeverría, L.E. Virtual commissioning of a robotic cell: An educational case study. In Proceedings of the 2019 24th IEEE International Conference on Emerging Technologies and Factory Automation (ETFA), Zaragoza, Spain, 10–13 September 2019; IEEE: New York, NY, USA, 2019; pp. 820–825.
35. Shen, W.; Hu, T.; Yin, Y.; He, J.; Tao, F.; Nee, A. Digital twin based virtual commissioning for computerized numerical control machine tools. In *Digital Twin Driven Smart Design*; Elsevier: Amsterdam, The Netherlands, 2020; pp. 289–307.
36. Wang, J.; Niu, X.; Gao, R.X.; Huang, Z.; Xue, R. Digital twin-driven virtual commissioning of machine tool. *Robot. Comput.-Integr. Manuf.* **2023**, *81*, 102499. [CrossRef]
37. Oppelt, M.; Urbas, L. Integrated virtual commissioning an essential activity in the automation engineering process: From virtual commissioning to simulation supported engineering. In Proceedings of the IECON 2014—40th Annual Conference of the IEEE Industrial Electronics Society, Dallas, TX, USA, 29 October–1 November 2014; IEEE: New York, NY, USA, 2014; pp. 2564–2570.
38. Oppelt, M.; Wolf, G.; Urbas, L. Towards an integrated use of simulation within the life-cycle of a process plant. In Proceedings of the 2015 IEEE 20th Conference on Emerging Technologies & Factory Automation (ETFA), Luxembourg, 8–11 September 2015; IEEE: New York, NY, USA, 2015; pp. 1–8.

engineering proceedings

MDPI

Proceeding Paper

Semantic Classification of Car Styling Using Machine Learning †

Hung-Hsiang Wang and Yen-Ting Lu *

Department of Industrial Design, National Taipei University of Technology, Taipei 10608, Taiwan; wanghh@ntut.edu.tw
* Correspondence: t112588012@ntut.org.tw

† Presented at the 2024 IEEE 7th International Conference on Knowledge Innovation and Invention, Nagoya, Japan, 16–18 August 2024.

Abstract: Product semantics is essential for car styling because it shapes how consumers perceive and interact with cars, influences user experiences, and allows for product differentiation. Although many AI tools are available to assist car designers, research on applying machine learning techniques to evaluate product semantics is rare. Therefore, we developed a classification model that helps designers identify and evaluate the semantics conveyed by car styling using the Waikato Environment for Knowledge Analysis (WEKA), a machine learning tool. We used Python web scraping to collect isometric drawings and introductory articles of 1320 SUV cars of various brands from 2009 to 2024 via websites such as Car Body Design and Car Design News. We also summarized four semantic types of car styling, namely "aggressive", "sporty", "clean", and "off-road", to create the dataset. We used WEKA image classification to randomly select 792 (60%) images from the dataset to train a classification model of car styling semantics. The remaining 528 images (40%) were used for verification. The classification model trained with the Binary Pattern Pyramid Filter and the Random Forest classifier achieved an accuracy of 84.6%. The model was evaluated in terms of whether 10 SUVs created by 10 graduate design students using AI conveyed the anticipated product semantics. Seven of the ten SUVs were correctly classified and the rest were not. All of the participants agreed that the predictions were satisfactory. However, it is necessary to improve the accuracy of each semantic classification, especially the "clean" type. The results of this study demonstrate the capability of machine learning to identify the semantics of car styling effectively, improve the communication and evaluation of product semantics by designers in the design process, and create a car styling with a good appearance that resonates with consumers.

Keywords: product semantics; car styling; industrial design; machine learning; image classification

1. Introduction

Among consumers' demands for car appearance, practicality, aesthetics, style, and individuality are critical. These demands drive car manufacturers to continuously update their designs to meet the market's requirements for beauty and style [1]. Therefore, designers must grasp the semantic design language that consumers associate with different cars. However, this largely relies on the personal experience of the designers. To date, there is still limited research on how to effectively identify the product semantics of car designs. Hence, we explored related literature utilizing machine learning to analyze product semantics and accurately identify car designs.

In contemporary product design, the development of deep learning technology has empowered designers to predict consumer preferences and identify product typology through image classification techniques. Researchers have utilized the Waikato Environment for Knowledge Analysis (WEKA) to predict the innovation threshold of product appearance [2]. This method enables designers to innovate product designs based on consumer preferences while maintaining the iconic features of the brand. With this technology, designers enhance the market appeal of products and effectively respond to market demands and consumer expectations.

Reference [3] explored the application potential of decision tree classifiers in WEKA software in architecture, particularly for the classification of cultural heritage images. The study aimed to evaluate the performance of different decision tree algorithms in handling small-scale architectural image sets. The research selected the Fast and Compact Transform-Histogram Filter (FCTH Filter), Edge Histogram Filter, and Discrete Cosine Transform (DCT) as feature extraction processing techniques and applied them to cultural heritage images such as archaeological sites, murals, and monasteries. The Random Forest (RF) algorithm exhibited significant superiority in these tasks. The effectiveness of WEKA in learning and predicting architectural styles was confirmed by demonstrating the wide applicability of this technology, providing a new technical approach for the automatic classification of architectural styles in the future.

Previous studies demonstrate the application achievements of WEKA in design, indicating that WEKA has significant development potential. From intricate product design to grand architectural modeling, WEKA provides solutions for high recognition. This finding reveals the effectiveness of machine learning technology in design innovation and opens up new possibilities for technical application in design.

Based on the previous results, we combine automotive styling and design semantics to develop a new evaluation method and enhance the judgment and cognition of automotive designers. Using this method, they can understand their design styles and provide feedback effectively through interactions with their instructors. We employed deep learning techniques to learn the semantics of automobiles, analyze the car designs of students, and map them to specific stylings. Then, a framework that supports students' creative thinking and provides a powerful teaching aid for instructors was created. Additionally, we explored the feedback from students regarding their experience with the system to assess its overall acceptance and practicality.

We categorized the collected sports utility vehicle (SUV) images into four semantic types: "aggressive", "sporty", "off-road", and "clean". WEKA machine learning was used to create a model that predicts the semantics of car styling. In automotive design, the discussion on adjectives describing styling and their semantic aspects is insufficient. Therefore, by applying machine learning technology, designers can analyze and evaluate the semantic features of car styling to provide precise semantic assessments during the design process.

2. Related Work

2.1. Semantics of Automotive Styling

In the automotive industry, aesthetics influence 60% of consumer purchasing decisions [4], demonstrating that automotive styling significantly affects buying intentions and the automotive industry at large. Therefore, automotive companies need to invest heavily in design to markedly impact market performance [5]. This underscores the critical role of automotive styling in the automotive sector. The significance of automotive semantics within semiotics cannot be overlooked. The positioning of automotive semantics in the semantic space is crucial for consumers to recognize different years and models of vehicles [6]. Even though vehicles' exteriors and technical specifications evolve, the continuity of semantics helps consumers identify and remember specific cars. Manufacturers adapt to market changes and consumer expectations by introducing new model semantics, changing existing names, and highlighting the key role of semantics in brand management. Overall, the role of automotive semantics in brand recognition, consumer perception, and market strategy is vital for further exploration and research. The importance of automotive styling and the critical impact of automotive semantics on consumers are emphasized. The significant influence of automotive semantics in the automotive industry is revealed, especially in highly technical and innovation-driven car design.

2.2. Semantics in Automotive Design Education

In product design education, particularly in automotive styling design education, applying semantics is important in the automotive design process [7]. Teaching helps students to understand design forms, convey complex shapes, and avoid conceptual compromises using 3D software through the appropriation of complex historical semantic elements. This pedagogical approach enhances students' abilities in terms of their understanding of design aesthetics and increases the number of sketch iterations and the overall aesthetic quality of the products [7]. Reference [8] demonstrated that the principles of product semantics significantly improve students' clarity in expressing design intentions, indicating that product semantics contributes to commercial success as an essential component of the educational process. These studies revealed the profound impact of product semantics in design practice and academic research.

2.3. Image Classification Using Machine Learning

Machine learning techniques are widely used in automotive styling and consumer preference-oriented research to predict aesthetic scores for car designs and automatically generate innovative and attractive designs. By training models, different styles of car designs are analyzed and generated to help designers improve efficiency in screening designs and enhance the innovativeness and marketability of designs [9]. WEKA is used to predict the threshold of product appearance innovation [2]. Designers have used this method to innovate product designs based on consumer preferences while maintaining brand characteristics. This technology increases the marketability of products and helps designers effectively meet market demands and consumer expectations. In classifying automotive and product styling, machine learning effectively recognizes differences between brands, serving as an expert system for brand style classification [10,11]. To improve the accuracy of machine learning, machine learning is used to investigate the car front style [12].

Machine learning learns automotive and product styling using image recognition. However, few studies have explored product semantics. Thus, we explored the accuracy of machine learning in identifying automotive styling semantics to develop a model to enhance communication efficiency and creative processes for creators, students, and related departments.

3. Methods

This research was conducted in two stages. In the first stage, we built and validated a model. First, the required images of automotive designs were collected and preprocessed. Then, these images were split into training and testing sets in a 6:4 ratio. Then, the WEKA image recognition functionality was used to train the model in the training set. Finally, the model generated from the training dataset was validated using the images in the testing dataset to evaluate accuracy. In the second stage, the trained model was applied to analyze and evaluate students' automotive design works. This process provided students with feedback on their design and guided their learning progress and improvement in innovation capabilities. The model was used to explore the practicality and challenges of machine learning technology in art design evaluation and provide a theoretical and empirical basis for future research and teaching on related technologies.

3.1. Data Collection

To conduct an in-depth image and text analysis, Python's web scraping techniques were used to collect 1320 SUV images from 2009 to 2024 (Figure 1) from two websites related to automotive design [13,14]. Adjectives related to the semantics of these images were determined. The semantics associated with each image were derived from car photos and articles on the websites and were filtered according to the advice of two experts. The finalized keywords included "aggressive", "sporty", "clean", and "off-road".

Figure 1. A total of 1320 SUV car images obtained from Car Body Design and Car Design News (placed the remaining images on the cloud drive).

3.2. Data Preprocessing

We processed the images to reduce the background interference and eliminate the effect of color differences on the analysis results. We utilized Python to preprocess the images. We applied the back removal process to improve the focus clarity and enhance the classification accuracy. The pixels of the images were set between 580 and 1100. Pixel normalization was conducted to ensure that all the images had the same resolution and size, thus enabling the algorithm to process and analyze them accurately. To minimize the interference of color image recognition, the images of cars with white bodies were purposely chosen for this study to minimize the interference of color. The surface property of the white color presents the details of the car body shadow clearly, which is important for highlighting the vehicle's shape and styling features. Finally, the scaling and deformation of the images were performed to increase data diversity, reduce over-simulation, simulate realistic conditions, and improve the robustness of the model. The process ensured the quality of the data and the accuracy and reliability of the subsequent analysis and the analysis of the visual features of the car.

3.3. Feature Extraction

We used WEKA machine learning software for image feature extraction and classification. We extracted features from the preprocessed images and tested four different image classifiers, namely the Binary Pattern Pyramid Filter (BPPF), Edge Histogram Filter, FCTH Filter, and Gabor Filter, which focus on the texture and feature analysis of images. In the experiment, 1320 images were used, of which 60 and 40% were allocated to the training and test datasets. According to the machine learning results of the WEKA software, BPPF showed superiority in image classification, with a classification accuracy of 85.8%. The high performance of BPPF in processing image texture features provides a reference for future research (Figure 2).

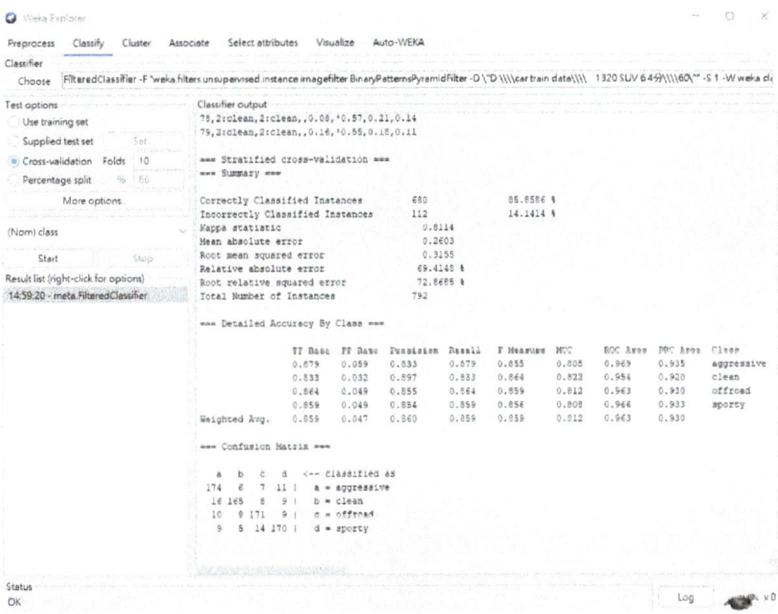

Figure 2. Prediction results of the style semantic model using WEKA.

3.4. Verification and Testing

To validate the generalization ability of the model, 40% of the 1320 images were allocated to the test dataset. We used 528 SUV images. These images and text descriptions were preprocessed for testing. Then, the 528 images were classified into styles using BPPF (Figure 3). A total of 447 images were correctly recognized, with an accuracy rate of 84.6% in the test dataset. The accuracy and effectiveness of the BPPF classifier were validated in recognizing the semantics of visual styles.

Figure 3. Car designs created by automotive design students.

3.5. Application and Evaluation

We applied the trained model to evaluate the SUV designs of 10 automotive design students using AI (Figure 3). The WEKA-trained automotive semantic model automatically categorized the style semantics of these students' works. Based on the categorization results, the students' designs were evaluated as different automotive semantic styles (Table 1, Figure 4). The results reflected the stylistic diversity of the students' designs and demonstrated the model's ability to identify and evaluate different semantic styles. Feedback on the categorization results was collected and summarized to verify the model's usefulness and applicability.

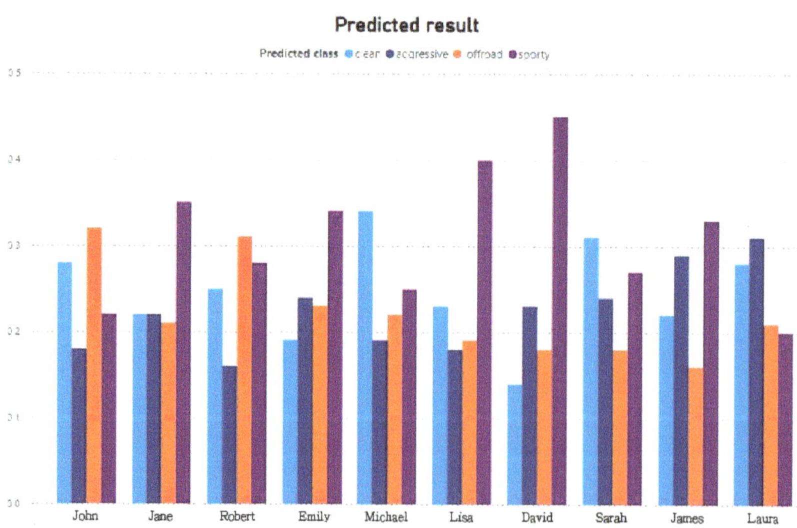

Figure 4. Prediction of semantics of car design works through WEKA and visualization with bar charts.

Table 1. Predicting semantics of car design works using WEKA.

Student	Aggressive	Clean	Off-Road	Sporty
John	0.18	0.28	**0.32**	0.22
Jane	0.22	0.22	0.21	**0.35**
Robert	0.16	0.25	**0.31**	0.28
Emily	0.24	0.19	0.23	**0.34**
Michael	0.19	**0.34**	0.22	0.25
Lisa	0.18	0.23	0.19	**0.40**

Table 1. *Cont.*

Student	Aggressive	Clean	Off-Road	Sporty
David	0.23	0.14	0.18	**0.45**
Sarah	0.24	**0.31**	0.18	0.27
James	0.29	0.22	0.16	**0.33**
Laura	**0.31**	0.28	0.21	0.20

The bolded values indicate the proportion of each car image classified under a particular style. The highest value is shown as Weka's final classification result.

We collected feedback from the students, including comments on the categorization and perceptions of their designs. Semi-structured interviews were conducted with 10 graduate students aged 18−24 years old with a background in automotive design. The following quotations were obtained from their feedback.

John: *"I designed an off-road and sporty SUV, but the model categorized it as off-road and partly sporty and simple. I think the model categorization is quite close to the styling style that I wanted to achieve, only that the simplicity component is higher than what I wanted to achieve"*.

Jane: *"I think this car is appropriately categorized as 'sporty', but as far as 'off-road' is concerned, I think only the width of the car matches this category. As for the rating of 0.22 for 'clean', I don't think it's quite accurate because the surface design is not particularly clean. Lastly, I think the rating of "aggressive" is similar to Weka's evaluation, and it is quite appropriate"*.

Robert: *"I think it is reasonable for my car to be categorized as Off-road by WEKA, including its clean and sporty data, which are very similar to what I wanted to design in the beginning." Student 4: "I think WEKA has a good idea to categorize my car as Off-road"*.

Emily: *"I think WEKA's judgment of sporty is very much in line with my creative concept, except that I think the semantic meaning of clean, off-road should be a little lower, and should not add up to more than 0.5, and the judgment of aggressive as 0.24 is quite in line with my thoughts"*.

Michael: *"The design direction of this car is off-road, but clean is the highest, although he has clean appearance line, but I think off-road will be the highest right, the other sporty and aggressive I am not too satisfied"*.

Lisa: *"The sporty is too much higher than I expected, I think it's about 0.25, off-road and clean together is more than 0.65, and aggressive can be reduced to 0.1, which will be better"*.

David: *"This car's predicted solution is very sporty and aggressive as I thought it would be, but I think the clean could be a little lower to 0.1 and the off-road to 0.22"*.

Sarah: *"I think it's quite inaccurate to classify this car as clean, it's more of an aggressive car, and the sporty value is quite consistent. The sporty value is quite consistent, and off-road can be lowered to 0.3, which would be better"*.

James: *"I think it's quite reasonable to classify this car as sporty, I think off-road needs to be higher. aggressive needs to be much lower. Besides, I think it's quite appropriate that clean is not classified"*.

Laura: *"I think it is very appropriate for this car to be classified as aggressive by WEKA, but the clean part can be lowered to 0.2, the off-road can be lowered to 0.3, and the sporty part can be raised even more"*.

The observations below are based on the comparison of each student's self-assessment of their design with the results of the WEKA model categorization.

3.5.1. Model Categorization

John felt that the model rated the "sportiness" and "simplicity" of his design higher, but he emphasized the off-road nature more. There was a difference between the students' understanding of "simplicity" and the model's. Jane thought that the width of the car was consistent with the "off-road" feature, but was surprised by the low rating of "clean" as the surface of the design was not simple enough. There was an interpretive difference in the definition of "clean" between the students and the model. Robert was satisfied with the scores of "off-road", "clean", and "sporty" given by the model, which indicated a high degree of agreement between his design intention and the model's understanding.

Emily was satisfied with the model's "sporty" rating and wondered why "clean" and "off-road" were rated lower, suggesting that less emphasis was placed on these elements than on the model's assessment. Michael, Lisa, and David expressed their dissatisfaction with the "clean" rating, which they felt did not match the off-road style of their design, highlighting the need for further clarification on how to balance design elements in teaching. Sarah thought that "clean" was over-rated and expected an "aggressive" style, which indicated that the students had a clearer perception of the aggressive features of the vehicle design. James and Laura were satisfied with or accepted the ratings of "sporty" and "aggressive" but were dissatisfied with the ratings of other elements such as "off-road", showing differences in the expectations of particular design elements.

3.5.2. Strengths and Weaknesses of WEKA Model

1. Participation and cognitive reflection

The students described their views on their car designs in detail, including their understanding of various design elements such as "sporty", "clean", "off-road", and "aggressive". Initiative and self-reflection are important in the design process as they prompt students to think about design intention and realization and make self-assessments and adjustments to the design.

2. Consistency

The WEKA model accurately reflected the students' design intentions, such as in the case of Robert, whose model ratings for "off-road", "clean", and "sporty" were highly consistent with the design objectives. In addition, the high rating of "sporty" for Emily showed the accuracy of the model in evaluating this design element. Laura's rating of "aggressive" showed that the model reasonably assessed the design. The model was validated in understanding and evaluating design elements and its ability to match the students' perceptions, which strengthens the students' trust in the use of the model and demonstrates the practical value of the model in real-world applications.

3. Inconsistency

Discrepancies between students' understandings of particular design elements and the model's ratings can lead to misunderstandings in the design intentions. For example, John, Jane, David, and Sarah were dissatisfied with the "clean" scale, indicating that there was an inconsistency between the students and the model in terms of the meaning of this term. This also suggests that "clean" is the main direction for the subsequent optimization of the training model.

4. Advantages of WEKA model

The WEKA model evaluated design elements to help students understand and master abstract concepts in design theories and apply these theories in practice. The students' feedback showed that the model effectively reflected the students' design intentions by evaluating design elements such as "sporty" and "aggressive" with consistency and accuracy. This enhanced the students' understanding of design and facilitated their self-assessment and improvement of the design effects. In addition, there were discrepancies between the model's scores and the students' self-assessments. Based on these, teachers can guide students to have in-depth discussions on the relationship between design concepts and results to understand design elements and the perception of design. This interactive learning process enhances students' critical thinking skills and strengthens their problem-solving skills.

The advantage of the WEKA model in design education is that it provides a real-time, specific, and accurate design assessment, which significantly impacts students' learning efficiency and design ability.

4. Conclusions

In this study, the WEKA model was established to categorize the design semantics of SUV cars for application in automotive design education. A high accuracy of 84.6% was achieved after analyzing 1320 SUV model images and model training, which demonstrates the ability of machine learning technology to recognize complex design semantics. In addition, the evaluation and application of the technology in students' design work enhanced their in-depth understanding of the design semantics and facilitated their effective communication and feedback on their work. The results emphasize the importance of digital tools in enhancing educational interactivity and student creativity and demonstrate the potential for broader application of technology in professional design education and practice.

A total of 7 out of 10 students recognized that the model helped them to better understand and apply the design concepts, and to evaluate and adjust their designs. However, there were challenges, especially in the ambiguity between students' perceptions and the model's assessment for several design elements such as "clean". The model needs to be optimized and adjusted in these aspects. When the model's assessment was inconsistent with the student's self-assessment, the students were prompted to think deeply, which provided the teacher with opportunities to guide discussion and critical thinking training. The WEKA model has shown pedagogical potential in design education and the potential to develop students' design skills and creativity. Future improvements are needed to improve the assessment accuracy and semantic richness of the model to more accurately distinguish semantic meanings to increase applicability.

Machine learning technology can be used to identify the semantics of automotive design. Through technological innovation and practical application, more innovation and progress can be achieved in design in the future.

Author Contributions: Conceptualization, H.-H.W.; method, H.-H.W.; resources, Y.-T.L.; data curation, Y.-T.L.; validation, H.-H.W.; writing—original draft preparation, Y.-T.L.; writing—review and editing, H.-H.W.; experiments, Y.-T.L.; supervision, H.-H.W. All authors have read and agreed to the published version of the manuscript.

Funding: This research was funded by NSTC, Taiwan, R.O.C., through grant 111-2410-H-027-019-MY2.

Institutional Review Board Statement: Not applicable.

Informed Consent Statement: Not applicable.

Data Availability Statement: Data are unavailable due to privacy reasons.

Acknowledgments: The authors gratefully acknowledge the financial support of NSTC, Taiwan, R.O.C., through grant 111-2410-H-027-019-MY2.

Conflicts of Interest: The authors declare no conflicts of interest.

References

1. Gartman, D. *Auto-Opium: A Social History of American Automobile Design*; Routledge: London, UK, 2013.
2. Wang, H.-H.; Liu, C.-K.; Yu, S.-B. Predicting Consumer Preferences by the Deformation Threshold of Product Appearance. *Eng. Proc.* **2023**, *55*, 67. [CrossRef]
3. Jankovic, R. Classifying cultural heritage images by using decision tree classifiers in WEKA. In Proceedings of the 1st International Workshop on Visual Pattern Extraction and Recognition for Cultural Heritage Understanding Co-Located with 15th Italian Research Conference on Digital Libraries (IRCDL 2019), Pisa, Italy, 30 January 2019; pp. 119–127.

4. Kreuzbauer, R.; Malter, A.J. Embodied cognition and new product design: Changing product form to influence brand categorization. *J. Prod. Innov. Manag.* **2005**, *22*, 165–176. [CrossRef]
5. Rubera, G. Design innovativeness and product sales' evolution. *Mark. Sci.* **2015**, *34*, 98–115. [CrossRef]
6. Aronoff, M. Automobile semantics. *Linguist. Inq.* **1981**, *12*, 329–347.
7. Berry, J. Overcoming the complexity of teaching form and detail aesthetics in product design education. In Proceedings of the DS 46: Proceedings of E&PDE 2008, the 10th International Conference on Engineering and Product Design Education, Barcelona, Spain, 4–5 September 2008; pp. 156–161.
8. Huang, C. Design for Ease of Use: Product Semantics and Design Education. Doctoral Dissertation, Manchester Metropolitan University, Manchester, UK, 1996.
9. Burnap, A.; Hauser, J.R.; Timoshenko, A. Product aesthetic design: A machine learning augmentation. *Mark. Sci.* **2023**, *42*, 1029–1056. [CrossRef]
10. Wang, H.-H.; Lin, Y.-Y.; Huang, H.-T. Application of Typicality in Predicting Product Appearance. *Eng. Proc.* **2023**, *55*, 66. [CrossRef]
11. Wang, H.H.; Chen, C.P. STYRULE-A Classification System of Car Styling. *J. Des.* **2009**, *9*, 107–121.
12. Ma, L.; Wu, Y.; Li, Q.; Yuan, X. Recognition of car front facing style for machine-learning data annotation: A quantitative approach. *Symmetry* **2022**, *14*, 1181. [CrossRef]
13. Car Body Design. Available online: https://www.carbodydesign.com/ (accessed on 1 May 2024).
14. Car Design News. Available online: https://www.cardesignnews.com/ (accessed on 1 May 2024).

engineering proceedings

MDPI

Proceeding Paper

Adopting Artificial Intelligence and Artificial Reality in an Interactive Sign Language Learning System: Acceptance of Interactive Technology †

Kuo-Feng Hwang [1], Yu-Han Wang [1,*], Tsui-Ying Tsao [2], Shu-Hui Chou [2], Shih-Ying Chueh [2], Cheng-Chun Wu [1] and Kuan-Yi Ho [1]

1 Department of Multimedia Design, National Taichung University of Science and Technology, Taichung 404336, Taiwan; kfhwang@nutc.edu.tw (K.-F.H.); aomi.yuan1120@gmail.com (C.-C.W.); dwvsmedia02@gmail.com (K.-Y.H.)
2 National Taiwan Library, Taipei 235066, Taiwan; tsauying@mail.ntl.edu.tw (T.-Y.T.); csh@mail.ntl.edu.tw (S.-H.C.); jill@mail.ntl.edu.tw (S.-Y.C.)
* Correspondence: han@gm.nutc.edu.tw
† Presented at the 2024 IEEE 7th International Conference on Knowledge Innovation and Invention, Nagoya, Japan, 16–18 August 2024.

Abstract: Growing awareness of inclusivity for people with disabilities has led the National Taiwan Library to promote digital and interactive approaches for improved communication between sign language users and the hearing community. The "Interactive System of Digital Sign Language Picture Books", developed with augmented reality (AR) and artificial intelligence (AI), enhances inclusive learning experiences. In this study, 82 participants with hearing disabilities aged 31–60 used the system and completed a technology acceptance model questionnaire. Results indicate that ease of use positively influences perceived usefulness, attitude toward use, and intention to use, supporting enriched learning and fostering social communication.

Keywords: AI recognition; sign language; augmented reality; technology acceptance model

1. Introduction

The awareness and acceptance of being inclusive toward individuals with disabilities have been increasing. Taiwanese Sign Language has been officially recognized as a tool for communication as it safeguards cultural and linguistic heritage and raises the awareness of culture and language for individuals with hearing disabilities. Beyond the essential support, the public must understand the value of inclusivity.

1.1. Technology in Education

The acceptance of educational technologies, particularly those to enhance language learning, has been a subject of research recently. Technology acceptance models (TAMs) and their applications provide frameworks for understanding users' attitudes and intentions toward new technologies. Perceived ease of use (PEOU) and perceived usefulness (PU) are key determinants of user acceptance [1]. When users perceive a technology as easy to use and useful, they are more likely to adopt it for educational purposes [2,3].

Innovations in sign language technology improve communication, accessibility, and inclusivity for the hearing-impaired community, underscoring the goal to enhance sign language learning and support the hearing-impaired [4]. Incorporating the perspectives of the hearing-impaired is crucial; their engagement with the community is beneficial to them [5].

 https://doi.org/10.3390/engproc2025089014

1.2. Technological Innovations in Sign Language Learning

Augmented reality (AR) and artificial intelligence (AI) are used to create innovative tools for learning sign language [6–9]. Such advanced technology offers unique opportunities to enhance the interactivity and effectiveness of educational resources, particularly for individuals with hearing impairment. For example, AR has been adopted in learning sign languages in English [10], Indonesian [7], and Peruvian [11]. Several applications have been developed to facilitate communication between the hearing and the hearing-impaired by adopting AI technologies and sign language capturing. Yet, it has been developed in English, Hindi, Thai, Greek, Korean, and Cantonese [12]. However, state-of-the-art technologies have not been adopted to fulfill the needs of the hearing-impaired in Taiwan.

Therefore, we developed an "Interactive System of Digital Sign Language Picture Books" that integrates cutting-edge technologies to enhance the learning experience for the hearing and the hearing-impaired. The system uses AR to enable users to interact with animated content more engagingly. AI technology focuses on sign language recognition to provide learners with immediate feedback on the accuracy of their sign language. This real-time feedback helps learners understand and correct their signs, thereby enhancing the learning experience. By combining AR and AI, the system improves the visual and interactive aspects of sign language education and supports an inclusive reading environment that caters to diverse user needs and abilities.

2. Materials and Methods

To foster a sustainable learning environment, the system was developed in Taiwanese Sign Language (Figure 1). For testing the system, 82 participants were recruited, including sign language instructors and students. The participants included the hearing and the hearing-impaired who had an interest in learning sign language.

Figure 1. Interactive system of digital sign language picture books (https://idl.nutc.edu.tw/, accessed on 1 February 2025).

Participants began by selecting a story of interest from the main page of the system. Upon selection, they proceeded to watch a video that featured storybooks accompanied by sign language representations embedded within the content. Following the viewing, participants engaged in sign language practice by signing in front of the camera, which was initiated by clicking a designated green button on the user interface. The AI-driven recognition system subsequently analyzed their signing performance and provided immediate feedback, presenting the accuracy of their signed vocabulary as a percentage (Figure 2).

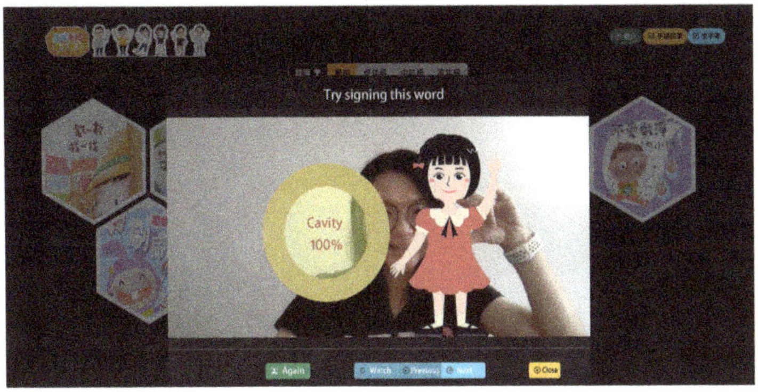

Figure 2. AI in recognizing the correctness of learners' sign language.

TAM was employed to examine learners' acceptance of the system and identify the perceived usefulness and perceived ease of use and their influence on users' attitudes toward and intentions. The results were used to predict users' behavior in the successful implementation of technology [1].

Before using the system, the participants were informed about the study's objectives and procedures. Each participant used the system for 10–15 min. Then, they completed a questionnaire to assess their acceptance of the system. The collected data were analyzed using structural equation modeling (SEM) to conduct factor and path analyses to predict the user acceptance of the system. A survey questionnaire was developed for this study, encompassing four constructs and 14 items. The items were tailored to align with the objectives of this research. The theoretical framework of TAM includes four constructs: perceived usefulness (PU), perceived ease of use (PEOU), attitude toward using (ATU), and behavioral intention to use (BIU), which are used to predict the actual system use (ASU) by the users.

In the questionnaire, the perceived usefulness (PU) construct included four items (1-1 to 1-4); the perceived ease of use (PEOU) construct included five items (2-1 to 2-5); the attitude toward using (ATU) construct included three items (3-1 to 3-3); and the behavioral intention to use (BIU) construct included two items (4-1 to 4-2). A five-point Likert scale ranging from (1) "Strongly Disagree" to (5) "Strongly Agree" was used to score the items. The constructs and items are detailed in Table 1. Data collection was conducted electronically through an online survey.

The participants completed questionnaires using a computer, tablet, or mobile phone. The collected data were analyzed using JASP and SmartPLS 4 for descriptive statistics and structural equation modeling (SEM) to explore relationships among latent variables of TAM. SmartPLS 4 was chosen for SEM analysis due to its availability for small sample sizes and its capability to handle complex causal models involving multiple constructs. Partial Least Squares (PLS), a method within SEM, was appropriate for this study as it has predictive power and model interpretability.

Table 1. Constructs and items in the questionnaire on this study.

TAM Construct		Items
Perceived usefulness	PU1	This sign language learning system helps me learn sign language better.
	PU2	This sign language learning system allows me to learn sign language more quickly.
	PU3	This sign language learning system makes it easier for me to learn sign language.
	PU4	I find this sign language learning system very helpful for learning sign language.
Perceived ease of use	PEOU1	Learning to operate this system is easy for me.
	PEOU2	Interacting with this sign language learning system is clear and easy to understand.
	PEOU3	I find this sign language learning system to be very versatile in its interactivity.
	PEOU4	Mastering the use of this sign language learning system is easy for me.
	PEOU5	Learning sign language using this system is easy for me.
Attitude toward using	ATU1	I think using this sign language system to learn sign language is a great way.
	ATU2	This system makes learning sign language more enjoyable.
	ATU3	I enjoy using this system to learn sign language.
Behavioral intention to use	BIU1	I plan to continue using this system for teaching.
	BIU2	I will continue using this system for teaching.

3. Results

In SmartPLS, the measurement model represents the first stage of testing. In this stage (the outer model), convergent validity and discriminant validity are determined to see whether the items in each construct reflect that construct. The second stage involves testing the inner model to verify the proposed hypotheses and evaluate the explanatory power of the constructs.

The descriptive statistics are presented in Table 2. Females outnumbered males at 74.4% and 25.6% of the total number of participants. The majority of participants were 31–60 years old. The ratio of the hearing and hearing-impaired was equal.

Table 2. Demographic Analysis.

		Participants (N = 82)	Percentage (%)
Gender	Male	21	25.6
	Female	61	74.4
Age	Under 18	3	3
	18–30	3	6
	31–40	19	23.2
	41–50	29	35.4
	51–60	24	29.3
	61+	4	4.9
Hearing Status	Deaf	41	50
	Hard of Hearing	1	1.2
	Hearing	40	48.8
Occupation	Sign Language Teacher	62	75.6
	General Public	20	24.4

Descriptive statistics for the four constructs in this study are presented in Table 3. The average scores for each construct are all above 4.5, with the highest average of 4.707 observed in "Attitude Toward Using" and the lowest average of 4.537 in "Behavioral Intention to Use".

Table 3. Descriptive statistics for each construct.

TAM Constructs	Mean	SD
Perceived usefulness	4.637	0.523
Perceived ease of use	4.593	0.566
Attitude toward using	4.707	0.504
Behavioral intention to use	4.537	0.659

3.1. Outer Model

In the outer model, internal consistency was assessed using Cronbach's alpha (Table 4). A reliability level of 0.7 or higher is recommended [13]. Cronbach's alpha of the constructs exceeded 0.7, ranging from 0.877 to 0.905. Composite reliability (CR represents the reliability of all measurements. The higher the CR value of a latent variable, the better the items measure the latent variable, indicating higher internal consistency of the construct [13]. In this study, the CR of all constructs exceeded 0.7, ranging from 0.822 to 0.908.

Table 4. Reliability and validity analysis values of construct items.

TAM Constructs	Mean	SD	Factor Loadings	Cronbach's Alpha (α)	CR	AVE
PU1	4.70	0.46	0.832	0.877	0.887	0.730
PU2	4.63	0.48	0.894			
PU3	4.61	0.56	0.846			
PU4	4.61	0.58	0.845			
PEOU1	4.59	0.60	0.892	0.905	0.908	0.726
PEOU2	4.54	0.61	0.862			
PEOU3	4.60	0.54	0.768			
PEOU4	4.59	0.56	0.859			
PEOU5	4.69	0.50	0.874			
ATU1	4.67	0.52	0.861	0.889	0.892	0.819
ATU2	4.77	0.42	0.936			
ATU3	4.68	0.56	0.915			
BIU1	4.57	0.63	0.924	0.822	0.822	0.849
BIU2	4.50	0.69	0.919			

CR and average variance extracted (AVE) were used to assess the reliability of constructs. If these values meet the standard thresholds, it indicates good construct reliability [14]. AVE measures the extent to which a latent variable explains the variance of its indicators. A higher AVE value indicates a greater degree of variance explained by the latent variable [15]. AVE must be greater than 0.5, with 0.36–0.5 being an acceptable range. A higher AVE value signifies a higher convergent validity of the construct [16]. In this study, AVEs were higher than 0.5, indicating sufficient convergent validity for each construct (Table 5).

Table 5. Analysis Values of Cronbach's α for each construct.

TAM Constructs	Numbers of Items	Cronbach's α Value
Perceived usefulness	4	0.872
Perceived ease of use	5	0.877
Attitude toward using	3	0.879
Behavioral intention to use	2	0.925

Discriminant validity is used to assess the extent to which measurement variables distinguish between different constructs, indicating low correlation between items from different constructs. According to the Fornell–Larcker criterion, the square root of the average variance extracted (AVE) for each latent construct should be greater than the correlation coefficients between that construct and other constructs, demonstrating discriminant validity between constructs [16,17]. The discriminant validity of the four constructs investigated in this study is shown in Table 6. The square root of AVE for each latent construct exceeds the correlation coefficients between that construct and other constructs, indicating discriminant validity among the constructs.

Table 6. Discriminant validity for each construct (Fornell–Larcker criterion).

	ATU	BI	PEOU	PU
ATU	0.905			
BI	0.815	0.921		
PEOU	0.839	0.726	0.852	
PU	0.659	0.620	0.687	0.854

However, Henseler et al. [18] argue that the Fornell–Larcker criterion may lack sensitivity, thus proposing the heterotrait–monotrait ratio of correlations (HTMT) as an alternative index to assess discriminant validity between constructs. According to their recommendation, the HTMT values between constructs should be less than 1 to indicate discriminant validity. In this study, all HTMT values between constructs are below 1, demonstrating discriminant validity (as shown in Table 7).

Table 7. Discriminant validity for each construct (HTMT criterion).

	ATU	BI	PEOU	PU
ATU				
BI	0.951			
PEOU	0.934	0.840		
PU	0.737	0.721	0.757	

3.2. *Structural Model/Inner Model: Hypothesis Verification and Analysis Test*

Structural model (inner model) analysis includes (1) model fit analysis, (2) overall model explained variance (R^2), (3) path coefficients, and (4) collinearity assessment.

3.2.1. Model Fit Analysis

Model fit analysis assesses the goodness of fit using the standardized root mean square residual (SRMR), where a good fit is indicated by an SRMR value below 0.08 or 0.1 [19]. In this study, the SRMR value for the model is 0.075 (see Table 8), indicating a good model fit.

Table 8. Model fit.

	Saturated Model	Estimated Model
SRMR	0.075	0.075
D_ULS	0.590	0.592
NFI	0.781	0.782

3.2.2. The Explained Variance (R^2) of the Overall Research Model

The explained variance (R^2), also known as the coefficient of determination, is used to explain the goodness of fit in a linear regression model. It indicates the causal relationships between the latent variables in the model, reflecting the extent to which the independent variables can explain the dependent variable. This metric can represent the in-sample predictive power and is commonly used as a criterion to judge the quality of a model. However, since R^2 can be influenced by the number of paths from other constructs, it is prone to bias. Therefore, the adjusted R^2 is often used for evaluation. The R^2 value ranges from 0 to 1, where a smaller value indicates poorer explanatory power, and a larger value indicates better explanatory power. A value of 0 means the independent variables are completely unrelated to the dependent variable, while a value of 1 means the independent variables can fully explain the dependent variable. The critical values for R^2 are 0.67, 0.33, and 0.19, which represent different levels of explanatory power for the model: substantial, moderate, and weak, respectively [20,21].

In the context of this research model, the constructs of attitude toward use (ATU) and behavioral intention (BI) exhibit substantial explanatory power, while the construct of perceived usefulness (PU) demonstrates moderate explanatory power (Table 9).

Table 9. Explained Variance of the Structural Model.

Construct	R^2	Adjusted R^2
ATU	0.717	0.710
BI	0.676	0.668
PU	0.473	0.466

3.2.3. Path Coefficient

Collinearity assessment utilizes the variance inflation factor (VIF) to evaluate collinearity among constructs. A VIF value should be less than 5 to indicate that there is no collinearity problem among the constructs. In this study, the VIF values range between 1 and 1.89, all of which are below 5, indicating no collinearity issues (see Table 10).

Table 10. Structural equation modeling results.

Hypothesis	Hypothesis Paths	Results
H1	PEOU → PU	Supported
H2	PEOU → ATU	Supported
H3	PU → ATU	Not supported
H4	PU → BIU	Not supported
H5	ATU → BI	Supported

3.2.4. Collinearity Assessment

Collinearity assessment utilizes the variance inflation factor (VIF) to evaluate collinearity among constructs. A VIF value should be less than 5 to indicate that there is no collinearity problem among the constructs. In this study, the VIF values range between 1 and 1.89, all of which are below 5, indicating no collinearity issues (see Table 11).

Table 11. The structural equation modeling results of this study.

Hypothesis	VIF	Path Coefficients	SD	T Statistics	*p* Value	Confidence Interval
H1	1.000	0.687	0.074	9.280	0.000	[0.536, 0.825]
H2	1.896	0.733	0.094	7.762	0.000	[0.531, 0.905]
H3	1.896	0.155	0.110	1.414	0.157	[−0.049, 0.380]
H4	1.767	0.147	0.086	1.720	0.086	[−0.020, 0.318]
H5	1.767	0.718	0.072	9.910	0.000	[0.566, 0.850]

The research hypotheses and their path coefficients are shown in Figure 3. The arrows on the figure represent the paths, and the numbers on the paths indicate the path coefficients. An asterisk following a path coefficient indicates that the path has a significant difference.

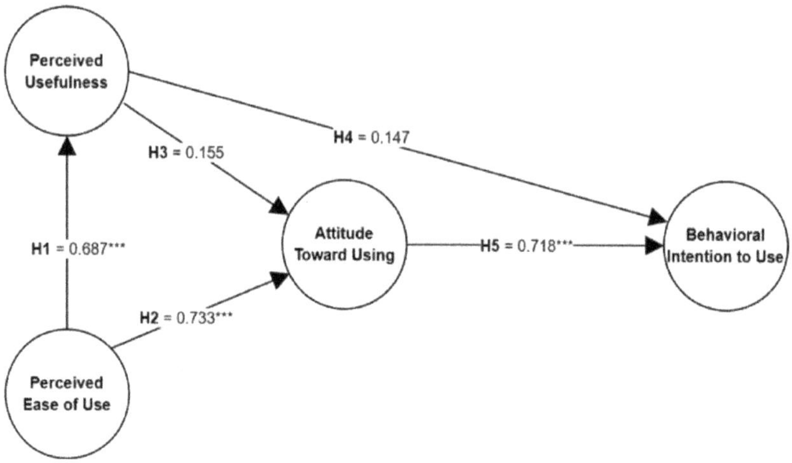

Figure 3. Hypotheses and their path coefficients (*** presents the significant differences in paths).

The research results show that the path coefficients for H1, H2, and H5 are positive. Perceived ease of use (PEOU) positively influences perceived usefulness (PU), as shown by path H1 ($p < 0.001$), and perceived ease of use (PEOU) positively influences attitude toward use (ATU), as shown by path H2 ($p < 0.001$). Thus, hypotheses H1 and H2 are supported. Additionally, attitude toward use (ATU) positively influences behavioral intention (BIU), as shown by path H5 ($p < 0.01$), supporting hypothesis H5.

However, perceived usefulness (PU) does not have a significant positive influence on attitude toward use (ATU), path H3 ($p > 0.01$), nor on behavioral intention (BIU), path H4 ($p > 0.01$). Therefore, hypotheses H3 and H4 are not supported. This indicates that perceived usefulness (PU) does not have a significant positive influence on attitude toward use or behavioral intention.

4. Discussion

The research results indicate multiple positive relationships regarding learners' acceptance of technology when interacting with the sign language picture book system developed in this study. Firstly, three hypotheses are supported: the perceived ease of use of the system for learning sign language positively influences perceived usefulness (H1), perceived ease of use positively influences attitude toward use (H2), and attitude toward use positively influences behavioral intention (H5). This suggests that if learners find the system easy to use, they will also perceive it as useful. Moreover, when learners consider the system useful, it positively affects their attitude toward using it and their behavioral intention to use it. However, the study results show that the system's perceived usefulness (PU) does not significantly positively impact attitude toward use (ATU) and behavioral intention (BIU). The possible reasons for this are as follows: (1) the sign language recognition accuracy of the system needs improvement, (2) the content of the sign language materials should provide thematic dialogue learning scenarios, and (3) a search function should be provided to allow users to look up content by entering keywords. Additionally, because the current teaching content is more suitable for learners with some foundational knowledge, it is recommended that the materials be categorized by different levels of difficulty. For beginners, the sign language actions and speed in the videos might be too fast, so it may be necessary to slow down the speed and provide more demonstrations in the video content. These suggestions aim to enhance users' perceived usefulness of the system.

5. Conclusions

The development of the Interactive System of Digital Sign Language Picture Books contributes to enriching the reading and learning experience of deaf and hearing-impaired communities. Furthermore, it could promote social inclusion by facilitating better communication between deaf and hearing individuals and raising awareness of sign language in broader society. Evaluating the interactive system using the technology acceptance model (TAM) has provided us with valuable insights into how target users perceive the system for learning and reading through Taiwanese Sign Language.

The limitations of this study include the lack of examination of external variables influencing the acceptance of the sign language picture book system. Examples of such variables could include instructional content design, information or system quality, self-efficacy, enjoyment, or gamification, among others. Therefore, future research is recommended to incorporate these external variables to further explore the system's acceptance.

Additionally, since participants were all first-time users of the system developed in this study, acceptance may vary over time with prolonged system use. Factors such as improvement in sign language proficiency or increased familiarity with system usage could influence acceptance. Therefore, future studies could also delve deeper into the impact of external variables on the acceptance of this system.

Author Contributions: Conceptualization, K.-F.H., T.-Y.T., S.-H.C., S.-Y.C. and Y.-H.W.; methodology, K.-F.H. and Y.-H.W.; software, K.-F.H.; validation, Y.-H.W.; formal analysis, Y.-H.W.; investigation, K.-F.H., Y.-H.W., C.-C.W. and K.-Y.H.; resources, S.-Y.C.; data curation, Y.-H.W.; writing—original draft preparation, Y.-H.W.; writing—review and editing, Y.-H.W.; visualization, Y.-H.W.; supervision, K.-F.H. and T.-Y.T.; project administration, K.-F.H.; funding acquisition, T.-Y.T., S.-H.C. and S.-Y.C. All authors have read and agreed to the published version of the manuscript.

Funding: This research was funded by the National Taiwan Library, grant number NTL-lsy1120028.

Institutional Review Board Statement: Not applicable.

Informed Consent Statement: Not applicable.

Data Availability Statement: Data are contained within the article.

Conflicts of Interest: The authors declare no conflicts of interest. The funders had no role in the design of the study; in the collection, analyses, or interpretation of data; in the writing of the manuscript; or in the decision to publish the results.

References

1. Eason, G.; Noble, B.; Sneddon, I.N. On certain integrals of Lipschitz-Hankel type involving products of Bessel functions. *Philos. Trans. R. Soc. Lond.* **1955**, *247*, 529–551.
2. Maxwell, J.C. *A Treatise on Electricity and Magnetism*, 3rd ed.; Clarendon: Oxford, UK, 1892; Volume 2, pp. 68–73.
3. Jacobs, I.S.; Bean, C.P. Fine particles, thin films and exchange anisotropy. In *Magnetism*; Rado, G.T., Suhl, H., Eds.; Academic: New York, NY, USA, 1963; Volume III, pp. 271–350.
4. Fox, N.; Woll, B.; Cormier, K. Best practices for sign language technology research. *Univers. Access Inf. Soc.* **2023**, 0123456789. [CrossRef]
5. Bragg, D.; Koller, O.; Bellard, M.; Berke, L.; Boudreault, P.; Braffort, A.; Caselli, N.; Huenerfauth, M.; Kacorri, H.; Verhoef, T.; et al. Sign language recognition, generation, and translation: An interdisciplinary perspective. In Proceedings of the ASSETS 2019—21st International ACM SIGACCESS Conference on Computers and Accessibility, Pittsburgh, PA, USA, 28–30 October 2019; pp. 16–31. [CrossRef]

6. Yorozu, Y.; Hirano, M.; Oka, K.; Tagawa, Y. Electron spectroscopy studies on magneto-optical media and plastic substrate interface. *IEEE Transl. J. Magn. Jpn.* **1987**, *2*, 740–741. [CrossRef]
7. Young, M. *The Technical Writer's Handbook*; University Science: Mill Valley, CA, USA, 1989.
8. Adeyanju, I.A.; Bello, O.O.; Adegboye, M.A. Machine learning methods for sign language recognition: A critical review and analysis. *Intell. Syst. Appl.* **2021**, *12*, 200056. [CrossRef]
9. Nedjar, I.; M'hamedi, M. Interactive system based on artificial intelligence and robotic arm to enhance arabic sign language learning in deaf children. *Educ. Inf. Technol.* **2024**, *29*, 24563–24580. [CrossRef]
10. Soogund, N.; Joseph, M. SignAR: A Sign Language Translator Application with Augmented Reality using Text and Image Recognition. In Proceedings of the 2019 IEEE International Conference on Intelligent Techniques in Control, Optimization and Signal Processing (INCOS), Tamilnadu, India, 11–13 April 2019; IEEE: Piscataway, NJ, USA, 2019. [CrossRef]
11. Cabanillas-Carbonell, M.; Cusi-Ruiz, P.; Prudencio-Galvez, D.; Salazar, J.L.H. Mobile Application with Augmented Reality to Improve the Process of Learning Sign Language. *Int. J. Interact. Mob. Technol.* **2022**, *16*, 51–64. [CrossRef]
12. Papastratis, I.; Chatzikonstantinou, C.; Konstantinidis, D.; Dimitropoulos, K.; Daras, P. Artificial intelligence technologies for sign language. *Sensors* **2021**, *21*, 5843. [CrossRef] [PubMed]
13. Nunnally, J.C. *Psychometric Theory*; McGraw Hill: New York, NY, USA, 1978.
14. Hair, J.; Anderson, R.; Tatham, R.; Black, W. *Multivariate Data Analysis*, 5th ed.; Prentice Hall: Hoboken, NJ, USA, 1998.
15. Chang, S. *Partial Least Squares Structural Equation Modeling (PLS-SEM): Application Using SmartPLS*; Wu-Nan: Taipei, Taiwan, 2021.
16. Fornell, C.; Larcker, D.F. Evaluating Structural Equation Models with Unobservable Variables and Measurement Error. *J. Mark. Res.* **1981**, *18*, 39–50. [CrossRef]
17. Segars, A.H.; Grover, V. Strategic Information Systems Planning Success: An Investigation of the Construct and Its Measurement. *MIS Q.* **1998**, *22*, 139–163. [CrossRef]
18. Henseler, J.; Ringle, C.M.; Sarstedt, M. A new criterion for assessing discriminant validity in variance-based structural equation modeling. *J. Acad. Mark. Sci.* **2015**, *43*, 115–135. [CrossRef]
19. Garson, G.D. *Partial Least Squares Regression and Structural Equation Models*; Statistical Associates Publishers: Raleigh, NC, USA, 2016.
20. Chin, W.W. The partial least squares approach for structural equation modeling. In *Modern Methods for Business Research*; Marcoulides, G.A., Ed.; Lawrence Erlbaum Associates Publishers: Mahwah, NJ, USA, 1998; chapter 10; pp. 295–336.
21. Höck, M.; Ringle, C.M. Local strategic networks in the software industry: An empirical analysis of the value continuum. *Int. J. Knowl. Manag. Stud.* **2010**, *4*, 132–151. [CrossRef]

engineering proceedings

MDPI

Proceeding Paper

Comparison of Armillary Sphere in Ancient China and Western World †

Jian-Liang Lin [1,*] and Kuo-Hung Hsiao [2]

1 Technology Education Division, National Science and Technology Muse, Kaohsiung 807, Taiwan
2 Exhibition Division, National Science and Technology Muse, Kaohsiung 807, Taiwan
* Correspondence: golduser007@gmail.com
† Presented at the 2024 IEEE 7th International Conference on Knowledge Innovation and Invention, Nagoya, Japan, 16–18 August 2024.

Abstract: Armillary spheres were developed in the East and the West for a long time. They independently developed various functions for astronomy. In this article, we discuss the differences in mechanical structures, appearance, and functions between the armillary spheres in ancient China and Europe. The earliest armillary sphere in ancient China was invented by Luo Xia Hong (落下閎) between 156 BC and 87 BC. Then, the armillary sphere in ancient China improved with the historical development of astronomy. The famous armillary sphere was built in an astronomical clock tower (水運儀象台) by Su Song (蘇頌) in the Song (宋) dynasty. This armillary sphere was an astronomical apparatus for the observation of celestial phenomena and the correction of time standards. However, the armillary sphere in Europe had different applications, even though the structures were similar. The armillary spheres in Europe simulated the sun's trajectory in one day to predict the sunrise and sunset positions. They adjusted the tilting angle of the celestial sphere with the altitude of observation to observe the path of the stars around the ecliptic. Through this review, the armillary spheres in ancient China and Europe are defined clearly.

Keywords: armillary sphere; ancient China; Europe

1. Introduction

Celestial bodies in the sky interested humankind to discover their secrets, whatever people located on the ground. People tried to explain the period phenomena and the cyclic motions of celestial bodies by constructing the appearance of the universe and the corresponding theories. Astrology and astronomy were successively developed with observation and records for a long time. Owing to the observation requirements, several astronomical mechanical devices were created, such as armillary spheres, quadrants, sundials, and astrolabes. Armillary spheres were invented by different ancient civilizations around the world. Even though the appearances of armillary spheres from various areas are different, their functions are almost the same. Such mechanical devices were applied to observe the positions of celestial bodies and even to correct time.

The armillary spheres are combined with several rings to form a sphere in ancient China or the ancient Western world, as shown in Figure 1. These rings with various scales were used to construct a coordinate system of the universe. The core of the formed sphere is the Earth. The design represents that the geocentric model was the essential theory in the period that the device was created. Though their appearances are similar, the sizes of the physical models have huge differences. The ancient Chinese model is heavy and huge, and it is a floor-standing type. The ancient Western model is smaller; it is a desktop or handheld type.

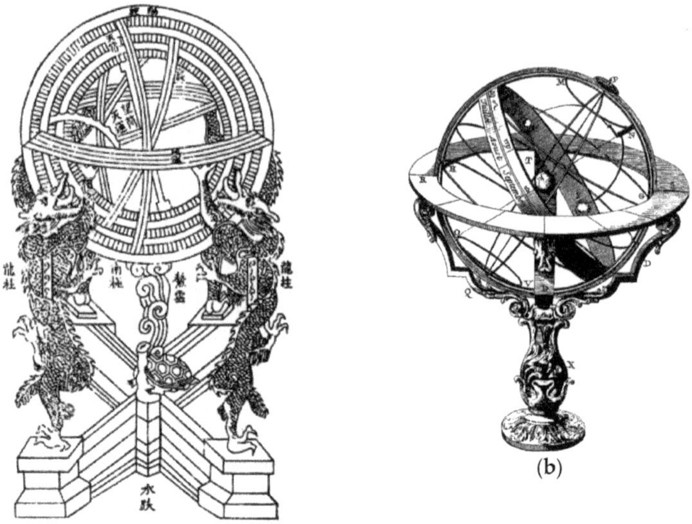

(**a**) (**b**)

Figure 1. Armillary spheres in (**a**) China and (**b**) Europe.

The origin of their invention in the ancient Western world is questionable. Eratosthenes (276–194 BC) and Anaximander of Miletus (611–547 BC) could be the inventors. In Ptolemy's *Almagest*, these instruments were much simpler, and the structure had three or four rings. They could be existing or simpler armillary spheres. There are no manuscripts to prove it. One known opinion is that the earliest complete armillary was invented by the meteōroskopion of the Alexandrine Greeks in about AD 140. This structure of the earliest Greek device included nine rings around the globe. The instrument had nine rings, including the horizon, meridian, Equator, tropics, polar circles, and ecliptic and a ring for supporting the skeleton of the sphere. The armillary sphere was an important astronomical device for observing and measuring the positions and motions of celestial bodies until the appearance of the telescope in the 17th century [1–3].

The armillary sphere was invented independently in ancient China. The earliest inventor is Luo Xia-Hong (落下閎) between 156 and 87 BC. However, people misunderstood that the first known armillary sphere in China was created by the astronomer Geng Shou-Chang (耿壽昌) around 52 BC. The early model had a simple design with several rings representing the celestial equator and meridian. Correctly understood, the astronomical model invented by Geng Shou-Chang (耿壽昌) was the celestial globe (Figure 2). The armillary sphere in ancient China improved with the historical development of astronomy. During the Eastern Han dynasty (漢朝), a famous astronomer Zhang Heng (張衡, 78–139 CE) improved the armillary sphere. He created a more complex design incorporating an equatorial ring, a meridian ring, and a horizon ring. However, the device did not have an element (tube) for observation. The astronomical device by Zhang Heng demonstrated celestial bodies. The most amazing armillary sphere was built in an astronomical clock tower (水運儀象台) by Su Song (蘇頌) in the Song dynasty (宋朝) (Figure 3). The clock tower included the armillary sphere, celestial bodies, a time chime device, and a mechanical clock. These devices were driven by a water wheel [4–6].

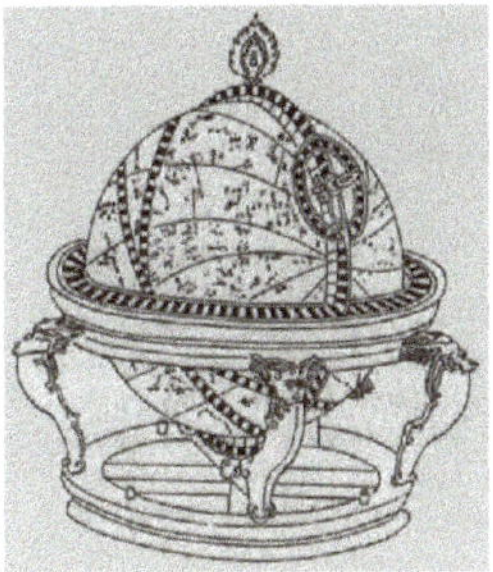

Figure 2. Celestial globe by Geng Shou-Chang (耿壽昌).

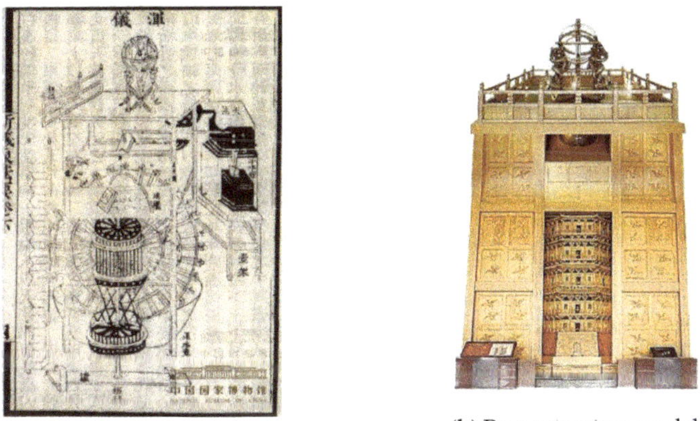

(**a**) Historical record (**b**) Reconstruction model

Figure 3. Astronomical clock tower (水運儀象台).

At present, people confuse the armillary sphere and celestial globes. There is incorrect information on websites. Armillary spheres independently developed in ancient China and the ancient Western world. Their functions and the structure are similar. Hence, we present the historical development of the armillary sphere by analyzing its structure and functions and identifying the armillary sphere and the celestial globe.

2. Armillary Sphere and Celestial Globe

Armillary spheres and celestial globes are different astronomical devices. Armillary spheres are used to observe, and celestial globes are used to demonstrate the positions of known stars in the sky at one moment in one night. However, owing to the description of the ancient literature, people misidentify these two mechanical devices without drawings or due to the misunderstanding of how to operate the devices recorded in the ancient manuscripts.

In many ancient transcripts, the appearances of armillary spheres and celestial globes are described as spheres. Armillary spheres are built by several rings to form a spherical structure, and the surface of the structure is not filled up. Several rings are fixed and some rings are rotatable. The rings support each other. The endpoints of fixed or rotatable inner rings are connected to the outer one, as shown in Figure 4. Because the armillary sphere is used for observation, this device has an element to improve the accuracy and convenience of observation, using a straight tube or a tube with a cross at the front sight. Furthermore, with the development of astronomical observation, the structures of armillary spheres became complicated. Rings were added to the device for measuring or simulating the planets' motions.

(a)

(b)

Figure 4. Ancient armillary spheres in (**a**) China and (**b**) Europe.

As to celestial globes, the common styles are constructed as hollow balls (Figure 5). The known stars are engraved or bulging on the surface of the ball. The ball constructed on the frame is rotatable. Therefore, users can rotate celestial globes to demonstrate the sky based on the time and day. Celestial globes were built based on observation records for a long time, which seem to constitute big data or statistics. Certainly, they were used to predict the celestial phenomena occurring in the next few days.

Figure 5. Ancient Chinese celestial globe.

The armillary sphere in ancient China is too complicated to conventionally operate. The structural designs of armillary spheres have more than nine rings. An astronomer (郭守敬) improved the device by separating its multiple functions and what was called the abridged armillary (簡儀), as shown in Figure 6. In the ancient Western world, a mechanical device assembled by several concentric rings was similar to the armillary sphere. The device aimed to simulate the periodic motions of the sun, the moon, and the planets. According to the literature survey, a device including 47 concentric rings survived [7].

Figure 6. Abridged armillary (簡儀).

3. Armillary Sphere in Ancient China

The armillary sphere, an ancient astronomical instrument representing the celestial sphere with rings and hoops, has a rich history in China, reflecting significant advancements in astronomy and instrumentation.

The theory of Sphere Heavens (渾天說) describes the evolution of the structure of the universe, which has become complete through continuous additions and expansions of concepts by scholars of various dynasties. When the theory of Sphere Heavens developed in the Eastern Han Dynasty, the famous astronomer Zhang Heng advocated that the sky and the Earth were spheres like an egg, with the sky being the eggshell and the earth being the yolk, the sky thus enclosing the Earth. Based on this, Zhang Heng improved the armillary sphere to create a complicated design incorporating an equatorial ring, a meridian ring, and a horizon ring. The more developed the astronomical theory became, the more complex the armillary sphere device became. In the Tang dynasty (唐朝, 618−907 AD), the astronomer Li Chunfeng (李淳風) created an astronomical observation device with three spherical layers in 633 AD called the "nest", which corrected astronomical observations from many aspects. He also designed a tube on the device for observing. The installation of the tube was finished, but the astronomer Yi Xing (一行, 683−727 AD) played a pivotal role in enhancing the instrument. He added more rings to represent the ecliptic and celestial poles, which increased its precision.

In ancient China, the most well-known form of the armillary sphere was presented by Su Song (蘇頌) in the Song Dynasty (宋朝, 960–1279 CE). This form, on a water-moving instrument platform, had three systems consisting of Liuhe Yi (六合儀), Sanchen Yi (三辰儀), and Siyou Yi (四游儀), combined with a viewing tube (窺管) and a stand (鰲雲柱), as shown in Figure 7 [4]. This armillary sphere, including nine rings, was built on top of an astronomical clock tower (水運儀象台). The base was a cross-shaped water supply mechanism (水趺). This special design of the base was to correct the horizontal level of the device. The mechanical devices in the tower were driven by a water wheel. Hence, the armillary sphere had a celestial ring running on water to track celestial bodies, observe the movements of stars, and record their coordinates. It is a symbol of the maturity of the development of China's armillary sphere.

Siyou Yi (四游儀) Sanchen Yi (三辰儀) Liuhe Yi (六合儀)

Figure 7. Structural analysis of ancient Chinese armillary sphere.

The viewing tube involving the Siyou Yi design is convenient to observe through because it can be rotated in three axial directions. Then, the astronomer added front sight on the viewing tube to improve measurement accuracy. In the Ming dynasty (明朝, 1368–1644 AD), Guo Shoujing (郭守敬) changed the style of the viewing tube so that the observation room would not be affected.

Throughout its development, the armillary sphere in ancient China exemplified the ingenuity and dedication of Chinese astronomers to understand the cosmos. The instrument's evolution reflects broader scientific and technological advancements and highlights China's significant contributions to the history of astronomy.

4. Armillary Sphere in Ancient West World

As far as armillary spheres in the ancient Western world are considered, the center of the early model is the Earth, known as Ptolemaic. The later models using the Sun as the center are known as Copernican. No matter whether the models are suspended or rested on a stand, their structures are similar. Only differences exist between these two conceptual models like the position and the tilting angle of the ecliptic.

The armillary sphere has a rich history in the Western world, tracing back to ancient Greek and Roman times and evolving through the Middle Ages into the Renaissance. In ancient Greece and Rome, the concept of the armillary sphere is attributed to the ancient Greeks. The earliest known references date back to 3rd century BCE. Hipparchus (190–120 BC) is known to have used an armillary sphere for his astronomical observations and calculations. Claudius Ptolemy (100–170 AD) described the operation of an armillary sphere in his work *Almagest*, which became a cornerstone of astronomical knowledge for centuries. During the early Middle Ages, much knowledge of armillary spheres was preserved and enhanced by Islamic scholars. Islamic astronomers such as Al-Battani (858–929 AD) and Al-Sufi (903–986 AD) used and refined armillary spheres, incorporating them into their astronomical studies.

Armillary spheres were reintroduced to Europe through translations of Arabic texts during the 12th and 13th centuries. Gerard of Cremona (1114–1187 AD) translated many of these works, reintroducing classical and Islamic astronomical knowledge to European scholars. Then, several revivals and innovations of armillary spheres were presented in the Renaissance. Johannes de Sacrobosco (1195–1256 AD) and Regiomontanus (1436–1476 AD) studied and wrote about the armillary sphere. Nicolas Copernicus (1473–1543 AD) utilized the armillary sphere to develop his heliocentric theory, challenging the geocentric model of Ptolemy. By the 16th and 17th centuries, armillary spheres were commonly used in universities for teaching astronomy. They also became valuable tools for navigation, aiding explorers during the Age of Discovery.

An armillary sphere represents the celestial sphere with rings and hoops centered on the Earth (geocentric model) or Sun (heliocentric model), as shown in Figure 8. The primary rings and hoops around the center are as follows [8].

Figure 8. Armillary spheres for the geocentric model and the heliocentric model.

- The equatorial ring represents the celestial equator.
- The ecliptic ring is tilted relative to the equatorial ring; it shows the path of the Sun (the ecliptic).
- The meridian ring represents the local meridian, passing through the celestial poles and the zenith.
- The horizon ring represents the observer's local horizon.
- The tropic rings represent the Tropic of Cancer and the Tropic of Capricorn.
- The Arctic and Antarctic Circles represent the Arctic and Antarctic Circles.

Except for the rings, there are additional components and adjustable features. Small circles, called almucantars, are parallel to the horizon in several armillary spheres. Rete is a mesh-like structure representing the starry sky, often used in conjunction with the ecliptic ring to show constellations and star positions. The rings rotate to align with the positions of celestial objects, helping to demonstrate celestial mechanics and navigation techniques. Between the sphere and the base, a slot is designed to adjust the tilting angle based on the altitude of observers, as shown in Figure 9.

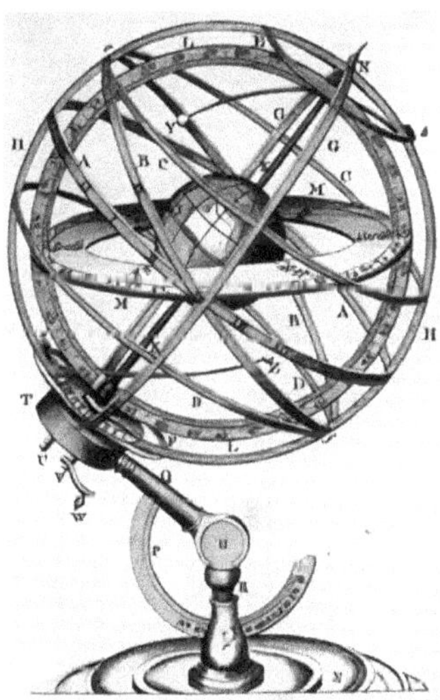

Figure 9. Adjustable design for altitude.

5. Conclusions

With the advent of the telescope and more advanced astronomical instruments, the practical use of armillary spheres declined. These devices played different roles. The armillary sphere was used for teaching, observational purposes, and as a tool for astronomers to understand the movements of celestial bodies. And they remain important historical artifacts displayed in museums as symbols of early scientific inquiry and discovery.

There is wrong information on the armillary sphere and the celestial globe. Hence, we introduce the origin of armillary spheres in ancient China and the ancient Western world. Then, the historical developments, the functions, and the structural analysis of armillary spheres in ancient China and the ancient Western world were presented. Although they are astronomical mechanical devices for observation and have similar appearances, they have different structures and applications.

Author Contributions: Conceptualization, J.-L.L. and K.-H.H.; formal analysis, J.-L.L.; investigation, J.-L.L. and K.-H.H.; resources, J.-L.L.; writing—original draft preparation, J.-L.L. and K.-H.H.; writing, J.-L.L. All authors have read and agreed to the published version of the manuscript.

Funding: This research was supported by the National Science and Technology Council (Taipei, Taiwan) under grant NSTC 113-2410-H-359-003, constituting the financial support of this work.

Institutional Review Board Statement: Not applicable.

Informed Consent Statement: Not applicable.

Data Availability Statement: All software code and data included in this study are available upon request to the corresponding author.

Conflicts of Interest: The authors declare no conflicts of interest.

References

1. Helmink, L. Armillary Sphere with Wind Heads. Available online: https://www.britannica.com/science/armillary-sphere (accessed on 1 November 2024).
2. Armillary Sphere. Available online: https://en.wikipedia.org/wiki/Armillary_sphere#External_links (accessed on 1 November 2024).
3. Armillary Spheres and Teaching Astronomy, Whipple Museum of the History of Science, University of Cambridge. Available online: https://www.whipplemuseum.cam.ac.uk/explore-whipple-collections/astronomy/armillary-spheres-and-teaching-astronomy (accessed on 1 November 2024).
4. Song, S. *Xin Yi Xiang Fa Yao (新儀象法要)*; The Commercial Press, Ltd. (台灣商務印書館): Taipei, Taiwan, 1968.
5. Zhenduo, W (王振鐸). *Song Dai Shui Yun Yi Xiang Tai De Fu Yuan (宋代水運儀象台的復原)*; Wen Wu Can Kao Zi Liao (文物參考資料); Wenwu: Beijing, China, 1958; Volume 9, pp. 1–9.
6. Junwei, Z. (張君瑋) Reconstruction and Innovative Design of Su Song's Armillary Sphere in the Northern Song Dynasty (北宋蘇頌渾儀的復原與創新設計). Master's Thesis, Southern Taiwan University of Science and Technology, Tainan, Taiwan, 2014.
7. Linton, C.M. *From Eudoxus to Einstein: A History of Mathematical Astronomy Annotated*; Cambridge University Press: Cambridge, UK, 2004.
8. Parkin, C. Armillary Sphere. Available online: https://www.hsm.ox.ac.uk/armillary-sphere (accessed on 1 November 2024).

engineering proceedings

MDPI

Proceeding Paper

Combining Interactive Technology and Visual Cognition—A Case Study on Preventing Dementia in Older Adults †

Chung-Shun Feng [1,2,*] and Chao-Ming Wang [3]

1 Graduate School of Design, National Yunlin University of Science and Technology, Yunlin 632, Taiwan
2 Department of Industrial Design, Chaoyang University of Technology, Taichung 413, Taiwan
3 Department of Digital Media Design, National Yunlin University of Science and Technology, Yunlin 632, Taiwan; wangcm@yuntech.edu.tw
* Correspondence: shun2704@gmail.com

† Presented at the 2024 IEEE 7th International Conference on Knowledge Innovation and Invention, Nagoya, Japan, 16–18 August 2024.

Abstract: According to the World Health Organization, the global population is aging, with cognitive and memory functions declining from the age of 40–50. Individuals aged 65 and older are particularly prone to dementia. Therefore, we developed an interactive system for visual cognitive training to prevent dementia and delay the onset of memory loss. The system comprises three "three-dimensional objects" with printed 2D barcodes and near-field communication (NFC) tags and operating software processing text, images, and multimedia content. Electroencephalography (EEG) data from a brainwave sensor were used to interpret brain signals. The system operates through interactive games combined with real-time feedback from EEG data to reduce the likelihood of dementia. The system provides feedback based on textual, visual, and multimedia information and offers a new form of entertainment. Thirty participants were invited to participate in a pre-test questionnaire survey. Different tasks were assigned to randomly selected participants with three-dimensional objects. Sensing technologies such as quick-response (QR) codes and near-field communication (NFC) were used to display information on smartphones. Visual content included text-image narratives and media playback. EEG was used for visual recognition and perception responses. The system was evaluated using the system usability scale (SUS). Finally, the data obtained from participants using the system were analyzed. The system improved hand-eye coordination and brain memory using interactive games. After receiving visual information, brain function was stimulated through brain stimulation and focused reading, which prevents dementia. This system could be introduced into the healthcare industry to accumulate long-term cognitive function data for the brain and personal health data to prevent the occurrence of dementia.

Keywords: dementia; interactive technology; visual cognition; experiential design

1. Introduction

With emerging technology and networks, the quality of medical care and medical research has been improved. The global birth rate is decreasing every year, and mortality rates are falling owing to the development of medical technology. Therefore, average life expectancy is prolonged, which is related to population aging. According to the United Nations, by 2050, more than 50% of countries will have entered the "aging society" stage [1].

The health of the elderly is the primary problem of an aging society. Elderly individuals face physical, mental, and emotional problems, and the need for prevention of dementia in this population is emphasized [1,2]. In the age group of individuals aged 65 years old and above, the brain's ability begins to decline, especially in terms of cognition, memory, attention, and real-time response [3]. According to the World Health Organization (WHO), in 2021, 55 million elderly people had dementia worldwide. In 2015, the Alzheimer's Disease Association (ADI) pointed out that dementia affected one person in the world every three seconds [3]. The National Development Commission of the Executive Yuan of Taiwan estimated that Taiwan will become an ultra-aging society by 2026 [4]. According to the Ministry of Health and Welfare, there will be more than five Taiwanese people with dementia in 2061 for every 100 individuals [4]. Today, with the advancement of internet technology and interactive technology, new technologies are being used to prevent dementia, which is a common problem in an aging society. Therefore, we developed a new system with interactive games to improve the hand–eye coordination and brain memory of elderly individuals to prevent dementia and to improve their health.

2. Literature Review

2.1. Dementia

Dementia is a progressive brain disorder syndrome, characterized by the deterioration of cognitive function, memory function, and behavioral abilities, impacting daily life [5]. With advancements in medicine and declining birth rates, many countries are experiencing aging populations, increasing the issue of dementia in elderly societies. According to a study by ADI, the number of individuals with dementia will reach 74.7 million by 2030 [5]. As this population ages, brain function will gradually decline through cognitive impairment, spatial awareness decline, judgment decline, and other functional declines associated with the chronic deterioration of brain memory function [6]. Epidemiological studies indicate that the likelihood of dementia increases with age, particularly after the age of 65 [7]. According to the Ministry of the Interior of Taiwan in 2018, 18.23% of elderly people aged 65 and above experienced mild cognitive impairment, while 7.86% had dementia. Table 1 lists the different stages and symptoms of dementia [8,9]. Low educational attainment is also a risk for dementia, as education level affects lifestyle and environmental health factors that may contribute to dementia [9,10].

Table 1. Dementia classification and symptoms.

Dementia Level	Symptoms of Dementia
Level 1: Mild cognitive impairment	Extremely mild cognitive decline.
Level 2: Mild dementia	Slight loss of memory and slight impairment of judgment.
Level 3: Moderate dementia	Moderate memory loss, disorientation, and memory impairment regarding people, times, and places.

2.2. Impact of Visual Cognition on Dementia

The human brain is composed of nerve cells in four regions: the frontal lobe, parietal lobe, occipital lobe, and temporal lobe (Figure 1). The frontal lobe is responsible for cognitive functions, motor control, social interaction, and problem-solving [10,11].

The brain produces various brain waves that are closely related to cognitive, sensory, and motor functions. In the brain, alpha brain waves play an important role in visual information processing [11,12]. Many studies have shown the relationship between brainwave presentation and activity state responses to visual and cognitive functions and behaviors [12]. People use vision to enhance brain function. Card games, table games, puzzle solving, and mathematical logic games to activate brain functions and emotional feelings and slow down cognitive decline in elderly individuals, thus preventing the occurrence of dementia [13]. Warren put forward the pyramid theory of the visual–perceptual hierarchy. He found that the visual–perceptual-related function is impaired in dementia, as shown in Table 2 [14,15].

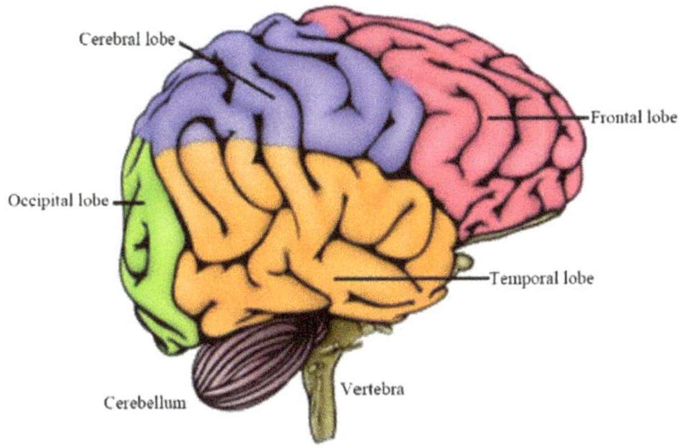

Figure 1. Brain structure [10].

Table 2. Dementia classification and symptoms visual symptoms of dementia [14].

Visual Transmission Paths	Visual Perception Problem	Visual Symptoms
Low-order transmission paths	Eye movement functions	Abnormal scanning and gazing
Low-order transmission paths	Basic visual functions	1. Contrast and visual field 2. Visually induced potentials
Higher-order transmission paths	Complex visual functions	1. Face and object recognition disorders 2. Spatial orientation disorder and dysarthria

Brain waves have different waveforms and frequency ranges, and each wavelength is distributed in the human brain and has biological significance [16]. The human brain has a frequency range of 1–50 Hz, and according to the International Brainwave Society (IBS), the frequency and amplitude of brainwave signals are classified as Delat (δ-wave), Theta (theta), Alpha (α), and Beta (β) wavelengths (Table 3) [17]. Different brainwaves affect the brain's awareness. For example, for the Alpha wave, the brain shows a relaxed and focused state and cognitive behavioral response, and brainwave activity state is shown by detecting the brainwaves using brainwave instrumentation [17]. The changes in brain waves can be used to understand the degree of cognitive ability, mental state, and brain deterioration in elderly dementia patients.

Table 3. Brainwave and characteristics.

Brainwave Type	Frequency Range	Characteristics and Physiological Status
Delat wave (δ-wave)	0.1~3 Hz	1. Unconscious wavelengths 2. Deep sleep
Theta wave (θ-Wave)	4~7 Hz	1. Physical relaxation and meditative states 2. Deep memories and sensations
Alpha wave (α-Wave)	8~13 Hz	1. Brain waves observed in a normal, relaxed adult 2. The brain is in a relaxed and focused state
Beta wave (β-wave)	13~30 Hz	1. The brain is awake and alert 2. The brain makes analytical decisions and deals with the senses

2.3. *Interactive Technology Experience*

In recent years, network technology and generative graphic artificial intelligence (AI) have been used to study graphic audio–visual and interactive behaviors and stimulate the cognitive ability of elderly individuals. The dynamic interaction of real-time feedback helps to prevent the occurrence of dementia [18]. Ascott et al. stated "New media art creation is mainly divided into five stages, linking, integration, interaction, transformation, emergence and other five basic elements" [19]. Berkeley said "New media art creation is mainly divided into five stages, linking, integrating, interacting, transforming and appearing" [19]. Alvin proposed that music can enhance the cognitive ability, communication and interaction, and self-expression of elderly individuals and can soothe their emotional state and comfort them [20]. PicMemory is an application of mobile technology in the treatment of dementia patients, combining digital video technology with a personalized narrative platform to create opportunities for family members to share, communicate, and interact with each other and share nostalgic memories to prevent the possibility of dementia [21]. WhizTouch, a two-way interactive call bell, provides sound and light changes to allow seniors to interact and give feedback [22]. WhizToys includes software that provides colors, music, images, sounds, and other interactive products to prevent dementia [22] (Table 4). The use of technology in daily life is becoming increasingly important for people to participate in society. However, the changes in the overall technological environment are not applied to the life and daily living of people with age-related dementia [22]. In dementia research, it has been found that the use of artistic messaging reduces the risk of dementia and promotes prevention. In the literature, it has been found that an emphasis on art or esthetics, painting, music, graphics, and interactive technology can be combined to create technological art and new media interactive experiences, such as virtual reality (VR) [23].

Table 4. Analysis of installation works of interactive technologies.

Item	Directions	Photographs
PicMemory	A sound interface for seniors with dementia to express their emotions and facilitate interaction.	
WhizTouch	Make a sound and change the light to green for interactive feedback from seniors.	
WhizToys	Use the color band for motor and cognitive training.	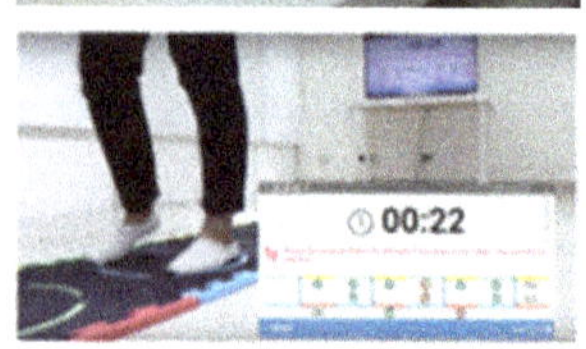

3. Methodology

In this study, we combined interactive technology and visual cognition to prevent dementia based on experimental methods and expert interviews. Using the framework of the interactive system, we designed an innovative interactive system. Meumann carried out an experiment that encompassed multiple empirical methods of conducting on-site systematic observation with oneself or others [24]. Through an expert interview, information on a two-way communication process was obtained [25]. Due to the rapid development of new technologies, interactive design has emphasized the integration of people, systems, and environments. Interactive systems are based on the concept of human-centered design, and the theoretical framework of interactive system design integrates people, activities, contexts, content, and technological components [26]. The structure of this study is shown in Figure 2.

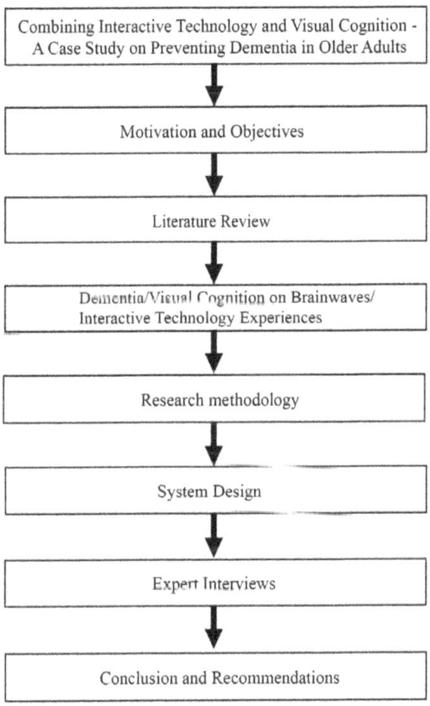

Figure 2. Research framework.

4. System Design

The participants were senior citizens. The experimental equipment included a MindWave Mobile 2 brainwave meter, three-dimensional objects, and visual interface design. When the participants carried out tasks, they wore the Cerebral Wave Meter on their heads and used their hands to grasp the "three-dimensional objects" and sense the information on the surface of the "three-dimensional objects" using the mobile phone. After receiving the information, the brain wave data were recorded using the Cerebral Wave Meter to show the concentration and relaxation of the brain, the cognitive ability for the images, and the emotional feelings, as shown in Figure 3.

Figure 3. System design.

4.1. Interactive Interface

The interactive system consisted of a three-dimensional square object and software on an interactive interface. The cube-shaped object has a 2D barcode and wireless communication function. The software on the visual interface was used to prevent dementia based on thinking and cognitive recognition. Thinking and cognitive recognition were displayed to show the brain's wavelength response. Increased brain stimulation for elderly individuals can delay the onset of dementia by training thinking, reading books, practicing regular exercise, and community interaction and communication.

The system was designed to include three different units of color, direction, and memory for the visual interface with different brain stimulation functions. The design elements of the visual interface were based on images, geometric shapes, and three-dimensional shapes, as shown in Table 5. The system can improve the hand–eye coordination and memory of middle-aged and senior citizens through the interactive process of games. After receiving the information content visually, an elderly individual can use their brain to read the information and stimulate brain function and activation for the prevention of dementia.

Table 5. Interactive interface design.

Item	Interface Visual Design
Color module	
Functional description	Using color recognition to detect and prevent one of the first signs of dementia.

Table 5. *Cont.*

Item	Interface Visual Design
Memory module	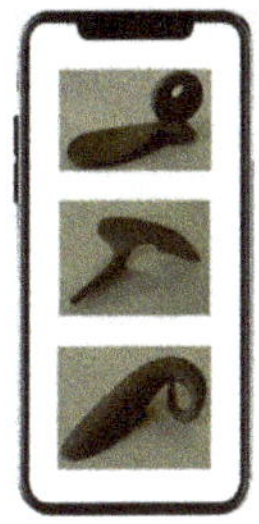
Functional description	Using three-dimensional modeling pictures to promote memory function and logical thinking, improve memory, and concentration.

4.2. System Architecture

Images, colors, and three-dimensional shapes from the obtained data were stored on a hard disk, as shown in Figure 4. The system was evaluated by middle-aged and elderly individuals using a pre-test questionnaire. In the second step, the tasks were selected for the participants to execute. Using sensing technology (QR-code and NFC), the information was displayed on a smartphone. The information was presented using two different interface presentation modes, namely, text–image narration and media–video playback. The brainwave meter was used to identify visual recognition and perceptual responses. In the third step, after the tasks, the system was evaluated for ease of use using SUS. Finally, the data were compiled and analyzed. SUS was assessed using a Likert scale consisting of ten sets of questions.

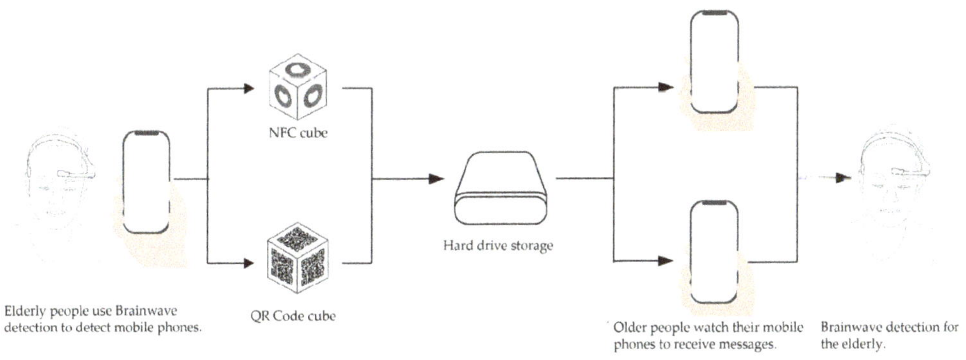

Figure 4. System architecture.

When the participants responded to the data through the brainwave meter, they could immediately understand the text and emotional responses. The system helped the participants prevent the occurrence of dementia.

5. Expert Interviews

Semi-structured interviews were conducted for invited experts in order to improve the interactive system. The interviewees included academic experts, scholars, professionals, or supervisors in related fields with experience and expertise, as shown in Table 6. Their suggestions and opinions on the development of the system provided the basis for the revision and adjustment of the system.

Table 6. Basic Information sheet for respondents.

Interviewees	Occupational Nature	Areas of Expertise
Academic experts	Associate professor	Human machine interface Human factors design
Industrial supervisor	Design manager	Product design. Interactive interface design
Industrial supervisor	Senior designer	Visual communication design Multimedia design

They suggested effectiveness, simplicity, enjoyment, and visual graphic design for brain activation. The prevention of dementia needs to be carried out under the principles of universal design. Sharp edges must be omitted to reduce risks during use. When using three-dimensional square objects, a layer of icons or other single colors on the surface of the electronic sensor can enhance the integrity of human–machine interface operations. In the interactive visual design, the visual image design must include analog icons to effectively help brain activation. The use of brainwave equipment must be more reliable and objective. When elderly individuals operate using visual images, they have to spend more time searching for specific images. Therefore, simple and bright or contrasting colors in the design are more helpful for the brain activation of elderly individuals with dementia.

6. Conclusions

We collected examples of products with interactive technology for dementia prevention. Michael and Chen proposed that serious games include three key elements: technological media, entertainment, and experience. In this study, audio effects were provided in the interactive system and interactive feedback was given. In the case of interactive products that provide lighting or changes in light and shadow, visual feedback was used. In the expert interviews, the issues related to the design of the interactive system were discussed to make suggestions regarding human–computer interface operating behavior; the indicators of interactive design, including effectiveness, simplicity, and enjoyment; and visual graphic design to help brain activation. The graphic–analog design helps elderly individuals' visual experience.

The literature on dementia, visual cognition, and interactive technology were explored in this study. Interactive products and software interface designs for dementia prevention were investigated in order to generalize the design principles. A new dementia prevention interactive system was designed accordingly. In the future, a survey to understand user experience and satisfaction using SUS is necessary to evaluate ease of use. The interactive system requires more feedback and experience for effective dementia prevention.

Author Contributions: Conceptualization, C.-M.W. and C.-S.F.; methodology, C.-M.W. and C.-S.F.; validation, C.-S.F.; investigation, C.-S.F.; data curation, C.-S.F.; writing—original draft preparation, C.-S.F.; writing—review and editing, C.-S.F.; visualization, C.-S.F.; supervision, C.-M.W. All authors have read and agreed to the published version of the manuscript.

Funding: This research received no external funding.

Institutional Review Board Statement: Not applicable.

Informed Consent Statement: All subjects gave their informed consent for inclusion before they participated in the study.

Data Availability Statement: The datasets used and/or analyzed during the current study are available from the corresponding author on reasonable request.

Conflicts of Interest: The authors declare no conflict of interest.

References

1. Gao, Y. Business Opportunities and Challenges of Global Aging. *Econ. Outlook* **2014**, *155*, 8–13.
2. Cheung, P.H.; Lam, F.S. A study on the design of dementia information APP interface for middle-aged and elderly dementia family carers. In Proceedings of the 2022 ICWVCD International Conference on Visual Communication Design and Basic Design Forum, Yunlin, Taiwan, 7 October 2022; pp. 94–103.
3. Frost, E.; Porat, T.; Malhotra, P.; Picinali, L. A novel auditory-cognitive training app for delaying or preventing the onset of dementia: Participatory design with stakeholders. *JMIR Hum. Factors* **2020**, *7*, e19880. [CrossRef] [PubMed]
4. Wong, F.S. Aging Society and Elderly Education. In *Adult and Lifelong Education Society of the Republic of China. Adult and Lifelong Education Society of the Republic of China*; Normal University Bookstore: Taipei City, Taiwan, 2004.
5. Wimo, A.; Jönsson, L.; Bond, J.; Prince, M.; Winblad, B.; International, A.D. The worldwide economic impact of dementia. *Alzheimer's Dement.* **2010**, *9*, 1–11. [CrossRef] [PubMed]
6. Leung, K.Y.; Ching, W.C.; Chan, R.H. A review of dementia highlights. *J. Intern. Med.* **2014**, *25*, 151–157.
7. Fu, C.L. Dementia in Taiwan. *Taiwan J. Geriatr. Gerontol.* **2008**, *3*, 169–181.
8. Rocha, I.C.; Arantes, M.; Moreira, A.; Vilaça, J.L.; Morais, P.; Matos, D.; Carvalho, V. Monitoring Wearable Devices for Elderly People with Dementia: A Review. *Designs* **2024**, *8*, 75. [CrossRef]
9. Pappadà, A.; Chattat, R.; Chirico, I.; Valente, M.; Ottoboni, G. Assistive technologies in dementia care: An updated analysis of the literature. *Front. Psychol.* **2021**, *12*, 644587. [CrossRef] [PubMed]
10. Lin, C.-Y. Mind's Eye Brainwave Emotional Reality Art Creation. Ph.D. Thesis, National Cheng Kung University, Tainan City, Taiwan, 2022.
11. Chen, Y.-M. Caring load of family carers of elderly dementia patients. *J. GaleMed* **2013**, *7*, 1–22.
12. Zeng, Z. The Causal Relationship Between Alpha Brainwave Phase and Visual Detection: An Investigation of Real-Time Brain Phase Locking. Master's thesis, Graduate Institute of Medical Electronics and Informatics, National Taiwan University, Taipei, Taiwan, 2021; pp. 1–75.
13. Zokaei, N.; MacKellar, C.; Čepukaitytė, G.; Patai, E.Z.; Nobre, A.C. Cognitive training in the elderly: Bottlenecks and new avenues. *J. Cogn. Neurosci.* **2017**, *29*, 1473–1482. [CrossRef] [PubMed]
14. Glosser, G.; Gallo, J.; Duda, N.; de Vries, J.J.; Clark, C.M.; Grossman, M. Visual perceptual functions predict instrumental activities of daily living in patients with dementia. *Cogn. Behav. Neurol.* **2002**, *15*, 198–206.
15. Warren, W.H., Jr. Self-motion: Visual perception and visual control. In *Perception of Space and Motion*; Academic Press: Cambridge, MA, USA, 1995; pp. 263–325.
16. Kang, C.S.; Shi, D.; Huang, S. Ontology-based detection of malicious emails. *J. Inf. Manag.* **2007**, *14*, 1–28.
17. Yau, S.M. A study of LEGO Wedo2.0 on the Concentration of First Year Students with Special Needs Using Brainwave Analysis. Master's Thesis, Tsinghua University, Beijing, China, 2021; pp. 1–106.
18. Lai, S.-T. A study of the Effectiveness of Interactive Games on Cognition, Problem Behaviour, and Depression in Nursing Home Residents with Dementia. Master's Thesis, Central Taiwan University of Science and Technology, Taichung City, Taiwan, 2020; pp. 1–80.
19. Shih, P.-T.; Wang, C.-M. Marine Contextual Interactive Sound Installation. In Proceedings of the 2019 Asian Society of Fundamental Architecture Conference, Taichung, Taiwan, 16 October 2019; pp. 271–278.
20. Zhang, Y. Intervention of Music Therapy in Dementia Care: A Case Study of Group Music Therapy to Promote Social Interaction and Self-Expression of Person with Dementia. Ph.D. Thesis, Department of Social Work, National Taiwan University, Taipei, Taiwan, 2023; pp. 1–83.
21. Fu, L.C.; Hsu, Y.J.; Chen, S.H.; Hung, Y.P.; Chen, C.H.; Liao, C.F.; Wu, C.H.; Chen, H.C.; Peng, H.N.; Chuang, W.L.; et al. Innovative Life Technology for Dementia Care. *J. Welf. Technol. Serv. Manag.* **2018**, *6*, 365–388.
22. Chandler, C.R.; Catanzaro, M.; Siette, J. "I know now that it's something that you can do something about": Deductive thematic analysis of experiences at an arts-based dementia risk reduction exhibit. *Arch. Gerontol. Geriatr.* **2024**, *123*, 105440. [CrossRef] [PubMed]
23. Sanders, D.; Scott, P. Literature review: Technological interventions and their impact on quality of life for people living with dementia. *BMJ Health Care Inform.* **2020**, *27*, e100064. [CrossRef] [PubMed]
24. Jeon, Y.H.; Foxe, D.; Suh, G.H.; Wang, H.; Dominguez, J.C.; Maukera, R.; Kounnavong, S.; Piguet, O. Post-diagnosis dementia care in the Western Pacific region: Assessment of needs and pathways to optimal care. *Lancet Reg. Health–West. Pac.* **2024**, *50*, 101182. [CrossRef] [PubMed]

25. Ren, P.; Liao, M.; Zhou, L.; Rong, H.; Zhong, H.; Li, L.; Xiao, T.; Li, F.; Zeng, J.; Chen, X. Quality of life, care problems and care needs of Chinese older patients with dementia under the perspective of long-term care insurance. *Geriatr. Nurs.* **2024**, *60*, 418–426. [CrossRef] [PubMed]
26. Haque, R.; Alam, K.; Gow, J.; Neville, C.; Keramat, S.A. Beyond the sum of their parts: The combined association of dementia and chronic pain with self-care limitations in older Australians. *J. Affect. Disord.* **2024**, *369*, 633–642. [CrossRef] [PubMed]

engineering proceedings

MDPI

Proceeding Paper

One-Node One-Edge Dimension-Balanced Hamiltonian Problem on Toroidal Mesh Graph †

Yancy Yu-Chen Chang and Justie Su-Tzu Juan *

Department of Computer Science & Information Engineering, National Chi Nan University, Nantou 54561, Taiwan; yancychang@gmail.com
* Correspondence: jsjuan@ncnu.edu.tw

† Presented at the 2024 IEEE 7th International Conference on Knowledge Innovation and Invention, Nagoya, Japan, 16–18 August 2024.

Abstract: Given a graph $G = (V, E)$, the edge set can be partitioned into k dimensions, for a positive integer k. The set of all i-dimensional edges of G is a subset of $E(G)$ denoted by E_i. A Hamiltonian cycle C on G contains all vertices on G. Let $E_i(C) = E(C) \cap E_i$. For any $1 \leq i \leq k$, C is called a dimension-balanced Hamiltonian cycle (DBH, for short) on G if $||E_i(C)| - |E_j(C)|| \leq 1$ for all $1 \leq i < j \leq k$. The dimension-balanced cycle problem is generated with the 3-D scanning problem. Graph G is called p-node q-edge dimension-balanced Hamiltonian (p-node q-edge DBH) if it has a DBH after removing any p nodes and any q edges. G is called h-fault dimension-balanced Hamiltonian (h-fault DBH, for short) if it remains Hamiltonian after removing any h node and/or edges. The design for the network-on-chip (NoC) problem is important. One of the most famous NoC is the toroidal mesh graph $T_{m,n}$. The DBC problem on toroidal mesh graph $T_{m,n}$ is appropriate for designing simple algorithms with low communication costs and avoiding congestion. Recently, the problem of a one-fault DBH on $T_{m,n}$ has been studied. This paper solves the one-node one-edge DBH problem in the two-fault DBH problem on $T_{m,n}$.

Keywords: Hamiltonian; dimension-balanced; toroidal mesh graph; faulty node; faulty edge

1. Introduction

With the popularity of the Internet, the topology of its network allows for the improvement of the performance of the entire system. According to the connection structure of this network, graphs are used to display its topology. Vertices represent devices, and edges between two vertices represent communication devices. The path between the vertices in the graph shows data transmission and communication between the two. A Hamiltonian cycle connects all vertices and returns to the original vertex to ensure the smoothness of the network [1]. However, when vertices or edges fail, the Hamilton cycle is broken, and data cannot be transferred within it. This causes significant inconvenience. It is important to ensure that there are new routes in the network to provide data transfer between them. Therefore, it has been studied whether the Hamiltonian cycle exists when there are faulty vertices or edges in the network graph [2,3].

For a graph $G = (V, E)$, the edge set E is partitioned into k dimensions $\{E_1, E_2, \ldots, E_k\}$ for a positive integer k. For any cycle C on G, the set of all i-dimensional edges of C, which is a subset of $E(C)$, is denoted by $E_i(C)$. If $||E_i(C)| - |E_j(C)|| \leq 1$ for $1 \leq i < j \leq k$, C is called a *dimension-balanced cycle* or DBC for short. If C is a Hamiltonian cycle on graph G and C is also a DBC, then C is called a DBH of G. The dimension-balanced cycle problem is generated with the 3-D scanning problem [4]. Research results on the issues of DBCs and DBHs have been obtained in recent years [5–7].

A graph is called a *toroidal mesh graph*, denoted by $T_{m,n}$, where $V(T_{m,n}) = \{(x, y) \mid 0 \leq x \leq m-1, 0 \leq y \leq n-1\}$ and the edge set $E(T_{m,n}) = \{(x_1, y_1)(x_2, y_2) \mid x_1 = x_2$ and $|y_1 - y_2| = (1$ or $n-1)$, or $|x_1 - x_2| = (1$ or $m-1)$ and $y_1 = y_2\}$. In other words, $T_{m,n}$ is viewed as the Cartesian product of cycles C_m and C_n. In any toroidal mesh graph $T_{m,n}$, let $E_1 = \{(x_1, y_1)(x_2, y_2) \mid |x_1 - x_2| = (1$ or $m-1)$ and $y_1 = y_2\}$ and $E_2 = \{(x_1, y_1)(x_2, y_2) \mid x_1 = x_2$ and $|y_1 - y_2| = (1$ or $n-1)\}$. Figure 1 shows an example of $T_{4,3}$. The design for the network-on-chip (NoC) problem has been an important issue recently. One of the most famous NoC problems is actually the toroidal mesh graph $T_{m,n}$.

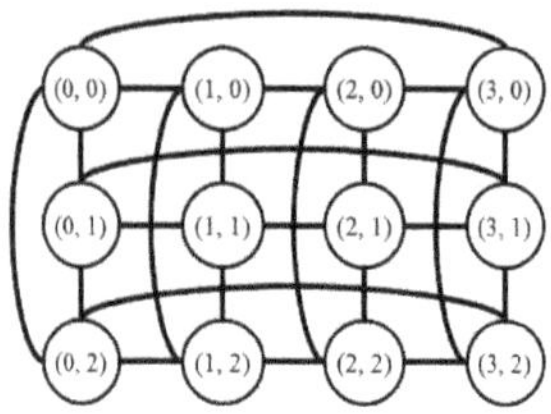

Figure 1. $T_{4,3}$.

For the NoC design problem, how to choose a suitable routing path to prevent congestion and good performance is important. The DBC problem on toroidal mesh graph $T_{m,n}$ is appropriate for designing simple algorithms with low communication costs, and it will avoid the situation of congestion. A graph G is called p-node q-edge *dimension-balanced Hamiltonian* (p-node q-edge DBH for short) if it has a DBH after removing any p nodes and any q edges. A graph G is called *h-fault dimension-balanced Hamiltonian* (*h-fault* DBH for short) if it remains Hamiltonian after removing any h node and/or edges [2]. Many researchers have studied the Hamiltonian problem on several topologies with faulty nodes or edges [3,8–10].

It has been proved that for any F with $|F| \leq 2$, $T_{m,n} - F$ has a Hamiltonian cycle if $m \geq 3$, $n \geq 3$, and n is odd, where F is a subset of the union of the node set and the edge set of the graph $T_{m,n}$ [2]. Recently, the problem of a one-fault DBH on $T_{m,n}$ has been studied [11]. Therefore, this study investigated the two-fault DBH problem on $T_{m,n}$, on the one-node one-edge DBH problem.

The rest of this paper is organized as follows: Section 2 gives preliminaries, Section 3 shows the main results, and Section 4 is the conclusion.

2. Preliminaries

In this section, we will first state some relevant theorems and definitions as a starting point for this study. For any vertex (i, j) in a toroidal mesh graph $T_{m,n}$, (i, j) is called *black* if $i + j$ is even; otherwise, (i, j) is called *white*. If m and n are both even, every edge joins two vertices with different colors. This means $T_{m,n}$ is a bipartite graph. Therefore, there is no odd cycle on $T_{m,n}$ if both m and n are even, so $T_{m,n} - F$ has no Hamiltonian cycle when F is a set of one node. We have the following theorems.

Theorem 1 [11]. *When m and n are both even, $T_{m,n}$ is not one-node DBH.*

This is because if there is no DBH on $T_{m,n} - F_1$ when F_1 is a set of one node, then there is also no DBH on $T_{m,n} - F_2$ when F_2 is a set of one node and one edge. The following theorem can be derived.

Theorem 2. *When m and n are both even, $T_{m,n}$ is not two-fault DBH because $T_{m,n} - F$ has no Hamiltonian cycle when F is a set of one node and one edge.*

A *vertex-transitive* graph, or a *node symmetric* graph, is a graph such that every pair of vertices is equivalent under some element of its automorphism group [12]. It is not difficult to see that a toroidal mesh graph $T_{m,n}$ is a vertex-transitive graph.

3. Main Results

This study explores the one-node one-edge DBH problem on $T_{m,n}$. Because of Theorem 2, we discuss this problem only for when m and n are not both even: (1) when one of m and n is odd and the other is even and (2) when m and n are both odd numbers. When one of m and n are is odd and the other is even, generality is decreased, assuming that m is an odd number and n is an even number. At first, Theorem 3 was used to analyze the situation when $m = 3$, and then Theorem 4 was used to analyze the situation when $m > 3$ based on the DBH on $T_{5,4}$ with one faulty node and one faulty edge.

Theorem 3. *There is a DBH on $T_{3,3} - F$ and $T_{3,4} - F$; however, when $n > 4$, there is no DBH on $T_{3,n} - F$, where F is a set of any one node and one edge.*

Proof. Figure 2 shows two DBHs on $T_{3,3} - F$ and $T_{3,4} - F$, respectively. Since $T_{m,n}$ is a vertex-transitive graph, the red vertex v is the faulty vertex ($v \in F$) without loss of generality. Due to the characteristics of vertex transitivity, the graphs of Figure 2 can be flipped and/or translated so that the faulty vertex is fixed at the original position. Six different extended DBHs are produced. Figure 3 shows those DBHs. Each edge does not exist on at least one DBH on $T_{3,3} - F$ and $T_{3,4} - F$.

$n > 4$ is considered in Figure 4. When the fault set $F = \{(1, 0), (0, 0)(2, 0)\}$, for vertices (0, 0) and (2, 0), only these four edges drawn in Figure 3 are chosen. If DBH C exists in $T_{3,n} - F$, at most $\lceil (3n - 1)/2 \rceil - 4$ horizontal edges can be selected in C. At least $2(n - 3)$ horizontal edges are needed in C to connect all vertices. This is because $2(n - 3) > \lceil (3n - 1)/2 \rceil - 4$ when $n > 5$. Therefore, it can be proven that when $n > 4$, there is no DBH on $T_{3,n} - F$. □

Lemma 1. *For any $T_{5,4} - F$, there exists a DBH, where F is a set of any one node and one edge.*

Proof. Since $T_{m,n}$ is a vertex-transitive graph, without loss of generality, the faulty node is (2, 0) (the red node in Figure 5). Next, all possible cases where edges are faulty edges are considered. Figure 5a proves that when the faulty edges are edges not used in the graph, $T_{5,4} - F$ exists DBH; the faulty edges that Figure 5a cannot solve are shown in Figure 5b. When a faulty edge occurs on the edges in Figure 5a, it is replaced with Figure 5b. □

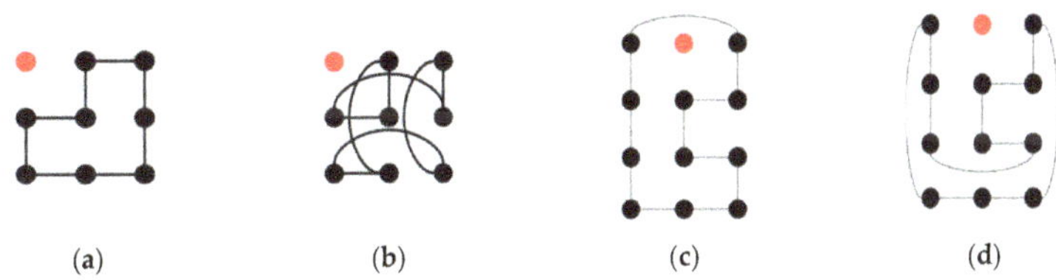

Figure 2. (**a**,**b**) Two DBHs on $T_{3,3} - F$; (**c**,**d**) two DBHs on $T_{3,4} - F$.

Since the graph is vertex-transitive, we remove the faulty edge at the positions of unused edges through rotation, reflection, and translation. When $m > 5$ is odd and/or $n > 4$ is even, we perform two operations on Figure 5 to obtain a DBH of $T_{m,n}$ − F recursively:

(1) expanding one column to both the left and right of $T_{m-2,4}$ and (2) expanding two rows downward of $T_{m,n-2}$.

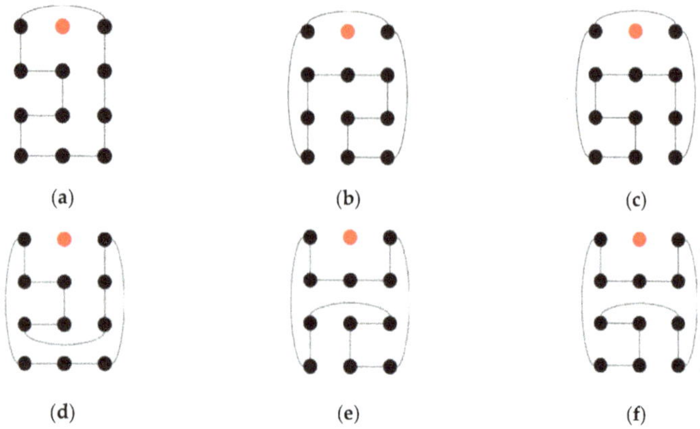

Figure 3. (**a–f**) Six different extended DBHs of Figure 2c,d.

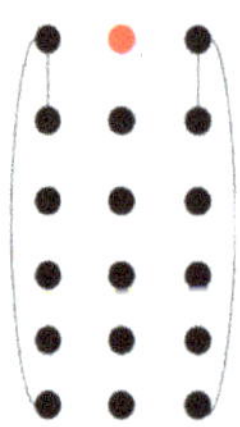

Figure 4. If $F = \{(1, 0), (0, 0)(2, 0)\}$, the edges must be used.

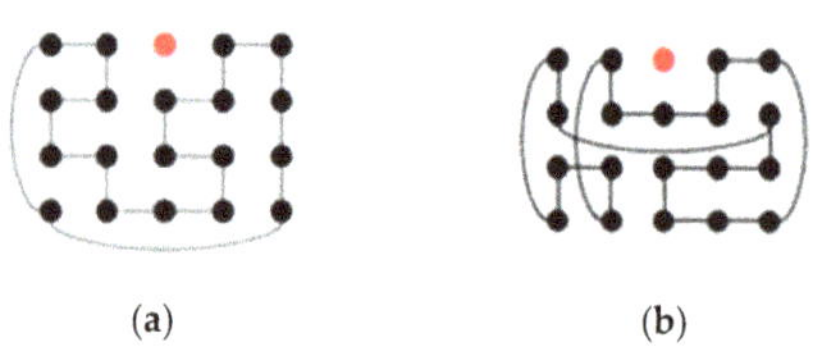

Figure 5. (**a**) A DBH of $T_{5,4} - F$; (**b**) another DBH of $T_{5,4} - F$.

The following theorem is proved according to these two operations. For convenience, in the following figures, the red dashed lines represent the deleted edges, and the blue lines represent the added edges.

Theorem 4. *When $m \geq 5$ is an odd number and $n \geq 4$ is an even number, there exists a DBH on $T_{m,n} - F$, where F is a set of any one node and one edge.*

Proof. This theorem is proved by dividing it into two steps as mentioned above. Similarly to Lemma 1, two Hamiltonian cycles C_1 and C_2 in $T_{m,n} - F$, as in Figure 5a,b (say type (a) and (b) for short), are constructed.

For any odd $m \geq 5$ and even $n \geq 4$, since $T_{m,n}$ is a vertex-transitive graph, the faulty node is $(\lfloor m/2 \rfloor, 0)$, without loss of generality. Lemma 1 shows that the theorem holds for $m = 5$ and $n = 4$. If $m > 5$ and $n = 4$, Step 1 proves that there exists a DBH on $T_{m,4} - F$ at first.

Step 1. One column is expanded to both the left and right of $T_{m-2,4}$: In this step, there exists a DBH on $T_{m,4} - F$, which is proved by induction. Claim 1 is proved as follows.

Claim 1: $|E_1(C_1)| = |E_1(C_2)| = \lfloor (mn-1)/2 \rfloor$ and $|E_2(C_1)| = |E_2(C_2)| = \lceil (mn-1)/2 \rceil$.

By Lemma 1, two DBHs C_1 and C_2 exist with a type (a) and (b), respectively, on $T_{5,4} - F$. Figure 5a,b prove Claim 1. If $m > 5$, there exist two DBHs C_1 and C_2 on $T_{m-2,4} - F$ with type (a) and (b), respectively, satisfying Claim 1. We divide this step into two cases according to the number of (m mod 4) for discussion.

Case 1. (m mod 4) = 3: There are two DBHs on $T_{m,4}$ to be constructed. Therefore, this case is divided into two parts and discussed separately depending on whether the constructed DBH is type (a) or type (b).

Case 1.1. To construct a DBH with type (a): In the hypothesis, there is a DBH C_1 on $T_{m-2,4}$ with type (a) satisfying Claim 1. First, C_1 is put in the middle of $T_{m,4}$ with edges {(1, 0)(1, 3), (1, 3)(m − 2, 3), (m − 2, 0)(m − 2, 1), (m − 2, 1)(m − 2, 2)}, (m − 2, 2)(m − 2, 3)} in C_1. Then, remove {(1, 3)(m − 2, 3), (m − 2, 1)(m − 2, 2)} from C_1 and add {(m − 2, 1)(m − 1, 1), (m − 1, 1)(m − 1, 2), (m − 2, 2)(m − 1, 2), (m − 2, 3)(m − 1, 3), (m − 1, 0)(m − 1, 3), (0, 0)(m − 1, 0), (0, 0)(0, 1), (0, 1)(0, 2), (0, 2)(0, 3), (0, 3)(1, 3)} into C_1. After calculation, the operation requires adding four first-dimensional edges and four second-dimensional edges. We removed one first-dimensional edge and one second-dimensional edge and added five first-dimensional edges and five second-dimensional edges. Therefore, the constructed Hamiltonian cycle is of type (a), satisfies Claim 1, and is a DBH. Figure 6 shows an example of $T_{7,4} - F$.

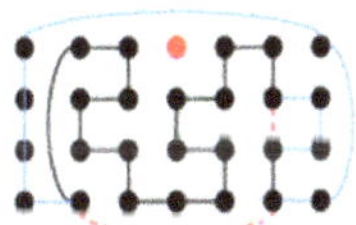

Figure 6. The constructed DBH C_1 on $T_{7,4} - F$.

Case 1.2. To construct a DBH with type (b): There is a DBH C_2 on $T_{m-2,4}$ with type (b) and that satisfies Claim 1. First, C_2 is put in the middle of $T_{m,4}$, with edge {(1, 1)(m − 2, 1)} in C_2. Then, {(1, 0)(1, 3), (1, 1)(m − 2, 1)} is removed to add {(0, 0)(1, 0), (0, 1)(1, 1), (0, 1)(0, 2), (0, 2)(m − 1, 2), (0, 3)(1, 3), (0, 0)(0, 3), (m − 2, 1)(m − 1, 1), (m − 1, 0)(m − 1, 1), (m − 1, 0)(m − 1, 3), (m − 1, 2)(m − 1, 3)}. After calculation, this operation requires adding four first-dimensional edges and four second-dimensional edges. We removed one first-dimensional and one second-dimensional edges and added five first-dimensional edges and five second-dimensional edges. Therefore, the constructed Hamiltonian cycle becomes type (b), satisfies Claim 1, and is a DBH. Figure 7 shows an example of the constructed C_2 on $T_{7,4} - F$.

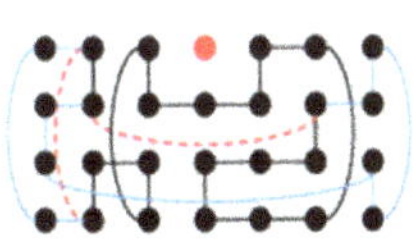

Figure 7. Constructed DBH C_2 on $T_{7,4} - F$.

Case 2. (m mod 4) = 1: There are also two DBHs on $T_{m,4}$ need to be constructed. Therefore, this case is divided into two parts and discussed separately depending on whether the constructed DBH is type (a) or (b).

Case 2.1. To construct a DBH with type (a), a DBH C_1 is used on $T_{m-2,4}$ with type (a) satisfying Claim 1. C_1 is similar to Figure 6 in this subcase. First, C_1 is put in the middle of $T_{m,4}$ with edges {(1, 0)(1, 1), (1, 1)(1, 2), (1, 2)(1, 3), (1, 0)(m − 2, 0), (m − 2, 0)(m − 2, 3)} in C_1. Then, remove {(1, 0)(m − 2, 0), (1, 1)(1, 2)} and add {(0, 0)(1, 0), (0, 1)(1, 1), (0, 1)(0, 2),

$(0, 2)(1, 2)$, $(0, 0)(0, 3)$, $(0, 3)(m-1, 3)$, $(m-2, 0)(m-1, 0)$, $(m-1, 0)(m-1, 1)$, $(m-1, 1)(m-1, 2)$, $(m-1, 2)(m-1, 3)\}$ into C_1. After calculation, this operation requires adding four first-dimensional edges and four second-dimensional edges. Similarly to case 1.1, we removed one first-dimensional edge and one second-dimensional edge and added five first-dimensional edges and five second-dimensional edges. Therefore, the constructed Hamiltonian cycle is of type (a), satisfies Claim 1, and is a DBH. Figure 8 shows an example of the constructed C_1 on $T_{9,4} - F$.

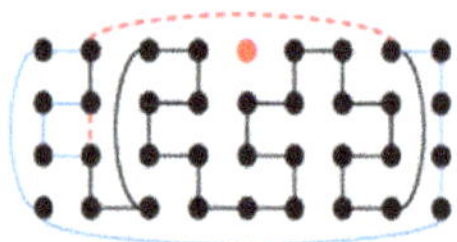

Figure 8. The constructed DBH C_1 of $T_{9,4} - F$.

Case 2.2. To construct a DBH with type (b): There is a DBH C_2 on $T_{m-2,4}$ with type (b) satisfying Claim 1. C_2 is similar to Figure 7 in this subcase. First, C_2 is put in the middle of $T_{m,4}$ with edge $\{(1, 2)(m-2, 2)\}$ in C_2. Then, remove $\{(1, 0)(1, 3), (1, 2)(m-2, 2)\}$ and add $\{(0, 0)(1, 0), (0, 1)(0, 2), (0, 1)(m-1, 1), (0, 2)(1, 2), (0, 3)(1, 3), (0, 0)(0, 3), (m-2, 2)(m-1, 2), (m-1, 0)(m-1, 1), (m-1, 0)(m-1, 3), (m-1, 2)(m-1, 3)\}$. Again, it can be determined that this operation requires adding four first-dimensional edges and four second-dimensional edges into C_2 to keep it dimension-balanced. We removed one first-dimensional edge and one second-dimensional edge and added five first-dimensional edges and five second-dimensional edges. Therefore, the constructed Hamiltonian cycle is of type (b) satisfying Claim 1 and a DBH. Figure 9 shows an example of the constructed C_2 on $T_{9,4} - F$.

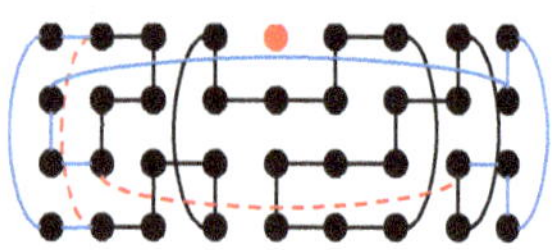

Figure 9. The constructed DBH C_2 of $T_{9,4} - F$.

Step 2. Two rows downward of $T_{m,n-2}$.

By Step 1, there exist two DBHs C_1 and C_2 with types (a) and (b), respectively, on $T_{m,4} - F$. $|E_1(C_1)| = |E_1(C_2)| = 2m - 1$, and $|E_2(C_1)| = |E_2(C_2)| = 2m$. Then, if $n > 4$, let $k = (n-4)/2$. We divided this step into two cases according to the number of (m mod 4) for discussion. Y_m is defined as the Cartesian product of C_m and P_2, called a prism graph, where P_n is a path with n vertices. The vertex set of Y_m is $V(Y_m) = \{(x, y) \mid 0 \le x \le m-1; 0 \le y \le 1\}$, and the edge set $E(Y_m) = \{(x_1, y_1)(x_2, y_2) \mid x_1 = x_2$ and $|y_1 - y_2| = 1$, or $|x_1 - x_2| = (1$ or $m-1)$ and $y_1 = y_2\}$.

Case 1. (m mod 4) = 1: There are two DBHs on $T_{m,n}$ to be constructed. Therefore, this case is divided into two parts and discussed separately depending on whether the constructed DBH is type (a) or (b).

Case 1.1. To construct a DBH with type (a): By Step 1, there is a DBH C_1 on $T_{m,4}$ with type (a). Copy the last two rows of C_1 to be a subgraph of Y_m named Z_m. Figure 10a,b show Z_5 and Z_9, respectively. Now, C_1 is put in the upper part of $T_{m,n}$, and k Z_ms in the lower part of $T_{m,n}$. Then, remove $\{(0, 3)(m-1, 3)\} \cup \{(2i, 0)(2i, 3), (2i+1, 3+2j)(2i+2, 3+2j) \mid 0 \le i < (m-1)/4, 0 \le j < k\} \cup \{(2i+1, 0)(2i+1, 3) \mid (m+3)/4 \le i < (m-1)/2\}$ and add $\{(2i, 0)(2i, n-1), (2i, 3+2j)(2i, 4+2j), (2i, 3+2j)(2i+1, 3+2j) \mid 0 \le i < (m-1)/4, 0 \le j < k\} \cup \{(2i+1, 0)(2i+1, n-1), (2i+1, 3+2j)(2i+1, 4+2j) \mid (m+3)/4 \le i < (m-1)/2,$

$0 \leq j < k\} \cup \{((m-1)/2, 3+2j)((m-1)/2, 4+2j), (m-1, 3+2j)(m-1, 4+2j) \mid 0 \leq j < k\}$. Figure 10c,d show examples for $T_{5,6} - F$, $T_{9,6} - F$. After calculation, the generated Hamiltonian cycle C satisfies $|E_1(C)| = |E_1(C_1)| + k(m-1) = nm/2 - (k+1)$ and $|E_2(C)| = |E_1(C_1)| + k(m+1) = nm/2 + k$. We added k or $k+1$ first-dimensional edges and removed k or $k+1$ second-dimensional edges from C to form a type (a) DBH of $T_{m,n} - F$. There are two sub-steps according to the value of n that can solve this situation.

Step 1.1.1. Let $x = \lfloor (n+2)/8 \rfloor = \lfloor (k+3)/4 \rfloor$. Find x smallest i with $\{(0, i)(0, i+1), (m-1, i)(m-1, i+1)\} \subseteq E(C)$. Then, remove $(0, i)(0, i+1)$, $(m-1, i)(m-1, i+1)$ and add $(0, i)(m-1, i)$, $(0, i+1)(m-1, i+1)$ to C. Figure 10e shows an example for $T_{5,6} - F$. This step adds $2\lfloor (k+3)/4 \rfloor$ first-dimensional edge and removes $2\lfloor (k+3)/4 \rfloor$ second-dimensional edges from C.

Step 1.1.2. Let $x = \lfloor (n-2)/8 \rfloor = \lfloor (k+1)/4 \rfloor$. Find x smallest i with $\{(m-2, i)(m-2, i+1), (m-1, i)(m-1, i+1)\} \subseteq E(C)$ and $\{(m-2, i)(m-1, i), (m-2, i+1)(m-1, i+1)\} \not\subseteq E(C)$. Then, $(m-2, i)(m-2, i+1)$, $(m-1, i)(m-1, i+1)$ are removed, and $(m-2, i)(m-1, i)$, $(m-2, i+1)(m-1, i+1)$ are added into C. Figure 10f shows an example. This step adds $2\lfloor (k+1)/4 \rfloor$ first-dimensional edges and removes $2\lfloor (k+1)/4 \rfloor$ second-dimensional edges from C.

Note that $2\lfloor (k+3)/4 \rfloor + 2\lfloor (k+1)/4 \rfloor = k$ or $k+1$. The created Hamiltonian cycle C is a DBH with type (a).

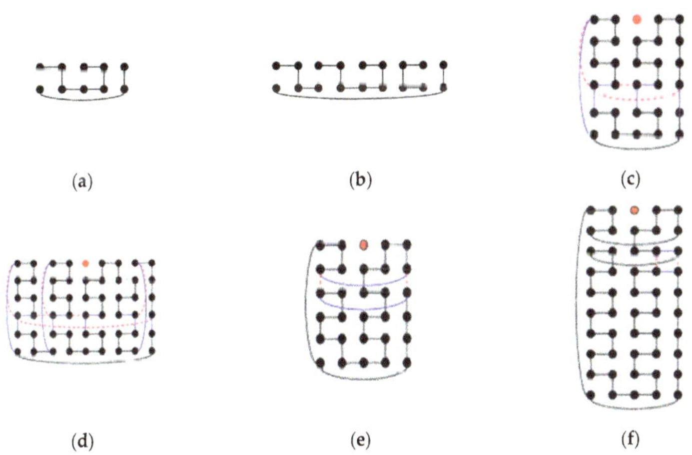

Figure 10. Examples for case 1.1 of Step 2 of Theorem 4. (**a**) Z_5, (**b**)Z_9, (**c**) a Hamiltonian cycle C on $T_{5,6} - F$, (**d**) a Hamiltonian cycle C on $T_{9,6} - F$, (**e**) the construct DBH of $T_{5,6} - F$, (**f**) the construct DBH of $T_{5,10} - F$.

Case 1.2. To construct a DBH with type (b): By Step 1, there is a DBH C_2 on $T_{m,4}$ with type (b). The last two rows of C_2 are copied to be a subgraph of Y_m named Z_m. Modify Z_m by adding edges $\{(0, 2i+1)(0, 2i+2) \mid 0 \leq i \leq (m-9)/4\}$ and deleting edges $\{(0, (m-5)/2)(0, (m-3)/2\} \cup \{(0, 2i)(0, 2i+1) \mid (m+5)/4 \leq i \leq (m-2)/2\}$. Figure 10a,b show Z_5 and Z_9, respectively. Then, C_2 is put in the upper part of $T_{m,n}$, and k Z_ms in the lower part of $T_{m,n}$. Then, for any $0 \leq i \leq m-1$, if edge $(0, i)(3, i)$ is in C_2, this edge is removed, and edges $\{(0, i)(n-1, i), (3+2j, i)(4+2j, i) \mid 0 \leq j < k\}$ are added to C_2 and k Z_ms to form a new Hamiltonian cycle C of $T_{m,n} - F$. Figure 11c,d show an example of $T_{5,6} - F$ and $T_{9,6} - F$. Similarly to case 1.1, after calculation, the generated Hamiltonian cycle C satisfies $|E_1(C)| = |E_1(C_1)| + k(m-1) = nm/2 - (k+1)$ and $|E_2(C)| = |E_1(C_1)| + k(m+1) = nm/2 + k$. We added k or $k+1$ first-dimensional edges and removed k or $k+1$ second-dimensional edges from C to form a type (b) DBH of $T_{m,n} - F$. There are two sub-steps according to the value of n to solve this situation. $m = 5$ is a special case and needs to be discussed separately.

Step 1.2.1. Let $x = \lfloor (n+2)/8 \rfloor = \lfloor (k+3)/4 \rfloor$. For $m = 5$, find x smallest i with $\{(0, i)(0, i+1), (m-1, i)(m-1, i+1)\} \subseteq E(C)$. Then, remove these two edges and add $(0, i)(m-1, i)$, $(0, i+1)(m-1, i+1)$ into C. Figure 11e shows an example. For $m > 5$, find x smallest i with $\{((m-1)/2+2, i)((m-1)/2+2, i+1), ((m-1)/2+3, i)((m-1)/2+3, i+1)\} \subseteq E(C)$. Then, these two edges are removed, and $((m-1)/2+2, i)((m-1)/2+3, i)$, $((m-1)/2+2, i+1)((m-1)/2+3, i+1)$ are added to C. Figure 11f shows an example for $T_{9,6} - F$.

Step 1.2.2. Let $x = \lfloor (n-2)/8 \rfloor = \lfloor (k+1)/4 \rfloor$. For $m = 5$, find the $2x$ edges $(0, i)(0, i+1)$, $(1, i)(1, i+1)$ such that $(0, i)(1, i)$, $(0, i+1)(1, i+1) \notin E(C)$ (after performing Step 1.2.1). Then, $(0, i)(0, i+1)$, $(1, i)(1, i+1)$ are removed, and $(0, i)(1, i)$, $(0, i+1)(1, i+1)$ are added to C. Figure 11g shows an example for $T_{5,10} - F$. For $m > 5$, find the $2x$ edges $((m-1)/2+3, i)((m-1)/2+3, i+1)$, $((m-1)/2+4, i)((m-1)/2+4, i+1)$ such that $((m-1)/2+3, i)((m-1)/2+4, i)$, $((m-1)/2+3, i+1)((m-1)/2+4, i+1) \notin E(C)$ and $i > 0$. Then, $((m-1)/2+3, i)((m-1)/2+3, i+1)$, $((m-1)/2+4, i)((m-1)/2+4, i+1)$ are removed, and $((m-1)/2+3, i)((m-1)/2+4, i)$, $((m-1)/2+3, i+1)((m-1)/2+4, i+1)$ are added to C. Figure 11f shows an example for $T_{9,10} - F$.

Note that $2\lfloor (k+3)/4 \rfloor + 2\lfloor (k+1)/4 \rfloor = k$ or $k+1$, so the created Hamiltonian cycle C is a DBH with type (b).

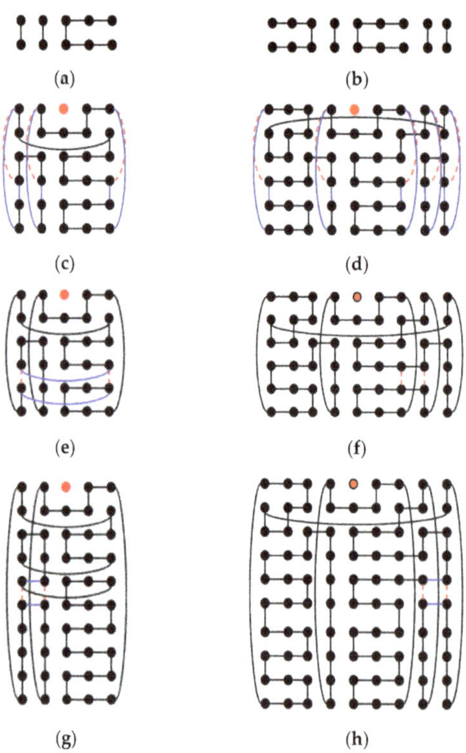

Figure 11. Examples for case 1.2 of Step 2 of Theorem 4. (**a**) Z_5, (**b**)Z_9, (**c**) a Hamiltonian cycle C on $T_{5,6} - F$, (**d**) a Hamiltonian cycle C on $T_{9,6} - F$, (**e**) the construct DBH of $T_{5,6} - F$, (**f**) the construct DBH of $T_{9,6} - F$, (**g**) the construct DBH of $T_{5,10} - F$, (**h**) the construct DBH of $T_{9,10} - F$.

Case 2. (m mod 4) = 3: There are two DBHs on $T_{m,n}$ constructed. Therefore, this case is divided into two parts and discussed separately depending on whether the constructed DBH is type (a) or type (b).

Case 2.1. To construct a DBH with type (a): By Step 1, there is a DBH C_1 on $T_{m,n-2}$ with type (a). The last two rows of C_1 are copied to be a subgraph of Y_m named Z_m. Figure 12a,b show that Z_7 and Z_{11}. C_1 are put in the upper part of $T_{m,n}$, and k Z_ms in the lower part of $T_{m,n}$. Then, $\{(1+2i,0)(1+2i,3), (m-1-2i,0)(m-1-2i,3) \mid 0 \le i < (m-3)/4\} \cup \{(2i,3+2j)(2i+1,3+2j) \mid 0 \le i < (m+1)/4, 0 \le j < k\}$ are removed, and $\{(1+2i,0)(1+2i,n-1), (m-1-2i,0)(m-1-2i,n-1), (1+2i,3+2j)(2+2i,3+2j), (1+2i,3+2j)(1+2i,4+2j), (m-1-2i,3+2j)(m-1-2i,4+2j) \mid 0 \le i < (m-3)/4, 0 \le j < k\} \cup \{((m-1)/2, 3+2j)((m-1)/2,4+2j), (0,3+2j)(0,4+2j) \mid 0 \le j < k\}$ are added. Figure 12c,d show $T_{7,6} - F$ and $T_{11,6} - F$. After calculation, the generated Hamiltonian cycle C will satisfy $|E_1(C)| = |E_1(C_1)| + k(m-1) = nm/2 - (k+1)$ and $|E_2(C)| = |E_2(C_1)| + k(m+1) = nm/2 + k$. We added k or $k+1$ first-dimensional edges and removed k or $k+1$ second-dimensional edges from C to form a type (a) DBH of $T_{m,n} - F$. There are two sub-steps according to the value of n that can solve this situation.

Step 2.1.1. Let $x = \lfloor (n+2)/8 \rfloor = \lfloor (k+3)/4 \rfloor$. First, $\{(0,4i+1)(0,4i+2), (m-1,4i+1)(m-1,4i+2) \mid 0 \le i < x\}$ are removed from C, and then $\{(0,4i+1)(m-1,4i+1), (0,4i+2)(m-1,4i+2) \mid 0 \le i < x\}$ are added to C. Figure 12e shows an example for $T_{7,6} - F$. In total, $2\lfloor (k+3)/4 \rfloor$ first-dimensional edges are added to remove $2\lfloor (k+3)/4 \rfloor$ second-dimensional edges from C.

Step 2.1.2. Let $x = \lfloor (n-2)/8 \rfloor = \lfloor (k+1)/4 \rfloor$. Then, $\{(0,4i+3)(0,4i+4), (1,4i+3)(1,4i+4) \mid 0 \le i < x\}$ are removed from C, and $\{(0,4i+3)(1,4i+3), (0,4i+4)(1,4i+4) \mid 0 \le i < x\}$ are added to C. Figure 12f shows an example of $T_{7,10} - F$. This step adds $2\lfloor (k+1)/4 \rfloor$ first-dimensional edges and removes $2\lfloor (k+1)/4 \rfloor$ second-dimensional edges from C.

Note that $2\lfloor (k+3)/4 \rfloor + 2\lfloor (k+1)/4 \rfloor = k$ or $k+1$, so the created Hamiltonian cycle C is a DBH with type (a).

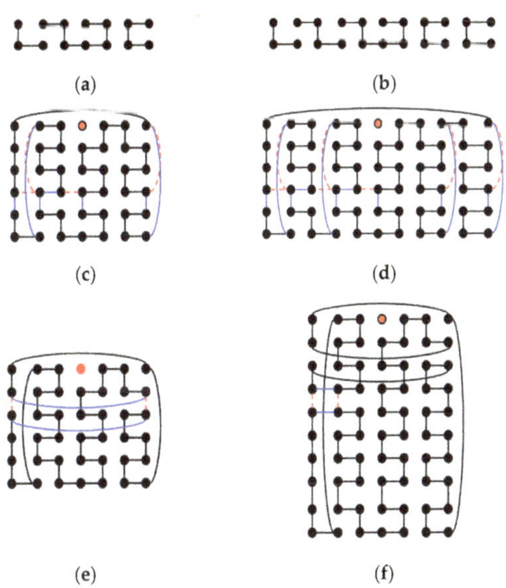

Figure 12. Examples for case 2.1 of Step 2 of Theorem 4. (**a**) Z_7, (**b**)Z_{11}, (**c**) a Hamiltonian cycle C on $T_{7,6} - F$, (**d**) a Hamiltonian cycle C on $T_{11,6} - F$, (**e**) the construct DBH of $T_{7,6} - F$, (**f**) the construct DBH of $T_{7,10} - F$.

Case 2.2. To construct a DBH with type (b): By Step 1, there is a DBH C_2 on $T_{m,4}$ with type (b). The last two rows of C_2 are copied to be a subgraph of Y_m named Z_m. Z_m is modified by adding edges $\{(0, 2i)(0, 2i + 1) \mid 0 \leq i \leq (m - 7)/4\}$ and deleting edges $\{(0, (m - 5)/2)(0, (m - 3)/2\} \cup \{(0, 2i)(0, 2i + 1) \mid (m + 5)/4 \leq i \leq (m - 3)/2\}$. Figure 13a,b show Z_7 and Z_{11}, respectively. Then, C_2 is put in the upper part of $T_{m,n}$, and k Z_ms in the lower part of $T_{m,n}$. Then, for any $0 \leq i \leq m - 1$, if edge $(0, i)(3, i)$ is in C_2, then this edge is removed to add edges $\{(0, i)(n - 1, i), (3 + 2j, i)(4 + 2j, i) \mid 0 \leq j < k\}$ into C_2 and k Z_ms to form a new Hamiltonian cycle C of $T_{m,n} - F$. Figure 13c,d show examples of $T_{7,6} - F$ and $T_{11,6} - F$. Similarly to case 2.1, after calculation, the generated Hamiltonian cycle C satisfies $\mid E_1(C) \mid = \mid E_1(C_1) \mid + k(m - 1) = nm/2 - (k + 1)$ and $\mid E_2(C) \mid = \mid E_1(C_1) \mid + k(m + 1) = nm/2 + k$. We added k or $k + 1$ first-dimensional edges and removed k or $k + 1$ second-dimensional edges from C to form a type (b) DBH of $T_{m,n} - F$. There are two sub-steps according to the value of n to solve this problem.

Step 2.2.1. Let $x = \lfloor (n + 2)/8 \rfloor = \lfloor (k + 3)/4 \rfloor$. Find the x smallest i with $\{(0, i)(0, i + 1), (m - 1, i)(m - 1, i + 1)\} \subseteq E(C)$. Then, these two edges are removed, and $(0, i)(m - 1, i)$, $(0, i + 1)(m - 1, i + 1)$ are added to C. Figure 13e shows an example of $T_{7,6} - F$.

Step 2.2.2. Let $x = \lfloor (n - 2)/8 \rfloor = \lfloor (k + 1)/4 \rfloor$. Find the x edge pairs $((m - 1)/2 - 2, i)((m - 1)/2 - 2, i + 1)$, $((m - 1)/2 - 1, i)((m - 1)/2 - 1, i + 1)$ such that $((m - 1)/2 - 2, i)((m - 1)/2 - 1, i)$, $((m - 1)/2 - 2, i + 1)((m - 1)/2 - 1, i + 1) \notin E(C)$, and $i > 3$. Then, $((m - 1)/2 - 2, i)((m - 1)/2 - 2, i + 1)$, $((m - 1)/2 - 1, i)((m - 1)/2 - 1, i + 1)$ are removed, and $((m - 1)/2 - 2, i)((m - 1)/2 - 1, i)$, $((m - 1)/2 - 2, i + 1)((m - 1)/2 - 1, i + 1)$ are added to C. Figure 13f shows an example of $T_{7,10} - F$.

Again, because of $2\lfloor (k + 3)/4 \rfloor + 2\lfloor (k + 1)/4 \rfloor = k$ or $k + 1$, so the created Hamiltonian cycle C is a DBH with type (b). □

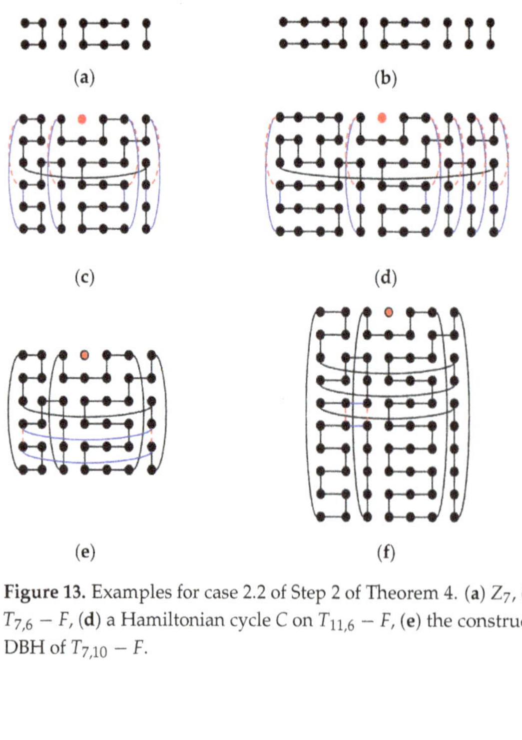

(a) (b) (c) (d) (e) (f)

Figure 13. Examples for case 2.2 of Step 2 of Theorem 4. (**a**) Z_7, (**b**)Z_{11}, (**c**) a Hamiltonian cycle C on $T_{7,6} - F$, (**d**) a Hamiltonian cycle C on $T_{11,6} - F$, (**e**) the construct DBH of $T_{7,6} - F$, (**f**) the construct DBH of $T_{7,10} - F$.

4. Conclusions

We discussed whether $T_{m,n}$ is a two-fault DBH or not, focusing on the one-node one-edge DBH problem. For any m and n, the conclusions of this study are summarized in Table 1. When both m and n are even, $T_{m,n}$ is not a one-node one-edge DBH; and when one of m and n is odd (>3) and the other is even, $T_{m,n}$ is a one-node one-edge DBH. It is proven that $T_{3,n}$ is not the one-node one-edge DBH for any $n \geq 5$. In the subsequent study where m and n are both odd, it is also proven that $T_{5,5}$ is not the one-node one-edge DBH. Due to space limitations, this part of the proof is omitted in this article. Finally, we provide a conjecture that when both m and n are odd and at least one is >5, $T_{m,n}$ is not the one-node one-edge DBH.

Conjecture 1. *There is no DBH on* $T_{m,n} - F$ *when m and* $n \geq 5$*, where F is a set of any one node and one edge.*

Table 1. Summary of one-node one-edge DBHs on $T_{m,n}$.

	$m = 3$	m **Is Odd (>3)**	$m = 4$	m **Is Even (>4)**
$n = 3$	**Yes** (Thm 3)	**No** (Thm 3)	**Yes** (Thm 3)	**No** (Thm 3)
n is odd (>3)	**No** (Thm 3)	**No** (Conjecture 1)	**Yes** (Thm 4)	
$n = 4$	**Yes** (Thm 3)	**Yes** (Thm 4)	**No** (Thm 1)	
n is even (>4)	**No** (Thm 3)			

Author Contributions: Conceptualization, J.S.-T.J.; methodology, J.S.-T.J. and Y.Y.-C.C.; validation, J.S.-T.J.; formal analysis, J.S.-T.J. and Y.Y.-C.C.; investigation, J.S.-T.J.; data curation, Y.Y.-C.C.; writing—original draft preparation, Y.Y.-C.C., writing—review and editing, J.S.-T.J.; visualization, Y.Y.-C.C.; supervision, J.S.-T.J.; project administration, J.S.-T.J.; funding acquisition, J.S.-T.J. All authors have read and agreed to the published version of the manuscript.

Funding: This work was partially supported by the National Science and Technology Council of Taiwan R.O.C. under grant NSTC 112-2115-M-260-001-MY2.

Institutional Review Board Statement: Not applicable.

Informed Consent Statement: Not applicable.

Data Availability Statement: Not applicable.

Conflicts of Interest: The authors declare no conflicts of interest.

References

1. Dimakopoulos, V.V.; Palios, L.; Poulakidas, A.S. On the Hamiltonicity of the Cartesian Product. *Inform. Process. Lett.* **2005**, *96*, 49–53. [CrossRef]
2. Kim, H.-C.; Park, J.-H. Fault Hamiltonicity of Two-Dimensional Torus Networks. In Proceedings of the Workshop on Algorithms and Computation WAAC'00, Tokyo, Japan, 21–22 July 2000; pp. 110–117.
3. Du, Z.-Z.; Xu, J.-M. A Note on Cycle Embedding in Hypercubes with Faulty Vertices. *Inf. Process. Lett.* **2011**, *111*, 557–560. [CrossRef]
4. Wang, H.-R. The Balanced Hamiltonian Cycle Problem. Master's Thesis, Department of Computer Science and Information Engineering, National Chi-Nan University, Nantou, Taiwan, 2011.
5. Peng, W.-F.; Juan, J.S.-T. The Balanced Hamiltonian Cycle on the Toroidal Mesh Graphs. World Academy of Science, Engineering and Technology, Open Science Index 65. *Int. J. Comput. Inf. Eng.* **2012**, *6*, 674–680.

6. Wu, R.-Y.; Lai, Z.-Y.; Juan, J.S.-T. The Weakly Dimension-Balanced Hamiltonian. In Proceedings of the International Conference on Parallel and Distributed Processing Techniques and Applications (PDPTA), Las Vegas, NV, USA, 30 July–2 August 2018; pp. 178–182.
7. Juan, J.S.-T.; Lai, Z.-Y. The Weakly Dimension-Balanced Pancyclicity on Toroidal Mesh Graph $T_{m,n}$ When Both m and n Are Odd. In *Lecture Notes in Computer Science*; Springer: Cham, Switzerland, 2021; Volume 13025, pp. 437–448.
8. Hsu, H.-C.; Chiang, L.-C.; Tan, J.J.; Hsu, L.-H. Fault hamiltonicity of augmented cubes. *Parallel Comput.* **2005**, *31*, 131–145. [CrossRef]
9. Cheng, C.-W.; Lee, C.-W.; Hsieh, S.-Y. Conditional Edge-Fault Hamiltonicity of Cartesian Product Graphs. *IEEE Trans. Parallel Distrib. Syst.* **2013**, *24*, 1951–1960. [CrossRef]
10. Abdallah, M.; Cheng, E. Fault-Tolerant Hamiltonian Connectivity of 2-Tree-Generated Networks. *Theor. Comput.* **2022**, *907*, 62–81. [CrossRef]
11. Juan, J.S.-T.; Ciou, H.-C.; Lin, M.-J. The One-Fault Dimension-Balanced Hamiltonian Problem on Toroidal Mesh Graph. *Symmetry* **2005**, *17*, 93. [CrossRef]
12. Xu, J. *Theory and Application of Graphs*; Kluwer Academic Publishers: Norwell, MA, USA, 2003.

engineering proceedings

MDPI

Proceeding Paper

Multiple Comparisons of the Influential Factors on the Quality of Government Information Systems †

Yi-Luen Lin * and Wei-Hsi Hung

Department of Management Information Systems, National Chengchi University, Taipei 116011, Taiwan; fhung@nccu.edu.tw
* Correspondence: 110356501@nccu.edu.tw

† Presented at the 2024 IEEE 7th International Conference on Knowledge Innovation and Invention, Nagoya, Japan, 16–18 August 2024.

Abstract: The satisfaction of users is important for the success of government information systems. According to a satisfaction survey of X city's Road Excavation Management Information System with 221 valid questionnaires in 2023 in Taiwan, however, the rate of those who were dissatisfied and felt a lack of systematical enhancement reached 20.81%. In this case, the reason for user dissatisfaction and the related aspects of user dissatisfaction must be discussed. Quality played an important role in influencing user satisfaction. Information, system, and service qualities have been ignored in the past. Yet, few studies have explored the application of these qualities to evaluate the context of government information systems. Government information systems in Taiwan have multiple stakeholders: the authorities concerned (i.e., government agencies), outsourcing contractors (i.e., IT vendors), and users (i.e., public infrastructure pipeline units). In this study, an e-service quality conceptual model and an information system success model were constructed to explore the differences in the quality of government information systems. We identified reasons for dissatisfaction. Using the Analytic Hierarchy Process, we analyzed the data of expert questionnaire surveys. The results showed that the authorities prioritized information quality, the outsourcing contractors prioritized system quality, and the public prioritized service quality.

Keywords: information quality; system quality; service quality; information system; analytic hierarchy process

1. Introduction

The government is supposed to introduce quasi-market mechanisms to provide public services centered on quality [1]. In other words, the government must provide citizens with appropriate public services. Farnham and Horton [2] proposed the concept of New Public Management by suggesting that the government must be responsive to the citizen. This means that government information systems must serve to provide rapid responses to citizens' needs. Opposite to the conventional government depending mostly on face-to-face services, an e-government, benefitting both from the rise of the Internet and the construction of government information systems, is geared toward information services more quickly and efficiently provided for government transparency and accountability [3,4].

Satisfaction is crucial for government information systems [5,6]. According to a 2023 satisfaction survey of the Road Excavation Management Information System in X city in Taiwan, the proportion of those who were dissatisfied and felt a lack of systematical enhancement reached 20.81%. In this case, the reason and aspects must be explored.

Therefore, we investigated factors affecting the satisfaction with government information systems from different perspectives (government agency authorities, IT vendors, and pipeline units). Based on the results, explanations and suggestions are provided.

2. Literature Review

DeLone and McLean [7] proposed the success model of information systems and suggested that the success of an information system must be examined from two perspectives: information quality and system quality. System quality refers to the measurement of technological success, while information quality measures semantic success. That is, the quality of the information is determined by the information system. System and information qualities influence the information system and user satisfaction. In using the information system, the users evaluate their satisfaction. Individual impact is classified into the understanding of information and the degree of information application. Individual impact is assessed using financial indicators and benefit evaluations.

Many countries have transformed service-oriented economic entities to service industries. DeLone and McLean [8] enhanced the initial information system success model by using the concept of service quality, defined as the support provided by the service provider (an IT department, a new organizational unit, or an outsourced contractor). So far, three key quality factors of the IS Success Model have been determined: system quality, information quality, and service quality. In this model, the use of the information system is differentiated depending on usage behavior.

Information system success models are applied to government information systems. Unlike commercial information systems that are profit-oriented, government information systems provide public services for the citizen. Hence, the Technology Acceptance Model (TAM) [9] can be applied to government information systems to assess the impact of quality, citizens' satisfaction, and benefits. For example, Chen [10] investigated the impact of quality on citizens' satisfaction using the online tax filing system. Fan and Yang [3] studied the impact of quality dimensions on citizens' perception of online government service quality. The latter combined its result with citizens' perception of offline government service quality to see the possible impacts on citizen satisfaction. By summarizing previous research results, we compiled the quality dimensions and their definitions as shown in Table 1.

Table 1. Quality factors.

Source	Chen, 2010 [10]	Xu et al., 2013 [11]
Information Quality	Relies on users' perceived value of information system outputs.	Capture the content of e-services, such as the level of completeness and the latest information.
System Quality	An evaluation of the information processing system itself.	Describe the structural characteristics of capturing e-commerce systems and the dynamics of management performance.
Service Quality	A multi-dimensional construct, usually measured by the SERVQUAL scale.	Users' overall subjective assessment of the quality of interaction with system service providers, including specific service levels that need to be satisfied.

Previous research has focused on the exploration of the dimensions and factors of information quality, system quality, and service quality. The importance of these factors has not been weighed prudently yet. Moreover, government information systems are closely related to people's daily life, yet few studies have explored the context of how these three quality dimensions are applied to government information systems. Also, the differences in the importance of quality factors recognized by government agencies, IT vendors, and users need to be explored.

In the past, discrepancies in the importance of quality factors perceived by different roles were investigated. Such discrepancies are divided into positive and negative ones. A positive discrepancy occurs when the provided quality exceeds the users' measured importance, which impacts user satisfaction positively but leads to unused resources and waste. Conversely, a negative discrepancy means that the provided quality falls short of the measured importance of users whose satisfaction is affects negatively. Therefore, it is necessary to study the quality that aligns with the measured importance perceived by different roles to minimize the possible discrepancy.

3. Method

The Analytic Hierarchy Process (AHP) is a hierarchical analysis method used for decision-making problems with multiple evaluation criteria. It is used to analyze and decompose a problem into hierarchical levels and then provide decision makers with quantitative judgments to choose appropriate solutions and reduce the risk of decision-making errors [12]. When applying the AHP method, a hierarchical structure is used to break down complex decision problems into concise hierarchical elements. This involves constructing pairwise comparison matrices for elements at each level to determine their eigenvalues and eigenvectors. Based on these calculations, weights are assigned to each element to facilitate further analysis and interpretation.

In this study, government agencies, IT vendors, and pipeline units were included. Government agencies act as the authorities concerned in the context of road excavation, specifically the X City government. IT vendors outsource contractors responsible for the management of the information system. Pipeline units represent the users of the information system regarding electricity, tap water, telecommunications, gas, and rainwater/sewage systems.

To investigate the quality discrepancy, we designed an AHP questionnaire to evaluate information quality, system quality, and service quality as hierarchical elements, and we conducted assessments on the aforementioned roles. In the AHP, a hierarchy structure was established. Then, pairwise comparison matrices were created to calculate eigenvalues and eigenvectors, perform consistency checks, and compute weights, as shown in Figure 1.

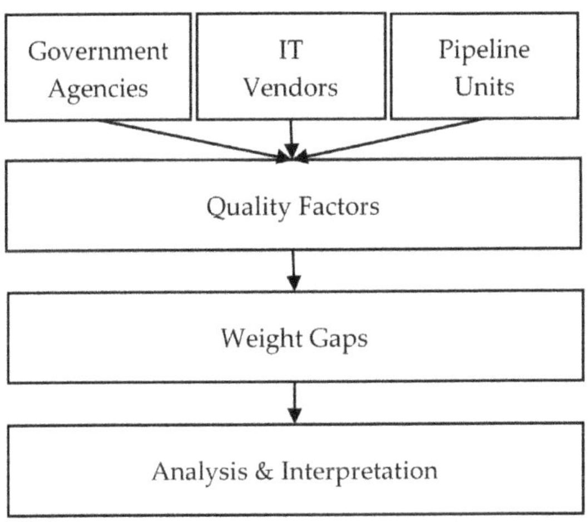

Figure 1. Research design.

4. Results

Based on a literature review, a questionnaire was constructed, and a pilot survey was conducted for scholars and experts. The questionnaire was revised according to the suggestions of scholars and experts before the official survey was administered. The AHP questionnaire was distributed to government agencies, IT vendors, and pipeline units. As a larger incentive for the interviewees to complete the questionnaire, a gift voucher was offered. After excluding the incomplete responses and those with a consistency index and consistency ratio greater than 0.1, 20 valid AHP questionnaires from government agencies, 26 from information vendors, and 18 from pipeline units were collected. The statistical results are shown in Table 2.

Table 2. AHP results.

Weight	Government Agencies	IT Vendors	Pipeline Units
Information Quality	54.30%	28.45%	30.16%
System Quality	24.30%	58.98%	14.65%
Service Quality	21.40%	12.57%	55.19%

In the Road Excavation Management Information System, the authorities prioritize information quality after the Kaohsiung gas explosion incident 2014 in Taiwan. The public officers must answer to the inquiry of higher-level agencies and the people. In an emergency that could occur regarding the tap water pipeline rupture shown in Figure 2, the authorities must retrieve an accurate and complete pipeline map immediately to make urgent repairs and prevent public discontent from occurring as soon as possible.

Figure 2. Emergency response site for tap water pipeline rupture.

For government agencies and pipeline units, a reliable information system is indispensable. To ensure the accuracy of the maps, pipeline units keep measuring pipelines. Then, the pipeline maps are uploaded in the road excavation management information system with its official review and approval. If errors are found in the maps, the authorities compel the pipeline units to make instant improvements and reupload the corrected maps promptly. The information system allows the authorities to review maps anytime and anywhere. An illustration of the pipeline maps in the information system of X city is shown in Figure 3.

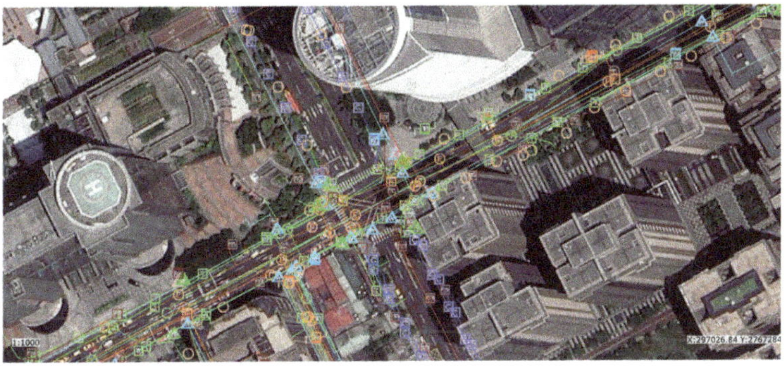

Figure 3. Pipeline maps.

System quality is the priority of IT vendors. Generally, government agencies allocate budgets for the information systems and technical support of IT vendors. IT vendors possibly take blame as they are responsible for the construction and maintenance of the information system. Due to the penalties for future contracts, the requirements of government agencies are taken seriously by IT vendors. Therefore, the primary focus of IT vendors is system quality. Once any issues or unreliability happens in the system, IT vendors must handle customer service calls and fix the problems. Government agencies require the Service Level Agreement (SLA) in outsourcing contracts with IT vendors. If the system quality is poor or fails to meet the SLA requirements, IT vendors may face penalties. To avoid such situations, IT vendors should prioritize system quality.

As outsourced contractors, IT vendors are obliged to make adjustments according to the government agencies' requirements to keep the maps accurate, up to date, and complete. Service quality is the third priority of IT vendors. When the government agencies' demands for system quality and information quality are satisfied, service quality can be recognized accordingly.

Pipeline units prioritize service quality the most. When issues arise while using the system, they need immediate technical support to resolve the problems. Therefore, they prioritize service quality, including the information vendors' technical capabilities and customer service responsiveness. Pipeline units also place importance on information quality. As previously mentioned, government agencies prioritize information quality. In X city, public infrastructure pipeline map maintenance guidelines have been established, specifying measurement methods and accuracy requirements for public infrastructure pipeline maps. Therefore, pipeline units must adhere to regulations and emphasize the information quality of public infrastructure pipeline maps.

5. Conclusions

The influential factors of the quality of a government information system indicate the major concern of three groups. Their pressures come from public condemnation or discredit, while penalties are given due to an untenable case resolution or a deferred payment. Essentially, public condemnation for disasters is a major source of pressure on government agencies; penalties for systematical deficits are major concerns for IT vendors;

and unprofessional assistance by IT vendors affects pipeline units. Each stakeholder has different concerns. If the difference in concerns of the three stakeholders is minimized, their cooperation can become efficient. For a smoother engineering procedure associated with information quality, system quality, and service quality, information quality can be enhanced with the data measured and uploaded precisely.

An IS Success Model is applied to government information systems to identify different perspectives. In a road excavation information system, a balance between the three stakeholders must be maintained. Information quality is constantly stressed by government agencies. A positive correlation between service quality and information quality needs to be considered for the cooperation of the government and IT vendors. As the main service supplier, IT vendors must cooperate with pipeline units and government agencies. Then, the negative discrepancy perceived by pipeline units can be minimized. Then, information quality can be improved. The positive discrepancy of system quality between IT vendors and government agencies needs to be mitigated. Given that the greatest concern is system quality, IT vendors must concentrate on system quality, which is related to the maintenance of information quality. Satisfaction with the information quality of government agencies can be enhanced through observation and analysis to determine quality factors of government information systems.

Author Contributions: All authors have the equal contribution to this article. All authors have read and agreed to the published version of the manuscript.

Funding: This research received no external funding.

Institutional Review Board Statement: Not applicable.

Informed Consent Statement: Not applicable.

Data Availability Statement: Data sharing is not applicable to this article.

Conflicts of Interest: The authors declare no conflict of interest.

References

1. Bellamy, C. Information and communications technology. In *Managing the New Public Services*; Farnham, D., Horton, S., Eds.; Palgrave: London, UK, 1996; pp. 137–152.
2. Farnham, D.; Horton, S. The political economy of public sector change. In *Managing the New Public Services*; Farnham, D., Horton, S., Eds.; Palgrave: London, UK, 1993; pp. 3–24.
3. Fan, J.; Yang, W. Study on e-government services quality: The integration of online and offline services. *J. Ind. Eng. Manag.* **2015**, *8*, 693–718. [CrossRef]
4. Yildiz, M. E-government research: Reviewing the literature, limitations, and ways forward. *Gov. Inf. Q.* **2007**, *24*, 646–665. [CrossRef]
5. Alawneh, A.; Al-Refai, H.; Batiha, K. Measuring user satisfaction from e-Government services: Lessons from Jordan. *Gov. Inf. Q.* **2013**, *30*, 277–288. [CrossRef]
6. Malodia, S.; Dhir, A.; Mishra, M.; Bhatti, Z.A. Future of e-Government: An integrated conceptual framework. *Technol. Forecast. Soc. Change* **2021**, *173*, 121102. [CrossRef]
7. DeLone, W.H.; McLean, E.R. Information systems success: The quest for the dependent variable. *Inf. Syst. Res.* **1992**, *3*, 60–95. [CrossRef]
8. DeLone, W.H.; McLean, E.R. The DeLone and McLean model of information systems success: A ten-year update. *J. Manag. Inf. Syst.* **2003**, *19*, 9–30.
9. Davis, F.D.; Bagozzi, R.P.; Warshaw, P.R. User acceptance of computer technology: A comparison of two theoretical models. *Manag. Sci.* **1989**, *35*, 982–1003. [CrossRef]
10. Chen, C.W. Impact of quality antecedents on taxpayer satisfaction with online tax-filing systems: An empirical study. *Inf. Manag.* **2010**, *47*, 308–315. [CrossRef]

11. Xu, J.; Benbasat, I.; Cenfetelli, R.T. Integrating service quality with system and information quality: An empirical test in the e-service context. *MIS Q.* **2013**, *37*, 777–794. [CrossRef]
12. Satty, T.L. How to make a decision: The analytic hierarchy process. *Eur. J. Oper. Res.* **1990**, *48*, 9–26. [CrossRef]

engineering proceedings

MDPI

Proceeding Paper

Role of Mathematics Teachers in Learner's Diversity Using AI Tools †

Wing-Kin Cheng

School of Education and Languages, Hong Kong Metropolitan University, Hong Kong, China; wikcheng@hkmu.edu.hk

† Presented at the 2024 IEEE 7th International Conference on Knowledge Innovation and Invention, Nagoya, Japan, 16–18 August 2024.

Abstract: The advancement of artificial intelligence (AI) has attracted attention across disciplines. Different research has revealed the role of AI and the outcomes of AI in education (AIEd). However, teachers need to use AI to cater to learner diversities in mathematics education, which needs exploration. Therefore, how different AI tools assist mathematics teachers in developing teaching materials for students was investigated in this study. Teachers were invited to utilize AI techniques to develop their teaching and learning materials. The findings can be used to enhance the remedial and enrichment measures in teaching secondary mathematics and construct a framework to help teachers with learner diversities with AI tools.

Keywords: artificial intelligence; secondary mathematics teachers; prompting techniques; remedial; enrichment

1. Introduction

Artificial intelligence (AI) offers opportunities and challenges to enhance teaching efficiency for teachers. To effectively harness AI's benefits, teachers' professional development is crucial. Teachers play a pivotal role in implementing AI-based curricula and understanding readiness and perceptions regarding AI usage. In the 21st century, equipping students with AI literacy is important. Therefore, this study aims to assess secondary mathematics teachers' use of AI integration in teaching secondary mathematics.

Yau et al. [1] studied how to design an AI curriculum. Through a collaborative effort of university professors and secondary school teachers, an innovative AI curriculum was co-designed, and teachers' technical knowledge gaps were identified. This study underscores the importance of collaboration among experts and educators in designing effective AI curricula tailored to students' contextual needs and improving learning outcomes. Therefore, teachers need to collaborate for the development of AI-based materials in secondary mathematics.

In this study, two focus groups of in-service local secondary mathematics teachers in Hong Kong were involved. The first group of teachers focused on how they used AI tools to prepare materials for remedial classes; the second group of teachers focused on how they used AI tools to prepare materials for enrichment classes. Also, how secondary mathematics teachers use AI tools was explored.

2. Literature Review

2.1. Readiness in Using AI

The successful implementation of generative AI in education relies on the readiness of teachers. The importance of teacher training is emphasized to integrate AI curricula for various subjects. Chiu and Li [2] emphasized the significance of professional training and support for teachers. Providing educators with the necessary tools and guidance is crucial for effective AI implementation. Ravi et al. [3] recommend that teacher support and professional development sessions precede the introduction of new AI curricula. This ensures that teachers, even those without prior experience, are equipped to confidently deliver foundational AI knowledge. Ravi et al. [3] delved into teachers' perceptions and experiences in implementing AI curricula in middle schools. Educators expressed a need for clearer directions and specifications in training materials to navigate novel concepts effectively. Additionally, they requested hands-on examples for incorporating AI into teaching practices. Loong and Herbert [4] explored the integration of digital technology by primary teachers in mathematics classrooms. Yet, the readiness of Hong Kong secondary mathematics teachers in this regard remains underexplored.

2.2. Prompting Techniques

Schorcht et al. [5] used four variants of prompting techniques in mathematical problem-solving: Zero-Shot, Chain-of-Thought, Ask-me-Anything, and Few-Shot. In Zero-Shot, a task is entered into AI without additional inputs. In Chain-of-Thought, the user provides sequential guides to AI and then AI offers an organized sequential output. These intermediate steps assist the generative AI in producing an accurate solution. It significantly improves the AI's performance in straightforward problem-solving tasks involving the calculation of basic computational steps.

In Ask-me-Anything (Arora et al. [6]), users avoid long answers. When AI is asked a question, the answer generated must be short. AI waits until its user gives it further instructions to proceed. In doing so, AI asks the user questions to confirm if it is performing what is expected. The process is deemed to be interactive. In Few-Shot, an additional task with a given solution is introduced to AI alongside the original problem. The user then inputs a similar mathematics problem with only small differences in details. Schorcht et al. [5] highlighted the effectiveness of Chain-of-Thought and Ask-me-Anything in enhancing AI's strategic approaches to problem-solving.

It is still necessary to research how secondary mathematics teachers use AI, particularly how they interact with various AI tools to tailor curricula to different learning needs.

3. Methodology

According to Williams and Katz [7], a focus group with individuals who share a common interest or characteristic provides information on a particular subject matter. Regarding the purposes of a focus group, Krueger and Casey [8] noted that a focus group promotes a comfortable atmosphere of disclosure among the members of the group to share their ideas, experiences, and attitudes about a topic. Participants play both an active and passive role as they influence and are influenced. The researcher, on the other hand, is the moderator, facilitator, observer, and investigator.

The concept of grounded theory was proposed by Glaser and Strauss [9] to build theory from data. Following this approach, researchers are encouraged to discover theories based on research data rather than testing existing hypotheses. According to Gibson and Hartman [10], the grounded theory is characterized by its openness. This notion of openness entails why the researcher shall not make use of preconceived concepts or ideas when conducting certain research. Since AI is an innovation in education, grounded theory can ensure the openness of the findings and implications drawn.

Grounded theory was applied by the two focus groups to elicit how teachers select AI tools, what teachers want to achieve with these selected AI tools, and how secondary mathematics teachers interact with the AI tools. Six secondary mathematics teachers were recruited, three of whom were assigned to the remedial group and three of them to the enrichment group. They designed materials that fit their teaching needs. They chose AI tools to evaluate the effectiveness of these tools.

3.1. Remedial Group

In this group, teachers used AI tools to prepare remedial materials and a remedial topic to determine the corresponding objectives. They captured all the prompts and responses from AI and recorded how they interacted with AI. They also compared the effectiveness of different AI tools and stated the benefits and the limitations of using these different AI tools.

3.2. Enrichment Group

In this group, teachers used AI tools to prepare enrichment materials for an enrichment topic and determined the corresponding objectives. They were given the same set of instructions as teachers in the remedial group.

Teachers from both groups chose AI tools and the topics in which they were interested. Teachers recorded the prompting questions and AI responses for data analysis.

4. Findings

4.1. Remedial Group

- Teacher A

Teacher A chose two questions: Q1. Plot a graph $x^2 - x - 1 = 0$; Q2. Solve $x^2 - x - 1 = 0$.

He used the mobile app version of the AI tools POE, Copilot, and Wolfram Alpha. He gave the same prompt to the three tools. He did not keep interacting with the tools. Teacher A's conclusion to Q1 is that *POE* could not plot the graph; it just showed the HTML code. It could not convert it to a graph or a picture. *Copilot* and *Wolfram Alpha* provided the correct graph. The graph on *Copilot* was downloadable. Teacher A's conclusion to Q2 was that *POE* and *Copilot* offered the correct answers in texts. Thus, it was not easy to read. *Wolfram Alpha* represented answers with the approximate value or exact value, and it could also give various solutions. The teacher described the benefit of AI tools as being fast and having lots of information. Several AI tools did not provide graphs. They only generated ideas with too much information.

- Teacher B

Teacher B wanted to use worksheet exercises designed by a textbook publisher. However, there were not enough exercises, and no solutions were provided. The topic chosen was factorization. Teacher B used two AI tools, *Magic School* and *Wolfram Alpha,* to generate more. Teacher B chose a YouTube video clip teaching factorization. He used the *YouTube Video Questions* tool in *Magic School.* He filled in the template with eighth-grade-level questions. It included 10 questions with multiple-choice answers and the URL of the YouTube video. Then, 10 multiple-choice questions were generated with the answer key. Teacher B used *Wolfram Alpha* to generate an exercise on quadratic factorization for students. He used *Wolfram Problem Generate* to conduct this. He set the prompt "Factor $w^2 + 13w + 42$" with the 'Beginner' level. Eight multiple-choice questions were generated with a set of answer keys. Teacher B also tried out the step-by-step function in *Wolfram Alpha.* The prompt was

"Factor $2x^2 - 3x - 5$"; a step-by-step solution was generated. AI tools saved him time in preparing the online remedial materials. Students could obtain hints from AI anytime and anywhere. However, the choice of words for the questions generated by AI was not the same as the teachers' choice of words.

- Teacher C

Teacher C had the same objectives as Teacher B, preparing a set of exercise questions on factoring quadratic polynomials. He used AI to prepare exercises on the factor method. He used *POE* and prompts, "Can you generate 10 questions for factorizing quadratic polynomial?" and "Once again, share the solution with me?". *POE* provided 10 questions with the correct solutions. He used the same prompts for *Copilot*. *Copilot* could also provide 10 questions with correct solutions. *POE* and *Copilot* provided text responses. Teacher C also wanted to know if AI could plot the graph of $y = 2x + 1$. He used *POE* and *Copilot* and asked "Can you use *Desmos* or *GeoGebra* to plot the graph of $y = 2x + 1$?". *Copilot* provided the steps of using *Desmos* and *GeoGebra* to plot the graph. *Desmos* gave a scatter plot, while *Copilot* looked for images and gave sketches of a graph that did not exactly show $y = 2x + 1$. AI tools generated questions and answers effectively and efficiently, particularly for those mechanical questions and topics. However, precise wording needed to be used for good AI responses. AI could not show the mathematics in the proper format. For example, *POE* did not show the answers in mathematics notation.

4.2. Enrichment Group

- Teacher D

Teacher D chose a question from the Hong Kong Mathematics Olympics Competition as shown in Figure 1.

From HKMO 2018 Heat Event:
Let $f(x)$ be a polynomial of degree 2, where $f(1) = \frac{1}{2}$, $f(2) = \frac{1}{6}$, $f(3) = \frac{1}{12}$.
Find the value of $f(6)$.

Figure 1. Prompting question 1 from Teacher D.

Teacher D used *POE*, *Thetawise*, and *Wolfram Alpha* to test whether these AI tools could give correct solutions. Teacher D used the same prompt for the 3 AI tools: "Let $f(x)$ be a polynomial of degree 2, where $f(1) = \frac{1}{2}, f(2) = \frac{1}{6}, f(3) = \frac{1}{12}$. Find the value of $f(6)$".

POE correctly set the system of three equations. However, it solved the system of equations incorrectly. Teacher D did not notice that *POE* gave the incorrect answer. *Thetawise* correctly set the system of three equations and solved the equations with numerical methods, which provided an estimate. He prompted *Thetawise* by adding the word "exact value" and the second prompt was "Let $f(x)$ be a polynomial of degree 2, where $f(1) = \frac{1}{2}$, $f(2) = \frac{1}{6}, f(3) = \frac{1}{12}$. Find the value of $f(6)$ with exact value". *Thetawise* gave what Teacher D wanted. Teacher D observed that *Wolfram Alpha* could not understand the questions. Teacher D did not revise the prompt and considered that *Wolfram Alpha* failed in this task. AI tools were time-saving and they generally gave clear and detailed explanations of the corresponding mathematics problems. It was difficult to identify the incorrect steps. Users need to make modifications when inputting the question.

- Teacher E

Teacher E captured the image of a DSE question—a high-stakes public examination for senior secondary students in Hong Kong (HKDSE 2024 paper 1 Q17) as shown in Figure 2. He imported the image of the question to three different AI tools: *Copilot*, *POE*, and *Thetawise*. He wanted to know if the AI tools could give the correct solutions within three prompts.

The coordinates of the points Q and R are $(10, -1)$ and $(-4, -9)$ respectively.

(a) Let P be a moving point in the rectangular coordinate plane such that $PQ = PR$. Denote the locus of P by Γ.

(i) Describe the geometric relationship between Γ and QR.

(ii) Find the equation of Γ.

(3 marks)

(b) Let C be the circle which passes through Q, R and the point $(4, 3)$.

(i) Find the equation of C.

(ii) The coordinates of the point U are $(10, 4)$. It is found that U lies outside C. UV and UW are the tangents to C at the points V and W respectively. Is the area of the circumcircle of ΔUVW greater than 100? Explain your answer.

(5 marks)

Figure 2. Prompting questions from Teacher E.

Copilot was able to read the question from the captured image. However, it missed the slope of Γ. Teacher E set the second prompt as "In (a)(i), the solution given by you, the slope of QR is incorrect, check again please". However, the slope of Γ generated by *Copilot* was still incorrect. Teacher E used *POE* with the same first prompt. *POE* was also able to read the question from the captured image and it was able to give the correct answer in (a)(i). However, *POE* missed the slope of QR and wrongly represented the sign. *POE* was almost correct in (b)(i). The sqrt (49 + 16) must be sqrt(65) instead of 7. It did not explain why the diameter of C is QR though. Teacher E set the second prompt as "In (a)(i), the solution given by you, the slope of QR is incorrect, check again please". *POE* corrected the slope of QR, but it failed to note that Γ is perpendicular to QR, so the slope of Γ was incorrect. Teacher E set the third prompt as "The new solution is not correct. Γ is a line perpendicular to line segment QR, if the slope of QR is 4/7, the slope of Γ should be −7/4". The AI made a mistake in this third and last step. Instead of $y + 5 = -(7/4)x + (21/4)$, it should be $y = -(7/4)x + ¼$. Teacher E concluded that *POE* could not successfully complete the task.

Teacher E used *Thetawise* with the first prompt as shown in Figure 2. *Thetawise* was also able to read the question from the captured image. It translated the math symbol into LaTeX form and gave the answers correctly in (a)(i) and (a)(ii). For part (b), the AI found the equation of C by setting up a system of simultaneous equations. *Thetawise* finished the calculation of (b)(i) correctly. In (b)(ii), it used a more complex approach than expected. The AI stopped before attempting to find the radius r of the circle. Teacher E then continued with the second prompt "Do you notice that the circumference UVW passes through the center of C in part (b)(i)? If yes, you should be able to find the diameter of the circumcircle UVW, hence the area of it". *Thetawise* responded and found the center of a circle from an equation that applied the distance formula. Yet, it seemed relatively inapt to comprehend deductive geometry in this situation. The distance between point U and the center of circle C should be the diameter of the required circle because its subtended angle is 90°.

Teacher E set the third prompt as "In your answer, the distance between point U and the center of C should be the diameter of the required circle because it's subtended angle is 90°". *Thetawise* then completed the question correctly. After using *Thetawise* to obtain the correct solution, Teacher E was interested in exploring if it could generate similar questions. From the above results, only *Thetawise* can generate the correct solution within a few prompts. Also, *Thetawise* was able to display easily readable solutions using LaTeX. Thus, Teacher E rated *Thetawise* over *POE* and *Copilot* to generate a question similar to the given one.

It is a handy tool to prepare questions and solutions with clear prompts and boosts self-directed learning because students learn at home using AI tools. AI provides instant feedback in learning; users can learn the background knowledge of the subject easily. Teacher E also observed that AI is a bit different from Google because AI organizes learning material for users, and teachers can even tailor-make the lesson with clear prompts such as "tell me about quadratic equation in 50 words". However, the solution given by AI may not be correct, so the effectiveness of AI depends on the depth of the users' knowledge about the subject matter.

- Teacher F

Teacher F selected a junior secondary competition question which was not a standard textbook-type or examination-type question. He was interested in exploring if AI provided the correct solution to the question. He used *Thetawise*, *MathGPT*, and *Julius AI* to solve this problem. Teacher F fed the image of Figure 3 to the three AI tools.

It is known that real numbers a, b and c satisfy $3a + 12b + 2c = 63$ and $2a + 7b + c = 35$. Find the value of $\frac{a+2b}{3b+c}$.

Figure 3. Prompting question by Teacher F.

Both *Thetawise* and *MathGPT* showed the step-by-step solution with correct answers. *Julius AI* provided the solutions with several steps, but it stopped before giving the full solutions and prompted Teacher F to say what he wanted. Teacher F set the second prompt as "Solve the whole question and give the full solution". *Julius AI* gave a wrong answer, and the third prompt was "It seems that the answer is 1/3. Did you make any mistake?". *Julius AI* finally gave the correct answer.

Besides using AI to develop enrichment materials, Teacher F shared his dialog with AI about how he prepared a set of remedial exercises. He used *Websim AI*. He used eight prompts to interact with AI. First prompt: "Create 5 factorization questions related to only taking out common factor, perfect square, and difference of two squares"; second prompt: "Put the answer at the end of all questions. Don't use Hide/Show Button"; third prompt: "Don't show "Factor the following expression" in each question, but show "Factorize the following expressions" at the beginning of all 5 questions"; fourth prompt: "Don't show "(common factor)" in each question. 1 row 1 question"; fifth prompt: "1 row 1 answer, no explanation needed"; sixth prompt: "Create 15 more questions, giving a total of 20 questions"; seventh prompt: "Answer of Q6 and Q12 can be further factorized using the difference of two squares"; eighth prompt: "Q16, Q18, and Q20 involve other skills. Delete them and create 3 others". Teacher F finally prepared a set of exercises with 20 questions on the factorization of quadratic polynomials with answers.

Using *Websim AI*, teachers can generate different types of questions to address the users' needs. It shortens the time to prepare when generating a lot of exercises with slight or large variations according to the user's input commands. The time to set questions, prepare solutions, and typeset is reduced. Another benefit shared by Teacher F is how *Wolfram Problem Generator* can generate different levels of questions within the scope on its website. The questions can be different as they are generated. This provides extra resources to teachers.

Its limitation is related to math questions with diagrams, in 2D or 3D, especially when the information is provided in the diagrams, instead of descriptions in the question. Mistakes made by AI are found in diagrams. Another limitation is related to math questions with cross-topics and high-order thinking. It is easy for AI to solve or create a difficult math question in a single specific topic, but it becomes problematic if several different topics are involved. For example, to work out a long question of an HKDSE Math Paper 1 which requires knowledge of arithmetic sequences and geometric sequences, the transformation of functions, quadratic equations, the centers of triangles, and coordinate geometry, the AI solver could not solve it successfully. When AI is asked to create a math question with different topics, these topics are dealt with in different parts and coherence among these different parts of the question is absent.

5. Analysis

5.1. Remedial Group

Teachers in the remedial group and enrichment group had different expectations of AI. For the remedial groups, teachers had the following expectations:

- Plot graphs (Teachers A, C);
- Give correct answers and solutions (Teachers A, B and C);
- Obtain step-by-step solutions (Teacher B);
- Generate exercises efficiently (Teachers B, C and F);
- Obtain more information on the topic (Teacher A);
- Generate questions with proper mathematics notation (Teacher C).

The AI tools that the teachers used in the remedial group are shown in Table 1. All teachers chose the AI tools according to their needs. For example, *Wolfram Alpha* had *Question Generator* that helped teachers prepare sets of questions swiftly. *Magic School* helped the teachers set questions based on a YouTube video. *Websim AI* helped teachers generate questions and save them on a website so that teachers could provide aids to students instantly.

Table 1. AI tools used in remedial group.

Teachers	AI Tools Used				
	POE	*Copilot*	*Wolfram Alpha*	*Magic School*	*Websim AI*
Teacher A	ü	ü	ü		
Teacher B			ü	ü	
Teacher C	ü	ü			
Teacher F			ü		ü

Teacher B chose *Wolfram Alpha* as they could generate lots of reliable questions and thus ensure efficiency in teaching and learning material development. Teachers A, C, and F criticized that AI tools could not draw graphs of functions and could not show proper mathematics notations. Better tools are needed to fulfill their needs.

5.2. Enrichment Group

Teachers in the enrichment group had different expectations of AI as follows:

- Solve structured questions (Teachers D, E and F);
- Give a prescriptive solution (Teachers D and F);
- Generate a similar exercise to a solved problem (Teachers E and F);
- Generate exercises with different levels of difficulty (Teacher E);
- Solve cross-topic questions (Teacher F).

The AI tools that the teachers chose and used in the enrichment group are shown in Table 2. Teachers D, E, and F chose *Thetawise* as it could solve different mathematics problems including questions at the university level. *POE* and *Copilot* were also popular.

Table 2. AI tools used in enrichment group.

Teachers	AI Tools Used					
	POE	*Copilot*	*Wolfram Alpha*	*Math GPT*	*Thetawise*	*Julius AI*
Teacher D	ü		ü		ü	
Teacher E	ü	ü			ü	
Teacher F				ü	ü	ü

Teachers in the enrichment group used different AI tools to solve structured questions chosen from public examinations or mathematics competitions. Teachers found it time-consuming to work on the solutions for such questions and seek help from AI. AI could provide them with lots of information and detailed steps. However, not all the solutions were correct. Teachers had to be careful to evaluate the solutions from AI and give AI the relevant hints. The three AI tools *Math GPT*, *Thetawise*, and *Julius AI* could solve lots of mathematics problems correctly. However, the solutions were sometimes too complicated and advanced and might not fit the educational purpose. Teachers also kept providing guidance and information to help AI correct the mistakes. Teacher feedback gathered from this study revealed that AI could not solve deductive geometry and trigonometry problems that involve geometrical diagrams.

Regarding what teachers expected to achieve with the selected AI tools, Teachers A, B, and C wanted to have well-structured exercises with solutions to help students consolidate particular mathematics knowledge and skills. It usually took the teachers much time to prepare remedial materials. Teachers D and F recognized that AI gave incorrect steps, and sometimes it was hard to identify the mistaken steps. Teachers E and F suggested that AI tools could not effectively solve some cross-topic mathematics problems.

5.3. Prompting Techniques

Schorcht et al. [5] suggested four variants of different prompting techniques including *Zero-Shot*, *Chain-of-Thought*, *Ask-me-Anything*, and *Few-Shot*. Teachers' general prompting techniques were matched into these four variants of prompting techniques as shown in Table 3.

Table 3. Four variants of different prompting techniques.

Teachers	*Zero-Shot*	*Chain-of-Thought*	*Ask-Me-Anything*	*Few-Shot*
Teacher A	ü			
Teacher B	ü			
Teacher C		ü		
Teacher D	ü			
Teacher E		ü		
Teacher F		ü		

Teachers were not trained with different prompting techniques. They used their own ways of interacting with the AI tools. Thus, teachers in this study used *Zero-Shot* and *Chain-of-Thought* prompting. Teachers A, B, and D used *Zero-Shot* prompting when dialoguing with AI. When the AI tools gave what they wanted (e.g., an exercise with solutions), they considered the task completed and evaluated that AI was useful. When it came to complex mathematics problems, *Zero-Shot* prompting was not useful in generating a correct solution. Teachers C, E, and F used *Chain-of-Thought* prompting to refine the questions, correct the mistakes, and provide more details to AI to amend the solutions. Teacher F used *Chain-of-Thought* prompting to generate a set of exercise questions in a particular format and re-generate appropriate questions.

Teachers A and D used *Zero-Shot* prompting when using their selected AI tools for all the questions and evaluated the effectiveness of AI from the answers generated. They did not keep interacting with AI to refine the answers generated. Teachers A, B, and D evaluated the effectiveness of the AI tools, concluding that the AI tools could not provide graphs and did not give proper mathematical notation. Teachers E and F interacted with AI by employing the *Chain-of-Thought* prompting technique. Their feedback revealed that AI could amend the solution if the guide given to AI was clear and relevant. The accuracy of the structured solutions depended largely on teachers' knowledge when feeding input to AI.

6. Conclusions and Implications

Different AI tools helped secondary mathematics teachers generate exercises on a particular content area, verify the answers to a problem, create exercises efficiently, and seek hints and information about complex mathematics problems. Mathematics teachers rated AI tools by evaluating whether the tools could (a) give a correct solution; (b) give a solution using a particular method; and (c) prepare exercises with solutions effectively. Teachers in the groups employed different prompting techniques when interacting with AI. One of the significant challenges faced by the teacher was how the *Zero-Shot* prompting technique failed to identify the mistakes made by AI and concluded that the AI gave the incorrect solution. Mathematics teachers used *Zero-Shot* and *Chain-of-Thought* prompting when interacting with their chosen AI tools. Teachers used *Zero-Shot* prompting techniques to generate remedial exercises efficiently. Teachers chose to use the default template in the AI tools to generate exercises to save time. Teachers who used *Chain-of-Thought* prompting techniques wanted to check whether AI tools could solve structured mathematics problems. Teachers also used the *Chain-of-Thought* technique to prepare similar questions with different levels of difficulty to the original problem.

In mathematics problem solving, there are solutions to the problem. Information and descriptions of various methods can be generated by AI. However, AI may (a) make calculation mistakes; (b) choose the wrong approach to tackle the problem; (c) develop different interpretations of the problem from the user's expectation. It is the teacher's role to correct mistakes from the answers generated by AI. Teachers need to interpret the questions and give information to AI. Teachers must be equipped with their content knowledge so that they can evaluate the solutions given by AI. In this study, *Chain-of-Thought* prompting demonstrated more accurate solutions in a better format and levels of difficulty that teachers intend to develop. Further study is necessary to explore the prompting framework with *Chain-of-Thought*. By differentiating the objectives in remedial and enrichment needs, secondary mathematics teachers can benefit from a well-structured prompting framework or prompting cycle. AI technology is advancing at an unprecedented rate. Different AI problem-solvers and geometry solvers are emerging, which necessitates further study on the collaboration of different AI tools.

Funding: This research was funded by a Hong Kong Metropolitan University Research Grant (No. RD/2024/1.2).

Institutional Review Board Statement: The study was reviewed, approved, and conducted by "Research Ethics Committee" of Hong Kong Metropolitan University (HKMU) with approval date: 14 May 2024; approval code: RD/2024/1.2. All the participants' related information collected in this study is treated with strict confidentiality and is reported in an aggregated manner and anonymized form to ensure data privacy and protection.

Informed Consent Statement: Informed consent was obtained from all the human subjects involved into this while conducting the activity.

Data Availability Statement: Due to the ethical, data privacy, and protection concerns, we cannot make the raw data publicly available.

Conflicts of Interest: The author declares no conflict of interest.

References

1. Yau, K.W.; Chai, C.S.; Chiu, T.K.; Meng, H.; King, I.; Yam, Y. A phenomenographic approach on teacher conceptions of teaching Artificial Intelligence (AI) in K-12 schools. *Educ. Inf. Technol.* **2023**, *28*, 1041–1064. [CrossRef]
2. Chiu, T.K.; Li, Y. How Can Emerging Technologies Impact STEM Education? *J. STEM Educ. Res.* **2023**, *6*, 375–384. [CrossRef]
3. Ravi, P.; Broski, A.; Stump, G.; Abelson, H.; Klopfer, E.; Breazeal, C. Understanding Teacher Perspectives and Experiences after Deployment of AI Literacy Curriculum in Middle-school Classrooms. *arXiv* **2023**, arXiv:2312.04839.
4. Loong, E.Y.; Herbert, S. Primary school teachers' use of digital technology in mathematics: The complexities. *Math. Educ. Res. J.* **2018**, *30*, 475–498. [CrossRef]
5. Schorcht, S.; Buchholtz, N.; Baumanns, L. Prompt the problem–investigating the mathematics educational quality of AI-supported problem solving by comparing prompt techniques. In *Frontiers in Education*; Frontiers Media SA: Lausanne, Switzerland, 2024; Volume 9, p. 1386075.
6. Arora, S.; Narayan, A.; Chen, M.F.; Orr, L.; Guha, N.; Bhatia, K.; Chami, I.; Re, C. Ask Me Anything: A Simple Strategy for Prompting Language Models. In Proceedings of the Eleventh International Conference on Learning Representations 2023, Kigali, Rwanda, 1–5 May 2023. Available online: https://openreview.net/pdf?id=bhUPJnS2g0X (accessed on 6 July 2024).
7. Williams, A.; Katz, L. The use of focus group methodology in education: Some theoretical and practical considerations. *IEJLL Int. Electron. J. Leadersh. Learn.* **2001**, *5*.
8. Krueger, R.; Casey, M. *Focus Groups: A Practical Guide for Applied Research*, 3rd ed.; Sage: Newbury Park, CA, USA, 2000.
9. Glaser, B.G.; Strauss, A.L. *The Discovery of Grounded Theory: Strategies for Qualitative Research*; Weidenfeld and Nicolson: London, UK, 1968.
10. Gibson, B.; Hartman, J. *Rediscovering Grounded Theory*; SAGE Publications Ltd.: Newbury Park, CA, USA, 2014.

engineering proceedings

MDPI

Proceeding Paper

Application of Artificial Intelligence to IELTS Learning †

Yu Hou *, Mohamad Ibrani Shahrimin Adam Assim and Shairil Izwan Taasim

Department of Social Science and Management, Universiti Putra Malaysia Bintulu Campus, Bintulu 97008, Malaysia; ibrani@upm.edu.my (M.I.S.A.A.); shairil@upm.edu.my (S.I.T.)
* Correspondence: beckyhouyu@gmail.com
† Presented at the 2024 IEEE 7th International Conference on Knowledge Innovation and Invention, Nagoya, Japan, 16–18 August 2024.

Abstract: In recent years, artificial intelligence (AI) has emerged as a transformative force in education, notably in language learning. We explored AI's integration into International English Language Testing System (IELTS) learning environments, focusing on psychological aspects that impact learning outcomes. AI-driven tools, mobile applications, and personalized platforms used in IELTS preparation were reviewed and discussed in terms of AI and educational psychology. Emphasis is placed on the benefits and challenges of AI in motivating learners, enhancing autonomy, providing feedback, and ultimately improving proficiency. Insights into how AI can be refined to address psychological and cultural challenges are also provided.

Keywords: AI; IELTS learning; psychological and cultural challenges

1. Introduction

AI technology has rapidly advanced and become instrumental in transportation, healthcare, and education. In language learning, AI-driven applications and systems have significantly transformed traditional methods, offering personalized, data-driven experiences that cater to individual learners' needs. In testing such as IELTS, AI is utilized to support preparation for listening, speaking, reading, and writing, bringing psychological and educational benefits to learners [1,2]. Considering IELTS's role in global academic and professional contexts, understanding the influence of AI on learning motivation, autonomy, and performance is crucial.

2. AI in Language Learning and IELTS Preparation

We examined the role of AI in IELTS preparation from an educational psychology perspective in this study. By exploring the psychological processes that AI technology influences, the results of this study provide a foundation for understanding the advantages and limitations of AI in language learning, ultimately contributing to improved educational practices and learner outcomes.

2.1. Introduction

AI applications in English education trace back to the early days of Computer-Assisted Language Learning (CALL), with its foundations set in the 1960s. The maturation of technologies of big data, cloud computing, and machine learning has enabled the development of AI-based platforms for personalized content generation, feedback, and virtual learning environments [2]. At present, AI provides interactive and engaging experiences and offers personalized learning trajectories that meet diverse needs and enhance engagement [3]. As shown in Figure 1, focusing on learner motivation, autonomy, feedback mechanisms, and skill development, we analyzed how AI can facilitate English language acquisition and test readiness.

Figure 1. The psychological process of language learning.

2.2. Personalized Learning and Motivation

One of the primary benefits of AI in IELTS learning is its ability to provide a personalized experience. AI-driven platforms analyze learners' data, customizing course content and difficulty levels to align with their strengths and weaknesses [4]. This personalization fosters a sense of engagement and motivation that is difficult to achieve in traditional classrooms. Students using AI-based tools for IELTS preparation demonstrate higher motivation levels and a sustained interest in improving their skills compared with those who rely solely on conventional methods [5]. Key aspects such as personalization, data analysis, engagement, motivation, and skill improvement are shown in Figure 2.

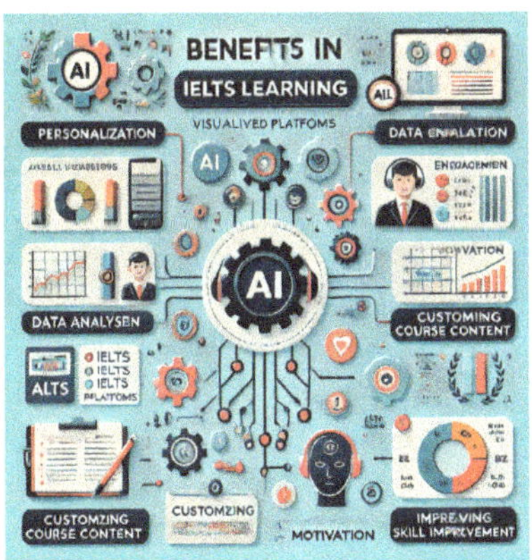

Figure 2. Summarizing the benefits of AI in IELTS learning.

AI offers multiple benefits in IELTS learning, including personalization, data analysis, interactive feedback, and skill development. By enhancing motivation and engagement, AI effectively supports self-directed learning and boosts exam preparation efficiency (Table 1).

Table 1. Benefits of AI in IELTS learning.

Feature	Description
Free Exploration	AI platforms enable learners to explore content at their own pace, choosing topics and areas of focus according to their interests. This approach promotes engagement, curiosity, and allows learners to connect with the material in a more meaningful way, fostering a deeper understanding and increasing motivation.
Gesture Interaction	Through AI and gesture-recognition technology, learners interact with virtual study environments by moving, selecting, or manipulating objects. This technology supports more immersive learning experiences, allowing students to interact with study materials in innovative ways that enhance understanding and engagement.
Physical Interaction	AI tools offer sensor-equipped interfaces that allow learners to engage in a physical manner, such as typing, speaking, or using digital pens. This interactive approach aids in kinesthetic learning, helping students to better absorb information by involving physical actions, thus enhancing retention and comprehension.
Multi-Person Collaboration	AI-facilitated virtual environments support collaborative learning, allowing multiple users to work together on shared tasks or exchange ideas. This setup encourages teamwork and peer learning, providing students with the social interaction that reinforces motivation and helps build communication skills necessary for collaborative work.
Visual Feedback	AI systems provide immediate visual feedback through animations, charts, and highlights, helping students understand their performance in real-time. This visual reinforcement is especially useful in language learning, where it helps learners identify areas for improvement and encourages iterative progress without needing to wait for teacher feedback.
Supporting Autonomous Learning	AI platforms empower learners with autonomy, allowing them to manage study schedules, select content, and focus on specific weaknesses. Tools like vocabulary apps adapt to individual performance, helping students take control of their learning journey, aligning with educational psychology principles that value self-motivation and independence.

2.3. Supporting Autonomous Learning

AI platforms offer learners autonomy by allowing them to manage their study schedules, select content, and focus on their specific weaknesses [6]. Autonomy is critical in tests such as IELTS, where self-directed study is essential for achieving desired outcomes. For example, AI applications such as mobile vocabulary tools track progress and adapt content to individual performance, encouraging students to take control of their learning journey. This aligns with educational psychology's emphasis on fostering learner independence and self-motivation in skill acquisition [7].

2.4. AI-Driven Feedback Mechanisms

AI plays a transformative role in providing timely feedback, particularly in writing and speaking components. Automated feedback systems in AI platforms assess grammar, coherence, and logical structure, enabling students to refine their skills with actionable insights, which can be demonstrated in Figure 3. While human teachers provide valuable contextual feedback, AI delivers immediate responses, allowing students to iteratively improve their performance [8]. However, several students report challenges in interpreting AI-generated feedback without human clarification, highlighting a need for more user-friendly and intuitive feedback designs.

Figure 3. Transformative role of AI in providing timely feedback for IELTS preparation.

3. Educational Psychology in AI-Enhanced IELTS Preparation

In IELTS preparation, maintaining motivation is a significant challenge. AI tools designed with gamified elements, goal tracking, and rewards enhance student motivation, creating a more engaging and less stressful learning environment. Educational psychology emphasizes that motivated students are more likely to succeed, and AI platforms leverage this by tracking progress and providing milestone achievements, which reinforces learners' commitment to achieving higher scores [9].

3.1. Autonomy and Learner Control

AI tools encourage students to exercise autonomy over their studies by allowing them to choose specific learning paths and materials. In IELTS, this autonomy helps students focus on weak areas without relying entirely on a teacher's direction. By fostering learner control, AI contributes to students' self-efficacy, a psychological concept that increases motivation and reduces test anxiety. Educational psychology suggests that autonomy can lead to a stronger sense of ownership and responsibility for learning outcomes, which is beneficial for IELTS preparation, as illustrated in Figures 4 and 5.

Figure 4. Student personalized learning paths.

Figure 5. Focus on building self-efficacy.

3.2. Psychological Impact of AI-Generated Feedback

AI feedback tools provide immediate corrective guidance in writing and speaking, helping students identify and address errors [5]. Educational psychology recognizes the role of timely feedback in promoting learning, yet learners can find AI feedback lacking the nuance or empathy of human feedback. This leads to frustration, especially for students unfamiliar with interpreting AI-generated suggestions. To improve this, AI systems integrate explanatory components or guided interpretations to better align with learners' psychological needs. The benefits and suggestions are described in Table 2.

Table 2. AI feedback in IELTS preparation.

Feature	Benefits	Suggestions for Improvement
Immediate Feedback	Enables quick correction and iterative learning.	Add context-based suggestions to enhance feedback depth.
Real-Time Responses	Promotes immediate learning and positive reinforcement.	Offer pacing options for feedback to suit student preferences.
Error Correction	Focuses on specific errors for targeted improvement.	Provide detailed explanations with examples for better understanding.
Empathy in Feedback	Encourages motivation and resilience.	Simulate empathetic responses to support student efforts.
Feedback Interpretation	Builds self-reliance in advanced learners.	Add guided interpretations for clearer understanding.
Skill Development	Supports frequent practice and self-assessment.	Balance with occasional human feedback for comprehensive learning.
Psychological Support	Enhances motivation and reduces test anxiety.	Include motivational components and adaptive learning paths.

4. Challenges in AI-Enhanced Language Learning

4.1. Reduced Human Interaction and Emotional Support

One of the limitations of AI in language learning is the absence of human interaction, which is essential for emotional support. Traditional teaching provides encouragement, empathy, and elements that foster a supportive learning environment. AI lacks this relational aspect, and while it can simulate interactions, it cannot replace the nuanced, motivational influence that teachers provide. Educational psychology indicates that emotional support is crucial in building confidence and reducing learning anxiety, especially for students facing rigorous tests, including IELTS [9].

4.2. Cultural Relevance in AI Learning

In using AI platforms, cultural diversity must be considered to cater to learners from different backgrounds. Since IELTS is taken by students worldwide, AI applications must be sensitive to varying learning styles and preferences. Cultural inclusivity enhances learning effectiveness by ensuring content is relatable and accessible to a broader audience. Continuous refinement of AI platforms to reflect cultural nuances is essential for maximizing AI's educational potential [3].

4.3. Over-Reliance on AI and Reduced Interpersonal Skills

While AI fosters independent learning, excessive reliance on it impairs interpersonal skills. IELTS preparation traditionally involves group studies or peer interactions, promoting social learning and communication skills. Overuse of AI can lead to isolation, depriving learners of collaborative opportunities crucial for comprehensive language acquisition. Educators need to balance AI applications with social learning activities to support the development of interpersonal skills.

5. Opportunities for Future Development in AI-Enhanced IELTS Preparation

5.1. Enhancing Feedback

To address students' challenges in understanding AI feedback, future AI systems need to incorporate more interactive feedback that explains errors with examples or suggests targeted exercises. Such enhancements align with educational psychology principles, making feedback more accessible and meaningful for learners [8].

5.2. Integration of Emotional AI

Advancements in emotional AI hold the potential to create a more empathetic learning environment. Incorporating emotional recognition features into AI platforms helps gauge learners' engagement and adjust content accordingly, providing a personalized and emotionally supportive experience [9].

5.3. Cultural Adaptation in Content and Interaction

AI needs to be refined to recognize and adapt to cultural differences in learning preferences. Through better data collection and adaptive algorithms, AI systems offer culturally relevant examples, dialogs, and scenarios, making learning inclusive and effective for a global audience.

6. Conclusions

AI offers transformative possibilities in IELTS preparation and English language learning, enhancing motivation, autonomy, and personalized feedback. From an educational psychology perspective, AI-driven tools address critical aspects of learner engagement and performance. However, challenges including limited emotional support, cultural sensitivity, and potential over-reliance on technology remain. Addressing these challenges through further research and development will allow AI to become a more robust and comprehensive tool in language education than at present. By balancing technological benefits with psychological insights, AI can contribute to a more effective and inclusive future for English language learning.

Author Contributions: Conceptualization, Y.H. and M.I.S.A.A.; methodology, Y.H.; software, Y.H.; validation, Y.H., M.I.S.A.A. and S.I.T.; formal analysis, Y.H.; investigation, Y.H.; resources, Y.H.; data curation, Y.H.; writing—original draft preparation, Y.H.; writing—review and editing, Y.H.; visualization, Y.H.; supervision, Y.H.; project administration, Y.H.; funding acquisition, M.I.S.A.A. All authors have read and agreed to the published version of the manuscript.

Funding: This research received no external funding.

Institutional Review Board Statement: Not applicable.

Informed Consent Statement: Informed consent was obtained from all subjects involved in the study.

Data Availability Statement: Data sharing is not applicable to this article.

Conflicts of Interest: The authors declare no conflicts of interest.

References

1. Żammit, J. Exploring the effectiveness of VR in teaching Maltese. *Comput. Educ. X Real.* **2023**, *3*, 100035.
2. Pearson, W.S. Student engagement with teacher written feedback on rehearsal essays undertaken in preparation for IELTS. *SAGE Open* **2022**, *12*, 3–8. [CrossRef]
3. Shemshack, A.; Spector, J.M. A systematic literature review of personalized learning terms. *Smart Learn. Environ.* **2020**, *7*, 33. [CrossRef]
4. Farsani, D. The name game: The role of cultural affordances in learning probability. *Learn. Math.* **2022**, *42*, 17–18.
5. Dai, W.; Lin, J.; Jin, H. Can Large Language Models Provide Feedback to Students? A Case Study on ChatGPT. In Proceedings of the 2023 IEEE International Conference on Advanced Learning Technologies (ICALT), Orem, UT, USA, 10–13 July 2023.
6. Koucheki, R.; Lex, J.R.; Morozova, A.; Ferri, D.; Hauer, T.M.; Mirzaie, S.; Ferguson, P.C.; Ballyk, B. Immersive VR and Cadaveric Bone are Equally Effective in Skeletal Anatomy Education: A Randomized Crossover Noninferiority Trial. *J. Surg. Educ.* **2023**, *80*, 1028–1038. [CrossRef]
7. Han, Y.; Xu, Y. Student feedback literacy and engagement with feedback: A case study of Chinese undergraduate students. *Teach. High. Educ.* **2021**, *26*, 181–196. [CrossRef]
8. Hu, R.; Trenkic, D. The effects of coaching and repeated test-taking on Chinese candidates' IELTS scores, their English profi. ciency, and subsequent academic achievement. *Int. J. Biling. Educ. Biling.* **2019**, *24*, 1486–1501. [CrossRef]
9. Chong, S.W. A research report: Theorizing ESL community college students' perception of written feedback. *Community Coll. J. Res. Pract.* **2020**, *44*, 463–467. [CrossRef]

engineering proceedings

MDPI

Proceeding Paper

Image Classification Models as a Balancer Between Product Typicality and Novelty †

Hung-Hsiang Wang and Hsueh-Kuan Chen *

Department of Industrial Design, National Taipei University of Technology, Taipei 10608, Taiwan; wanghh@mail.ntut.edu.tw
* Correspondence: t112588026@ntut.org.tw

† Presented at the 2024 IEEE 7th International Conference on Knowledge Innovation and Invention, Nagoya, Japan, 16–18 August 2024.

Abstract: Car styling is crucial for consumer acceptance and market success. Since vehicle manufacturers produce electric vehicles, they have faced the challenge of maintaining the typicality of their original products and presenting the innovation of new technologies. We propose a method that integrates artificial intelligence (AI)-generated images and image classification technology to help designers effectively balance between typicality and novelty. We collected 118 pictures of electric vehicles and 122 pictures of fuel vehicles in 2024 from the BMW official website. Focusing on seven key visual features of the vehicles, we used the Waikato environment for knowledge analysis (WEKA) to train an image classification model on the dataset through three separate training and testing sessions. First, we used the prompts that described typical BMW design to generate images of new BMW electric vehicles in Stable Diffusion. The images consisted of 21 front views, 20 side views, and 20 rear views. The accuracy of the model of front views trained with the pyramid histogram of oriented gradients filter (PHOG)-Filter and random forest classifier was 78.5%, and the test accuracy reached 95%. The accuracy of the model of rear views trained with BinaryPatternsPyramid-Filter and random forest classifier was 80.5%, and the test accuracy was 90%. However, the accuracy of the model of side views did not reach 70%. That implies the distinction between BMW fuel vehicles and its electric vehicles is mainly based on the front and rear views, rather than the side view. The results of this study showed that integrating image classification and AI-generated images can be used to examine the balance between product typicality and novelty, and the application of machine learning and AI tools to study car style.

Keywords: AI-generated image; image classification; car styling; electric vehicle; product typicality

1. Introduction

Image recognition technology has become mature and is widely used in healthcare [1], transportation [2], and agriculture [3]. This technology has demonstrated its maturity and effectiveness. However, its practical application in product design is less common. With the rapid development of artificial intelligence (AI) in recent years, many related tools have emerged. Stable diffusion has become a tool for generating digital art and modifying images [4]. These advancements have brought fundamental changes to the operational models of the design industry. Many designers with limited programming skills use AI-generated images for the initial stages of design ideation to produce a large number of creative concepts. This approach enhances creative efficiency and promotes diversity in design thinking. To verify whether AI-generated images reflect the typical characteristics of a brand, we employ machine learning and image classification techniques to provide a solid analytical foundation for innovative design. Through this method, designers can find an appropriate balance between innovation and brand typicity.

2. Related Work

2.1. Waikato Environment for Knowledge Analysis (WEKA)

Deep learning has become ubiquitous using tools such as TensorFlow and PyTorch 1.8.1. However, these methods require a foundational knowledge of programming. Industrial designers or automotive designers can have limited programming skills, and a simple machine learning tool, WEKA, is an alternative for image classification. WEKA is used without programming, offering significant ease of use for conducting machine learning (ML) and deep learning (DL) image analysis. Convolutional neural networks (CNNs) are an important application in deep learning (DL). CNNs consist of convolutional layers, pooling layers, and fully connected layers, which segment feature images, extract features, and perform recognition [5]. Image recognition in machine learning is common, and WEKA image recognition has been applied in analyzing car shapes and distinguishing between fuel vehicles and electric vehicles [6]. Therefore, this method was used in this study to integrate machine learning into product design.

2.2. AI in Design

With significant advancements in AI in recent years, AI technology has become a necessary trend. Sbai et al. emphasized the potential of generative networks in providing design inspiration for human designers [7]. AI image generation is categorized into text-to-image generation and image-to-image generation. Automotive designers need to create ideas before developing innovative prototypes. However, this stage is time-consuming. The inclusion of AI breaks through creative bottlenecks and significantly reduces time and costs [8]. In education, AI image generation has shown significant benefits. AI-generated images effectively present the context of the words, thereby enhancing learning efficiency [9].

2.3. Analysis of Automotive Typicality

There are four design elements of a car to help identify its brand. These are (1) front badge, (2) headlights, (3) taillights, and (4) rear bumper [10] (Figure 1). In initial sketching, the side view of the vehicle serves as the defining standard for the vehicle's stance. The side view is divided into six features: body, bodyside, cabin (greenhouse), daylight opening (DLO), wheel, and tire [11] (Figure 2).

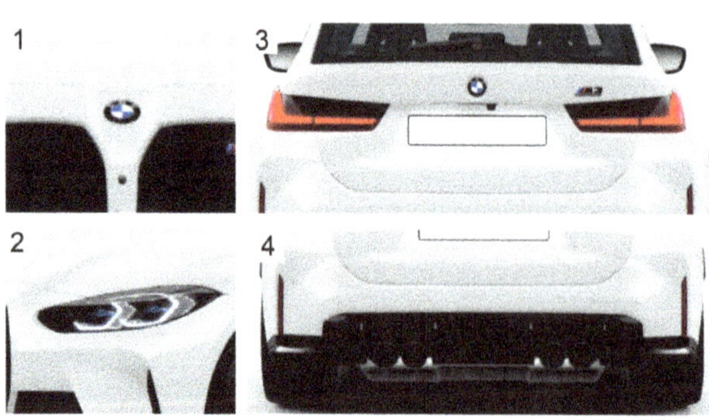

Figure 1. Headlights, taillights, and rear bumper.

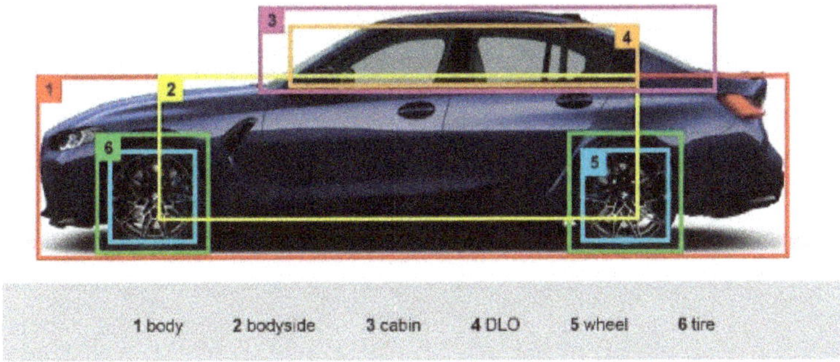

Figure 2. Image source: [11] Redrawn by the author.

Each brand has a unique design DNA. For example, Lexus has its spindle grille, BMW features a kidney grille, and Lamborghini's exterior is muscular with Y-shaped and hexagonal accents. These five design elements are the key to helping us recognize the vehicles.

3. Methods

We used open-source software throughout, including WEKA for machine learning and Stable Diffusion for AI image generation. We used DL methods through ML for image classification. The study is divided into the following three stages:

- In the first stage, WEKA is chosen as the machine learning tool. Using its image classification feature, we classify images and train models, adjusting parameters to obtain the most suitable and effective model samples.
- In the second stage, Stable Diffusion is used to generate images that will be used as the test dataset for the next stage.
- The third stage is the validation phase, where the accuracy of the AI-generated images is assessed by testing the classification model trained with AI-generated images.

3.1. Vehicle Features and Brand Style

Based on the application of typicality in predicting product appearance using WEKA [12], we collected images of electric and fuel vehicles from BMW's official website in 2024. The collected images include visual elements, such as the full vehicle, body, bodyside, headlights, taillights, front bumper, and rear bumper.

These images were divided into the aforementioned individual components, with each part undergoing subsequent image processing to facilitate model training to identify the design features and key elements. The designs of electric vehicles and fuel vehicles are compared to provide data to support visual differences. Backgrounds were removed to ensure all images were clean and focused on the vehicle. Using the post-processed image data, we trained models with machine learning techniques to automatically recognize and classify the design elements of both types of vehicles and understand the relationship between brand style and vehicle type characteristics. Through training and validation based on a deep learning (DL) framework, we achieved optimal accuracy in distinguishing the vehicles.

3.2. AI-Generated Images

AI tools present a new creative approach for designers with limited programming skills. According to Liu et al.'s tests on Stable Diffusion in industrial design, product styles were successfully transformed, demonstrating that AI has a high capability to judge color, material, and style [4]. Thus, we used Stable Diffusion to generate images of new electric vehicles, utilizing the text-to-image generation feature (Figure 3). This process involves both positive and negative prompts.

Figure 3. AI text-to-image generation.

For the vehicle body, the positive prompts include "clean exterior, BMW style, electric vehicle, future electric concept vehicle, concept, creative, car design, smooth and soft lines". Additional prompts were added based on vehicle type ("sedan, hatchback, wagon, SUV") and view angle ("front view, side view, rear view"). The negative prompts for the vehicle body include "old car design, old design language, boxy cars, air intake, radiator grille, exhaust". The generated images underwent image processing to be used for evaluation in the third stage.

3.3. Evaluation of Generated Image

As the application of AI-generated images in design continues to increase, ensuring the quality and standard of these images becomes critically important. Before evaluating the electric vehicle images generated by Stable Diffusion, the images were processed and converted to be consistent with the training model. Subsequently, the preprocessed images were fed into the previously trained deep learning model to test accuracy, using CNN to identify and classify vehicle images. By observing whether the model classified the Stable Diffusion-generated images as electric vehicles, we assessed the usefulness of AI-generated images for designers.

4. Results and Discussions

4.1. Usability of Machine Learning

We divided the test dataset into three parts: front three-quarter view and front detail images, side view and side detail images, and rear three-quarter view and rear detail images. The images were converted in grayscale (Figure 4). For training the model, we used Python 3 to apply a fixed amount of stretching deformation to some images, using deformed images as training dataset materials. The dataset was split into 80% of the collected images for training and 20% for testing.

Figure 4. Grayscale images from different perspectives.

In the selection of classifiers, we used sequential minimal optimization (SMO) and Random Forest. SMO is appropriate for high-dimensional data classification, capable of finding the optimal classification hyperplane to distinguish different categories for classification and regression analysis and efficient handling of large-scale datasets. Random Forest is known for its high robustness and accuracy for classification and regression tasks, effectively dealing with high-dimensional data and complex problems. Both are effective in image classification. In the feature extraction phase, we used various texture image filters in WEKA, including the pyramid histogram of oriented gradients filter (PHOG)-Filter, the BinaryPatternsPyramid-Filter, the EdgeHistogram-Filter, the Gabor-Filter, and the fast and compact transform-histogram filter (FCTH)-Filter.

By pairing classifiers with filters, all training sets were used through 10-fold cross-validation. The results of image training and testing showed that the random forest classifier performed the best for the classification of front three-quarter view and front detail images. When using the EdgeHistogram-Filter, the training dataset achieved an accuracy of 72.6%, and the test dataset achieved an accuracy of 90.4%. When using the PHOG-Filter, the training dataset achieved an accuracy of 78.5%, and the test dataset accuracy was 95% (Table 1). For the classification of rear three-quarter view and rear detail images, the combination of the random forest classifier and the BinaryPatternsPyramid-Filter achieved an accuracy of 80.5% in the training dataset and 90% in the test dataset. The combination of the sequential minimal optimization (SMO) classifier and the BinaryPatternsPyramid-Filter achieved an accuracy of 70% in the training dataset and 100% in the test dataset (Table 2). In the classification of side view and detail images, both the Random Forest and SMO classifiers were used with an accuracy of lower than 70%.

Table 1. Model for front three-quarter view and front detail images.

Classifier Name	Filter Name	Filter Function	Training Model Accuracy	Test Accuracy
Random Forest	EdgeHistogram-Filter	Quantifying edge information into histograms based on MPEG-7	72.6%	90.4%
Random Forest	PHOG-Filter	Extracting histograms of oriented gradients from images	78.5%	95%

Table 2. Model for rear three-quarter view and rear detail images.

Classifier Name	Filter Name	Filter Function	Training Model Accuracy	Test Accuracy
Random Forest	BinaryPatterns Pyramid-Filter	Extracting multi-scale local binary pattern (LBP) features	80.5%	90%
SMO	BinaryPatterns Pyramid-Filter	Extracting histograms of oriented gradients from images	70%	100%

When selecting models, the model's generalization ability was considered to ensure that the training dataset and test dataset performance were relatively consistent. This helped guarantee that the model maintains stable classification capabilities when encountering new data. The random forest classifier paired with the PHOG-Filter and BinaryPatternsPyramid-Filter showed better training and testing accuracy across different perspectives, indicating generalization ability. The SMO classifier, with a test dataset accuracy of 100%, performed ideally, but its training dataset accuracy was relatively low due to overfitting. Therefore, it is not recommended to rely solely on such results.

4.2. Distinguishment of Electric and Fuel Vehicles

To understand the differences between these two types of vehicles, we categorized the images into electric vehicles and fuel vehicles and experimented with different feature extraction methods with classifiers. The results indicated that machine learning analyzed AI-generated images effectively. Front and rear view images provide relevant feature information about the vehicle's power system, resulting in better classification. In contrast, side view features are less effective in determining the vehicle's power system. Side view features and shapes are less significant for judging automotive designs.

5. Conclusions

We used the machine learning tool, WEKA, to successfully distinguish AI-generated images of electric vehicles and fuel vehicles using image classification from different perspectives. The random forest classifier performed best in classifying front and rear view images, with an accuracy of exceeding 90%. However, the accuracy for classifying side-view images was relatively low. This indicates that electric vehicles and fuel vehicles have more distinct characteristics in front and rear views. In using WEKA for image classification, various image filters and classifiers were paired using the 10-fold cross-validation method to ensure the accuracy and stability of the model.

The developed method enables designers to quickly and accurately identify and distinguish different types of automotive features using machine learning techniques. This helps companies precisely convey product semantics during the design and development process for innovative or iterative designs. By combining AI image generation, designers can generate a large number of concept images in the early stages of the design process to enhance and innovate design efficiency. Machine learning effectively assesses whether the AI-generated images meet expectations. Future research is necessary to expand the dataset size, explore other machine learning algorithms and image filters, and apply this technology to home design, consumer electronics design, and others to improve the automation of the design process.

Author Contributions: Conceptualization, H.-H.W. and H.-K.C.; methodology, H.-K.C.; software, H.-K.C.; validation, H.-H.W. and H.-K.C.; formal analysis, H.-K.C.; investigation, H.-K.C.; resources, H.-K.C.; data curation, H.-K.C.; writing—original draft preparation, H.-K.C.; writing—review and editing, H.-H.W.; visualization, H.-K.C.; supervision, H.-H.W.; project administration, H.-H.W.; funding acquisition, H.-H.W. All authors have read and agreed to the published version of the manuscript.

Funding: This research was funded by NSTC, Taiwan, through grant 111-2410-H-027-019-MY2.

Institutional Review Board Statement: Not applicable.

Informed Consent Statement: Not applicable.

Data Availability Statement: Data are unavailable due to privacy reasons.

Acknowledgments: The authors gratefully acknowledge the financial support of NSTC, Taiwan, through grant 111-2410-H-027-019-MY2.

Conflicts of Interest: The authors declare no conflicts of interest.

References

1. Shen, D.; Wu, G.; Suk, H.-I. Deep learning in medical image analysis. *Annu. Rev. Biomed. Eng.* **2017**, *19*, 221–248. [CrossRef]
2. Hu, W.; Wang, W.; Ai, C.; Wang, J.; Wang, W.; Meng, X.; Liu, J.; Tao, H.; Qiu, S. Machine vision-based surface crack analysis for transportation infrastructure. *Autom. Constr.* **2021**, *132*, 103973. [CrossRef]
3. Chiu, M.T.; Xu, X.; Wei, Y.; Huang, Z.; Schwing, A.G.; Brunner, R.; Khachatrian, H.; Karapetyan, H.; Karapetyan, H.; Dozier, I.; et al. Agriculture-vision: A large aerial image database for agricultural pattern analysis. In Proceedings of the IEEE/CVF Conference on Computer Vision and Pattern Recognition, Seattle, WA, USA, 13–19 June 2020; pp. 2828–2838.
4. Liu, M.; Hu, Y. Application potential of stable diffusion in different stages of industrial design. In Proceedings of the International Conference on Human-Computer Interaction, Copenhagen, Denmark, 23–28 July 2023; Springer: Berlin/Heidelberg, Germany, 2023; pp. 590–609.
5. Yang, X.; Ye, Y.; Li, X.; Lau, R.Y.; Zhang, X.; Huang, X. Hyperspectral image classification with deep learning models. *IEEE Trans. Geosci. Remote Sens.* **2018**, *56*, 5408–5423. [CrossRef]
6. Wang, H.-H.; Shen, Y.-Y.; Hung, Y.-Y. Car Types and Semantics Classification Using Weka. In Proceedings of the 2022 5th Artificial Intelligence and Cloud Computing Conference, Osaka, Japan, 17–19 December 2022; pp. 34–41.
7. Sbai, O.; Elhoseiny, M.; Bordes, A.; LeCun, Y.; Couprie, C. Design: Design inspiration from generative networks. In Proceedings of the European Conference on Computer Vision (ECCV) Workshops, Munich, Germany, 8–14 September 2018.
8. Radhakrishnan, S.; Bharadwaj, V.; Manjunath, V.; Srinath, R. Creative intelligence-automating car design studio with generative adversarial networks (GAN). In Proceedings of the International Cross-Domain Conference for Machine Learning and Knowledge Extraction, Hamburg, Germany, 27–30 August 2018; Springer: Berlin/Heidelberg, Germany, 2018; pp. 160–175.
9. Aktay, S. The usability of images generated by artificial intelligence (AI) in education. *Int. Technol. Educ. J.* **2022**, *6*, 51–62.
10. Liem, A.; Abidin, S.; Warell, A. Designers' perceptions of typical characteristics of form treatment in automobile styling. In Proceedings of the 5th International Workshop on Design & Semantics of Form and Movement, DesForm, Taipei, Taiwan, 26–27 October 2009; pp. 144–155.
11. Lee, G.; Kim, T.; Suk, H.-J. GP22: A Car Styling Dataset for Automotive Designers. In Proceedings of the IEEE/CVF Conference on Computer Vision and Pattern Recognition, New Orleans, LA, USA, 19–20 June 2022; pp. 2268–2272.
12. Wang, H.-H.; Lin, Y.-Y.; Huang, H.-T. Application of Typicality in Predicting Product Appearance. *Eng. Proc.* **2023**, *55*, 66. [CrossRef]

engineering proceedings

MDPI

Proceeding Paper

Research on Optimal Fabrication of High-Quality Ceramic by Process Analysis Technology †

Jianhua Cheng * and Minhui Tong

School of Management and Economics, Jingdezhen Ceramic University, Jingdezhen 333000, China; tongminhui@jcu.edu.cn
* Correspondence: 241005@jcu.edu.cn

† Presented at the 2024 IEEE 7th International Conference on Knowledge Innovation and Invention, Nagoya, Japan, 16–18 August 2024.

Abstract: We developed a composite recognition method combining hybrid sensor recognition and artificial intelligence image recognition techniques and optimized quality monitoring of the ceramic production process. By integrating deep learning models and image preprocessing techniques, appearance defects (such as cracks and color differences) were detected in ceramic products, and key parameters in the sintering process using temperature, pressure, and gas-sensitive sensors were monitored in real-time. The developed composite recognition method significantly improved the quality control level of ceramic production, reduced production energy consumption and wastes, and improved production efficiency.

Keywords: laser-induced breakdown spectroscopy (LIBS); artificial intelligence recognition; ceramic production; quality optimization

1. Introduction

In ceramic production, traditional quality monitoring methods are inefficient and inaccurate. To solve this problem, we introduced laser-induced breakdown spectroscopy (LIBS) technology and combined it with artificial intelligence (AI) image recognition technology [1]. A deep learning model was constructed to detect the appearance defects of ceramic products in real-time and improve the detection accuracy. Temperature, pressure, and gas-sensitive sensors were introduced to monitor key parameters in the sintering process in real-time. Based on the data, a new and efficient quality monitoring solution was developed for ceramic production [2].

2. Methods

2.1. Process Analytical Techniques

Process Analytical Technology (PAT) is an advanced manufacturing technology used in the pharmaceutical, chemical, and food industries. PAT is issued for designing, analyzing, and controlling the production by real-time monitoring of the quality and performance of raw materials, intermediates, and processes [1]. The technology ensures the high quality of the final product while increasing productivity and reducing costs. The PAT technology reduces the uncertainty in the production process and improves the controllability of the process, leading to a stable and predictable production process. It consists of online, at-line, and in-line analytical methods that monitor Critical Quality Attributes (CQAs) and Critical Process Parameters (CPPs) in the production process.

2.2. LIBS

LIBS is an atomic spectroscopy technique for elemental analysis using high-power laser pulses focused on the surface of a sample, ablating the sample, and forming a short-lived, high-temperature plasma [2]. The LIBS technique allows for the rapid and portable measurement of major and trace element concentrations in solid, liquid, or gaseous samples to record a material's chemical signature (fingerprinting). Each LIBS spectrum contains information about the concentrations of all naturally occurring elements, isotopic ratios, and information on the atomic structure of the material. LIBS ablates atoms from the surface of the sample to form a high-temperature plasma using high-power laser pulses as the energy source. The plasma temperature is high above 10,000 K and has enough energy to excite electrons in outer orbitals. As the plasma cools, the excited electrons decay to lower energy orbitals, emitting photons that are inversely proportional to the energy difference between the excited and base orbitals. Each element has multiple emission wavelengths due to the existence of multiple possible excited states.

2.3. Partial Least Squares Regression

Partial Least Squares Regression (PLS) is used to model the relationship between multiple independent and dependent variables, especially in the presence of multicollinearity. PLS combines the ideas of Principal Component Analysis (PCA) and Typical Correlation Analysis (CCA) by extracting independent and dependent variables to establish a linear relationship between them. These main components are called PLS components or factors, which present linear combinations of the original variables and maximum covariance with the dependent variable. The goal of PLS is to find a new set of variables (PLS components) that are linear to the original independent variables and have maximum correlation with the dependent variable.

To extract PLS components, the first set of PLS components, which reflect both the trend of the independent variable and the trend of the dependent variable, are extracted by an iterative algorithm (e.g., the NIPALS algorithm). Specifically, PLS looks for linear combinations of X and Y that maximize their covariance (1).

$$\mathrm{maxCov}(\mathrm{Xp},\mathrm{Yq}) = \sqrt{\mathrm{Var}(\mathrm{Xp})\mathrm{Var}(\mathrm{Yq})}\mathrm{R}(\mathrm{Xp},\mathrm{Yq}) \quad (1)$$

3. Model Construction

3.1. Regression Modeling

In the analysis of the chemical composition of ceramic materials based on LIBS, PLS is an effective statistical method for establishing a linear relationship between the LIBS spectral data (independent variable) and the chemical composition of the ceramic material (dependent variable). PLS is used to construct a prediction model by extracting the common variance information between the two sets of data, thus realizing the accurate prediction of chemical composition [3,4]. In the basic form of a PLS model, the chemical composition is predicted by establishing a linear relationship between the independent variable (LIBS spectral data, X) and the dependent variable (chemical composition of the ceramic material, Y). The model is expressed as (2).

$$\mathrm{Y} = \mathrm{XB} + \mathrm{E} \quad (2)$$

where Y represents the dependent variable, i.e., the chemical composition of the ceramic material, X represents the independent variable, i.e., LIBS spectral data, B represents the matrix of regression coefficients for the PLS model, and E represents the error term, indicating the portion of variation in Y that the model fails to explain.

The core of the PLS model is to find the optimal projection between X and Y. In this study, we used the score vector (T) and the load vector (P) to establish a linear relationship between them by using (3).

$$\begin{aligned} X &\approx TPT^T + E \\ Y &\approx TTC + F \end{aligned} \tag{3}$$

where T is the score matrix that contains a projection of the X and Y co-variance information, P is the loading matrix representing the coefficients of X in the score space, r is the number of components selected, i.e., PLS components retained, t_i and p_i are the ith score vector and load vector, respectively, E is the residual matrix of X, indicating the fraction of the score space that is not explained by the model, and C and F are the coefficients of the score matrix and load matrix of Y, respectively, which are used to describe the structure of Y in the score space.

In this way, the PLS model effectively extracts key information related to chemical composition from high-dimensional LIBS spectral data, and then accurately predicts the chemical composition of ceramic materials [5]. Specifically, the modeling is screened by a computer (Figure 1).

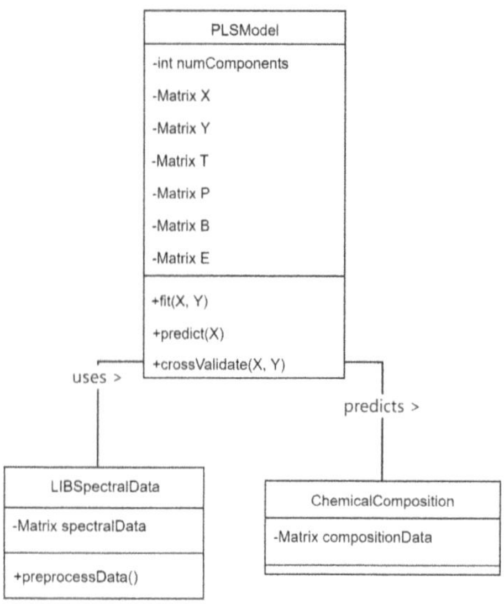

Figure 1. Computer modeling class diagram for constructing regression models underlying process analysis in ceramic production.

Within the framework of PLS, the selection of an appropriate number of components (r) is crucial for the construction of the model, which usually relies on the cross-validation technique. The number is used to evaluate the predictive efficacy of the model and effectively avoid the overfitting phenomenon to ensure the accuracy of the model prediction. Optimization and validation of the model are important for guaranteeing the predictive accuracy and reliability of the PLS model [6]. The process of model optimization and validation is of great significance in ensuring the accuracy and reliability of PLS models [6].

3.2. Data Fusion and PAT in Ceramics Production

The application of PAT is crucial in the study of optimization of high-quality ceramic production, and multiple data source correction plays an important role. In this study, data from different sensors, such as thermometers, manometers, flow meters, and spectral analyzers, were integrated to improve the accuracy and consistency of the data. The purpose of the integration of multiple data sources is to ensure that data collected from different sensors provide comprehensive and accurate information about the production process. This involves data preprocessing, feature extraction, data fusion, and result validation. In preprocessing, the data need to be filtered, denoised, and normalized [7]. Feature extraction, on the other hand, extracts key information from the raw data that has an impact on the production process. Data fusion is the process of integrating data from different sensors to form a comprehensive and consistent dataset, and result validation is an important step to ensure that the corrected data meet the production requirements.

In ceramic production, dry-press molding technology significantly improves the production efficiency of silicon carbide ceramic micro-airborne disks while maintaining high product consistency. By optimizing the dry-press molding process, the dimensional accuracy of SiC micro-empty discs is significantly improved to meet the application requirements of precision devices. The sintered SiC micro-empty discs show excellent mechanical properties, including high flexural strength and wear resistance. The application of PAT is not limited to the analysis of single-sensor data, but through the integration of a wide range of sensor data, such as temperature and pressure, the production process is monitored. These technologies are used to analyze key quality attributes of ceramic materials such as chemical composition, crystal shape, and particle size distribution. By using PCA and PLS, invalid information is effectively filtered out and the problem of covariance is solved for complex analysis [3,7]. The multi-data source correction technique plays a crucial role in PAT-based optimization studies of high-quality ceramic production as it provides an accurate view of the production process by integrating and correcting data from different sensors, thus helping manufacturers optimize the production process, improve product quality, reduce costs, and ultimately achieve automated and intelligent production [8] (Figure 2).

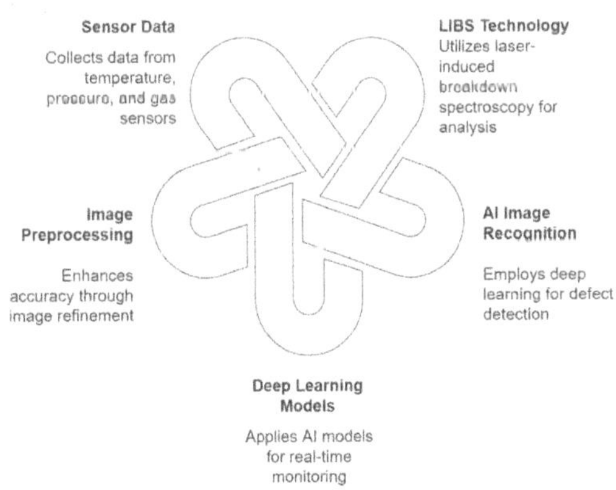

Figure 2. Schematic framework of the application of the data fusion language PAT in ceramic production.

3.3. *Optimization of AI in Process Analysis*

PAT plays a vital role in the optimization of high-quality ceramic production. The core of PAT lies in the real-time monitoring and control of CQAs and CPPs in the production process and ensuring the consistency and quality of the final product. In ceramic production, this involves precise control of all aspects of raw material proportioning, molding, drying, and sintering [9]. CQAs and CPPs are important in ceramic products and real-time monitoring of these parameters allows timely adjustment of the production process to ensure consistency and predictability of product quality. In ceramic production, sintering temperature and time are the key parameters for ceramic strength and density, while the strength of ceramics is a key quality measure. With PAT technology, the temperature and time can be monitored in real-time in the sintering process to ensure that the ceramic products produced meet the required strength criteria [10] (Table 1).

Table 1. Indicators for optimizing CQAs and CPPs for high-quality ceramic production.

CQAs	Key attributes affecting product quality	Compressive strength of ceramics
		Surface finish
		Dimensional accuracy
		Color uniformity
CPPs	Process parameters affecting CQAs	Sintering temperature
		Sintering time
		Raw material mixing uniformity
		Drying rate

AI image recognition technology in the optimization of ceramic production is used to analyze image data for automated quality inspection and defect identification. Through deep learning algorithms such as convolutional neural networks (CNNs), features are automatically extracted from image data to identify defects such as cracks, bubbles, and color differences on the surface of the product. Then, real-time monitoring and quality control of the production process are achieved. LIBS is a fast and non-destructive detection means for the chemical composition analysis of ceramic materials. LIBS technology provides the chemical composition of materials in real-time by analyzing the plasma emission spectra formed on the surface of the samples after they are ablated by high-energy laser pulses. Combined with statistical methods such as Principal Component Analysis (PCA) and Linear Discriminant Analysis (LDA), the ceramic fragments are quickly identified. By integrating data from different sensors, such as LIBS, machine vision, weight monitoring, and acoustic wave detection, the production process is monitored and the accuracy, and reliability of the data analysis are improved. This multi-sensor fusion approach provides information to optimize the production process and improve product quality and productivity.

4. LIBS and PLS

4.1. Simulation

Numerical simulation was conducted to investigate the propagation characteristics of high-power laser pulses in ceramic materials by constructing a mathematical model. We analyzed the kinetic process of the interaction of laser pulses with ceramic materials and the spatial and temporal distributions of the laser energy inside the materials [11]. A series of key laser parameters was defined as follows.

Wavelength ($\lambda = 1064 \times 10^{-9}$ meters)
Transmission distance ($L = 1 \times 10^{-3}$ meters)
Time step ($dt = 1 \times 10^{-14}$ seconds)
Beam radius of the laser beam ($w_0 = 1 \times 10^{-3}$ meters)
Amplitude of the laser pulse ($A = 1$)
pulse width ($\tau = 50 \times dt$)
peak time ($t_0 = 200 \times dt$)

These parameters presented the spatiotemporal characteristics of the laser pulse for subsequent numerical simulations. In this study, a mathematical model based on Gaussian distribution was simulated to describe the spatiotemporal distribution of the laser pulse. The Gaussian distribution of the laser pulse in the time dimension and the attenuation characteristics along the transmission distance were simulated by using this model. A three-dimensional grid was created using the meshgrid function in MATLAB R2021b software, which covers the entire spatial and temporal range of the simulation, and the distribution of the electric field intensity was visualized using three-dimensional mapping techniques [12] (Figure 3).

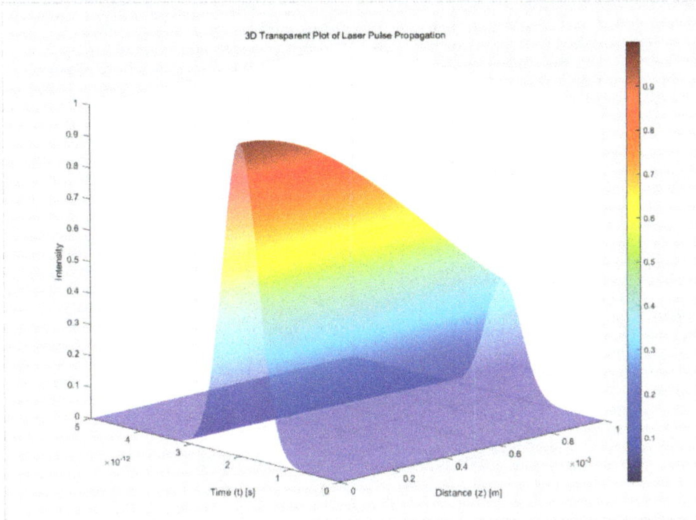

Figure 3. Experimental simulation of propagation characteristics of high-power laser pulses in ceramic materials.

4.2. PLS

PLS analysis was performed to establish a linear relationship between LIBS data and the chemical composition of ceramic materials. In the LIBS analysis, the spectral data (usually denoted as V1, V2, . . ., Vn) represent the emission intensity of a ceramic material at a specific wavelength. There is a complex nonlinear relationship between these spectral data and the chemical composition of the ceramic material. The linear relationship between these spectral data and the chemical composition was established by using a PLS model to understand and predict the chemical properties of the materials. The LIBS spectral data and the chemical composition data were normalized to eliminate the effect of different magnitudes and used to model using the plsr function. The PLS model was established by the leave-one-out (LOO) method. Cross-validation of the different numbers of principal components was carried out to assess their effect on the prediction performance. The results of the experiment are shown in Table 2.

Comp 1 is positively correlated with V4 and V8 and negatively correlated with V3. Spectral data for V4 and V8 are important in characterizing certain chemical components, while V3 may be negatively correlated with these chemical components. Comp 2 is positively correlated with V5 and negatively correlated with V3 and V6. V5 represents the emission characteristics of a specific element in the ceramic material, while V3 and V6 are negatively correlated with the content of that element. Comp 3 is positively correlated with V1 and V9 and negatively correlated with V6. V1 and V9 are more critical in describing certain chemical properties of ceramic materials, while V6 is negatively correlated with these properties.

Table 2. PLS analysis results.

	Comp 1	Comp 2	Comp 3
V1	0.225805958	−0.252367019	0.551012434
V2	−0.086603498	0.081155404	0.329024037
V3	−0.393079975	−0.279781825	−0.047168851
V4	0.55781294	−0.194465185	−0.221014594
V5	−0.012045793	0.711815864	−0.208958142
V6	−0.129256317	0.237695426	−0.517497336
V7	0.154309764	0.485083531	0.202877961
V8	0.417147893	−0.207197416	0.125774642
V9	−0.262474818	0.108665398	0.515956505
V10	0.462480944	−0.011035023	−0.019325213

The experimental results showed that the PLS model performed well in explaining the variance of the LIBS spectral data. The model's ability to explain the variance of the independent variable (LIBS spectral data) improved with the increase in the number of principal components, from 13.62 to 34.12% on the training set. For the dependent variable (chemical composition), the model's variance explaining ability increased from 10.42 to 10.72%.

5. Conclusions

We developed and validated a composite recognition model based on LIBS and AI image recognition techniques for monitoring the quality of ceramic products. The appearance defects of ceramic products were detected in real-time, and the key parameters during the sintering process were monitored, which significantly improved the quality control of ceramic production. By reducing energy consumption and waste in production, production efficiency increased and production costs were reduced. Further composite recognition research is required for other industrial productions, promoting the intelligent and automated production process.

Author Contributions: Conceptualization, J.C. and M.T.; methodology, J.C.; software, J.C.; validation, J.C. and M.T.; formal analysis, J.C. and M.T.; investigation, J.C.; resources, J.C.; data curation, J.C.; writing—original draft preparation, J.C.; writing—review and editing, J.C. and M.T.; visualization, J.C.; supervision, J.C.; project administration, J.C.; funding acquisition, M.T. All authors have read and agreed to the published version of the manuscript.

Funding: This paper is a research outcome of the Social Science Fund Project of Jiangxi Province (Research on the Internal and Practical Logic of New Quality Productivity Promoting the High-Quality Development of the Jingdezhen Ceramic Industry, Project No.: 24YJ38).

Institutional Review Board Statement: Not applicable.

Informed Consent Statement: Informed consent was obtained from all subjects involved in the study.

Data Availability Statement: Data sharing is not applicable to this article.

Conflicts of Interest: The authors declare no conflict of interest.

References

1. Jing, Y.; Yang, Y.; Wang, Q. Mechanical analysis and MATLAB simulation of parallel beam weighing system for chemical continuous equipment. *Tianjin Chem. Ind.* **2024**, *38*, 85–88.
2. Cai, X. Design of fuzzy PID temperature control system for high temperature furnace based on swarm intelligent optimization algorithm and MATLAB simulation. *Autom. Instrum.* **2024**, 216–219+224. [CrossRef]
3. Tu, W.; Liu, Y.; Lu, X.; Zhang, L.; Li, X. Multi-indicator synthesis and optimization of the extraction process of liver protection and antidepressant combinations based on the concept of quality by design. *Chin. J. Hosp. Pharm.* **2023**, *43*, 651–656. [CrossRef]

4. Liu, L.; Tang, L.; Yang, Y.; Chen, S.; Chen, B. Green recycling supplier selection method for shared motorcycle under hybrid information fusion strategy. *Comput. Integr. Manuf. Syst.* **2023**, *29*, 3869–3886. [CrossRef]
5. Li, D.; Sun, G.; Mu, S.; Zhang, K.; Tan, Y.; Zhang, Y.; He, S. A short-term power load forecasting method based on multidimensional time series information fusion. *Chin. J. Electr. Eng.* **2023**, *43*, 94–106.
6. Magno, F.; Dossena, G. The effects of chatbots' attributes on customer relationships with brands: Pls-sem and importance-performance map analysis. *TQM J.* **2023**, *35*, 1156–1169. [CrossRef]
7. Huang, H.; Qiu, H.; Huang, H.; Song, X. Dilemmas and countermeasures for high-quality development of building ceramics industry--Based on the perspective of supply and demand. *China Ceram.* **2024**, *60*, 53–59.
8. Srinivas, S.S.; Sarkar, R.K.; Runkana, V. Battery GraphNets: Relational Learning for Lithium-ion Batteries (LiBs) Life Estimation. *arXiv* **2024**, arXiv:2408.07624.
9. Zhang, P.; Sun, L.; Qi, L.; Yu, H. On-line quantitative analysis of major elements in phosphate slurry using LIBS assisted by plasma information from orthogonal directions imaging. *Measurement* **2024**, *234*, 114827. [CrossRef]
10. Wang, L.; Yang, Y.; Pan, H. Study on the trajectory of extraction process of honeysuckle in feverfew injection. *World Sci. Technol.-Mod. Tradit. Chin. Med.* **2017**, *19*, 1277–1282.
11. Wang, Z.; Zhang, Q.U.S. FDA Guidance for Industry: An Innovative Framework System for Drug Development, Manufacturing, and Quality Assurance—PAT. *Chin. J. Tradit. Chin. Med.* **2009**, *34*, 3304–3309. [CrossRef]
12. Xie, S.; Huang, Y.; Sun, X.; Zheng, J.; Tao, Q. Regulation Tools and Application Progress of Process Analysis Technology in Pharmaceutical Industry. *Chin. J. Pharm.* **2022**, *57*, 1589–1595. [CrossRef]

engineering proceedings

MDPI

Proceeding Paper

Tool Object Analysis and ChatGPT Integration for Creative Problem-Solving Algorithm Development †

Jung-Suk Hyun [1] and Chan-Jung Park [2,*]

1 Department of Management Information Systems, Jeju National University, Jeju 63243, Republic of Korea; jshyun@jejunu.ac.kr
2 Department of Computer Education, Jeju National University, Jeju 63243, Republic of Korea
* Correspondence: cjpark@jejunu.ac.kr
† Presented at the 2024 IEEE 7th International Conference on Knowledge Innovation and Invention, Nagoya, Japan, 16–18 August 2024.

Abstract: TRIZ, the theory of inventive problem solving, can be used to solve contradictions in high-level problems in inventions and patents. Contradictions arise when two contradictory requirements or conditions cannot be satisfied simultaneously and resolving them can lead to innovative ideas. TRIZ experts have developed 76 standard solutions for applying TRIZ. Though TRIZ's 76 standard solutions can be applied to a wide range of problem types, they are too complex. We developed a simplified tool object analysis algorithm using TRIZ's 76 standard solutions. The analysis was used to analyze the interaction between components of a system to identify where problems occur. The interaction between a tool and an object was categorized into beneficial actions (positive relationship), no effect (zero relationship), and harmful actions (negative relationship). When beneficial actions, and gaps or beneficial actions and harmful actions occur in the interaction between a tool and an object, conflicting relationships arise in problem-solving. We classified the interaction between tools and objects into beneficial actions and gaps and beneficial actions and harmful actions to provide appropriate problem-solving strategies for each situation. The analysis has the advantage of specifying the causes and solutions of problem-solving. The analysis was used in creative problem-solving to train ChatGPT. By training ChatGPT for it, specific problem-solving solutions were obtained.

Keywords: creative problem solving; TRIZ 76 standard solutions; tool object analysis; ChatGPT

1. Introduction

There are many instances in the history of science and technology where contradictions have been used to achieve innovation. For example, James Watt solved the problem of a cylinder needing to be both hot and cold [1], Siemens resolved the issue of heating and cooling in the continuous production of plate glass [2], Florey and Chain solved Fleming's problem of failing to extract penicillin by heating it, Auguste Piccard was the first to reach the stratosphere in a closed gondola, and Henry Bessemer developed a blast furnace that could introduce iron ore while blocking external air. All of these individuals resolved contradictions [3,4].

Despite the importance of contradiction problems, Western scholarship, which emphasizes rationality, has regarded contradictions as wrong or impossible to achieve [3]. On the other hand, Heinrich Altshuller and his colleagues in Russia, who were familiar with dialectics, easily accepted contradictions. They noticed that innovation occurred when contradictions were resolved, based on their analysis of creative patent specifications. They established TRIZ, an acronym derived from the Russian words for the theory of inventive problem-solving. Despite TRIZ's unique perspective in viewing innovative problem-solving from the standpoint of contradiction resolution, TRIZ has developed its theories inductively rather than deductively [5,6].

By introducing symbols into the problem-solving process, problems are resolved easily without trial and error, as they can be succinctly represented. Symbols help to prevent differing interpretations and overcome language ambiguity. For example, the Pythagorean theorem ($a^2 + b^2 = c^2$) clearly defines relationships in a right-angled triangle [7].

We used propositional logic in symbolic form to derive generalized solution strategies for various problems. Using this, solution strategies are identified appropriately for contradiction problems. By determining the correct problem-solving strategy, trials and errors can be minimized [7]. The butterfly model identifies and addresses contradiction problems using symbolic logic, offering specific solution strategies. This model reduces trial and error by accurately determining a problem-solving strategy and exploring available resources through the tool object analysis, leading to efficient problem resolution [8,9].

We developed a simplified tool object analysis algorithm. The analysis examines interactions between system components, classifying them as beneficial (positive), neutral (independent), or harmful (negative). By identifying these interactions and their conflicts, we provided tailored problem-solving strategies. This method is applied to creative problem-solving cases and is used to trained on ChatGPT 4.0 to offer specific solutions.

2. The Tool Object Analysis

In TRIZ's tool object analysis, a system's tool affects the attributes of the object, and the object is influenced by the tool. The interaction between the tool and the object involves the generation, absorption, and transfer of energy (field) [10,11]. The interactions between the tool and the object are categorized into (1) performing a desirable action, (2) performing an undesirable action, and (3) failing to perform a desirable action. When the tool performs a desirable action on the object, it is defined as a useful action. When the tool performs an undesirable action on the object, it is defined as a harmful action. Finally, when the tool fails to perform a desirable action on the object, it is defined as a null action. The interactions between the tool and the object are represented in Figure 1, with useful actions represented by solid arrows, harmful actions represented by wavy arrows, and null actions represented by dotted arrows.

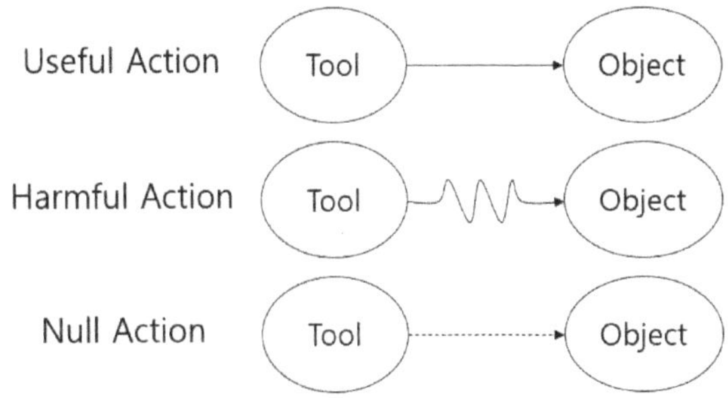

Figure 1. Interactions between tools and objects.

In the interaction between the tool and the object, the function performed by the tool is analyzed in terms of syntax by combining the object and the transitive verb [12]—that is, it is expressed as 'Who (tool) does what (object) how (action)'. When applying the components of Kipling's Six Honest Serving Men to the analysis, 'Who' refers to the system user who operates the tool. Since the tool is the means and method by which the object is acted upon, 'How' corresponds to the tool. The object is the entity upon which the tool acts, so 'What' corresponds to the object. Therefore, the problem of toilet water wastage can be expressed as 'The toilet (Who) blocks odors (What) because the pipe is S-shaped (How)' or using the instrumental case marker 'by', 'The toilet blocks odors by the S-shaped pipe'. Here, the noun corresponding to the tool is 'pipe'. The nouns corresponding to the object are 'odor' and 'amount of water'. 'Why (cause and goal)' is analyzed as preventing sewer odors from coming up to maintain cleanliness and not requiring a large amount of water for flushing (see also Figure 2). The analysis enables the application of the 76 standard solutions based on the substance-field structure and serves as a key to selecting the necessary solutions from the 40 inventive principles [6].

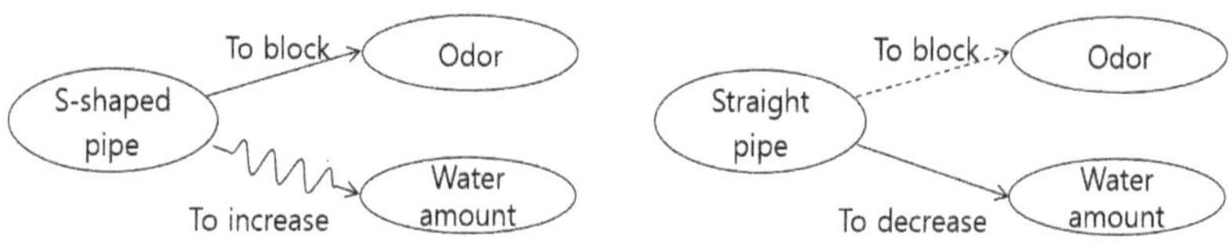

Figure 2. The tool object analysis of the toilet water waste problem.

In the analysis, the S-shaped bent pipe contains water that blocks odors, which is a beneficial action. However, to flush waste, a large amount of water is needed to push the waste over the S-trap, which is a harmful action. A straight pipe (1-shaped) does not require a large amount of water for flushing waste due to the lack of height difference in the pipe, which is a beneficial action. However, a straight pipe cannot hold enough water to block odors, resulting in a null action where the system is exposed to odors.

3. Tool Object Analysis and ChatGPT Integration

We defined the types of interactions that occur between tools and objects in order to train for ChatGPT to perform the tool object analysis and develop solutions as follows:

- When the tool performs beneficial and neutral actions on the object.
- When the tool performs beneficial and harmful actions on the object.
- When the tool performs beneficial actions on the object, but the object performs harmful actions on the tool.
- When the tool performs beneficial actions on the object but performs harmful actions on itself.

The first case is exemplified by the relationship between a pencil and paper, the second case by the relationship between a tree support and a tree, the third case by the relationship between a paper cup and water, and the fourth case by the relationship between a rocket fuel tank and a rocket (Figure 3). We created a PDF file containing the above content to train ChatGPT.

The following two problems were then presented, and then ChatGPT was requested to perform the tool object analysis and provide specific solutions. As a result, the outcomes were obtained as shown in Figures 4 and 5.

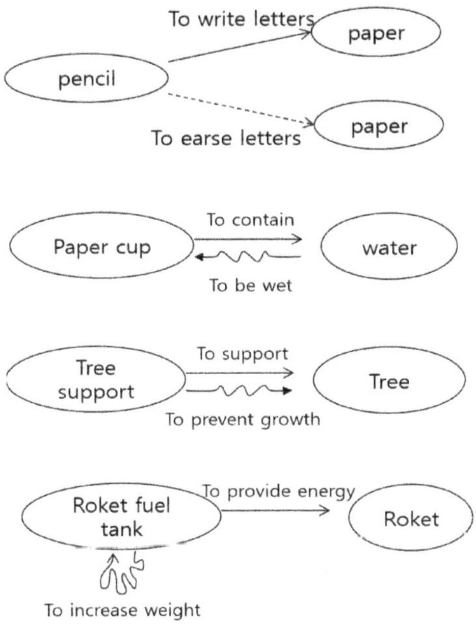

Figure 3. Training examples.

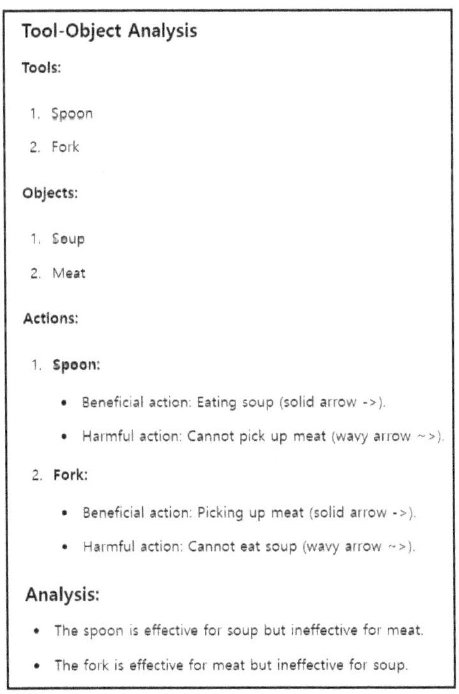

Tool-Object Analysis

Tools:

1. Spoon
2. Fork

Objects:

1. Soup
2. Meat

Actions:

1. **Spoon:**
   - Beneficial action: Eating soup (solid arrow ->).
   - Harmful action: Cannot pick up meat (wavy arrow ~>).
2. **Fork:**
   - Beneficial action: Picking up meat (solid arrow ->).
   - Harmful action: Cannot eat soup (wavy arrow ~>).

Analysis:

- The spoon is effective for soup but ineffective for meat.
- The fork is effective for meat but ineffective for soup.

Figure 4. Result from ChatGPT for the first problem.

Tool-Object Analysis

Tools:

1. Control rope
2. Ceiling plate of the gondola

Objects:

1. Hermetically sealed gondola

Actions:

1. **Control Rope:**
    - **Beneficial Action:** Controls the release of hydrogen (solid arrow ->).
    - **Harmful Action:** Creates a gap in the seal, allowing air to escape (wavy arrow ~>).
2. **Ceiling Plate:**
    - **Beneficial Action:** Maintains hermetic seal (solid arrow ->).
    - **Harmful Action:** Restricts the motion of the control rope if the gap is eliminated (wavy arrow ~>).

Analysis:

- The control rope must pass through the ceiling to function but creates a gap in the seal.
- The ceiling plate must maintain the hermetic seal but restricts the rope's movement if the gap is eliminated.

Figure 5. Training example results from ChatGPT for the second problem.

The first problem was defined as follows: "Now I will present a problem". Based on the contents of the above PDF file, please conduct a tool object analysis and provide a specific solution. "A spoon can be used to eat soup, but it cannot pick up meat. Conversely, a fork can be used to pick up meat, but it cannot be used to eat soup. In this case, please suggest a way to enable both eating soup and picking up meat. Also, please conduct a tool object analysis". The following solution is generated from ChatGPT: "To solve the problem of needing to eat both soup and meat with the same tool, we need to combine the functionalities of both the spoon and the fork. One practical solution is to create a hybrid tool commonly known as a "spork", which is a combination of a spoon and a fork". We confirmed that the tool object analysis and the solutions generated by ChatGPT were appropriate.

The second problem was as follows, "In 1930, Auguste Piccard designed a sealed, pressurized gondola that could safely carry him into the upper atmosphere to conduct scientific studies. The primary purpose of Piccard's first ascent was cosmic ray investigation. In 1931, he ascended over 51,775 feet. Piccard faced the gondola problem as follow: To rise, the balloon canopy was inflated with lighter-than-air hydrogen. To descend, Piccard pulled a rope to release hydrogen from the top of the canopy, making the whole balloon heavier. The gondola of the balloon was hermetically sealed. A control rope for the outlet hydrogen-release valve of the balloon had to pass through the ceiling of the gondola. In order to keep the air inside the gondola, the gap between the rope and the ceiling plate should not have existed. But this would have prevented free motion of the rope. How did Piccard solve the rope problem? Perform the tool-and-object analysis and find the concrete solution of this problem" [13].

The solution generated by ChatGPT was as follows: "To solve this problem, Piccard needed a mechanism that maintained the hermetic seal of the gondola while allowing the rope to move freely. A possible concrete solution is to use a gland seal (stuffing box). A gland seal or stuffing box allows the control rope to pass through the ceiling plate. The gland seal consists of a flexible, airtight material (such as rubber or silicone) that can compress around the rope, creating a tight seal while allowing the rope to move freely".

The analysis results and solutions generated by ChatGPT were appropriate. However, while ChatGPT suggested flexible materials such as rubber or silicone, the actual solution required mercury. This difference aside, we confirmed that ChatGPT performed the analysis accurately in generating specific solutions.

4. Conclusions

TRIZ suggests the separation principle and 40 inventive principles as methods for resolving contradictions in problems. However, beginners in problem-solving often find it difficult to apply TRIZ due to its complexity and vast content. To address this issue, this research developed an algorithm that defines contradiction problems and derives solutions using the tool object analysis of TRIZ. The analysis has the advantage of analyzing the problem's goals, means of resolution, and the time and space where the problem occurs. During this process, it identifies the goal of the problem (Why), the object to be achieved (What), and the means to achieve the goal (How). It also helps in understanding the interactions between the components of the system. Based on this, this study proposed a tool object analysis method using ChatGPT to assist in resolving contradiction problems. The developed simplified tool object analysis algorithm analyzes the interactions between system components, classifying them as beneficial, neutral, or harmful, and identifies these interactions and conflicts to provide tailored problem-solving strategies. This method was applied to creative problem-solving cases and was used to be trained on ChatGPT, enabling it to generate specific solutions, which is the significance of this research.

Author Contributions: Conceptualization, J.-S.H.; methodology and software, C.-J.P.; validation, J.-S.H.; writing-original draft preparation, C.-J.P.; writing-review and editing, J.-S.H. and C.-J.P. All authors have read and agreed to the published version of the manuscript.

Funding: This research was supported by the 2025 scientific promotion program funded by Jeju National University.

Institutional Review Board Statement: Not applicable.

Informed Consent Statement: Not applicable.

Data Availability Statement: The original contributions presented in this study are included in the article. Further inquiries can be directed to the corresponding author.

Conflicts of Interest: The authors declare no conflict of interest.

References

1. Ikenson, B.B. *Patents: Ingenious Inventions How They Work and How They Came to Be*; Running Press: Jenkintown, PA, USA, 2012.
2. Utterback, J.M. *Mastering the Dynamics of Innovation, Cambridge*; Harvard Business School Press: Brighton, MA, USA, 1994.
3. Watkins, P. *From Hell Island to Hay Fever*; Brown Dog Books: Hinesburg, VT, USA, 2018.
4. Bessemer, H. *Sir Henry Bessemer, F.R.S.: An Autobiography*; Offices of Engineering: London, UK, 1905.
5. Altshuller, G.; Zlotin, B.; Zusman, A.; Philatov, V. *Tools of Classical TRIZ*; Ideation International Inc.: Southfield, MI, USA, 1999.
6. Savransky, S.D. *Engineering of Creativity: Introduction to TRIZ Methodology of Inventive Problem Solving*; CRC Press: Boca Raton, FL, USA, 2002.
7. Polya, G. *How to Solve It: A New Aspect of Mathematical Method*; Princeton University Press: Princeton, NJ, USA, 1971.
8. Hyun, J.S.; Park, C.J. The butterfly algorithm: A contradiction solving algorithm based on propositional logic for TRIZ. *Int. J. Softw. Eng. Its Appl.* **2016**, *10*, 27–34.
9. Hyun, J.S.; Park, C.J. Butterfly chatbot: Finding a concrete solution strategy to solve contradiction problems. *J. Adv. Inf. Technol. Converg.* **2019**, *9*, 77–87.

10. Altshuller, G. *Creativity as an Exact Science: The Theory of the Solution of Inventive Problems*; CRC Press: Boca Raton, FL, USA, 1984.
11. Fey, V.; Rivin, E. *Innovation on Demand: New Product Development Using TRIZ*; Cambridge University Press: Cambridge, UK, 2005.
12. Shin, W.S.; Hyun, J.S.; Ha, H.H. TRIZ contradiction resolution algorithm for new product development: Focusing on the linkage process of Kipling's six honest serving men and tool-object analysis. *J. Manag. Consult.* **2022**, *22*, 445–459.
13. Piccard, A. *Earth, Sky, and Sea*; Oxford University Press: Oxford, UK, 1956.

engineering proceedings

MDPI

Proceeding Paper

Investigating the Effect of Performing Secondary Tasks on Reaction Time While Driving by Computer Analysis †

Chia-Wen Tsai [1,2,*] and Dengchuan Cai [1]

1 Graduate School of Design, National Yunlin University of Science & Technology, Douliu City 640301, Taiwan; caidc@yuntech.edu.tw
2 Department of Industrial Design, National Taipei University of Technology, Taipei City,10608, Taiwan
* Correspondence: evelyn.tsai04@hotmail.com
† Presented at the 2024 IEEE 7th International Conference on Knowledge Innovation and Invention, Nagoya, Japan, 16–18 August 2024.

Abstract: Performing secondary tasks affects the performance of primary tasks. Engaging in secondary tasks while driving often leads to accidents. Therefore, this study aimed to investigate the impact on reaction time when drivers perform secondary tasks while driving. The participants of this study were 30 participants, including 11 males and 19 females. During driving, participants watched a driving video. When a warning triangle appeared in the center of the road, participants immediately pressed the Enter button. When the participant pressed the Enter button, the computer automatically recorded the time of the button press. In a within-subjects design, all participants took part in secondary tasks during the primary driving task, and outcomes were compared between secondary tasks (no task, conversation with the passenger, and listening to the radio). The results show that the reaction time during conversation with the passenger in the primary task was significantly longer than that during listening to the radio or having no task. However, there was no significant difference in reaction time between having no task and listening to the radio. The average reaction time with no task was longer than while listening to the radio. Fatigue increased the reaction time for no task. Having a conversation with the passenger did not affect the average reaction time for both tests due to the conversation with the passenger mitigating the effects of fatigue. The results of this study provide a reference for further research on driving behavior.

Keywords: task performance; reaction time; driving behavior; cognitive load

1. Introduction

The components of transportation are people, vehicles, roads, and the environment. Traffic accidents are related to human error [1]. The World Health Organization's Global Status Report on Road Safety 2018 stated that approximately 1.35 million people die each year due to traffic accidents worldwide [2]. When drivers are exposed to external stimuli or their attention is diverted from the driving task, the risk of accidents increases [3]. Distraction leads to unsafe outcomes, especially in complex driving situations as it reduces risk awareness and increases the likelihood of accidents [4]. Therefore, inattentiveness is a major cause of accidents or rear-end collisions [5].

Most of the risks from driver distraction result from the cognitive effects of diverted attention from the current task, although inattentiveness stems from external stimuli [3]. Distraction leads to unsafe outcomes during driving, especially in complex driving situations, as interference reduces risk awareness and increases driver response times [4].

Vision, as the primary tool for sensing and receiving, leads to direct sensory responses to external stimuli [6,7]. It is also the most important sense for drivers to obtain information about their surroundings while driving. Therefore, vision has a considerable impact on driving safety [8]. In driving accidents, many drivers have experienced visual fatigue during long driving or when driving in monotonous conditions [9–12]. Visual fatigue refers to the excessive strain on the visual organs, surpassing compensatory ability [13]. According to the American Optometric Association, prolonged use of the visual system reduces efficiency in visual processing functions [14]. Fatigue impairs the ability to maintain speed and stay in the lane, increases reaction times, and leads to decision-making errors, thereby increasing the risk of traffic accidents [15]. Perception–response time (PRT) refers to the time interval between the moment a driver perceives a danger on the road and takes action in response [16]. Driver distraction is caused by diverted attention from the current task. According to data from the National Highway Traffic Safety Administration (NHTSA), distractions include engaging in deep conversations with passengers, talking on the phone, holding objects, singing, dancing, smoking, adjusting the stereo or air conditioning, and eating [4]. Therefore, we investigated the impact of drivers performing secondary tasks while engaged in primary driving tasks on their response times.

2. Materials and Methods

2.1. Experiment

The participants in this study searched for the warning triangle signs in videos. The independent variables were the participant's ethnic group and the secondary task (no task, conversation with the passenger, and listening to the radio), while the dependent variable was the participant's reaction time. The test was conducted by watching a pre-recorded driving video on the road. When a warning triangle appeared on the road, the participant pressed the confirmation button immediately. The reaction time was defined as the time taken for the participant to press the Confirm key after noticing the warning triangle. We used a within-subjects design, with each participant performing the following three secondary tasks.

1. No task

The driver had no specific tasks but driving. The participants focused on the primary task and did not perform secondary tasks.

2. Conversation with the passenger

The conversation with questions comprised a series of general knowledge and arithmetic problems in simple and complex categories [17]. The participants performed addition for each arithmetic problem and classified the result as either odd or even [18]. The conversation samples included general knowledge and personal topics [17]. Due to the difficulty in controlling the influence of intermediate variables, casual conversation was not used [19].

3. Listening to the radio

The participants listened to the radio while driving. The radio content included information such as music, traffic conditions, road construction, natural disasters, and emergency rescue updates. Ear Pods were used for 30 min at 56−64 dB.

The height of the car warning triangle was 38.5 cm (Figure 1). The triangle was placed 150 min in front of the car. To make the simulation environment look realistic [20], a GoPro camera was used to record driving conditions on the road. The warning sign was placed 150 m ahead of the vehicle (Figures 2 and 3). The camera recorded the visual transition of the triangular warning sign from far to near distance. Non-linear video production software was used to edit the video. To provide the participants with visually immersive and naturally realistic interactive means [21], a mobile phone with GoVR Player software and Google Cardboard VR headwear glasses were used. The Google Cardboard VR headwear glasses were composed of two 45 mm focal length lenses and formed a virtual distance in the optical structure. The edited video was processed into the GoVR Player app to split the screen into left and right frames, corresponding to each eye. This allowed viewers to experience a sense of immersion, as the brain overlays the images through the convex lenses of the VR headset, creating a three-dimensional visual effect.

Figure 1. Height of warning triangle sign.

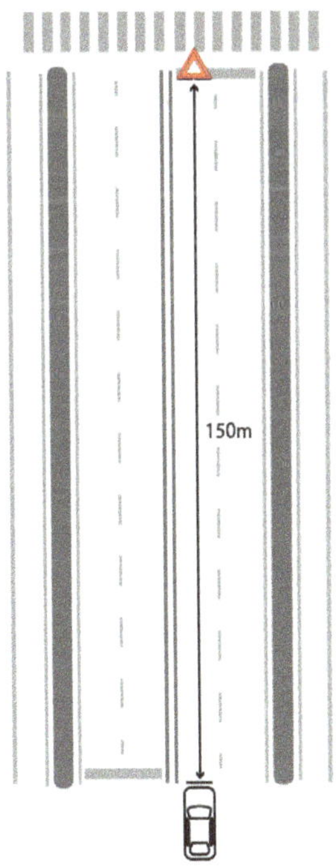

Figure 2. Distance of 150 m from starting point.

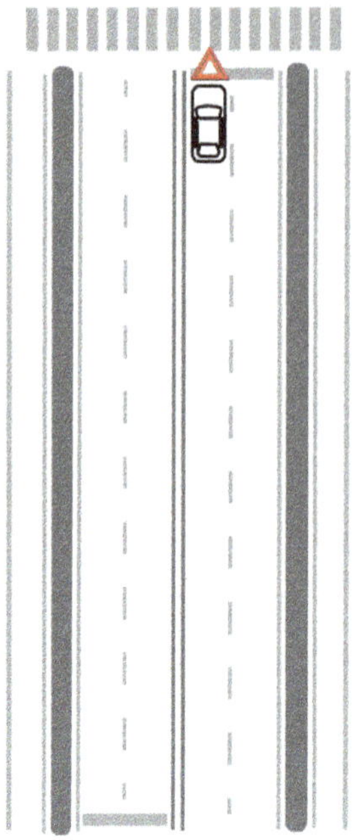

Figure 3. Driving forward to front of triangular warning sign.

2.2. Experiment

There were 30 participants in this study (11 males and 19 females). The average age was 29.50 years old (Standard Deviation (SD) = 11.89) (Table 1). All participants had a regular driver's license and driving experience with an average of 7.7 years. The participants had normal vision or corrected vision of 0.8 (20/25) [22].

Table 1. Descriptive statistics of participants.

Gender	N	Min.	Max.	Mean	SD
Male	11	20	65	30.64	13.46
Female	19	20	65	28.84	10.84
Total	30	20	65	29.50	11.89

3. Results and Discussion

We investigated the effects of no task, a conversation with the passenger, and listening to the radio on the reaction time of the participants. The reaction time of the participants was measured whenever triangular warning signs appeared. The triangular warning sign appeared 16 times, and the Confirm key was pressed 480 times. The average reaction time in each task was 3.9 (SD = 5.2), 3.7 (SD = 4.9), and 6.7 (SD = 13.9) (Table 2).

Table 2. Data of secondary tasks.

Secondary Task	N	Min	Max	Mean	SD
No task	480	0.0	46.0	3.9	5.2
Listening to the radio	480	0.0	47.0	3.7	4.9
Conversation with the passenger	480	0.0	80.0	6.7	13.9

The average reaction time in conversations was significantly higher than that for listening to the radio (t = 4.687, $p = 0.00$) and no task (t = −4.274, $p = 0.00$). However, there was no significant difference between the average reaction time when listening to the radio and the average reaction time when performing no task (t = −1.041, $p = 0.298$) (Table 3).

Table 3. Paired sample *t*-test results for secondary tasks.

Paired Sample *t*-Test	t	*p*	T Test
No task–Conversation with the passenger	−4.274	0.000	No task < Conversation with the passenger
Listening to the radio–No task	−1.041	0.298	Listening to the radio = No task
Conversation with the passenger–Listening to the radio	4.687	0.000	Conversation with the passenger > Listening to the radio

The reaction time for the 9th time was longer for no task (mean = 12.62) than listening to the radio (mean = 11.26) and conversation with the passenger (mean = 19.65). Therefore, the response time of the 9th time was eliminated, and the triangle warning label was checked for differences in response time (Table 4). Alhaag and Ramadan noted that muscle fatigue caused by muscle activity occurs 10 min after participants view the images [23]. There were significant differences in average reaction times for listening to the radio (t = −2.274, $p < 0.05$) and no task (t = −2.573, $p < 0.05$). Therefore, holding a conversation during the driving task mitigated participants' fatigue, resulting in no significant difference in reaction time (Table 5).

Table 4. Descriptive statistics related to the number of appearances of warning signs.

Descriptive Statistics					
No task					
Times	N	Min	Max	Mean	SD
1	30	1.49	8.00	4.47	1.88
2	30	1.15	7.00	2.86	1.48
3	30	1.49	7.00	3.16	1.25
4	30	1.00	7.00	2.74	1.47
5	30	1.00	7.00	3.21	1.62
6	30	1.00	5.00	2.85	1.11
7	30	0.00	6.00	3.44	1.55
8	30	1.15	6.00	3.09	1.33
9	30	1.29	46.00	12.62	18.39
10	30	1.00	6.00	2.83	1.26
11	30	1.62	7.00	4.55	1.40
12	30	0.00	6.73	3.84	1.69
13	30	1.70	7.00	3.67	1.77
14	30	1.47	9.00	3.64	1.94
15	30	1.49	8.00	3.55	1.72
16	30	1.00	6.40	3.29	1.54
Listening to the radio					
Times	N	Min	Max	Mean	SD
1	30	1.85	7.00	3.74	1.73
2	30	1.00	5.00	2.73	0.99
3	30	1.84	6.00	3.12	1.22
4	30	1.00	6.00	2.63	1.14
5	30	0.00	6.00	2.88	1.44
6	30	1.00	7.00	2.79	1.50
7	30	1.00	9.00	3.79	1.92

Table 4. *Cont.*

Descriptive Statistics					
Listening to the radio					
Times	N	Min	Max	Mean	SD
8	30	0.00	6.77	2.96	1.58
9	30	1.00	47.00	11.26	17.62
10	30	0.00	7.00	2.74	1.51
11	30	2.00	7.00	4.20	1.35
12	30	1.75	7.00	3.74	1.51
13	30	1.86	8.00	3.37	1.66
14	30	0.00	6.00	3.09	1.56
15	30	1.00	8.00	3.48	1.77
16	30	1.24	7.00	3.29	1.58
Conversation with the passenger					
Times	N	min	max	Mean	SD
1	30	0.00	79.38	6.41	13.91
2	30	0.00	39.45	3.98	6.87
3	30	1.00	79.29	5.80	13.96
4	30	1.00	78.92	4.91	14.03
5	30	0.00	79.96	5.41	14.17
6	30	1.00	38.84	4.23	6.71
7	30	1.00	79.53	5.61	14.04
8	30	1.00	78.81	5.63	13.90
9	30	1.00	78.79	19.65	23.59
10	30	0.00	79.32	5.32	14.04
11	30	0.00	80.07	8.56	15.94
12	30	0.00	39.09	5.80	6.62
13	30	1.78	79.36	6.54	13.89
14	30	0.00	79.03	6.36	13.90
15	30	0.00	79.82	8.08	16.01
16	30	1.89	38.99	5.48	6.59

Table 5. Independent sample *t*-test for reaction times for different tasks.

Secondary Tasks	Tasks	N	Mean	t	*p*	*t*-Test
Conversation with the passenger	Former	240	5.2466	−1.128	0.260	Former = Latter
	Latter	210	6.5931			
Listening to the radio	Former	240	3.0809	−2.274	0.023	Latter > Former
	Latter	210	3.4146			
No task	Former	240	3.2320	−2.573	0.010	Latter > Former
	Latter	210	3.6255			

4. Conclusions

In three different tasks, the participants' reaction times during conversations while driving were significantly longer than during radio listening and while performing no task. Reaction times for no task and radio listening were not significantly different. For radio listening and no task, reaction times in the latter part were longer than in the former part, indicating significant fatigue. However, there were no significant differences in the conversation tasks in the former and latter parts. This may be because the first part of the conversation involved arithmetic problems, while the latter involved common sense, offsetting fatigue effects. Talking to a real person or through a digital mode while driving reduced drivers' risk awareness, increasing reaction times due to distraction. These results provide important insights into driving behavior.

Author Contributions: Conceptualization, D.C. and C.-W.T.; methodology, D.C. and C.-W.T.; validation, C.-W.T.; investigation, C.-W.T.; data curation, C.-W.T.; writing—original draft preparation, C.-W.T.; writing—review and editing, C.-W.T.; visualization, C.-W.T.; supervision, D.C. All authors have read and agreed to the published version of the manuscript.

Funding: This research did not receive any specific grant from funding agencies in the public, commercial, or not-for-profit sectors.

Institutional Review Board Statement: Not applicable.

Informed Consent Statement: All subjects gave their informed consent for inclusion before they participated in the study.

Data Availability Statement: The datasets used and/or analyzed during the current study are available from the corresponding author on reasonable request.

Conflicts of Interest: The authors declare no conflict of interest.

References

1. Bartley, G.P. (Ed.) Causes of road traffic accidents in developing countries. In *Traffic Accidents: Causes and Outcomes*; Nova Publishers: New York, NY, USA, 2008.
2. World Health Organization. *Global Status Report on Road Safety 2018*; World Health Organization: Geneva, Switzerland, 2019.
3. Geden, M.; Staicu, A.M.; Feng, J. The impacts of perceptual load and driving duration on mind wandering in driving. *Transp. Res. Part F Traffic Psychol. Behav.* **2018**, *57*, 75–83. [CrossRef]
4. Choudhary, P.; Velaga, N.R. Effects of phone use on driving performance: A comparative analysis of young and professional drivers. *Saf. Sci.* **2019**, *111*, 179–187. [CrossRef]
5. Lundqvist, L.M.; Eriksson, L. Age, cognitive load, and multimodal effects on driver response to directional warning. *Appl. Ergon.* **2019**, *76*, 147–154. [CrossRef]
6. Gobe, M. *Emotional Branding: The New Paradigm for Connecting Brands to People*; Simon and Schuster: New York, NY, USA, 2010.
7. Kroemer, K.H. *'Extra-Ordinary' Ergonomics: How to Accommodate Small and Big Persons, the Disabled and Elderly, Expectant Mothers, and Children (Vol. 4)*; CRC Press: Boca Raton, FL, USA, 2005.
8. Waard, D.D.; Jessurun, M.; Steyvers, F.J.; Reggatt, P.T.; Brookhuis, K.A. Effect of road layout and road environment on driving performance, drivers' physiology and road appreciation. *Ergonomics* **1995**, *38*, 1395–1407. [CrossRef] [PubMed]
9. Liu, T.; Liu, Y.; He, W.; He, W.; Yu, X.; Guo, S.; Zhang, G. A passenger reduces sleepy driver's activation in the right prefrontal cortex: A laboratory study using near-infrared spectroscopy. *Accid. Anal. Prev.* **2016**, *95*, 358–361. [CrossRef]
10. Zhang, T.; Chan, A.H. Sleepiness and the risk of road accidents for professional drivers: A systematic review and meta-analysis of retrospective studies. *Saf. Sci.* **2014**, *70*, 180–188. [CrossRef]
11. Gall, B. A guide to improving alertness through effective management of fatigue. *Pet. Rev.* **2006**, *60*, 24–25.
12. Caldwell, J.A. Methodological issues associated with measuring the operational impact of fatigue in aviation. In Proceedings of the International Conference on Fatigue Management in Transport Operations: A Framework for Progress, Boston, MA, USA, 23–25 March 2009; US Department of Transportation: Washington, DC, USA, 2009.
13. Simmerman, H. Visual fatigue. *Optom. Vis. Sci.* **1950**, *27*, 554–561. [CrossRef] [PubMed]
14. American Optometric Association. *Guide to the Clinical Aspects of Computer Vision Syndrome*; American Optometric Association: St. Louis, MO, USA, 1995.
15. Williamson, A.; Friswell, R.; Olivier, J.; Grzebieta, R. Are drivers aware of sleepiness and increasing crash risk while driving? *Accid. Anal. Prev.* **2014**, *70*, 225–234. [CrossRef]
16. Olson, P.L.; Sivak, M. Perception-response time to unexpected roadway hazards. *Hum. Factors* **1986**, *28*, 91–96. [CrossRef] [PubMed]
17. Amado, S.; Ulupınar, P. The effects of conversation on attention and peripheral detection: Is talking with a passenger and talking on the cell phone different? *Transp. Res. Part F Traffic Psychol. Behav.* **2005**, *8*, 383–395. [CrossRef]
18. Mohebbi, R.; Gray, R.; Tan, H.Z. Driver reaction time to tactile and auditory rear-end collision warnings while talking on a cell phone. *Hum. Factors* **2009**, *51*, 102–110. [CrossRef]
19. Yan, W.; Xiang, W.; Wong, S.C.; Yan, X.; Li, Y.C.; Hao, W. Effects of hands-free cellular phone conversational cognitive tasks on driving stability based on driving simulation experiment. *Transp. Res. Part F Traffic Psychol. Behav.* **2018**, *58*, 264–281. [CrossRef]
20. Engström, J.; Johansson, E.; Östlund, J. Effects of visual and cognitive load in real and simulated motorway driving. *Transp. Res. Part F Traffic Psychol. Behav.* **2005**, *8*, 97–120. [CrossRef]
21. Bayarri, S.; Fernandez, M.; Perez, M. Virtual reality for driving simulation. *Commun. ACM* **1996**, *39*, 72–76. [CrossRef]

22. Colenbrander, A. Aspects and ranges of vision loss. In Proceedings of the Internacional Council of Ophthalmology at the 29th International congress of Ophthalmology, Sydney, Australia, 21–25 April 2002.
23. Alhaag, M.H.; Ramadan, M.Z. Using electromyography responses to investigate the effects of the display type, viewing distance, and viewing time on visual fatigue. *Displays* **2017**, *49*, 51–58. [CrossRef]

engineering proceedings

MDPI

Proceeding Paper

Internet of Things and Autonomous Robots to Develop Intelligent Solutions for Sterilization and Disease Prevention †

Ling-Hsiang Hung [1], Zong-Jie Wu [2], Chu-Hwa Yan [3] and Chien-Liang Chen [4,5,*]

1 Department of Computer Science, Santa Monica College, Santa Monica, CA 90405, USA; seanis190@gmail.com
2 Department of Industrial Engineering and Management, National Yang Ming Chiao Tung University, Hsinchu 30010, Taiwan; zongjie13@gmail.com
3 Department of Information Management, University of Kang Ning, Taipei 70970, Taiwan; echoyan@ukn.edu.tw
4 Department of Innovative Living Design, Overseas Chinese University, Taichung 40721, Taiwan
5 Department of Industrial Engineering and Management, National Quemoy University, Kinmen County 892, Taiwan
* Correspondence: clchen@nqu.edu.tw
† Presented at the 2024 IEEE 7th International Conference on Knowledge Innovation and Invention, Nagoya, Japan, 16–18 August 2024.

Abstract: As the epidemic affected everyone across the world, the solution to the epidemic was developed globally. Many applications adopt Internet of Things (IoT) technology to detect epidemics, and effective monitoring systems are developed to monitor air pollution, personal transmission, early detection of serious cases, and remote assessment. However, care facilities in an aging society require effective disinfection and sterilization to prevent viral transmission. We integrated the interactive and real-time features of the Internet of Things (IoT) to design and build an intelligent self-propelled sterilization robot for sterilization. Intelligent sterilization and disinfection planning and task allocation mechanisms were designed for sterilization in clinics. For healthcare facilities, the developed robot can reduce the burden on healthcare professionals, help to manage the disinfection and sterilization process, and ensure patient safety. At the same time, robots promote the development of epidemic prevention industries and prepares for future attacks from harmful air pollutants or new infections.

Keywords: Internet of Things; autonomous robots; sterilization; disease prevention

1. Introduction

The World Health Organization (WHO) declared the global outbreak of a new coronavirus (COVID-19) on the 11 March 2020. The rapid spread of the virus put the world in danger and resulted in the death of many people. According to the WHO, most of the deaths caused by the virus occurred in patients with underlying health conditions, such as diabetes or cardiovascular disease. In Canada, 80% of the deaths from the epidemic occurred in long-term care facilities for six months. These deaths occurred in long-term care facilities, which were susceptible to and severely affected by the epidemic due to congregate living environments and multiple comorbidities [1,2].

Despite extensive research, the full transmission routes of COVID-19 remain unclear. Infection has a significant impact on the human respiratory system. The respiratory system is affected by air pollution caused by suspended particles such as PM2.5, PM10, and gases emitted by industries. These particles block airways and exacerbate the damage caused by the virus [3]. In 2020, Wu et al. [4] collected data from more than 3000 counties in the United States from the Johns Hopkins University Center for Systems Science and Engineering Coronavirus Resource Center and conducted a study on the causes of death during the pandemic. The study revealed that a marginal increase in PM2.5 levels resulted in a considerable increase in the mortality rate of COVID-19. In the same year, Bashir et al. studied the correlation between environmental pollution indicators and the main patterns of the pandemic. The state of California experienced the most severe pandemic outbreak in the United States, resulting in more than 1800 deaths. Kendall and Spearman [5] conducted a correlation analysis using data from the Environmental Protection Agency and causes of death investigations. Their findings indicated that PM10, PM2.5, sulfur dioxide (SO_2), volatile organic compounds (TVOC), carbon monoxide (CO), and nitrogen dioxide (NO_2) in the air are significantly related to the mortality rate of cases of COVID-19.

The progressive integration of information technology, the Internet of Things (IoT), big data, and artificial intelligence is conducted in engineering design, precision medicine, and home healthcare. The IoT facilitates the interconnection of all computations, machinery, and data, thereby enabling seamless network transmission without the need for human intervention. Throughout the ongoing pandemic, technology has shown particular utility in monitoring healthcare [6]. Although the IoT has been used in epidemic prevention measures, its applications are predominantly limited to detecting and sensing whether individuals are infected. A notable deficiency is the lack of attention to preventive detection and preventive measures, particularly concerning the safety of long-term care facilities. Consequently, in addition to the fact that the Taiwanese population is increasingly elderly, it is paramount to ensure the personal safety of patients, residents, and caregivers in long-term care facilities and to provide a secure working environment.

It is essential to detect and clean harmful substances in the air and to continuously monitor air quality. In healthcare facilities, the efficient and rapid cleaning and disinfection of noxious substances is necessary. Therefore, this study aims to use the IoT to introduce long-term care services. The system contains air quality sensors, wireless transmission technology, and communication gateways to scan the environment. Based on air quality indicators, different levels of disinfection are determined. Additionally, a self-propelled robot with sterilization capabilities is designed for automatic cleaning to reduce the workload of nursing staff. The robot creates a smart decontamination service for long-term care facilities, ensuring a healthy environment for all residents.

2. Related Work

2.1. IoT for Epidemiological Support and Assistance

The remote monitoring technology of the IoT represents a highly effective solution to avoid face-to-face contact in critical situations and continuously monitor patient conditions. In addition to providing real-time environmental monitoring, the IoT also enables data notification and transmission through the interaction of connected devices. This technology encompasses various areas, including big data, cloud computing, smart sensors, and artificial intelligence. During the pandemic, the IoT demonstrated considerable potential and provided significant support in the healthcare sector (Table 1). Currently, in situations where increased healthcare demand leads to a shortage of manpower, the IoT is used in healthcare settings, addressing numerous challenges.

Table 1. Internet of Things in healthcare.

Applications	Descriptions
Patient Treatment [7]	Internet of Things with real-time location services for optimal treatment of COVID-19 patients
Smart Hospital [8]	Use of connected IoT devices and private networks, and provision of automated monitoring in hospitals
Patient Data Access [9]	Using IoT devices to store and analyze patient physiological data to provide better treatments
Smart Bed [10]	Designing a smart IoT bed that changes height according to the patient's requirements
Disease Alert [11]	Alerts patients to disease warnings in critical situations using real-time tracking technology

During a pandemic, intelligent robots can be used to improve patient care and alleviate the burden on healthcare systems. Robots represent an advanced application of IoT technology, and their mobility, individuality, versatility, and automation capabilities can be used with greater precision. Reference [12] developed autonomous mobile robots embedded with navigation and positioning technology for mapping and using ultraviolet light for environmental disinfection. These robots were designed for use in closed areas as they were remotely controlled over Wi-Fi using a smartphone or tablet. Although the use of robots in various environments continues to expand, there are concerns about the effectiveness of their disinfection capabilities. Furthermore, the deployment of robots in healthcare facilities, care institutions, and schools costs a lot. The use of ultraviolet (UV)-C needs to be complemented by manual cleaning for optimal results. Consequently, an integrated and cost-effective smart disinfection system using the IoT, artificial intelligence, and robotics helps care facilities, medical institutions, and even home environments create a safe and hygienic living environment.

2.2. Sterilization and Disinfection Scheduling

We assessed the efficacy of a sterilization robot for disinfection in specific locations or floors. In addition to evaluating the efficacy of the sterilization equipment itself, it was essential to consider a multitude of variables, including the time required for each disinfection, the interval between each disinfection, and the number of robots. We identified the most effective scheduling of sterilization robots to enhance the cleanliness of the area. Process planning is a widely used technique in numerous industries, including hospital operating rooms and airport boarding arrangements. The efficacy of process planning is not limited to the success of the task in question; rather, it also has an impact on the operation of the entire organization.

Effective process planning significantly reduces time costs [13]. Park [14] proposed a mixed planning and scheduling model for physicians and operating rooms, considering factors such as the number of tasks, operating rooms, physicians, and preferences. This improves the quality of medical care. The random generation of combinations effectively reduces costs compared to manual scheduling. Additionally, a number of medical equipment can be reused during surgery. Consequently, Coban [15] developed a model that uses mixed integer linear programming (MILP) to minimize time costs, including the use of sterilization equipment, surgical delays, and surgical execution time. The outbreak of the new coronavirus (2019-nCoV) resulted in a global shortage of medical personnel, highlighting the critical importance of effective staff allocation in hospitals. Hasan [16] proposed an algorithm, designated HSSAGA, which combines the Salp Swarm Algorithm (SSA) and the Genetic Algorithm (GA). First, the GA randomly generates combinations that meet various constraints. Second, the SSA is used to select the most optimal individuals for iterative calculations. Finally, the optimal combination within a specified temporal framework is selected to enhance scheduling and assignment challenges for nursing personnel.

3. Methodology

The global pandemic caused by COVID-19 had a profound impact on the lives and safety of people. This issue cannot be overlooked, particularly in densely populated housing facilities. Air pollution is a persistent threat to human health and is also a significant pathway for virus transmission. The elderly are adversely affected by air pollution, having increased vulnerability to infection. Therefore, it is necessary to monitor air quality and design a smart epidemic prevention and disinfection system. The optimal air quality required for human activity needs to be monitored. Smart robots can perform disinfection to prevent bacterial reproduction and spread. In addition, warning messages are issued to remind management personnel to pay attention to the current environment and to prevent rapid airborne transmission and the inhalation of harmful substances.

At the site of the experiment (Figure 1), sensors and robots were deployed to ensure the complete collection of air values within the organization. The sensors analyzed the number of sensors required for deployment within the organization. Air sensors and smart robots are used in hospitals to maintain the health conditions of residents and the quality of the facilities. When a device's data are sent through a gateway, a vertical connection to the information platform is established so that the organization's caregivers can manage the environment in a unified way. This approach is consistent with the basic concepts of protecting residents, caregivers, and the environment.

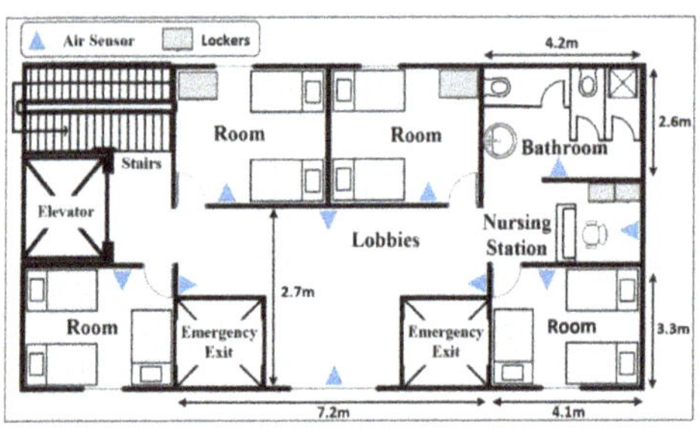

Figure 1. Experimental environment.

3.1. Environmental Design for Air Quality Monitoring and Wireless Transmission

It is significant to achieve the desired effect of prevention and disinfection using information technology. Given the prevalence of care institutions, it is imperative to consider the impact of air quality on human health in various environments. For example, in the bathroom, it is crucial to monitor oxygen content, as a deficiency can cause dizziness in the elderly. Consequently, the environmental architecture process enables the organization to work in a structured manner, with alert messages for air quality. This approach allows for the simultaneous improvement of environmental air quality and the assurance of the safety of the elderly. Figure 2 illustrates the environmental architecture of the IoT infrastructure in the following three layers: the perception layer (devices), the network layer (communication protocols), and the application layer (applications).

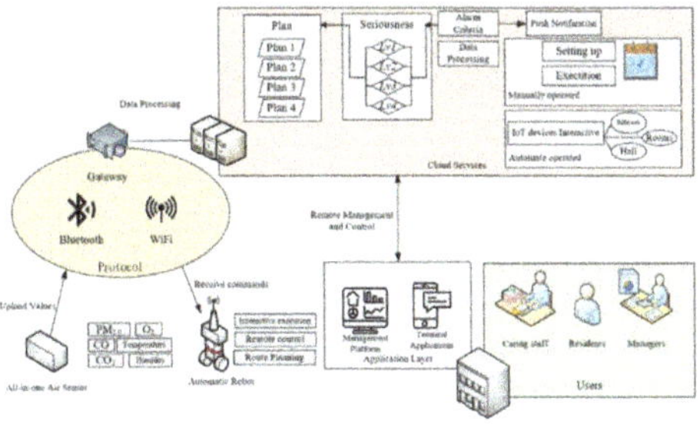

Figure 2. Environmental architecture diagram.

3.2. *Air-Sensing Specification*

To monitor air quality in the environment, we used an all-in-one air sensor. Real-time monitoring, interactive response, and remote operation services were performed by an automated robot to meet the demand for environmental monitoring in the sensor layer of the IoT devices. After integrating the collected data and transmitting them through the network layer, the continuous presentation of the data was conducted to meet the needs for real-time air monitoring.

3.3. *Wireless Transmission*

Bluetooth and Wi-Fi were used as transmission media at the network level for the data collected by the all-in-one air sensor. The data were interpreted by uploading them to the gateway. These two transmission methods are capable of satisfying different transmission requirements and achieving high research efficiency. The Bluetooth modules use Bluetooth 4.0 and Bluetooth 5.0, with a maximum transmission efficiency of 48 Mbit/s and a transmission distance of up to 300 m. This is an ideal solution for care facilities with short transmission distances and multiple transmitting devices, as well as for achieving low-latency transmissions. Wi-Fi facilitates an overall environment. The 802.11 b/g/n transmission standard was used, operating within the 2.4 GHz band. Support was provided for bandwidths of 20 MHz and 40 MHz, thus meeting the transmission requirements of various Wi-Fi devices. The Wi-Fi transmission range was 300–400 m, which is appropriate for the transmission range of devices in organizations. Therefore, these two transmission methods were chosen as the network layer architecture for this study.

3.4. *Test Data*

It is important to recognize the importance of the application layer in the IoT architecture. This allows managers to view relevant records at a glance, from the development of end applications to the back-end management platform. The data collected by the gateway are stored in a cloud database. Through the display function of the application and the management platform, managers can monitor changes in air quality in each environment. To ensure the correct functioning of the air sensors and understand environmental variability, tests are conducted to collect air quality data from air sensors. These data are used to verify the sensor operation and provide information on air quality in different environments.

3.5. *Edge Computing and Cloud Services*

To provide a highly efficient and stable transmission environment, the edge computing approach is used. The sensors transmit air quality data to each edge computing center for alarm reading and data processing. If air quality is poor and exceeds a set threshold, an alarm message is sent to the back-end management to alert them to the cause of air quality degradation. In cases where life is at risk, treatment and elimination are performed to meet the standards for ambient air quality. The data from the marginal calculation are stored in the cloud service and the thresholds of the marginal calculation can be controlled by operating commands from the cloud service.

The cloud service encompasses a multitude of processing elements, including definitional value calculation algorithms, severity levels, execution scenarios, sterilization, and automated IoT equipment. To provide optimal solutions within a conducive processing environment and space, all execution processes are configured through the cloud according to the time slots, various processing scenarios, and severity rankings based on threshold criteria. Various automation devices are unified and connected to edge computing and application services, thus facilitating the efficiency of cloud transfer as required by administrators during operation.

4. Discussion

Emergency preparedness is important in long-term care organizations. The immediate elimination of crises by front-line staff is crucial for incident management. However, the impact of air quality on human health is not uniform. Therefore, assessing air quality is essential to establish a staged threshold. The level of risk of contamination at different levels of sterilization must be considered, including the length, scope, and means of disinfection. Additionally, each sterilization process must be accurately recorded for further investigation and comprehension by caregivers.

4.1. Emergency Event Design

Air-sensing modules are installed at various locations to detect the air quality condition of a site in real time. Each floor is equipped with intelligent sterilization robots. The efficient operation of the robot is monitored. We planned intelligent sterilization scenarios including three main types: special events, temporary tasks, and cyclical processes. Each scenario is described in the following text.

4.1.1. Special Events

When the air sensor module detects an abnormal air value, an emergency is determined. In this case, the robot stops and proceed to the area of concern for air purification. Even if the robot is engaged in other tasks when an abnormality occurs, it stores the current sterilization status and transmits it to the cloud server. Upon removal of the emergency, the robot resumes the progress of the unfinished task, ensuring the integrity and performance of the sterilization process. At the same time, the robot communicates with the operator in real time through the cloud server and reports the status and progress of abnormal events in real time, thus allowing the remote operator to monitor the robot's operation. All sterilization data and event logs are recorded and saved for future data analysis and improvement.

4.1.2. Temporary Tasks

This scenario is classified as a Level 2 event. The trigger condition is a task event manually assigned by the staff within the organization. In addition to providing residential care services, the organization is responsible for maintaining the residential environment. Previously, cleaning staff used disinfectants to clean the premises. This method proved to be highly effective but it resulted in the air being filled with the smell of disinfectants. Although this is not a health hazard, it reduces the quality of residents' stay and the willingness of residents and their families to stay in the organization. However, by sending robots to clean the air immediately after the venue is manually cleaned, the odor can be quickly eliminated. This type of air purification quickly eliminates odors and improves the quality. Robots are efficient and have no adverse effects on the health of residents or staff, thus increasing the willingness of residents and their families to stay in the facility.

The robot can conduct regular air monitoring to ensure that indoor air quality is optimal and to adjust the operation of the air purification system on time. Consequently, robots improve the efficacy of cleaning and mitigates the olfactory impact of disinfectants. It facilitates the creation of a more hygienic and agreeable living environment for residents.

4.1.3. Cyclical Processes

Given the variability in site size, layout, and furnishings between different organizations, a fixed process arrangement for decontamination is viable. A lot of research has shown that effective process planning can improve efficiency and reduce costs. We use the Genetic Algorithm (GA) proposed by Holland to automatically calculate the optimal decontamination flow arrangement for each field. The Genetic Algorithm enables automatic calculation and optimization of the optimal decontamination process for each site based on its specific needs and constraints, ensuring that the effectiveness of the decontamination is maximized and the cost is minimized. This approach improves the efficiency of the decontamination process and allows for a more responsive approach to varying environmental conditions.

When applying the GA method, the general flow chart of operations in Figure 3 is used. This method is used to identify the optimal combination through a process of selection, crossover, and mutation across multiple generations. The fitness value of each individual is calculated to verify the overall number of overlaps. This process provides a powerful and efficient method to identify the optimal solution. If a population is created, the best gene for the crossover is identified. To avoid too similar a selection, mutations of the selection are taken into account and repeated iterative calculations are carried out until the conditions are met to obtain the best selection as the final result.

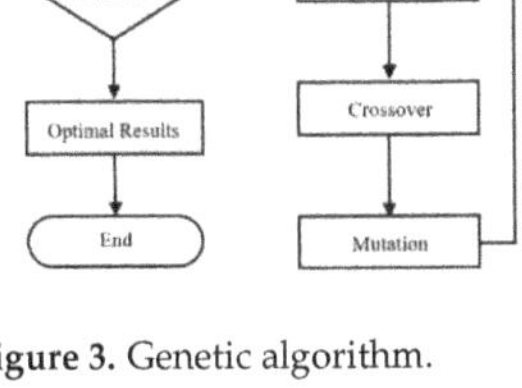

Figure 3. Genetic algorithm.

1. Initialize Population: First, it is necessary to generate n permutation samples for subsequent operations. However, since each room must be sterilized, it is essential to ensure that the randomly generated samples meet two requirements: (1) each room can only be visited once and cannot be repeated, and (2) we must return to the starting point after completing the sterilization operation. To satisfy these two requirements, a method of generating permutation samples using backtracking can be employed. Visiting rooms is the first thing to be performed. Then, a selection set is constructed. This set is constructed by randomly selecting the next room to be

visited. This process is repeated for all rooms. To ensure that each room is sterilized and cannot be repeated, a selection must be created. At the same time, the set of choices must be updated when backtracking to select the next room until all rooms are sterilized. This backtracking method generates a permutation sample to satisfy the requirements of the sterilization operation. When performing a sterilization operation, rooms are sorted according to priority or other criteria to ensure that urgent rooms or rooms special with needs are sterilized first. This approach allows for more efficient completion of the sterilization task and ensures that each room is treated according to the requisite standards.

2. Fitness Calculation: This study aims to achieve the maximum sterilization effect in the shortest time. In addition to the scenarios mentioned, there are special events and temporary tasks. If sterilization is performed at the same time or too close to the actions of a scenario, it results in a waste of time and little benefit. To address this problem, we consider the overall time and intervals between sterilization in each room to improve the effectiveness of air cleaning. When the overall time and interval of the sterilization operation are taken into account, the overall efficiency of air purification is improved. When sterilization operations are performed at the optimal time and the most appropriate location, the effectiveness of sterilization is maximized to obtain the best results.

   The total time required to reach each room is used as a parameter in the evaluation. Although the actual situation may be complex, one departure point and three rooms are included in this study, as shown in Figure 4. The exit point is represented by A. It is observed that when there are fewer and simpler parameters, all possible permutations can be exhaustively considered and the total time cost can be calculated. The green routes in Figure 5 offer the greatest savings in time. After calculating the sterilization time at each site, the total time is as follows: 2 + 20 + 4 + 30 + 2 + 10 + 6 is 74. Under the above conditions, the optimal path for the specified time is 74 for A > B > C > D > A.

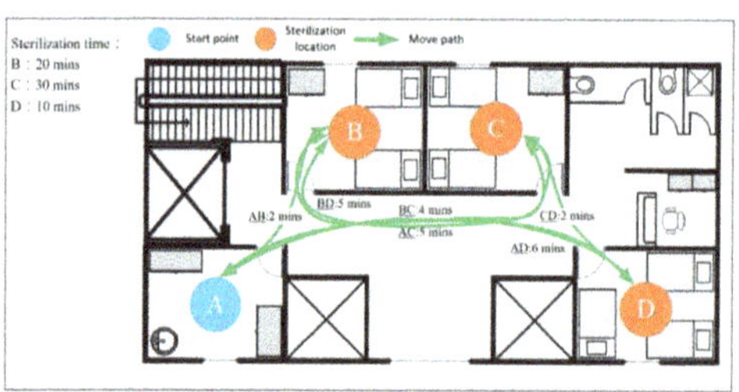

Figure 4. Simulation of the sterilization path: general case.

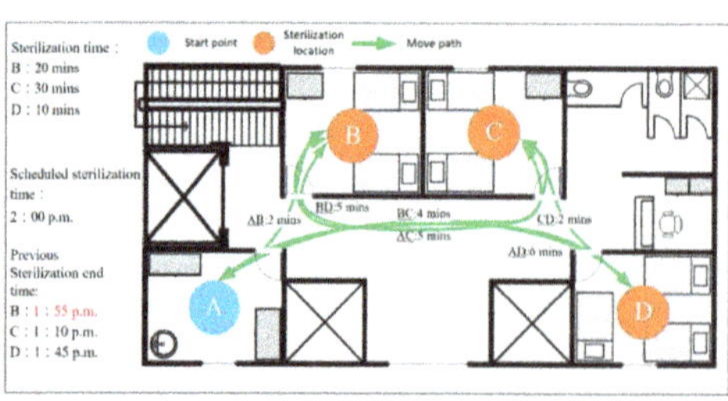

Figure 5. Simulation of the sterilization path: special case.

In a special case, we show a scenario in which the scheduled sterilization is currently at 2:00 p.m. and is about to begin (Figure 5). However, due to an anomalous value observed in room B, the sterilization process is finished at 1:55 p.m. When adapting the path parameters, it becomes clear that recommending room B shows a time interval that is not sufficient. In addition to considering the path parameters, we introduce a time interval penalty parameter. At each room, the interval between the time of arrival and completion of the previous sterilization are calculated. This interval is deducted after the complete process is calculated. This ensures enough time between sterilization, preventing a situation where we have just decontaminated and are straightaway heading to the next execution.

3. Selection: Once the fitness functions of all the combinations are obtained, they are ranked. When calculating the fitness function, it is possible to identify combinations with smaller fitness values as representing better results. These combinations require less time and longer intervals. In simpler cases, the optimal combination is found by sorting. However, in most cases, the scenarios are complex, so it is necessary to find more precise combinations of paths by means of mating and mutation. This iterative approach allows for the gradual improvement of the initial result until the optimal solution is reached. The use of mating and mutation in optimization problems is justified by the ability to refine the initial result until the optimal solution is identified. By employing these techniques, the precise combinations of paths can be identified to minimize the time spent and the time intervals.
4. Crossover: The top 10% of the sorted populations are selected for crossover. The purpose of crossover is to identify the segments of the superior pathway within superior-ranked combinations. Next, two sets of paths are randomly selected from the filtered populations for crossover. Considering two paths, where the path σ_1 is A > B > E > D > C > E and path σ_2 is A > D > E > B > C > A, a randomly selected segment of path σ_1, e.g., D > C, is combined with path σ_2 as a reserved path. However, due to the constraints mentioned above, it is necessary to delete D and C from path σ_2. Finally, the remaining paths are randomly assigned before or after the reserved paths, resulting in the combination path A > E > D > C > B > A. This obtained combination is added to the scenario, and its fitness function is calculated to produce a better combination of paths.
5. Mutation: It is still necessary to find the best path by matching the best genes and creating a better combination. If the two paths are similar, the algorithm may find the best solution. This can be solved by incorporating mutation processes. For example, say the selected path is A > D > E > B > C > A. In this case, a random exchange is conducted. If the replacement rooms are C and E, the path is A > D > C > B > E > A. This is performed to address the issue of path similarity. Finally, it is necessary to add the new combination path to the population and calculate its fitness function.

5. Conclusions

To monitor air quality, disinfect, and track hazardous substances in nursing homes, a fast and effective method to clean and disinfect harmful substances in the air is developed in this study. Considering the different needs of various rooms, IoT technology is used to implement a single air quality sensor, wireless transmission technology, and communication door control in long-term care facilities. The method allows for environmental detection based on air sensors and adjusts disinfection levels according to air quality indicators. An autonomous robot with disinfection capabilities is designed to provide intelligent sterilization and ensure the health and living conditions of nursing homes. The robot effectively addresses health risks related to air quality in nursing facilities and provides reliable sterilization solutions, creating a comfortable and safe environment for caregivers while improving the overall quality of service.

Author Contributions: Conceptualization, L.-H.H. and Z.-J.W.; methodology, L.-H.H. and C.-L.C.; writing—original draft preparation, L.-H.H. and Z.-J.W., C.-H.Y. and C.-L.C.; writing—review and editing, C.-H.Y. and C.-L.C.; visualization, Z.-J.W.; supervision, C.-L.C.; All authors have read and agreed to the published version of the manuscript.

Funding: This research is supported by the National Science and Technology Council of Taiwan, under research Project NSTC 112-2221-E-240-003.

Institutional Review Board Statement: Not applicable.

Informed Consent Statement: Not applicable.

Data Availability Statement: Data sharing is not applicable to this article.

Conflicts of Interest: The authors declare no conflict of interest.

References

1. Rose, W.J.; Ralston, P.M.; Autry, C.W. Urbanness and Its Implications for Logistics Strategy: A Revised Perspective. *Transp. J.* **2020**, *59*, 165–199. [CrossRef]
2. Yau, B.; Vijh, R.; Prairie, J.; McKee, G.; Schwandt, M. Lived experiences of frontline workers and leaders during COVID-19 outbreaks in long-term care: A qualitative study. *Am. J. Infect.* **2021**, *49*, 978–984. [CrossRef]
3. Espejo, W.; Celis, J.E.; Chiang, G.; Bahamonde, P. Environment and COVID-19: Pollutants, impacts, dissemination, management and recommendations for facing future epidemic threats. *Sci. Total Environ.* **2020**, *747*, 141314. [CrossRef] [PubMed]
4. Wu, X.; Nethery, R.C.; Sabath, M.B.; Braun, D.; Dominici, F. Exposure to air pollution and COVID-19 mortality in the United States: A nationwide cross-sectional study. *medRxiv* **2020**. [CrossRef]
5. Bashir, M.F.; Jiang, B.; Komal, B.; Bashir, M.A.; Farooq, T.H.; Iqbal, N.; Bashir, M. Correlation between environmental pollution indicators and COVID-19 pandemic: A brief study in Californian context. *Environ. Res.* **2020**, *187*, 109652. [CrossRef] [PubMed]
6. Rahman, M.S.; Peeri, N.C.; Shrestha, N.; Zaki, R.; Haque, U.; Ab, H. Defending against the Novel Coronavirus (COVID-19) outbreak: How can IoT (IoT) help to save the world? *Health Policy Technol.* **2020**, *9*, 136–138. [CrossRef] [PubMed]
7. Takabayashi, K.; Tanaka, H.; Sakakibara, K. Integrated Performance Evaluation of the Smart Body Area Networks Physical Layer for Future Medical and Healthcare IoT. *Sensors* **2018**, *19*, 30. [CrossRef] [PubMed]
8. Ndibanje, B.; Lee, H.-J.; Lee, S.-G. Security Analysis and Improvements of Authentication and Access Control in IoT. *Sensors* **2014**, *14*, 14786–14805. [CrossRef]
9. Hossain, M.; Islam, S.M.R.; Ali, F.; Kwak, K.-S.; Hasan, R. An Internet of Things-based health prescription assistant and its security system design. *Future Gener. Comput. Syst.* **2018**, *82*, 422–439. [CrossRef]
10. Vedaei, S.S.; Fotovvat, A.; Mohebbian, M.R.; Rahman, G.M.; Wahid, K.A.; Babyn, P.; Marateb, H.R.; Mansourian, M.; Sami, R. COVID-SAFE: An IoT-Based System for Automated Health Monitoring and Surveillance in Post-Pandemic Life. *IEEE Access* **2020**, *8*, 188538–188551. [CrossRef]
11. Pasluosta, C.F.; Gassner, H.; Winkler, J.; Klucken, J.; Eskofier, B.M. An Emerging Era in the Management of Parkinson's Disease: Wearable Technologies and IoT. *IEEE J. Biomed. Health Inform.* **2015**, *19*, 1873–1881. [CrossRef] [PubMed]
12. Bačík, J.; Tkáč, P.; Hric, L.; Alexovič, S.; Kyslan, K.; Olexa, R.; Perduková, D. Phollower—The Universal Autonomous Mobile Robot for Industry and Civil Environments with COVID-19 Germicide Addon Meeting Safety Requirements. *Appl. Sci.* **2020**, *10*, 7682. [CrossRef]
13. Sang, P.; Begen, M.A.; Cao, J. Appointment scheduling with a quantile objective. *Comput. Oper. Res.* **2021**, *132*, 105295. [CrossRef]
14. Park, J.; Kim, B.-I.; Eom, M.; Choi, B.K. Operating room scheduling considering surgeons' preferences and cooperative operations. *Comput. Ind. Eng.* **2021**, *157*, 107306. [CrossRef]
15. Coban, E. The effect of multiple operating room scheduling on the sterilization schedule of reusable medical devices. *Comput. Ind. Eng.* **2020**, *147*, 106618. [CrossRef]
16. Hasan, Q.; Rahmati, S.; Sharifi, A.; Ahmadi, M. HSSAGA: Designation and scheduling of nurses for taking care of COVID-19 patients using novel method of Hybrid Salp Swarm Algorithm and Genetic Algorithm. *Appl. Soft Comput.* **2021**, *108*, 107449.

engineering proceedings

MDPI

Proceeding Paper

Image Classification to Identify Style Composition Ratios in Crossover Cars †

Hung-Hsiang Wang and Hung-Jui Su *

Department of Industrial Design, National Taipei University of Technology, Taipei 10608, Taiwan; wanghh@mail.ntut.edu.tw
* Correspondence: howard9155@gmail.com
† Presented at the 2024 IEEE 7th International Conference on Knowledge Innovation and Invention, Nagoya, Japan, 16–18 August 2024.

Abstract: As the global demand for multifunctional and high-performance vehicles increases, automotive manufacturers face significant challenges in designing new crossover models. Consumers expect vehicles to blend features from various car models, which pushes the industry to adopt innovative design tools and methods. We explored the use of Waikato Environment for Knowledge Analysis (WEKA) image classification to predict the style composition ratios of sedans, hatchbacks, multi-purpose vehicles (MPVs), and sport utility vehicles (SUVs) as crossover vehicles. We collected 240 high-resolution side-view images of luxury vehicles from brands including Mercedes-Benz, BMW, and Lexus, and preprocessed the data using format unification and feature enhancement. We employed WEKA to extract image features and train a classification model using the edge histogram filter and sequential minimal optimization (SMO) classifier, which achieved an 86% classification accuracy. Subsequently, we used Vizcom, a generative Artificial Intelligence(AI) tool, to simulate realistic designs for new crossover cars and predict their style composition ratios. The proposed designs were evaluated by five experts, who found that the model accurately identified style composition ratios and helped designers create new car styles with market potential. The novel application of image classification can be used for analyzing blended styles in automotive design and enables designers to identify, evaluate, and control styles to meet market demands.

Keywords: car styling; car model; image classification; WEKA

1. Introduction

The automotive industry has shifted from a manufacturer-driven market to a consumer-driven market, where consumer needs are prioritized in selecting specific car models [1]. In the contemporary automotive industry, developing innovative vehicle types, especially electric vehicles and hybrid models, has become a trend, following the growing market demand for diversity and environmental protection [2]. Therefore, it is important to develop new vehicles that align with market trends and consumer expectations by applying image recognition technology and machine learning. This promotes design innovation and aligns with the future development direction of the automotive industry.

We explored the application of image recognition technology in automotive design, particularly in the development of new vehicles. With the increasing market demand for innovative designs and environmentally friendly models, artificial intelligence (AI) techniques need to be used for automotive development [3].

2. Related Work

2.1. AI in Industry

The application of AI in the industry has garnered widespread attention. By integrating digital sensors, network technologies, and automation systems [4], AI is redefining the economy and creating a new industrial era. Existing applications of AI are categorized into the following two types: "strong AI", which simulates human behavior and reasoning, and "weak AI", which uses machine learning to perform simple tasks traditionally carried out by humans, such as image recognition or natural language processing [3,4].

The core of AI lies in decision-making, and its greatest contribution to humanity is decision-making efficiency. When applied to business forecasting or industrial production, AI provides data-driven, objective decision-making recommendations. Similarly, AI in design thinking plays a crucial role in implementing core innovative decisions [5]. Therefore, AI is significant in promoting innovation and improving decision-making efficiency.

2.2. AI-Driven Automotive Styling

Applications of AI in automotive-related fields have primarily focused on autonomous driving systems [6], automotive manufacturing processes [7], intelligent traffic management [8], and vehicle recognition [9,10]. Previous studies have demonstrated the application of machine learning tools in car styling classification and consumer preference prediction. Hong et al. used AI to differentiate between fuel and electric vehicle types and product semantics, achieving high accuracy through image classification [11]. Wang et al. utilized deep learning and WEKA to predict consumer preferences for the front view of the BMW M3 series models, accurately explaining consumer preferences for innovative appearances [12]. Additionally, Wang and Chen analyzed the evolution of car styling and brand consistency for Dodge and Jaguar using deep learning methods, which accurately identified brand styles and design features [13]. However, these studies have not explored the application of machine learning in the design of different car types. To fill this gap, we developed an image recognition model to identify and select vehicle designs with market potential, thereby promoting design innovation in the automotive industry.

2.3. Image Recognition Technology

2.3.1. Applications of Machine Learning

Machine learning technology has been widely applied in the classification of automotive styling. Methods such as support vector machines (SVMs) and convolutional neural networks (CNNs) have shown excellent performance in image classification. These technologies are used in design and commercial trend forecasting. For instance, Wang and Chen used CNNs and heatmap analysis to classify and compare the styles of the Dodge and Jaguar brands. They demonstrated that deep learning is highly effective in identifying automotive design features and brand consistency [13]. Hong et al. used WEKA, a machine learning tool, to distinguish between the types and product semantics of fuel and electric vehicles. By analyzing 100 vehicle images, they showed that even small datasets can be used to train high-accuracy models, effectively distinguishing different vehicle types and semantics [11].

2.3.2. Image Feature Extraction Techniques

In automotive design, image feature extraction is a crucial step. The edge histogram, as a feature extractor, can effectively identify edge variations in vehicles, which is particularly important for distinguishing different vehicle models [14]. Analyzing edge histogram features in images allows machine learning models to accurately classify vehicle models. This technology enhances vehicle identification, which is crucial for intelligent transportation applications, and thereby improves classification performance.

3. Methods

We used the WEKA toolkit for image recognition to analyze styling components of sedans, hatchbacks, MPVs, and SUVs from brands including Mercedes-Benz, BMW, and Lexus and explore their application in designing new models. The research process consisted of the following three stages: data collection and preprocessing, model training, and application development.

3.1. Procedure

3.1.1. Data Collection and Preprocessing

We collected 60 high-resolution side-view images for each vehicle type from Mercedes-Benz, BMW, and Lexus, totaling 240 images. These images were standardized in format, resized for uniform resolution and dimensions, and enhanced using contrast and brightness adjustments and noise reduction to improve quality and subsequent analysis accuracy.

3.1.2. Feature Extraction and Model Training

Using WEKA's image processing tools, we extracted key features from vehicle images, including edge detection, texture analysis, and color histograms (specifically edge histogram). These features were used to train machine learning models. Based on preliminary trials, the best-performing classifiers, including SVM, random forest (RF), deep learning models, and optimized SVM, were selected. Cross-validation was employed to evaluate the models' generalization ability and accuracy, ensuring their effective application in practical styling recognition.

3.1.3. Application and Expert Evaluation

We proposed new hybrid vehicle designs based on model analysis results. We used AI technology to generate and optimize design sketches to determine market trends and consumer preferences. Using 2023 European preferences for mini cars (A-Segment) and crossovers, we employed generative AI (specifically Vizcom) to simulate designs of the Fiat 500e and Dacia Spring [2]. We developed a robust image recognition model to identify and select marketable vehicle designs, promoting innovation in the automotive industry.

We trained the WEKA classification model to analyze AI-generated design proposals and submit the results to industry experts for evaluation. We selected five experts with extensive automotive exterior design experience to provide comprehensive feedback, offering a wide range of professional insights. In the evaluation, each expert spent 20 min in a one-on-one assessment session and compared the AI analysis with their intuitive judgments to assess consistency and application value.

3.2. Data Collection

We applied image recognition technology in automotive design for the development of innovative models. To conduct in-depth image and text analysis, we collected a total of 240 images of sedans, hatchbacks, MPVs, and SUVs from the official sales websites of Mercedes-Benz, BMW, and Lexus between 2009 and 2024, as shown in Figure 1. Semantic terms related to "style" were selected. These curated data served as the primary data source for the study, supporting subsequent analyses, especially in image recognition and styling composition analysis.

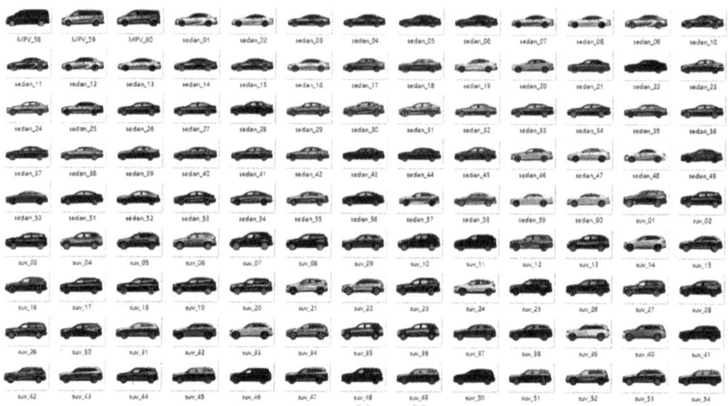

Figure 1. Examples of car design images.

3.3. Data Preprocessing

We standardized images collected from the official websites of Mercedes-Benz, BMW, and Lexus, as well as those acquired through the Copilot assistant integrated within Internet Explorer (IE). This involved unifying image formats and adjusting dimensions to ensure consistent resolution and size. To enhance automation and efficiency, we used Python 3 scripts to traverse images in Portable Network Graphic (PNG) in the designated folders and applied the rembg library for background removal. Using the deep learning-based library, we separated the foreground from the background to highlight the main features of the vehicles. All images were resized to fixed dimensions: 16 cm in width and 10 cm in height. Image enhancement processes, including contrast adjustment, brightness adjustment, and noise reduction, were performed to improve analysis precision. These steps were executed using Photoshop, and the saturation of the photos was adjusted to black and white grayscale using Microsoft tools to ensure consistency in visual analysis.

When the processing was complete, the images were renamed and saved in the output folder, according to WEKA's naming conventions. For this, we used Python 3's counter and string formatting functions, which significantly sped up processing and ensured consistency in the naming format. We enhanced the visual quality of the images effectively, ensuring optimal conditions for subsequent image recognition and feature extraction. This, in turn, supported the training of machine learning models and styling recognition analysis. Through these carefully designed preprocessing steps, the model laid a solid foundation for improving the overall quality and reliability of the research.

3.4. Feature Extraction and Classification Techniques

We used various image processing tools of the WEKA platform to analyze and extract key image features (Figure 2), including edge detection, texture analysis, and color histograms. These features are particularly important for identifying and classifying vehicle styles. Additionally, we utilized the SMO classifier in WEKA to train the SVM model and effectively classify vehicle types by combining different kernel functions and regularization parameters [15]. Extracting features such as edges, textures, and colors was essential for transforming vehicle image data into inputs for machine-learning models. This process captured the stylistic characteristics of different vehicle models for the development of predictive models for automatic vehicle style recognition. By using the filters, the models' predictive capability was enhanced, and a foundation for understanding various vehicle design characteristics was provided. This approach aids in accurately identifying vehicle styles and offers designers valuable insights, supporting new directions and innovations in vehicle design.

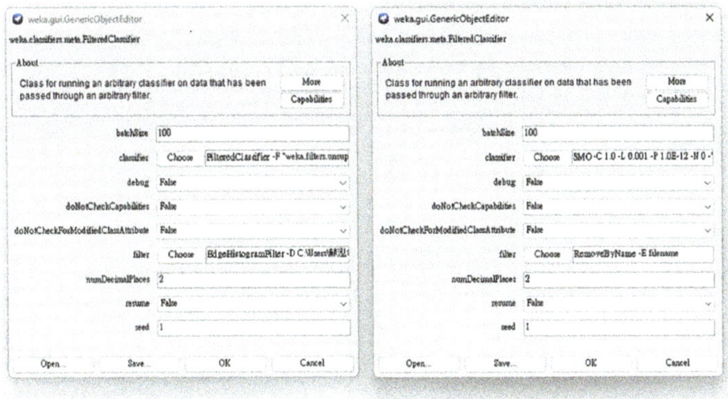

Figure 2. Example of WEKA filters and classifiers.

3.5. Validation and Testing

To verify the usability and accuracy of the database, we experimented with 240 images. Of these, 60% (144 images) were used as the training set, and the remaining 40% (96 images) were used as the testing set. The images included four different vehicle types to evaluate the model's performance in vehicle-type recognition. The experimental results (Figure 3) indicated that the model achieved an accuracy of 86.1% on the test set. To confirm reliability, we manually compared the classifications of the 96 images in the test set. After careful examination, 15 images were incorrectly classified, resulting in an actual accuracy of 84.4%. Although the accuracy was lower than the initial automatic accuracy, the model remained reliable and effective.

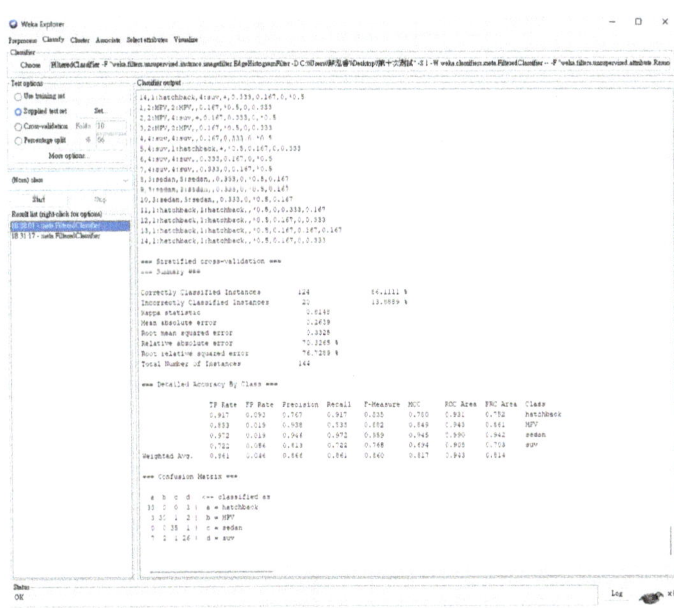

Figure 3. Accuracy of model on the testing set was up to 86.1%.

We expanded the training dataset from 144 to 240 images to improve the model's generalization ability and accuracy. After training, the model's accuracy increased to 86.6%, improving its performance with the increased data volume. To verify the model's effectiveness in recognizing complex vehicle styles, we tested two different crossover models from different manufacturers (Figures 4 and 5). When classifying models from the same series of the same manufacturer, the model accurately identified style compositions and proportions. The style compositions and proportions of the same model types were similar, indicating that the model could identify the style composition of different vehicles (Tables 1 and 2).

Figure 4. Six test samples of crossover designs from BMW.

Figure 5. Three test samples of crossover designs from Lexus.

Table 1. Analysis results of six crossover models from BMW.

	Hatchback	MPV	Sedan	SUV
Unknown01	16.7%	33.3%	0%	50%
Unknown02	33.3%	16.7%	0%	50%
Unknown03	33.3%	16.7%	0%	50%
Unknown04	33.3%	0%	16.7%	50%
Unknown05	33.3%	0%	16.7%	50%
Unknown06	33.3%	0%	16.7%	50%

Table 2. Analysis results of three crossover models from Lexus.

	Hatchback	MPV	Sedan	SUV
Unknown01	33.3%	0%	33.3%	33.3%
Unknown02	33.3%	0%	33.3%	33.3%
Unknown03	50%	0%	16.7%	33.3%

The model performed satisfactorily in basic vehicle type recognition and in analyzing complex style combinations. The model's effectiveness was also verified in basic vehicle type recognition and complex style combinations, providing a solid foundation for future research. Additionally, they offer information on automotive design and market analysis, helping the industry understand and predict consumer preferences for various vehicle styles.

Considering European consumer preferences for mini cars (Segment A) and crossovers from 2023 to the present, Fiat 500e and Dacia Spring models were analyzed by the invited experts [2]. Generative AI technology (Vizcom) was used to create these proposals (Figure 6). The model analysis results of the simulated proposals were compiled into a detailed chart, presenting the style composition data for each vehicle model (Table 3). The experts were able to make well-informed evaluations and provide valuable feedback for improving the model and its application in vehicle design.

Figure 6. Two test samples of crossover designs generated using Vizcom.

Table 3. Analysis results of two crossover models generated using Vizcom.

	Hatchback	MPV	Sedan	SUV
Unknown01	16.7%	33.3%	0%	50%
Unknown02	33.3%	0%	33.3%	33.3%

We collected and analyzed expert opinions and suggestions on the style results, focusing on areas where the insights of the experts differed from the model's output. After the meeting, we integrated their feedback to assess the AI model's accuracy and reliability. Based on the experts' feedback, we made necessary adjustments to enhance the model's performance and practical value.

We invited five experts, including two women and three men, with more than 4 years of industrial design experience to participate in the validation via an interview. From the interview, the reasonableness of the analyzed vehicle style proportions was evaluated (Table 4), and the model's value in predicting sales trends and designing new vehicle models was estimated (Table 5). In the 20 min interview, data presentation, initial insights, discrepancies, and potential improvement measures were discussed.

Table 4. Five experts' answers to Q1.

Answer	Reasons and Suggestions
Expert A (agree)	Car model one has the highest proportion of SUVs, which aligns with my understanding. The proportion of hatchbacks and MPVs is also appropriate. Car model two appears to be evenly split among hatchbacks, sedans, and SUVs. In reality, it combines the front of an SUV, the chassis of a sedan, and the rear of a hatchback, making it highly valuable for reference.
Expert B (agree)	In model one, SUVs have the highest proportion, matching technical and design logic. The proportions of hatchbacks and MPVs are also appropriate. Model two seems to be evenly split among hatchbacks, sedans, and SUVs. In reality, it combines the front of an SUV, the chassis of a sedan, and the rear of a hatchback. This innovative design shows the potential of technical fusion and is valuable for reference.

Table 4. *Cont.*

Answer	Reasons and Suggestions
Expert C (agree)	Model one shows SUVs as the highest proportion, matching my judgment. Hatchbacks and MPVs are also reasonable. This composition combines an SUV front, sedan chassis, and hatchback rear. The analysis matches my observations and is valuable for reference.
Expert D (agree)	The model for car one shows the highest proportion of SUVs, slightly lower for MPVs, and an appropriate proportion of hatchbacks, reflecting actual hatchback features. For car two, the analysis shows an even split among hatchbacks, sedans, and SUVs, but I believe the hatchback proportion should be higher and the SUV and sedan proportions lower. Despite this discrepancy, the analysis method is still valuable for reference.
Expert E (agree)	The model for car one shows the actual car is composed of various car features. For car two, the analysis results show an even split among hatchbacks, sedans, and SUVs. However, I believe hatchbacks should be 60% and sedans close to 0%. Despite the model fitting proportions from a features perspective, it has blind spots in length and wheelbase. This analysis method is still valuable for reference.

Table 5. Five experts' answers to Q2.

Answer	Reasons and Suggestions
Expert A (valuable)	The classification model, through analyzing sales data, can accurately predict market reactions to new car models and develop effective strategies. It is expected to help designers create more popular models, increasing the success rate.
Expert B (valuable)	Using the classification model to analyze sales data of best-selling car models helps designers understand market demands, improving the acceptance and sales success rate of new car models.
Expert C (valuable)	From a product development perspective, the classification model can refine design directions, reduce risks, improve efficiency, and ensure that new car models better meet market demands, increasing their chances of success.
Expert D (valuable)	The application of the classification model can provide designers with market insights, assisting in the creation of more popular car models. If applied in design education, it can help educational institutions adjust their curriculum to train competitive design talent, promoting the development of the automotive industry.
Expert E (valuable)	I believe the application of the classification model is very helpful. It demonstrates the potential of data analysis in business decisions and provides valuable market insights, helping me better understand consumer needs.

These discussions leveraged the experts' knowledge to optimize the research model and advance the project. Integrating expert feedback was crucial for refining the model and ensuring its practical application in vehicle design and market analysis.

3.5.1. Reasonableness of Analyzed Car Model

We asked the experts two questions: Q1: "Do you agree with the analyzed car model proportions?" and Q2: "Please explain your reasons and discuss whether you think these proportions are reasonable". Table 4 presents the answers of five experts to these questions. All experts agreed with the analysis results and provided detailed reasons and suggestions. Experts A and B noted that the proportion of SUVs in car model one was the highest, aligning with technical and design logic, and they considered the hybrid design of car model two to be highly valuable. Experts C and D believed that the proportion of hatchbacks in car model 2 should be higher, but they still found the analysis valuable. Expert E suggested that the proportion of hatchbacks had to reach 60%. These opinions indicated that the analysis in this study gained expert recognition and provided a reference for the improvement of our analysis methods, supporting designers in the decision-making process for innovative hybrid car models.

3.5.2. Use of the Classification Model in Predicting Car Sales Trends and New Car Model Design

We explored the possibility of the application of the classification model in predicting car sales trends and designing new car models. The experts answered the following question: "How do you assess the practical application value of the classification model in these areas? Please explain your reasoning". We analyzed the sales data of best-selling car models to predict whether new design proposals align with market trends, helping designers understand market demands. The results assist car manufacturers in conducting competitive analysis by providing data on competitors' car model compositions to improve or emulate them. Table 5 presents the responses of five experts, all of whom found the method valuable and provided detailed reasons and suggestions. They emphasized that the classification model helps to optimize design directions, reduce risks, improve efficiency, and develop the automotive industry. These opinions indicate that the analysis method can be used for predicting car sales trends and designing new car models, providing information for further improvement.

In summary, the experts found the model's classification to be reasonable and valuable. The high proportion of SUVs in model one was evaluated to be appropriate, while model two demonstrated innovative design with the potential for technical integration. All experts agreed that the classification model has significant potential in predicting car sales trends and designing new car models. By analyzing sales data, designers can better understand market demands, creating more consumer-friendly models and increasing the success rate of new designs. Additionally, the classification model refines design directions early in the development process to reduce risks, improve design efficiency, and enhance market competitiveness.

The experts' feedback highlights the importance of the classification model in car design and market analysis, confirming its effectiveness and reliability in practical applications.

4. Conclusions

We explored the use of the WEKA toolkit and image recognition technology by analyzing the styling compositions of sedans, hatchbacks, MPVs, and SUVs sold by Mercedes-Benz, BMW, and Lexus. We developed machine learning models to classify stylistic features of these vehicle types and applied generative AI to design new vehicle models. The usefulness of the developed model was validated by the invited experts and demonstrated its effectiveness in automotive design and development. The classification model helps designers better understand market demands by aligning vehicle designs with consumer preferences and focusing on design directions to effectively shorten development time and reduce risks. Suggestions were made for how the automotive industry and design education can use AI tools more effectively. Future research is needed to develop advanced techniques for and apply them to style composition.

Author Contributions: Conceptualization, H.-H.W.; method, H.-H.W.; resources, H.-J.S.; data curation, H.-J.S.; validation, H.-H.W.; writing—original draft preparation, H.-J.S.; writing—review and editing, H.-H.W.; experiments, H.-J.S.; supervision, H.-H.W. All authors have read and agreed to the published version of the manuscript.

Funding: This research was funded by NSTC, Taiwan, through grant 111-2410-H-027-019-MY2.

Institutional Review Board Statement: Not applicable.

Informed Consent Statement: Not applicable.

Data Availability Statement: Data are unavailable due to privacy reasons.

Conflicts of Interest: The authors declare no conflicts of interest.

References

1. Agarwal, T.; Saroha, N.; Kumar, G. Consumer Behaviour Analysis for Purchasing a Passenger Car in Indian Context. In Proceedings of the Mathematical Analysis II: Optimisation, Differential Equations and Graph Theory: ICRAPAM 2018, New Delhi, India, 23–25 October 2020; pp. 165–177.
2. Guan-Chun, Z. 2023 Global Electric Vehicle Sales Ranking: PHEVs and Mini Cars Become Market Favorites. Available online: https://www.artc.org.tw/tw/knowledge/articles/13733 (accessed on 10 July 2024).
3. Verganti, R.; Vendraminelli, L.; Iansiti, M. Innovation and Design in the Age of Artificial Intelligence. *J. Prod. Innov. Manag.* **2020**, *37*, 212–227. [CrossRef]
4. Iansiti, M.; Lakhani, K.R. From Disruption to Collision: The New Competitive Dynamics. *MIT Sloan Manag. Rev.* **2020**, *61*, 34–39.
5. Liedtka, J. Perspective: Linking Design Thinking with Innovation Outcomes through Cognitive Bias Reduction. *J. Prod. Innov. Manag.* **2014**, *32*, 925–938. [CrossRef]
6. Bayyou, D. Artificially intelligent self-driving vehicle technologies, benefits and challenges. *Int. J. Emerg. Technol. Comput. Sci. Electron.* **2019**, *26*, 5–13.
7. Wang, J.; Ma, Y.; Zhang, L.; Gao, R.X.; Wu, D. Deep learning for smart manufacturing: Methods and applications. *J. Manuf. Syst.* **2018**, *48*, 144–156. [CrossRef]
8. Dhingra, S.; Madda, R.B.; Patan, R.; Jiao, P.; Barri, K.; Alavi, A.H. Internet of things-based fog and cloud computing technology for smart traffic monitoring. *Internet Things* **2021**, *14*, 100175. [CrossRef]
9. Kamencay, P.; Markovic, M.; Dubovan, J.; Dado, M.; Benedikovic, D. Neural-Network-Optimized Vehicle Classification Using Clustered Image and Fiber-Sensor Datasets. *IEEE Access* **2023**, *11*, 41315–41324. [CrossRef]
10. Hicham, B.; Ahmed, A.; Mohammed, M. Vehicle Type Classification Using Convolutional Neural Network. In Proceedings of the 2018 IEEE 5th International Congress on Information Science and Technology (CiSt), Marrakech, Morocco, 21–27 October 2018; pp. 313–316.
11. Wang, H.-H.; Shen, Y.-Y.; Hung, Y.-Y. Car Types and Semantics Classification Using Weka. In Proceedings of the 2022 5th Artificial Intelligence and Cloud Computing Conference, Osaka, Japan, 17–19 December 2022; pp. 34–41.
12. Wang, H.-H.; Liu, C.-K.; Yu, S.-B. Predicting Consumer Preferences by the Deformation Threshold of Product Appearance. In Proceedings of the 2023 IEEE 5th Eurasia Conference on Biomedical Engineering, Healthcare and Sustainability, Tainan, Taiwan, 2–4 June 2023.
13. Wang, H.-H.; Chen, C.-P. A case study on evolution of car styling and brand consistency using deep learning. *Symmetry* **2020**, *12*, 2074. [CrossRef]
14. Tan, S.H.; Chuah, J.H.; Chow, C.-O.; Kanesan, J.; Leong, H.Y. Artificial intelligent systems for vehicle classification: A survey. *Eng. Appl. Artif. Intell.* **2024**, *129*, 107497. [CrossRef]
15. Asry, N.I.M.; Mustapha, A.; Mostafa, S.A. Comparative Analysis of Machine Learning Algorithms in Vehicle Image Classification. In *Methods and Applications for Modeling and Simulation of Complex Systems*; Communications in Computer and Information Science; Springer: Singapore, 2024; pp. 426–436.

engineering proceedings

MDPI

Proceeding Paper

Response Measures to Climate Change to Maintain Resilient Rice Production in Taiwan †

Chin-Song Wu and Tzu-Che Lin *

Department of Plant Industry, National Pingtung University of Science and Technology, Shuefu Road, Neipu, Pingtung 912, Taiwan; wuchinsung@mail.npust.edu.tw
* Correspondence: tclin@mail.npust.edu.tw

† Presented at the 2024 IEEE 7th International Conference on Knowledge Innovation and Invention, Nagoya, Japan, 16–18 August 2024.

Abstract: Recently, the intensification of the contrast between Taiwan's wet and dry seasons due to climate change has led to stringent water use restrictions in agriculture, significantly impacting Taiwan's primary staple crop, rice. This necessitates the development of effective agricultural management strategies to ensure the resilience of Taiwan's agriculture. Therefore, we simulated potential challenges in rice cultivation due to climate change under specific conditions: flooded cultivation and soil water tension levels of −20 and −40 kPa. At −40 kPa, the soil becomes excessively dry, causing severe soil surface cracking. This results in a 20% reduction in plant height and a 30% decrease in yield compared to flooded cultivation. At −20 kPa, plant height and yield are comparable to those under flooded conditions. In resource efficiency, flooded cultivation demonstrates low irrigation water use efficiency (0.22 and 0.42 kg/m^3) due to sustained high water levels. Conversely, the condition with −20 kPa shows the highest irrigation water use efficiency (2.15 and 2.16 kg/m^3) with no significant difference in nitrogen use efficiency compared to flooded management. Although the irrigation water use efficiency under −40 kPa is better than flooded management (0.87 and 1.02 kg/m^3), the nitrogen utilization efficiency is significantly low. Irrigation under the condition of −20 kPa climate change does not reduce yields and offers additional benefits. This strategy ensures stable crop production and conserves water resources for crop cultivation during the dry season, providing an effective means to stabilize production and mitigate the impacts of climate change on Taiwan's agriculture.

Keywords: climate change; rice; soil water tension; irrigation water use efficiency; nitrogen use efficiency

1. Introduction

Rice (*Oryza sativa* L.) is the key crop in Asia, and it is eaten by half of the world's population. The global total rice cultivation area is 168,356,566 ha, and the annual output is 799,999,504 metric tons [1]. The repercussions of global warming, a direct result of changes in the international environment, have changed the rice production landscape. Shifts in rainfall and temperatures have directly affected water storage capacity and soil moisture, while other climatic factors such as cloud cover, humidity, and evapotranspiration have further complicated the utilization and allocation of water resources. This presents a pressing challenge to rice production, necessitating immediate attention and action [2].

Taiwan has a rich environment with high temperatures and rainy weather. It has monsoon rain, orographic rain, thermal thunderstorms, typhoon rain, and tropical depression rainfall, resulting in uneven space and time distribution. Seasonal uneven distribution

 https://doi.org/10.3390/engproc2025089027

and long-term distribution are observed. The annual rainfall is unevenly distributed, leading to a disparity in rainfall during high and dry periods, short rivers, and poor water storage rates.

With the recent constraints on agricultural water use due to freshwater scarcity, the necessity for water-saving measures in rice cultivation has become pronounced. For instance, the implementation of the first-phase rice fallow policy in the Chianan region in 2023 and the promotion of comprehensive water-saving measures for crop rotation have far-reaching effects on issues such as rice seedling supply across Taiwan. These measures are not beneficial as they are imperative for the sustainability of rice production.

In Taiwan's traditional paddy fields, water use progresses through three stages, from land preparation to harvesting. The first stage involves water delivery, distribution, and division, which transport water to the fields and convert it into soil moisture. The second stage is the absorption of soil moisture by crop roots, transforming it into the water the crops need. In the final stage, water enters the plant through diffusion or penetration, facilitating transport to various parts for growth and metabolism. However, significant water losses occur at all three stages, including 16 to 18% due to evaporation and 50 to 72% due to runoff and leakage, resulting in only 10 to 12% of the irrigation water being effectively utilized by the rice plants [3].

Recently, the development and implementation of climate change adaptation strategies have gained traction. In the growth stage of rice, the timing of water shortages significantly influences various yield components. For instance, water stress during the tillering stage can reduce the number of panicles per plant, while water shortages during the panicle primordium differentiation stage can decrease the number of grains per panicle. Insufficient water uptake during critical physiological stages reduces leaf water potential and photosynthesis rates, leading to a decrease in photosynthetic products, which ultimately affects both the yield and quality of rice [4].

The impacts of continuous flooding and water-saving irrigation were investigated on rice growth, yield, water use efficiency, and nitrogen use efficiency to identify irrigation practices that conserve water while maintaining stable yields.

2. Materials and Methods

2.1. Study Site and Methodology

This experiment was conducted in September 2022 (2022—2nd cropping season) and February 2023 (2023—1st cropping season) at the National Pingtung University of Science and Technology. The rice cultivar is Tainong No. 71, and the experimental site is divided into three scenario-setting areas: traditional cultivation (continuous flooding, CF) as the control group; soil water tension meter (kilopascal, kPa) readings −20 and −40 kPa were set for treatment groups; each test area was about 155 m^2; and the interval between each scenario setting area was 1.5 m.

In each treatment, the water content was controlled to 2 to 3 cm on the soil surface in the first 10 days after transplanting to facilitate the stable growth of the seedlings to the survival stage. Afterward, soil moisture tensiometers (Spectrum Technologies- Irrometer Tensiometers 6420, Spectrum Technologies, Inc. Aurora, IL, USA) were embedded in the water-saving scenario test areas. Continuous flooding occurs when the water level of rice is maintained at 3 to 5 cm during the vegetative growth period. After the highest tillering period, a one-week drying period is carried out. At the beginning of the reproductive growth period, the water level is maintained at 3 cm. After the plant is stable, the water level is raised to 5 to 10 cm. The water level is lowered to 3 cm during the early stage, raised from 5 to 10 cm after earing begins, and reduced to 3 cm during the yellow ripening stage until the water is cut off 5 days before harvesting.

Water-saving cultivation includes −20 and −40 kPa in the soil water tension meter. During the vegetative growth period, it is monitored whether it has reached −20 or −40 kPa daily. If it reaches −20 kPa or −40 kPa, water is supplied. If it reaches −5–−8 kPa, the water return stops to let the value reach 0 kPa. Fertilization is provided before planting (base fertilizer), 15 days after planting, and 30 days after planting; ear fertilizer is also provided. The nitrogen ratio is 3:2:2:3 (base fertilizer: first top dressing: second top dressing: ear fertilizer) based on a nitrogen content of 120 kg/ha. The prevention and control of pests, diseases, and weeds are based on conventional agricultural methods. Water meters were installed on irrigation water pipes in each scenario setting field to record water consumption.

2.2. Sampling

The first sampling was carried out after the seedlings survived. In each setting area, 40 samples were taken for measurement at the opportunity. The plant height and relative chlorophyll content were measured every Sunday morning. After harvesting after the maturity stage, the plants in each scenario setting area were cut into a 1 × 1 m area, and two areas were randomly selected to measure parameters, such as effective tiller, grain number, fertility rate, thousand-grain weight, and yield per unit area.

2.3. Evaluation of Values

Under climate change, the stability, advantages, and disadvantages of growth and production were assessed when using water-saving or other methods for cultivation under climate change. After determining if cultivation is stable, the value of cultivation under climate change is evaluated. During the experiment, irrigation water use efficiency (IWUE) and nitrogen use efficiency (NUE) were measured. IWUE refers to the yield divided by irrigation water consumption, which is used to calculate irrigation water efficiency [5]. NUE is the yield divided by the total amount of nitrogen applied to understand the economics of nitrogen application.

2.4. Statistical Methods

All the data analysis was performed using SAS 9.4. Analysis of variance (ANOVA) was used to assess the significant differences among the data. Fisher's Least Significant Difference (LSD) test was used to determine the significant differences with a significance level of $\alpha = 0.05$.

3. Results and Discussion

3.1. Rice Growth in Different Scenarios

We recorded the effects of various water supply modes on growth traits during rice cultivation to assess the need for adjustments in cultivation practices under different scenario settings (Figures 1 and 2). The control group (CF) and treatment groups with soil water tensions of −20 and −40 kPa were analyzed for differences in plant height and relative chlorophyll content. During the 2022 second cropping season (49 days after transplanting, DAT) and the 2023 first cropping season (28 DAT), plant height in the CF group was larger than in the −20 kPa and −40 kPa groups in all cultivation stages (Figure 1). The SPAD value, an indicator of chlorophyll content and photosynthetic efficiency, was lower in CF than in other treatments from 7 to 42 DAT but surpassed the scenarios of −20 and −40 kPa during the mid-growth period (49 to 56 DAT) (Figure 2). In the heading stage, the rice plant height must be between 90 and 110 cm. Too high plants show easy lodging, and too short plants yield poorly [6]. SPAD is a major indicator of plant photosynthetic efficiency. The CF scenario presents the risk of easy lodging under the −40 kPa scenario with a risk of poor yield. The rapid decline in SPAD during the reproductive growth period is crucial to carefully select the appropriate water supply pattern to balance the difficulty of cultivation with the performance of the final yield.

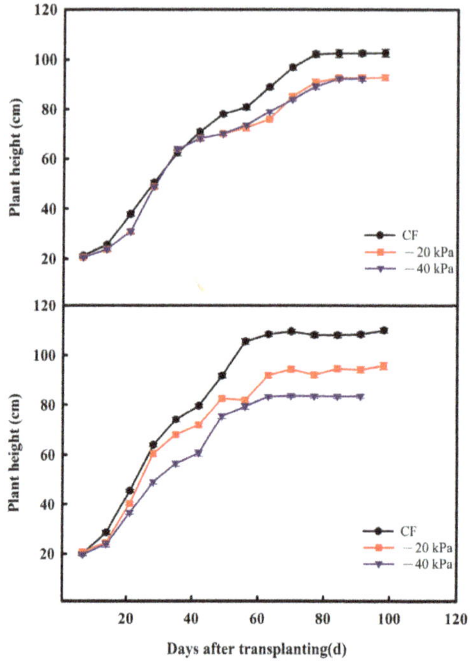

Figure 1. Effects of different situations on rice growth.

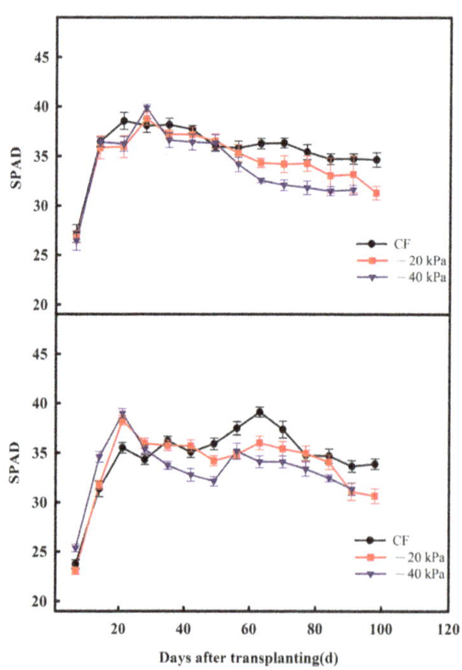

Figure 2. Effects of different scenarios on SPAD of rice.

3.2. Post-Harvest Yield Survey

A post-harvest survey was conducted after the rice maturity period (Table 1). If different water supply modes were used in the two seasons of cropping, regardless of the number of effective tillers, grain number per plant, yield, and thousand-grain weight, the CF and −20 kPa scenarios had more yields than the −40 kPa scenarios. If the soil water tension remains below −40 kPa for a long time, the effective tiller number of rice reduces, thereby decreasing the yield, ultimately leading to a decrease in yield. In addition, −20 kPa allows for maintaining yield performance similar to that of CF, which is related to root biomass [7].

Table 1. The impact of different scenarios on the components and yield of rice yield.

	ET (no.)	GN (no.)	FR (%)	TGW (g)	Yield (kg/ha)
2022—2nd cropping season					
CF	17.8 ^{a}	1071.0 ^{a}	70.2 ^{a}	22.6 ^{a}	3626.2 ^{a}
−20 kPa	17.3 ^{a}	1051.4 ^{a}	75.0 ^{a}	19.3 ^{ab}	3204.4 ^{a}
−40 kPa	16.1 ^{a}	604.9 ^{b}	54.9 ^{b}	18.0 ^{b}	1851.3 ^{b}
2023—1st cropping season					
CF	28.2 ^{a}	1277.9 ^{a}	72.3 ^{a}	26.4 ^{ab}	4454.7 ^{a}
−20 kPa	21.9 ^{a}	1060.6 ^{b}	75.4 ^{a}	27.2 ^{a}	4709.7 ^{a}
−40 kPa	16.6 ^{c}	874.1 ^{b}	57.4 ^{b}	25.9 ^{b}	4110.9 ^{b}

Means followed by different letters within columns differ significantly at $p < 0.05$ according to Fisher's LSD test. ET: effective tiller, GN: grain number, FR: fertility rate, TGW: thousand-grain weight.

3.3. Feasibility Assessment

While adequate irrigation enhances rice yields, it may also lead to unproductive water loss. Reducing water supply generally decreases plant height, particularly due to restricted growth during the vegetative stage. However, the −20 kPa scenario demonstrated that appropriate water-saving irrigation can maintain yields, as there was no significant difference in yield during the second cropping season. In addition, CF increases the number of rice tillers, growth consistency, excellent growth potential, and grain number per plant, thereby increasing yield. Water also participates in important physiological functions, such as photosynthesis and protein synthesis. The CF scenario increases tillering, consistent growth, superior growth potential, and a higher grain number per plant, contributing to higher yields. Water plays a crucial role in physiological processes, including photosynthesis and protein synthesis (Table 2).

Table 2. Assessment of added value in different situations.

	Irrigation Water Consumption (m^3)	IWUE (kg/m^3)	NUE (kg/m^3)
2022—2nd cropping season			
CF	135.00	0.42 ^{c}	30.22 ^{a}
−20 kPa	23.00	2.16 ^{a}	26.70 ^{a}
−40 kPa	33.00	0.87 ^{b}	15.43 ^{b}
2023—1st cropping season			
CF	316.00	0.22 ^{c}	37.12 ^{a}
−20 kPa	34.00	2.15 ^{a}	39.24 ^{a}
−40 kPa	62.00	1.02 ^{b}	34.26 ^{b}

Means followed by different letters within columns differ significantly at $p < 0.05$ according to Fisher's LSD test.

Regarding added value under different scenarios (Table 2), traditional CF cultivation exhibits a high nitrogen use efficiency but poor irrigation water use efficiency (IWUE).

Given the concentration of rainfall in Taiwan, traditional CF may not be sustainable under climate change [8]. The −20 kPa scenario offers better IWUE and stable nitrogen use efficiency (NUE), making it a viable cultivation method under changing climate conditions. In contrast, the −40 kPa scenario, despite having good IWUE, has significantly poor NUE, leading to lower yields and making it a less favorable option. The −40 kPa scenario has good IWUE and significantly poor NUE, indicating that the yield is worse than in other scenarios, so it is not a better cultivation situation.

4. Conclusions

In addition to maintaining stable yields and reducing the impact of climate change on agriculture, planting methods under climate change have advantages in terms of added value. Therefore, in the face of the challenge of climate change, irrigation in the −20 kPa reduces yields and has the advantage of stabilizing production and increasing agricultural resilience under climate change.

Author Contributions: Conceptualization, C.-S.W. and T.-C.L.; methodology, C.-S.W.; software, C.-S.W.; validation, C.-S.W. and T.-C.L.; formal analysis, C.-S.W.; investigation, C.-S.W.; resources, T.-C.L.; data curation, C.-S.W.; writing—original draft preparation, C.-S.W.; writing—review and editing, T.-C.L.; visualization, T.-C.L.; supervision, T.-C.L.; project administration, T.-C.L.; funding acquisition, T.-C.L. All authors have read and agreed to the published version of the manuscript.

Funding: This research received no external funding.

Institutional Review Board Statement: Not applicable.

Informed Consent Statement: Not applicable.

Data Availability Statement: The raw data supporting the conclusions of this article will be made available by the authors upon request.

Conflicts of Interest: The authors declare no conflicts of interest.

References

1. FAOSTAT. Production/Yield quantities of Rice. 2023. Available online: https://www.fao.org/faostat/en/#data/QCL/visualize (accessed on 26 February 2025).
2. Maraseni, T.N.; Deo, R.C.; Qu, J.; Gentle, P.; Neupane, P.R. An international comparison of rice consumption behaviours and greenhouse gas emissions from rice production. *J. Clean. Prod.* **2018**, *172*, 2288–2300. [CrossRef]
3. Bouman, B.A.M.; Tuong, T.P. Field water management to save water and increase its productivity in irrigated lowland rice. *Agric. Water Manag.* **2001**, *49*, 11–30. [CrossRef]
4. Fageria, N.K. Yield physiology of rice. *J. Plant Nutr.* **2007**, *30*, 843–879. [CrossRef]
5. Haonan, Q.; Jie, W.; Shihong, Y.; Zewei, J.; Yi, X. Current status of global rice water use efficiency and water-saving irrigation technology recommendations. *J. Agron. Crop Sci.* **2023**, *209*, 734–746. [CrossRef]
6. Zhang, Y.; Yu, C.; Lin, J.; Liu, J.; Liu, B.; Wang, J.; Huang, A.; Zhao, T. *OsMPH1* regulates plant height and improves grain yield in rice. *PLoS ONE* **2017**, *12*, e0180825. [CrossRef] [PubMed]
7. Yan, J.; Wu, Q.; Qi, D.; Zhu, J. Rice yield, water productivity, and nitrogen use efficiency responses to nitrogen management strategies under supplementary irrigation for rain-fed rice cultivation. *Agric. Water Manag.* **2022**, *263*, 107486. [CrossRef]
8. Wu, J.; Lin, T.; Chen, Z.; Yeh, Y. Effect of Different Irrigation Regimes on Rice Yield and Water Use Efficiency. *J. Agric. Assoc. Taiwan* **2022**, 264–275. [CrossRef]

engineering proceedings

MDPI

Proceeding Paper

Evaluating Feasibility of Pose Detection with Image Rotation for Monitoring Elderly People at Home †

Sinan Chen [1,*] and Masahide Nakamura [1,2]

1 Center for Mathematical and Data Sciences, Kobe University, 1-1 Rokkodai-cho, Nada-ku, Kobe 657-8501, Japan; masa-n@cs.kobe-u.ac.jp
2 RIKEN Center for Advanced Intelligence Project, 1-4-1 Nihonbashi, Chuo-ku, Tokyo 103-0027, Japan
* Correspondence: chensinan@gold.kobe-u.ac.jp
† Presented at the 2024 IEEE 7th International Conference on Knowledge Innovation and Invention, Nagoya, Japan, 16–18 August 2024.

Abstract: This study aims to enhance human pose detection performance through image preprocessing. With the growing global elderly population, supporting in-home elderly individuals has become increasingly crucial, especially in Japan, where approximately 90% of the elderly live at home. In this study, abnormal behaviors in in-home elderly individuals were detected using high-precision pose detection technologies, such as the MoveNet model, based on machine learning. However, there was a discrepancy in pose detection accuracy between the standing and lying positions. Hence, preprocessing techniques such as image rotation, flipping, resizing, and noise reduction were applied, and their effects were analyzed in detail. As a result, an improvement in pose detection accuracy for most body parts, except for specific regions, was observed. By constructing an optimal preprocessing pipeline, it is expected to reduce false detection and missed detection, contributing to the practical application of pose detection technology.

Keywords: human pose detection; image preprocessing; elderly monitoring; assistive technology

1. Introduction

As of 2021, approximately 80% of the world's elderly population aged 65 and over live at home [1]. The proportion of elderly individuals living at home is particularly high in developed countries; for instance, in Japan, about 90% of the elderly population resides at home [2]. The number of dementia patients worldwide is estimated to reach 152 million by 2050 [3]. There is an increasing need to monitor dementia patients who are at higher risk of going missing or experiencing accidents [4]. The weakening of community ties and the decline in the caregiving capacity of families are contributing to the isolation of the elderly. Ensuring their safety and providing life support for isolated elderly individuals has become a critical issue. Consequently, the demand for monitoring services that support home care is rising [5].

Several systems and services utilizing engineering technology are currently available for monitoring elderly individuals living at home. Systems that detect the behavior and daily rhythms of the elderly using sensors and notify caregivers of any abnormalities have become widespread [6].

Services that employ artificial intelligence (AI) to learn the behavior patterns of the elderly and detect anomalies have also emerged [7]. Remote communication tools such as video calls and chat allow elderly individuals to connect with their families and caregivers [8]. Additionally, services that utilize mobile robots and assistive robots to support and monitor the daily lives of the elderly are now available [9].

We have developed monitoring support services for elderly individuals living at home by utilizing image-based human pose detection. We have proposed technologies for the quantification and measurement of physical activity quality within the home [10], as well as local sensing, using pre-trained pose detection models based on machine learning. Recently, we incorporated the high-precision pose detection model MoveNet [11] to predict and detect irregular behaviors of elderly individuals at home. Figure 1 shows the overview of human pose detection technology. Additionally, we have utilized pre-trained, image-based models that can be executed locally, such as facial expression emotion recognition models, hand movement detection models, and in-home object detection models.

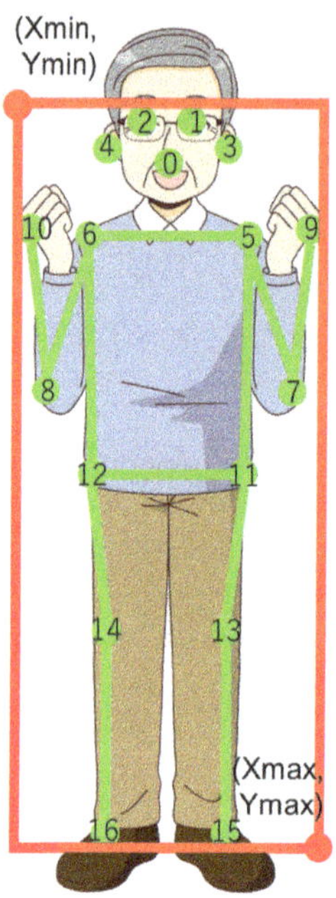

ID No.	Part
0	nose
1	left_eye
2	right_eye
3	left_ear
4	right_ear
5	left_shoulder
6	right_shoulder
7	left_elbow
8	right_elbow
9	left_wrist
10	right_wrist
11	left_hip
12	right_hip
13	left_knee
14	right_knee
15	left_ankle
16	right_ankle


```
# Sample of Output Results
{
  "score" : 0.955,
  "keypoints" : [
    {
      "score" : 0.994,
      "part" : "nose",
      "position" : {
        "x" : 192.798,
        "y" : 25.926
      }
    },
    ...
  ],
  "boundingBox" : {
    "maxX" : 223.567,
    "maxY" : 145.361,
    "minX" : 171.044,
    "minY" : 22.202
  }
}
```


Figure 1. Overview of human pose detection technology.

This study aims to explore the potential for enhancing pose detection performance through the preprocessing of acquired images. The key idea involves the rotation of images using Python and OpenCV. Specifically, we apply preprocessing techniques such as rotation, flipping, resizing, and noise reduction to the images and analyze how these techniques affect the accuracy and efficiency of the pose detection algorithms. By improving the image quality, false detections and missed detections are reduced. Additionally, we examine various combinations of preprocessing methods to establish an optimal preprocessing pipeline. Ultimately, we evaluate the effectiveness of these preprocessing techniques in practical applications and contribute to the practical implementation of pose detection technology.

2. Preliminaries

2.1. Technical Challenges

Technological challenges include posture, viewpoint, image resolution, and occlusion which can significantly affect the accuracy of pose detection. Consequently, the accuracy of pose detection varies between the standing and lying humans due to these differing conditions (Figure 2).

(a) (b)

Figure 2. Person lying on a sofa is drawn by human pose detection: (**a**) Full view (**b**) Magnified view.

- Differences in Posture: For standing humans, the positions of the joints are identifiable, resulting in higher accuracy of pose detection. In contrast, for lying humans, the positions of the joints may be obscured, making pose detection more challenging.
- Differences in Viewpoint: Standing humans are typically captured from the front, allowing for clear identification of the joint positions. Lying humans, however, are often captured from the side, making it more difficult to discern joint positions.
- Image Resolution: Standing humans appear larger within the image, leading to higher resolution. Lying humans appear smaller, resulting in lower resolution. Lower resolution decreases the accuracy of pose detection.
- Effect of Occlusion: Standing humans are less likely to be affected by occlusion, whereas lying humans may have parts of their body concealed by obstructions, thereby reducing the accuracy of pose detection.

2.2. Image Rotation with OpenCV

OpenCV is an open-source software library for computer vision and machine learning, providing a wide range of functionalities. Its newest version is 4.11.0. This library supports a variety of tasks, including image processing, video analysis, object detection, face recognition, and training of machine learning models, making it a powerful tool for researchers and developers. Furthermore, OpenCV is accessible through multiple programming languages, such as C++, Python, Java, and MATLAB, catering to a broad user base. Notably, Python is a preferred language among many researchers and engineers due to its high readability and low learning curve. As an example of OpenCV implementation in Python, Listing 1 demonstrates a sample code for rotating an image. This code implements the functionality of rotating an image by a specified angle, which is highly useful for image preprocessing tasks.

Listing 1. A sample code for rotating an image using Python.

```
import cv2
import numpy as np
img = cv2 . imread ( ' image . j p g ' ) rows , c o l s = img . shape [ : 2 ]
c e n t e r = ( c o l s / 2, rows / 2)

a n g l e = 45 s c a l e = 1 . 0
r o t a t i o n m a t r i x = cv2
. g e t R o t a t i o n M a t r i x 2 D ( c e n t e r, angle, s c a l e )

r o t a t e d i m g = cv2
. w a r p A f f i n e ( img, r o t a t i o n m a t r i x , ( c o l s , rows ) )

cv2 . imshow ( ' R o t a t e d Image ' , r o t a t e d i m g ) cv2 . wait Key ( 0 )
cv2 . destroy All Windows ( )
```

3. Proposed Method

3.1. Goal and Key Idea

The purpose of this study is to investigate the potential for improving human pose detection performance through the preprocessing of acquired images. The key idea involves the use of image transformations with OpenCV. Specifically, we apply preprocessing techniques such as image rotation, flipping, resizing, and noise reduction, analyzing in detail how these affect the accuracy and efficiency of human pose detection algorithms. By enhancing image quality, we expect to reduce false detection and missed detection. Additionally, this study explores various combinations of preprocessing techniques to construct an optimal preprocessing pipeline. Ultimately, we aim to evaluate the effectiveness of these preprocessing methods in practical applications and contribute to the practical implementation of human pose detection technology.

3.2. Overall Architecture

Figure 3 illustrates the overall architecture of the proposed method, which consists of the following five steps: (Step 1) Input the image into the pose detection model. (Step 2) Determine the rotation parts and angles based on the detection results. (Step 3) Rotate the input image according to the determined rotation angles. (Step 4) Input the rotated image into the pose detection model. (Step 5) Compare and analyze the output results before and after the image rotation.

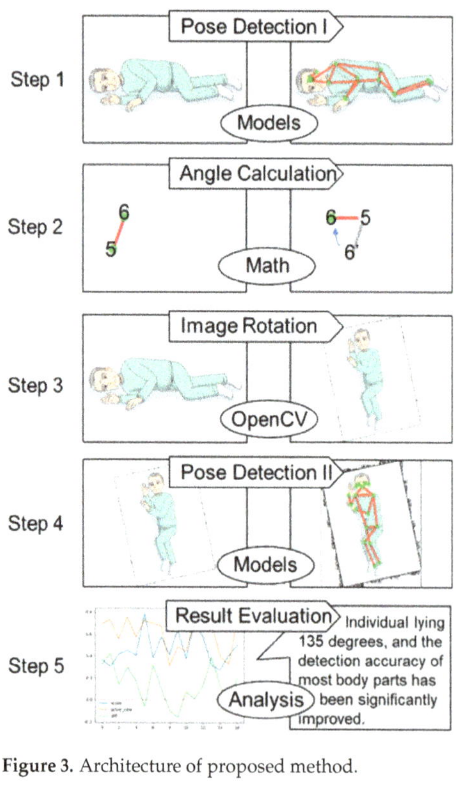

Figure 3. Architecture of proposed method.

3.3. Method Description of Key Steps

Listing 2 presents three methods for calculating slopes using Python, while Listing 3 demonstrates one method for converting slopes to angles using Python. Furthermore,

Listing 4 illustrates a method for rotating images using Python. By appropriately rotating the images based on the original output results and re-detecting them, it is expected that the accuracy of the results will improve, thereby standardizing the human postures.

Listing 2. Three ways to calculate slope using Python.

```
# Method 1
from scipy.stats import linregress

x_ = [keypoints_xlist[5],
keypoints_xlist[6]]
y_ = [keypoints_ylist[5],
keypoints_ylist[6]]

slope, intercept, r_value, p_value,
std_err = linregress(x_, y_)
print(slope)

# Method 2
x1_ = keypoints_xlist[5]
y1_ = keypoints_ylist[5]
x2_ = keypoints_xlist[6]
y2_ = keypoints_ylist[6]

def slopee(x1, y1, x2, y2):
x = (y1 - y2) / (x1 - x2)
return x

print(slopee(x1 , y1_, x2_, y2_))

# Method 3
import numpy as np

slope_, intercept_ = np.polyfit(x_, y_, 1)
print(slope_)
```

Listing 3. One way to convert ratio to angle using Python.

```
import math

angle = math.degrees(math.atan(slope))
print(angle)
```

Listing 4. One way to rotate an image using Python.

```
from PIL import Image

im = Image.open(file_path)
im_rotate = im.rotate(angle, expand=True)

plt.figure(figsize=(16, 12), dpi=20)
plt.imshow(im_rotate)
plt.gca().set_axis_off()
plt.subplots_adjust(top = 1, bottom = 0,
right = 1, left = 0,
hspace = 0, wspace = 0)
plt.margins(0,0)
plt.savefig(file_path + "_rotate.jpg")
```

4. Case Study and Discussion

We use a human pose detection model and input an image of an elderly person lying on their left side into the proposed method. The image is rotated to an upright position before performing pose detection, as shown in Figure 4. Specifically, the detection accuracy for body parts such as the left shoulder, right elbow, left wrist, and left ankle decreases, while the accuracy for the nose, left eye, left ear, right shoulder, and left knee significantly improves.

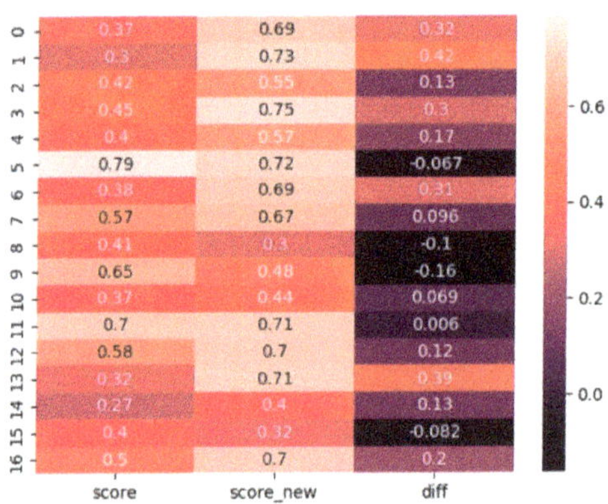

Figure 4. Results of case study with heatmap.

Image rotation as a preprocessing method improves the pose detection accuracy for joints while decreasing it for others. In light of these findings, we consider the following four points:

- Reconsideration of Preprocessing Methods: Since image rotation is effective for some joints but not for others, it is necessary to explore preprocessing methods beyond rotation, such as interpolation using augmented reality technology or image completion from other viewpoints.
- Model Improvement: Current pose detection models, including the MoveNet model, may be optimized for specific postures. To maintain high accuracy for lying positions as well, it is necessary to improve the model and introduce additional training data.
- Utilization of Multi-view Data: Integrating data from multiple viewpoints, rather than relying solely on images from a single viewpoint, has the potential to improve

overall detection accuracy. Especially for lying postures, combining information from different angles can mitigate the effects of occlusion.

- Challenges Toward Practical Application: To realize a monitoring system for the elderly, a flexible detection system that can accommodate various postures and environments is required. Based on the current results, it is necessary to conduct further validation in a wider range of environments and postures to clarify the challenges toward practical application.

5. Conclusions

In this study, we enhance pose detection performance through image preprocessing. Specifically, we apply preprocessing techniques such as rotation, flipping, resizing, and noise reduction using OpenCV and analyze in detail how these techniques affect the accuracy and efficiency of pose detection algorithms. As a result, we observe a significant improvement in pose accuracy for body parts other than 5, 8, 9, and 15. This indicates that improving image quality can reduce false detection and missed detection. Furthermore, by examining combinations of different preprocessing methods, we suggest the potential for further improving the accuracy and efficiency of pose detection technology by constructing an optimal preprocessing pipeline.

It is still necessary to address the following challenges:

- Validation with Diverse Datasets: This study was conducted under specific conditions, so it is necessary to examine the generalizability of the preprocessing methods using more diverse datasets.
- Real-Time Processing: It is crucial to evaluate whether the implementation of preprocessing methods is suitable for real-time processing and to ensure practical applicability.
- Optimization of Preprocessing Methods: There is a need to optimize combinations of different preprocessing methods to develop a more accurate preprocessing pipeline.
- Comparison with Other Pose Detection Models: It is important to compare the proposed method with other pose detection models besides MoveNet to verify its generality and effectiveness. Addressing these challenges is expected to lead to the development of more accurate pose detection technologies and represent a significant step toward practical implementation.

Author Contributions: Writing—original draft preparation, S.C.; writing—review and editing, S.C. and M.N.; supervision, M.N.; validation, S.C. All authors have read and agreed to the published version of the manuscript.

Funding: This research received no external funding.

Institutional Review Board Statement: Not applicable.

Informed Consent Statement: Not applicable.

Data Availability Statement: Data are contained within the article.

Acknowledgments: This research was partially supported by JSPS KAKENHI Grant Numbers JP20H05706, JP22H03699, JP22K19653, JP23H03401, JP23H03694, 24K02765, JP23K17006, and JST SICORP Grant Number JPMJKB2312.

Conflicts of Interest: The authors declare no conflict of interest.

References

1. World Health Organization. Available online: https://www.who.int/data/gho (accessed on 25 February 2025).
2. Ministry of Health, Labour and Welfare. Available online: https://www.mhlw.go.jp/toukei/saikin/hw/kaigo/service18/index.html (accessed on 25 February 2025).

3. Senior Care Around the World: Innovation and Bright Ideas from Countries Across the Globe. Available online: https://arosacare.com/senior-care-around-the-world/ (accessed on 25 February 2025).
4. Why is Geriatric Care Services in Such High Demand? Available online: https://www.springhills.com/resources/why-is-geriatric-care-in-high-demand (accessed on 25 February 2025).
5. Mobasseri, K.; Allahverdipour, H.; Matlabi, H.; Kousha, A. Experiences regarding home care for older adults from the viewpoint of family, paid caregivers, nurses, and experts: A qualitative study. *Health Soc. Care Community* **2024**, *2024*, 4591881. [CrossRef]
6. Rodrigues, M.J.; Postolache, O.; Cercas, F. Physiological and behavior monitoring systems for smart healthcare environments: A review. *Sensors* **2020**, *20*, 2186. [CrossRef] [PubMed]
7. Deep, S.; Zheng, X.; Karmakar, C.; Yu, D.; Hamey, L.G.; Jin, J. A survey on anomalous behavior detection for elderly care using dense- sensing networks. *IEEE Commun. Surv. Tutor.* **2019**, *22*, 352–370. [CrossRef]
8. Zamir, S.; Hennessy, C.H.; Taylor, A.H.; Jones, R.B. Video-calls to reduce loneliness and social isolation within care environments for older people: An implementation study using collaborative action research. *BMC Geriatr.* **2018**, *18*, 62. [CrossRef] [PubMed]
9. Fiorini, L.; De Mul, M.; Fabbricotti, I.; Limosani, R.; Vitanza, A.; D'Onofrio, G.; Tsui, M.; Sancarlo, D.; Giuliani, F.; Greco, A.; et al. Assistive robots to improve the independent living of older persons: Results from a needs study. *Disabil. Rehabil. Assist. Technol.* **2021**, *16*, 92–102. [CrossRef] [PubMed]
10. Chen, S.; Saiki, S.; Nakamura, M. Nonintrusive fine-grained home care monitoring: Characterizing quality of in-home postural changes using bone-based human sensing. *Sensors* **2020**, *20*, 5894. [CrossRef] [PubMed]
11. Goyal, G.; Di Pietro, F.; Carissimi, N.; Glover, A.; Bartolozzi, C. Moveenet: Online high-frequency human pose estimation with an event camera. In Proceedings of the IEEE/CVF Conference on Computer Vision and Pattern Recognition Workshops (CVPRW), Vancouver, BC, Canada, 17–24 June 2023; pp. 4024–4033.

engineering proceedings

MDPI

Proceeding Paper

Impact of Customer Perceptions of Virtual Influencers on Value Co-Creation in Streaming Platforms †

Jing-Wen Huang [1,*], Shan-Lin Yang [2] and Shao-Lan Chang [3]

1 Graduate Institute of Educational Administration, National Pingtung University, Pingtung 90003, Taiwan
2 Pingtung County Fang Liao High School, Pingtung 90003, Taiwan; youngshanlin@gmail.com
3 Kaohsiung Municipal Sha-Luen Elementary School, Kaohsiung 813031, Taiwan; elizabeth690108@yahoo.com.tw
* Correspondence: jwhuang@mail.nptu.edu.tw
† Presented at the 2024 IEEE 7th International Conference on Knowledge Innovation and Invention, Nagoya, Japan, 16–18 August 2024.

Abstract: The continuous advancement and application of generative artificial intelligence (AI) allows for virtual influencers to be active. Therefore, how virtual influencers respond to customer needs and create value has attracted the interest of streaming platforms. Using the stimulus–organism–response model, we explored how customer perceptions of virtual influencers affect their involvement in streaming platforms and value co-creation behavior. Customer perceptions of virtual influencers are considered as the stimulus, including customer preferences, interactivity, and functionality. Enduring engagement is viewed as the psychological organism, and value co-creation is regarded as the behavioral response. A questionnaire survey was used to investigate the customer perceptions of virtual influencers on streaming platforms or applications to link their experiences into a virtual world and enhance value co-creation. Customer perceptions, including preferences, interactivity, and functionality, have positive effects on enduring involvement and value co-creation. Enduring involvement affects the relationship between customer perceptions and value co-creation. Such results help to understand customer perceptions and behaviors toward virtual influencers to enhance the service and user experience of streaming platforms.

Keywords: generative artificial intelligence; virtual influencer; value co-creation; streaming platform

1. Introduction

The continuous advancement and application of generative artificial intelligence (AI) helps streaming platforms deliver personalized, timely, and relevant communications based on viewers' preferences, thereby enhancing customer experience, engagement, and loyalty [1,2]. Generative AI supports the ability of virtual influencers to enrich their forms, functionality, and interactivity. Virtual influencers on streaming platforms generate personalized content, user comments, and repeat visits [3]. Personalized content and user experiences provide virtual influencers with information on customer participation and involvement, increasing their platform stickiness and sustainable consumption behavior [4,5]. Therefore, how virtual influencers respond to customer needs and create value is important for streaming platforms.

The experience of enduring involvement enhances trust through psychological connection and relational interaction that go beyond purchasing [6,7]. Value co-creation is enabled by the importance of users' active participation and engagement in the creation of business value [8,9]. Customers increase involvement through collaboration and communication with other customers, which provides solutions and content creation [1,10,11]. For example, Netflix utilizes generative AI (e.g., virtual influencers, avatars, AI intelligent customer service) to facilitate customers' continuous involvement. Streaming media promotes connectedness by gathering different parties to jointly collaborate, adopt solutions, and generate value creation within and outside the media.

Related issues in generative AI regarding consumer behavior in marketing have been researched [1]. However, the relationship between customer perception of virtual influencers and value co-creation is still lacking. Thus, we explored the effects of customer perception of virtual influencers (including regarding innovativeness, interactivity, and functionality) provided by streaming platforms on continuous involvement and examined the effect of continuous involvement on value co-creation. The mediating role of continuous involvement between customer perception and value co-creation was also explored.

2. Literature Review

2.1. Stimulus–Organism–Response (SOR) Model

The SOR model comprises a stimulus, organism, and response. The model states that the online environment stimulates individuals to respond to the online environment through all explicit behaviors, such as discovery, stay, affiliation, and connection or desire to inactively act [12]. Stimuli evoke customers' emotional states. Organisms mediate between stimuli and final actions or responses. Responses reflect behavioral outcomes such as subscription behavior or purchasing behavior [13]. Stimuli trigger cognitive and emotional reactions (organism), which in turn affect behaviors (response) [12,13]. Customer perceptions of virtual influencers are considered as a stimulus, including innovativeness, interactivity, and functionality; continuous involvement is viewed as a psychological organism and value co-creation is regarded as the behavioral response.

2.2. Customer Perceptions of Virtual Influencers

Prior research has indicated how cognitive appraisal influences a user's intention to use social media, as well as subsequent usage behavior [14,15]. Factors affecting the customer's cognitive appraisal include video content, text layout, visual imagery, and user interface [13]. Thus, we examine how customers evaluate the stimulus of generative AI applications in terms of innovativeness, interactivity, and functionality.

Innovativeness refers to unique, innovative, and engaging content created by virtual influencers on streaming platforms [4]. Generative AI application generates human-like and easily customized text, images, and videos, offering customers content creation processes and unique experiences [1]. Customers inspired by innovativeness enhance their willingness to engage in community interaction and build long-term relationships with other members [16]. They have an interest in involvement in community activities and enhance their behavior with regard to recommendations and paid subscriptions [2,4].

Interactivity represents a perception of connection to other customers, which increases active participation and exploration, and influences the content [4,17]. The value co-creation perspective emphasizes interaction as the nature of co-creation [8,18]. When customers perceive the interactivity of virtual influencers, they are inclined to spend time and enhance their involvement in the streaming platform for favorable value creation [11,19].

Functionality indicates the extent to which customers' judgment about a given site's overall excellence and fitness for use in assisting with the task and making an online decision [14,20]. Customers evaluate functionality based on their perceptions of usefulness, accessibility, and entertainment. Virtual influencers enhance customers' intention to give feedback about entertaining content or make online decisions. Perceived functionality keeps customers' curiosity and motivates them to participate in streaming platforms [14,21].

According to the above, we formulated the following hypotheses.

- H1a: Customers' perceived innovativeness of virtual influencers has a significantly positive impact on continuous involvement.
- H1b: Customers' perceived interactivity of virtual influencers has a significantly positive impact on continuous involvement.
- H1c: Customers' perceived functionality of virtual influencers has a significantly positive impact on continuous involvement.

2.3. Continuous Involvement

Involvement reflects the personal relevance and importance of the target firm's service to customers [6,22]. The strength of customer involvement depends on their intrinsic interests, needs, and values. Continuous involvement refers to the ongoing concern with the product class association with the customers' self-concepts, values, or egos [7]. When customers have high levels of involvement, they tend to receive information related to products and services, and thus pay more attention to information [7,23]. Reference [10] explored the perceived value of a sharing platform adoption. The interactive discussions and activities on social media platforms allow for consumers' resource integration into co-creating brand meaning and value. Reference [11] revealed that engagement in Pokémon Go facilitates customer value creation comprising hedonic, achievement, and nostalgia value. The experience offered by continuous involvement enhances perceived trust through psychological connection and relational interaction that go beyond purchase, facilitating brand loyalty and value co-creation [6]. When customers are motivated to process the content of virtual influencers, they are involved in the community, giving suggestions on service improvement and participating actively the in co-creation of services [6,24]. Thus, we proposed the following hypothesis:

- H2: Continuous involvement has a significantly positive impact on value co-creation.

2.4. Mediating Effects

H1a, H1b, and H1c are related to customer perceptions (innovativeness, interactivity, and functionality) of virtual influencers on streaming platforms with continuous involvement. H2 links continuous involvement with value co-creation. Continuous involvement plays a mediating role in the relationship between customer perceptions (independent variable) and value co-creation (dependent variable).

In the case of Netflix, virtual influencers refer to popular actors, directors, or content creators whose work influences customer viewing choices and preferences. Customers form opinions about virtual influencers on Netflix based on their perceptions of content innovativeness, character interactivity, and the overall viewing experience [3,4]. Continuous involvement on Netflix reflects customers' sustained interest and emotional connection to specific shows, actors, or genres over time [5]. Customers with strong continuous involvement are inclined to create fan art, discussion forums, or social media content related to their favorite shows, contributing to the value co-creation process.

Continuous involvement is a mediator because it influences how customer perceptions of virtual influencers translate into value co-creation. Customers with continuous involvement are likely to have a stronger emotional bond with virtual influencers, leading to a profound impact on their perceptions and the co-creation of value. Customer perceptions of virtual influencers are translated into favorable value co-creation via the intermediate effect of continuous involvement. Thus, this study proposes the following hypothesis.

- H3: Continuous involvement mediates the relationship between customer perceptions of virtual influencers and value co-creation.

3. Method

A questionnaire survey was conducted to collect data. The research subjects were audiences and users of streaming platforms, who were familiar with the video streaming platforms, and who were 18 years old or older. The questionnaire was used to investigate the application of generative AI into the streams of virtual influencers in streaming platforms, allowing customers to link their use experiences into a virtual metaverse. A Likert-type scale was used in the survey. A total of 500 questionnaires were distributed, and 486 were recovered, with a valid response rate of 97.2%.

The calculated t-statistics for gender, age, and viewing time were statistically insignificant ($p > 0.1$). In behavioral research, the method of Ref. [25] was used to test the common method biases problem. The Harman's one-factor test result showed that a single factor did not emerge, and one general factor did not account for most of the variance. Thus, common method bias was unlikely to be a serious problem in the data [25].

4. Results

The Cronbach alpha of the survey exceeded 0.70, showing acceptable reliability and internal consistency [26]. The average variance extracted (AVE) was higher than 0.61, exceeding the threshold of 0.50 [27]. The fit index of confirmatory factor analysis for the proposed model was adequate. The standardized loadings of all the measurement items were significant on their posited underlying construct, indicating convergent validity and discriminant validity [28].

Customer perceptions of virtual influencers, including innovativeness, interactivity, and functionality, showed positive effects on continuous involvement ($p < 0.05$). Thus, H1a, H1b, and H1c were supported. The continuous involvement in online streaming platforms had a positive effect on value co-creation ($p < 0.05$). Thus, H2 was supported. The mediating role of continuous involvement was verified in affecting the relationship between customer perceptions of virtual influencers and value co-creation. Thus, H3 was supported.

5. Conclusions and Implication

Customer perceptions of virtual influencers, including innovativeness, interactivity, and functionality, facilitate their continuous involvement in streaming platforms. The continuous involvement helps co-create value among customers or between customers and streaming platforms. The mediating role of continuous involvement in the relationship between customer perceptions of virtual influencers and value co-creation was verified.

The SOR model shows that customer perceptions of virtual influencers play a significant role in their continuous involvement with streaming platforms and their involvement in value co-creation activities. Factors such as innovativeness, interactivity, and functionality of virtual influencers positively impact customer continuous involvement, which in turn leads to enhanced value co-creation behavior.

Customer perceptions of virtual influencers are presented in live streaming, too [1,3]. Understanding and leveraging customer perceptions of virtual influencers improves the service quality and user experience of streaming platforms. To improve customer perceptions, streaming platform operators need to design and promote the innovativeness, interactivity, and functionality of virtual influencers. By recognizing the mediating role of continuous involvement in the relationship between customer perceptions and value co-creation, streaming platform operators can tailor their strategies to enhance customer engagement and promote collaborative value creation within the virtual influencer ecosystem.

By considering value co-creation, value creation highlights the importance of relation and user value, and regards customers as co-creators [8,9,18]. The collaborative nature of value creation in influencer–consumer interactions suggests that it is not solely driven by the influencer, but involves active participation from customers. Continuous involvement mediates customer perceptions and value co-creation, providing value co-creation in sustainable competitive advantages [10,29]. The results of this study contribute to consumer behavior literature [6,10,11] by highlighting the importance of customer engagement in influencer marketing and the subsequent impact on value creation. The importance of customer involvement enhances value co-creation for the media-streaming industry.

Author Contributions: Conceptualization, all authors; methodology, all authors; investigation, all authors; writing—original draft preparation, J.-W.H.; writing—review and editing, J.-W.H. All authors have read and agreed to the published version of the manuscript.

Funding: This research received no external funding.

Institutional Review Board Statement: Not applicable.

Informed Consent Statement: Not applicable.

Data Availability Statement: No new data were created.

Acknowledgments: The authors thank the National Science and Technology Council for its support, and the authors are solely liable and responsible for this derived work.

Conflicts of Interest: The authors declare no conflicts of interest.

References

1. Kshetri, N.; Dwivedi, Y.K.; Davenport, T.H.; Panteli, N. Generative artificial intelligence in marketing: Applications, opportunities, challenges, and research agenda. *Int. J. Inf. Manag.* **2023**, *75*, 102716. [CrossRef]
2. Rahman, M.S.; Bag, S.; Hossain, M.A.; Abdelfattah, F.; Gani, M.O.; Rana, N. The new wave of AI-powered luxury brands online shopping experience: The role of digital multisensory cues and customers' engagement. *J. Retail. Consum. Serv.* **2023**, *72*, 103273. [CrossRef]
3. Chen, H.; Shao, B.; Yang, X.; Kang, W.; Fan, W. Avatars in live streaming commerce: The influence of anthropomorphism on consumers' willingness to accept virtual live streamers. *Comput. Hum. Behav.* **2024**, *156*, 108216. [CrossRef]
4. Hussain, K.; Khan, L.M.; Malik, A.A. Exploring audience engagement with Chat GPT-related content on YouTube: Implications for content creators and AI tool developers. *Digit. Bus.* **2024**, *4*, 100071. [CrossRef]
5. Xu, Y.; Ruan, Y. AI and human broadcasters: Relative impact on consumer engagement in live streaming commerce. *Electron. Commer. Res. Appl.* **2023**, *62*, 101335. [CrossRef]
6. Cheung, M.F.Y.; To, W.M. Effect of customer involvement on co-creation of services: A moderated mediation model. *J. Retail. Consum. Serv.* **2021**, *63*, 102660. [CrossRef]
7. Sharma, P.; Roy, R.; Rabbanee, F.K. Interactive effects of situational and continuous involvement with perceived crowding and time pressure in pay-what-you-want (PWYW) pricing. *J. Bus. Res.* **2020**, *109*, 882100. [CrossRef]
8. Prahalad, C.K.; Ramaswamy, V. Co-creation experience: The next practice in value creation. *J. Interact. Mark.* **2004**, *18*, 5–14. [CrossRef]
9. Vargo, S.L.; Lusch, R.F. From repeat patronage to value co-creation in service ecosystems: A transcending conceptualization of relationship. *J. Bus. Mark. Manag.* **2010**, *4*, 169–179. [CrossRef]
10. Liang, T.P.; Lin, Y.L.; Hou, H.C. What drives consumers to adopt a sharing platform: An integrated model of value-based and transaction cost theories. *Inf. Manag.* **2021**, *58*, 103471. [CrossRef]
11. Wu, J.H.; Robinson, S.; Li, Q.; Huang, T.Y.; Chen, Y.C. Online gaming: Mediation of situational engagement in facilitating player value. *J. Electron. Commer. Res.* **2021**, *22*, 382–402.
12. Jacoby, J. Stimulus-Organism-Response reconsidered: An evolutionary step in modeling (consumer) behavior. *J. Consum. Psychol.* **2002**, *12*, 51–57. [CrossRef]
13. Perez-Vega, R.; Kaartemo, V.; Lages, C.R.; Razavi, N.B.; Mannist, J. Reshaping the contexts of online customer engagement behavior via artificial intelligence: A conceptual framework. *J. Bus. Res.* **2021**, *129*, 902–910. [CrossRef]
14. Friedrich, T.; Schlauderer, S.; Overhage, S. The impact of social commerce feature richness on website stickiness through cognitive and affective factors: An experimental study. *Electron. Commer. Res. Appl.* **2019**, *36*, 100861. [CrossRef]

15. Lee, D.; Moon, J.; Kim, Y.J.; Yi, M.Y. Antecedents and consequences of mobile phone usability: Linking simplicity and interactivity to satisfaction, trust, and brand loyalty. *Inf. Manag.* **2015**, *52*, 295–304. [CrossRef]
16. Kim, E.; Tang, L.R.; Bosselman, R. Measuring customer perceptions of restaurant innovativeness: Developing and validating a scale. *Int. J. Hosp. Manag.* **2018**, *74*, 85–98. [CrossRef]
17. Pinto, R.D.; Peixoto, B.; Melo, M.; Cabral, L.; Bessa, M. Foreign language learning gamification using virtual reality—A systematic review of empirical research. *Educ. Sci.* **2021**, *11*, 5222. [CrossRef]
18. Priem, R.L. A consumer perspective on value creation. *Acad. Manag. Rev.* **2007**, *32*, 219–235. [CrossRef]
19. Cheung, M.L.; Pires, G.D.; Rosenberger, P.J.; Leung, W.K.S.; Ting, H. Investigating the role of social media marketing on value co-creation and engagement: An empirical study in China and Hong Kong. *Australas. Mark. J.* **2020**, *29*, 118–131. [CrossRef]
20. Lee, J.; Kim, J.; Choi, J.Y. The adoption of virtual reality devices: The technology acceptance model integrating enjoyment, social interaction, and strength of the social ties. *Telemat. Inform.* **2019**, *39*, 37–48. [CrossRef]
21. Xu, Y.; Chen, Z.; Peng, M.Y.; Anser, M.K. Enhancing consumer online purchase intention through gamification in China: Perspective of cognitive evaluation theory. *Front. Psychol.* **2020**, *11*, 581200. [CrossRef] [PubMed]
22. Liu, J.; Jo, W. Value co-creation behaviors and hotel loyalty program member satisfaction based on engagement and involvement: Moderating effect of company support. *J. Hosp. Tour. Manag.* **2020**, *43*, 23–31. [CrossRef]
23. Zhang, Y.; Hua, L.; Jiao, Y.; Zhang, J.; Saini, R. More than watching: An empirical and experimental examination on the impacts of live streaming user-generated video consumption. *Inf. Manag.* **2023**, *60*, 103771. [CrossRef]
24. García-Jurado, A.; Torres-Jiménez, M.; Leal-Rodríguez, A.L.; Castro-González, P. Does gamification engage users in online shopping? *Electron. Commer. Res. Appl.* **2021**, *48*, 101076. [CrossRef]
25. Podsakoff, P.M.; MacKenzie, S.B.; Lee, J.Y. Common method biases in behavioral research: A critical review of the literature and recommended remedies. *J. Appl. Psychol.* **2003**, *88*, 897–903. [CrossRef]
26. Hair, J.F.; Black, W.C.; Babin, B.J. *Multivariate Data Analysis: A Global Perspective*; Prentice-Hall: Upper Saddle River, NJ, USA, 2010.
27. Barclay, D.W.; Thompson, R.; Higgins, C. The Partial Least Squares (PLS) approach to causal modeling: Personal computer adoption and use an illustration. *Technol. Stud.* **1995**, *2*, 285–309.
28. Anderson, J.C.; Gerbing, D.W. Structural equation modeling in practice: A review and recommended two-step approach. *Psychol. Bull.* **1988**, *103*, 411–423. [CrossRef]
29. Wang, Y.; Li, C.; Zhang, D.; Wu, J.; Liu, Y. A deeper investigation of different types of core users and their contributions for sustainable innovation in a company-hosted online co-creation community. *J. Clean. Prod.* **2020**, *256*, 120397. [CrossRef]

engineering proceedings

MDPI

Proceeding Paper

Synthetic Image Data Generation for Wiring Harness Component Detection Using Machine Learning †

Huong Giang Nguyen *, Patrick Bründl and Jörg Franke

Institute for Factory Automation and Production Systems, Friedrich-Alexander-Universität Erlangen-Nürnberg, Fürther Str. 246b, 90429 Nürnberg, Germany; patrick.bruendl@faps.fau.de (P.B.); joerg.franke@faps.fau.de (J.F.)

* Correspondence: huong.nguyen@faps.fau.de

† Presented at the 2024 IEEE 7th International Conference on Knowledge Innovation and Invention, Nagoya, Japan, 16–18 August 2024.

Abstract: Machine learning is a powerful tool for computer vision tasks in manufacturing, as features are automatically extracted and a high variety of components or failures are reliably detected. A focal prerequisite for high-performing machine learning models is a database that is large in quantity as well as quality, and representative for the computer vision task in the manufacturing environment. In addition, manufacturing applications require a domain-specific dataset. Thus, we generated and integrated synthetic data for object detection using convolutional neural networks, specifically for wiring harness component detection. A synthetic data generation pipeline for images was developed and implemented. Experiments were conducted to assess the domain gap between synthetic and real images and to determine factors that are beneficial to synthetic data generation. The experimental findings demonstrate relevant training approaches to integrate synthetic data, factors that have a positive impact on training, and high-performance results comparable to using real data only.

Keywords: machine vision; optical inspection; machine learning; synthetic data generation; wiring harness

1. Introduction

The potential of machine learning for manufacturing applications has been widely addressed and proven by researchers [1]. Machine learning models automatically extract relevant features from Euclidian and non-Euclidian data. Convolutional neural networks (CNNs), a type of machine learning model, are well known for processing image data as they can exploit shift-invariance and local connectivity in image data [2]. However, manufacturing applications are highly specific, meaning that off-the-shelf datasets lack relevance for CNN training. Thus, a new computer vision task for a manufacturing application results in the need to create a dataset with a latent feature space specific to the application.

The challenge of data scarcity and high resource investment in database creation is also found in the manufacturing application of wiring harness component detection. Researchers addressing this application generate CNNs by investing a high amount of time and manpower into generating and labeling images. To increase the efficiency of data generation, synthetic data are generated and integrated into model training. This facilitates the rapid development of robust CNN models and the adaptation of existing models to data drift in real-world production environments. Synthetic data are created in a virtual environment by, e.g., reusing computer-aided design (CAD) models or utilizing generative models. Therefore, using synthetic data reduces the amount of time and the costs involved because physical setups, manual image capturing, and manual labeling are no longer necessary. We present the synthetic image data generated and processed in this study.

The remainder of this article is structured as follows. First, an overview of the state of the art of traditional and machine-learning approaches for wiring harness component detection is provided. Second, the data generation and processing pipeline for wiring harness images is presented in detail. Experiments are conducted to evaluate the value-added through synthetic data in contrast to using real data only. Third, the last section concludes this article and gives an outlook on future research endeavors.

2. Machine Vision for Wiring Harness Component Detection

The wiring harness consists of thousands of components such as wires, connectors, clips, tubes, tapes, and grommets [3]. Assembled into the final product, wiring harnesses serve to connect sensors, actuators, and control units, e.g., in vehicles, planes, or electronic devices. The detection of the components in wiring harness manufacturing is difficult to realize, due to the presence of high component variety, deformable and rigid objects, and product complexity [4]. A door wiring harness is shown in Figure 1.

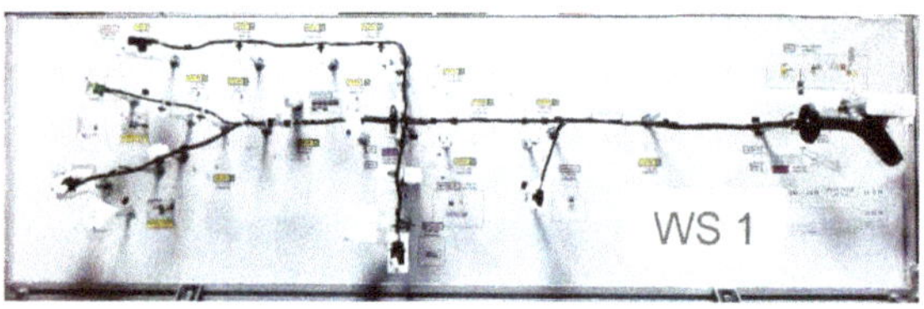

Figure 1. A wiring harness assembled on an assembly board.

2.1. Traditional Machine Vision

A variety of proposed solutions for wiring harness component detection stem from traditional computer vision algorithms. Focus is often directed towards assembled wires due to their deformable character and high occlusion between wires. Thus, a combination of multiple algorithms is proposed. Algorithms for thresholding [5], manual feature extraction [6], and edge detection [7] are often applied for component detection. The addition of algorithms for template matching facilitates the determination of the presence and position of tape, clips, wire bundles, and wheel-shaped components [8]. While the proposed solutions are successfully applied to the individual research use cases, the robustness to data drift, e.g., inducted by environmental changes and novel components, is not given. Thus, adapting algorithms to data drift necessitates the adjustment of a multitude of parameters. Matching algorithms focus on wiring harness localization for wire manipulation. Matching algorithms rely on a topological representation form of the wiring harness to determine wire branches and components [9].

2.2. Machine Learning Approaches

Wiring harness branch detection is realized with a CNN processing image as well as depth data, and the experimental findings show the benefit of processing data from several modalities to achieve a high CNN performance [10]. Reference [11] focuses on connector detection and pose estimation using a single-shot detector. These works rely on manual image data collection and labeling. Only a few researchers rely on synthetic data as the primary database for CNN training. The real-time deformable linear objects instance segmentation (RT-DLO) approach addresses instance segmentation for wires using synthetic data [12,13]. A CNN is proposed for semantic segmentation to generate a binary mask for additional algorithms for instance extraction. Synthetic data generation is based on the copy-and-paste approach. Therefore, wires are extracted from existing images and pasted on a variety of complex backgrounds. The results show the successful application of RT-DLO to real data.

Synthetic data can be generated without relying on real data for feature extraction. Wiring harness simulation and rendering allow data generation using digital artifacts. Wire instance segmentation of control cabinet wires and high-voltage wires is accomplished using models trained with synthetic data [14]. The synthetic data generation process encompasses a model generator to generate geometric models of wires considering their deformable character; scene generation, as the wires to be detected are placed in small load carriers; and scene rendering using Blender. A similar approach was applied in [15] to generate synthetic point clouds for point cloud segmentation. The simulation involved individual wires and wire bundles in an assembled state, the scene composition entailed wire assembly on an assembly board, and the rendering was implemented using Blender. The researchers investigated different training approaches to achieve a high performance. In [16,17], semantic part segmentation was conducted using geometric deep learning trained with meshes deducted from CAD data.

3. Synthetic Data Generation

Synthetic data generation is used to train a CNN for semantic segmentation of images containing wire harness assemblies. Following our previous works, we implemented and assessed semantic segmentation for wiring harness assembly states [18]. Previous research investigated the feasibility of machine learning-based component detection. We investigated the added value through synthetic data generation for semantic segmentation. The classes are (1) 'connector', (2) 'clip type 1', (3) 'clip type 2', (4) 'clip type 3', (5) 'grommet', (6) 'fully taped wire bundles', (7) 'spiral taped wire bundles', (8) 'untaped wire bundles', and (9) 'background'. Synthetic data generation, the ideal approach to synthetic data integration in the training process, and the determination of impacting factors for synthetic data generation are explored in this study. The real dataset for wire harness assembly states has been introduced in previous works [18].

3.1. Data Generation

The synthetic data generation approach and required input are visualized in Figure 2. The main steps are 'wiring harness model generation', 'scene composition', and 'rendering'. The starting point for wiring harness model generation was a structured bill of material (SBOM) with the assembly sequence and component material number. Furthermore, a CAD database with a geometric representation of the components was mandatory. The components possess meta information pertaining to the color, material, and label. As wiring harnesses are customizable products, a configuration list provided information on which components need to be retrieved from the database for specific wiring harness configurations. In conjunction with the SBOM, different assembly states, specifically work in progress and finished assembly, were derived. The wiring harness models were generated in accordance with the proposed steps using the tools 'Siemens NX' [19] as well as 'IPS Cable Simulation' [20]. The latter was required to simulate the deformable characteristics of wires and wire bundles.

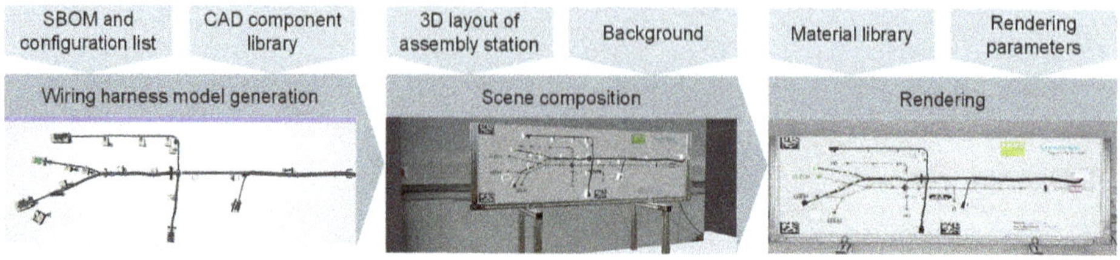

Figure 2. Synthetic data generation process for wiring harness assembly states.

Next, the manufacturing scene was created. The wiring harness model required an expansion to include the manufacturing equipment and environment. The background of the scenes can be composed of digital models of the manufacturing or background images. We used CAD models of the manufacturing equipment and a variety of images of manufacturing backgrounds. The final model, consisting of wiring harnesses and the scene, was then enriched by rendering relevant parameters. Therefore, a material library was required to assign material and texture information to individual components according to component meta information. Scene composition and rendering were realized using the tool 'Blender' [21].

For image generation, a parameter space for lighting, camera position, and background was created in this study. Lighting sources were located in the scene and varying lighting intensities were defined. Furthermore, the camera for image synthesis was set to match the camera of the physically existing setup. A parameter space for the camera location was set to capture the scene from different perspectives and to capture images of different light intensities. The light intensity was set in Blender and parametrized with 5, 25, and 50 W. Defined parameter spaces were implemented since the field of view was known for manufacturing applications. Lighting conditions did not vary significantly in the manufacturing environment. Therefore, a random function was implemented to select the camera location within the parameter space and for background retrieval. The background library consisted of a real image of the physically existing laboratory setup and background images of manufacturing environments retrieved from the internet. Thus, the background possessed contextual relevance. We used the rendering engine 'Cycles' provided by Blender. The output was images with a size of 1120 × 640 pixels. As the images contained metadata on the components and label annotations, the masks for semantic segmentation were automatically extracted. Masks were generated simultaneously with the image and saved as .jpg according to a naming convention to maintain an unambiguous assignment of image and mask for training. The images and masks generated for a wire harness assembly state are shown in Figure 3.

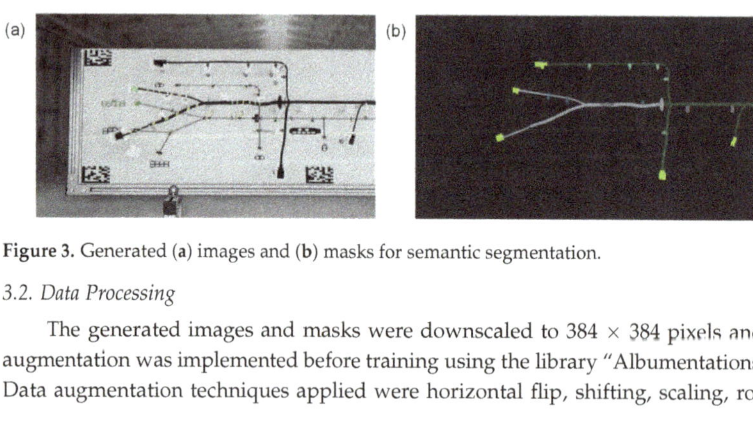

Figure 3. Generated (**a**) images and (**b**) masks for semantic segmentation.

3.2. Data Processing

The generated images and masks were downscaled to 384 × 384 pixels and data augmentation was implemented before training using the library "Albumentations" [22]. Data augmentation techniques applied were horizontal flip, shifting, scaling, rotation, image cropping, and padding. Moreover, contrast, brightness, gamma, blur, and hue saturation were randomly applied. The hold-out method was applied by splitting the dataset into a training dataset for training and validation according to a ratio of 80:20. The size of the test dataset was limited to the amount of real data available. A total of 166 real images with associated masks were available. The resulting number of images according to experiments conducted is shown in Table 1. Semantic image segmentation relied on the U-Net and LinkNet architecture with the backbone models SEResNet50 and EfficientNetB3, according to [23]. Training was performed for 50 epochs, using the Adam optimizer, adaptive learning rate adaptation, and early stopping. Hyperparameters were optimized using grid search for the parameters batch size (2, 4, 8, 12), learning rate (0.001 and 0.0001), architecture, and model backbone. The hardware used was Intel Core i7-8565U CPU and NVIDIA GeForce RTX 2080TI.

Table 1. Overview of experiments and datasets.

No	Experiments	Training and Validation		Testing	
		Domain	Images	Domain	Images
1	Benchmark	Real	132	Real	34
2	Domain gap	Synthetic	10,728	Real	166
3	Parameter study	Synthetic	445	Real	166
4	Mixing	Synthetic Real	747 83	Real	83
5	Fine-tuning	Synthetic Real	10,728 83	Real	83

4. Results

Experiments were conducted to understand the domain gap between synthetic and real data, as well as the added value provided by synthetic data for training. The goal for synthetic data implementation is to rely on as little real data as possible while achieving an equally high or higher performance. The results were obtained on the evaluation metrics of intersection over union (IoU) and F1 score. The IoU evaluates the spatial accuracy. The F1 score allows the precision and recall to be concluded. The first experiment was designed to establish a benchmark for comparative analysis when utilizing the generated synthetic data. This benchmark serves as a reference for evaluating the performance and generalizability of the following experiments. Training resulted in a mean IoU score of 0.7976 and a mean F1 score of 0.8375.

4.1. Domain Gap and Parameter Study

The experiment 'Domain gap' was conducted to identify the domain gap between real and synthetic data. CNNs were trained with synthetic data only and tested on real data. The training curves for the Unet architecture and EfficientNetb3 backbone trained with solely synthetic data reached an accuracy of 99.08% and a mean IoU of 0.6248. Contrastingly, testing with real data resulted in a mean IoU of 0.5531. A domain gap between the real dataset and the synthetic dataset was observed. The overview of IoU by classes is depicted in Table 2. The classes (1) connector, (4) clip type 2, (5) clip type 3, and (9) background achieved an IoU over 0.5. The class background displayed a superior performance because of the underlying class imbalance. The classes (5) grommet and (6) untapped wire bundle achieved an IoU score below 0.25. A comparison between ground truth and prediction mask shows the misclassification of background objects as a grommet and incomplete segmentation of untaped wires, especially at the edges of the objects (Figure 4). Thus, the model trained with synthetic data could not entirely detect all wire harness components of the real dataset.

Table 2. Evaluation of IoU by class.

No	Experiments	Classes								
		1	2	3	4	5	6	7	8	9
1	Benchmark	0.900	0.794	0.794	0.647	0.840	0.633	0.746	0.782	0.993
2	Domain gap	0.692	0.484	0.777	0.675	0.220	0.248	0.452	0.482	0.975
3	Parameter study	0.671	0.618	0.235	0.265	0.378	0.281	0.466	0.125	0.982
4	Mixing	0.874	0.181	0.664	0.613	0.845	0.543	0.748	0.515	0.991
5	Fine-tuning	0.894	0.759	0.747	0.687	0.880	0.628	0.787	0.506	0.991

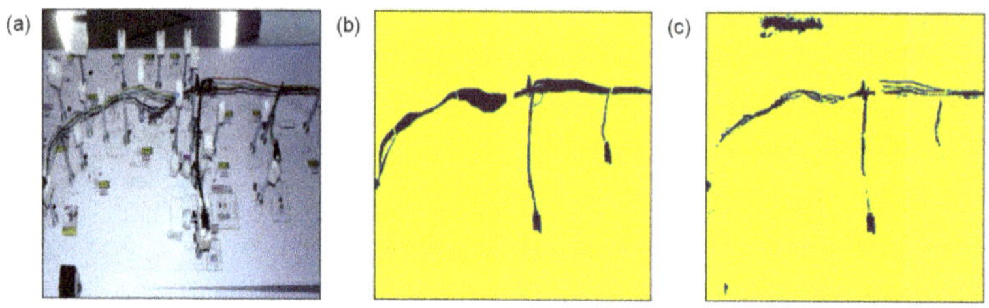

Figure 4. Prediction results for the experiment 'domain gap' (**a**) image from the real dataset, (**b**) ground truth mask, and (**c**) prediction mask.

To create a more in-depth understanding of the parameters applied during synthetic data generation, the parameters were determined according to the 'Design of Experiment' methodology. Parameters of the process steps 'scene composition' and 'rendering' were selected for the experiment, specifically light intensity, color, texture, and background. The parameter for light intensity was defined at three levels: 5, 25, and 50 W. The light intensity was chosen because this factor significantly impacts image generation time. Whereas an image with 5 W light intensity was generated in approx. 27 s, the 50 W image took 179 s to generate. The background was differentiated to distinguish between the real background and a background retrieved from the internet. Furthermore, images were generated with and without color and texture information. A full factorial design was employed. Thus, 24 datasets were created for all possible combinations of factors and levels. In the experiments, 14 blocks were tested, whereby the number of blocks and combinations reflected the hyperparameter optimization runs. Each dataset consisted of 355 images for training and 90 images for validation. The experiments were conducted with the LinkNet architecture.

Regression analysis and main effect analysis results showed the relevance of the factors of light intensity and color information. Figure 5 illustrates the main effect plot, showing that light intensity has a slight impact on IoU, while color has a significant impact on the performance. Texture and background have a minor impact on the performance. Consequently, the texture information can be neglected during rendering and the background selection does not require a systematic approach. The highest performance was a mean IoU of 0.2645 and a mean F1 score of 0.3337 with a parameter setting of 5 W, color, texture, and random background.

4.2. Mixing and Transfer Learning

Mixing datasets and transfer learning were implemented to overcome the domain gap between synthetic and real data. First, the amount of real data was reduced by substituting real data with synthetic data in the training and validation dataset. A mixing ratio of 10% real data and 90% synthetic data for training and validation was applied. The best-performing model resulted in a mean IoU of 0.6440 and a mean F1 score of 0.7000. While the IoU scores of most classes were slightly below the benchmark experiment, Class (2) clip type 1 yielded poor results, since the IoU was 0.181.

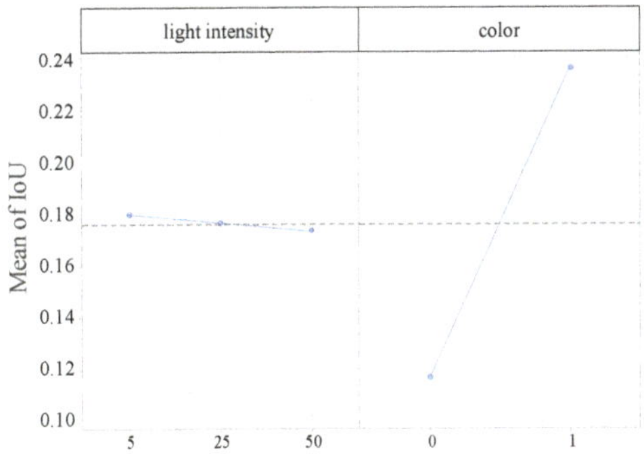

Figure 5. Main effect plot for intersection over union.

Fine-tuning was conducted to transfer the model pretrained with synthetic data to the domain of real data. The best-performing model generated in the experiment 'domain gap' was fine-tuned with real data through training with small learning rates. The mean IoU was 0.7496 and the mean F1 score of 0.7849. While classes (4) clip type 3, (5) grommet, and (7) fully taped wire bundles achieved higher IoU scores compared to the benchmark experiment. The class (8) spiral-taped bundles exhibited lower performance. As shown in Figure 6, thin wire bundles pose a challenge for some images in the testing dataset. The performance of the remaining classes is comparable to the performance achieved during the benchmark experiment. The model trained through fine-tuning achieved results equivalent to those of the benchmark model, despite utilizing only half the amount of real data for training. The experiment results indicated the importance of synthetic data for CNN pretraining and the model performance for fine-tuning with real data.

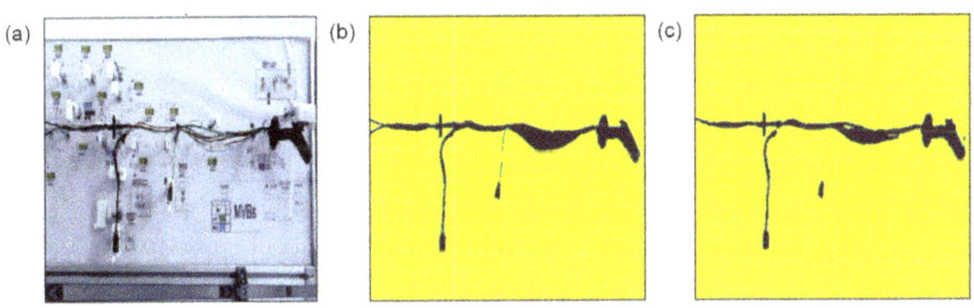

Figure 6. Prediction results for the experiment 'fine-tuning' (**a**) image from the real dataset, (**b**) ground truth mask, and (**c**) prediction mask.

5. Conclusions and Outlook

The potential of synthetic data to overcome the data scarcity for domain-specific machine learning applications was validated specifically in the detection of wire harness components. The proposed synthetic data generation approach was implemented. The experimental results present factors in the steps of 'scene composition' and 'rendering' that are critical to object detection. The texture and type of background can be neglected, and color information is critical to performance when the CNN is transferred to a real dataset. The identification of critical factors allows faster synthetic data generation because factors without impact are eliminated. Even though a domain gap has been identified, different dataset compositions and training approaches show that the need for real data is significantly reduced for this approach, but it can still achieve a high CNN performance equal to that achieved by training with real data only. By fine-tuning a CNN that has been trained with synthetic data, the use of real data leads to higher results compared to mixing datasets. However, further research is needed to explore other mixing strategies beyond a 1:9 ratio.

The limitation of the proposed synthetic data generation approach and the experimental findings is its applicability to the specific use of wire harness component detection and the dataset used for evaluation. The dataset's specific features, such as data distribution and sample size, significantly influence the experimental results. It is necessary to improve and extend the approach to synthetic data generation for wire harness component detection. Testing the approach on a bigger dataset validates its robustness and allows for the identification of optimization measures. The generalizability of the approach to other use cases and datasets also needs to be addressed. Extensive evaluation that includes diverse use cases is important to holistically understand the method's strengths and weaknesses with regard to its implementation in manufacturing applications.

Author Contributions: Conceptualization, methodology, validation, investigation, writing—original draft preparation, H.G.N.; writing—review and editing, P.B. and J.F.; supervision, J.F. All authors have read and agreed to the published version of the manuscript.

Funding: This research was supported by the project 'Next2OEM' of the Federal Ministry for Economic Affairs and Climate Action (BMWK) based on a decision by the German Bundestag and funded by the European Unions.

Institutional Review Board Statement: Not applicable.

Informed Consent Statement: Not applicable.

Data Availability Statement: No new data were created or analyzed in this study. Data sharing is not applicable to this article.

Conflicts of Interest: The authors declare no conflicts of interest.

References

1. Dogan, A.; Birant, D. Machine learning and data mining in manufacturing. *Expert Syst. Appl.* **2021**, *166*, 114060. [CrossRef]
2. Gu, J.; Wang, Z.; Kuen, J.; Ma, L.; Shahroudy, A.; Shuai, B.; Liu, T.; Wang, X.; Wang, G.; Cai, J.; et al. Recent advances in convolutional neural networks. *Pattern Recognit.* **2018**, *77*, 354–377. [CrossRef]
3. Tilindis, J.; Kleiza, V. The Effect of Learning Factors due to Low Volume Order Fluctuations in the Automotive Wiring Harness Production. *Procedia CIRP* **2014**, *19*, 129–134. [CrossRef]
4. Nguyen, H.G.; Kuhn, M.; Franke, J. Attributes for open innovation with automotive suppliers in new product development: A case study of the electrical and electronic system. *Int. J. Ayurveda Tradit. Med.* **2023**, *1*, 1. [CrossRef]
5. Supriya, S.K.; Ashwini, A.K. Automatic Optical Inspection System for wiring harness using Computer Vision. In Proceedings of the 2021 IEEE International Conference on Electronics, Computing and Communication Technologies (CONECCT), Bangalore, India, 9–11 July 2021; pp. 1–5.
6. Le, M.-H.; Hoang, V.-D.; Ngo, V.-T. Electrical colored wires inspection algorithm for automatic connector producing machines. In Proceedings of the 2016 International Conference on System Science and Engineering (ICSSE), Puli, Taiwan, 7–9 July 2016; pp. 1–4.
7. Ghidoni, S.; Finotto, M.; Menegatti, E. Automatic Color Inspection for Colored Wires in Electric Cables. *IEEE Trans. Autom. Sci. Eng.* **2015**, *12*, 596–607. [CrossRef]
8. Lee, W.; Cao, K. Application of Machine Vision to Inspect a Wiring Harness. In Proceedings of the 2019 IEEE International Conference on Industrial Cyber Physical Systems (ICPS), Taipei, Taiwan, 6–9 May 2019; pp. 457–460.
9. Zürn, M.; Wnuk, M.; Lechler, A.; Verl, A. Topology Matching of Branched Deformable Linear Objects. In Proceedings of the 2023 IEEE International Conference on Robotics and Automation (ICRA), London, UK, 29 May–2 June 2023; pp. 7097–7103.
10. Kicki, P.; Bednarek, M.; Lembicz, P.; Mierzwiak, G.; Szymko, A.; Kraft, M.; Walas, K. Tell Me, What Do You See?—Interpretable Classification of Wiring Harness Branches with Deep Neural Networks. *Sensors* **2021**, *21*, 4327. [CrossRef] [PubMed]
11. Zhou, H.; Li, S.; Lu, Q.; Qian, J. A Practical Solution to Deformable Linear Object Manipulation: A Case Study on Cable Harness Connection. In Proceedings of the 2020 5th International Conference on Advanced Robotics and Mechatronics (ICARM), Shenzhen, China, 18–21 December 2020; pp. 329–333.

12. Caporali, A.; Galassi, K.; Zagar, B.L.; Zanella, R.; Palli, G.; Knoll, A.C. RT-DLO: Real-Time Deformable Linear Objects Instance Segmentation. *IEEE Trans. Ind. Inform.* **2023**, *19*, 11333–11342. [CrossRef]
13. Zanella, R.; Caporali, A.; Tadaka, K.; Gregorio, D.D.; Palli, G. Auto-generated Wires Dataset for Semantic Segmentation with Domain-Independence. In Proceedings of the 2021 International Conference on Computer, Control and Robotics (ICCCR), Shanghai, China, 8–10 January 2021; pp. 292–298.
14. Dirr, J.; Gebauer, D.; Yao, J.; Daub, R. Automatic Image Generation Pipeline for Instance Segmentation of Deformable Linear Objects. *Sensors* **2023**, *23*, 3013. [CrossRef] [PubMed]
15. Nguyen, H.G.; Habiboglu, R.; Franke, J. Enabling deep learning using synthetic data: A case study for the automotive wiring harness manufacturing. *Procedia CIRP* **2022**, *107*, 1263–1268. [CrossRef]
16. Scheffler, B.; Bründl, P.; Nguyen, H.G.; Stoidner, M.; Franke, J. A Dataset of Electrical Components for Mesh Segmentation and Computational Geometry Research. *Sci. Data* **2024**, *11*, 309. [CrossRef] [PubMed]
17. Bründl, P.; Scheffler, B.; Stoidner, M.; Nguyen, H.; Baechler, A.; Abrass, A.; Franke, J. Semantic part segmentation of spatial features via geometric deep learning for automated control cabinet assembly. *J. Intell. Manuf.* **2023**, *35*, 3681–3695. [CrossRef]
18. Nguyen, H.G.; Franke, J. Deep learning-based optical inspection of rigid and deformable linear objects in wiring harnesses. *Procedia CIRP* **2021**, *104*, 1765–1770. [CrossRef]
19. Siemens Software, NX. Available online: https://www.plm.automation.siemens.com/global/en/products/nx/ (accessed on 17 February 2024).
20. fleXstructures, IPS Cable Simulation. Available online: https://flexstructures.com/products/ips-cable-simulation/ (accessed on 17 February 2024).
21. Blender Foundation, Blender—A 3D Modelling and Rendering Package. Available online: http://www.blender.org/ (accessed on 17 February 2024).
22. Buslaev, A.; Iglovikov, V.I.; Khvedchenya, E.; Parinov, A.; Druzhinin, M.; Kalinin, A.A. Albumentations: Fast and Flexible Image Augmentations. *Information* **2020**, *11*, 125. [CrossRef]
23. Yakubovskiy, Segmentation Models. GitHub Repository, 2019. Available online: https://github.com/qubvel/segmentation_models (accessed on 10 September 2023).

engineering proceedings

MDPI

Proceeding Paper

Chip Design of Multithreaded and Pipelined RISC-V Microcontroller Unit †

Mao-Hsu Yen [1], Yih-Hsia Lin [2], Tzu-Feng Lin [1], Yu-Hui Chen [1,*], Yuan-Fu Ku [3] and Chien-Ting Kao [4]

1 Department of Computer Science and Engineering, National Taiwan Ocean University, Keelung 202301, Taiwan; ymh@mail.ntou.edu.tw (M.-H.Y.); 11157002@email.ntou.edu.tw (T.-F.L.)
2 Department of Electronic Engineering, Ming Chuan University, Taoyuan 333321, Taiwan; yslin@mail.mcu.edu.tw
3 Taiwan Testing and Certification Center, Taoyuan 333011, Taiwan; kuyuanfu@etc.org.tw
4 Bureau of Standards, Metrology and Inspection (BSMI), Taipei 100026, Taiwan; teddy.kao@bsmi.gov.tw
* Correspondence: 11257045@email.ntou.edu.tw
† Presented at the 2024 IEEE 7th International Conference on Knowledge Innovation and Invention, Nagoya, Japan, 16–18 August 2024.

Abstract: Multithreading is widely used in microcontroller unit (MCU) chips. Multithreaded hardware is composed of multiple identical single threads and provides instructions to different threads. Using the concept of thread-level parallelism (TLP), pauses are compensated for during single-thread operation to increase the throughput at the same unit. The principle of pipelined management is to use instruction-level parallelism (ILP) to split the MCU into multiple stages. When an instruction is given in a certain stage, other instructions are provided to operate in other idle stages and improve their execution efficiency. Based on the four-thread and pipelined RISC-V MCU architecture, we analyzed the instruction types of three benchmarks, i.e., Coremark, SHA, and Dijkstra. A total of 94% of the instructions use the arithmetic logic unit (ALU). Based on the executable four-thread architecture, we developed two to four RISC-V architectures with different numbers of ALUs and a dispatch algorithm. This architecture allows for the simultaneous delivery of multiple instructions, enabling parallel processing of instructions and increasing efficiency. Compared to the traditional RISC-V architecture with only one ALU, the test results showed that the instructions per clock (IPCs) of RISC-V architectures with two, three, and four ALUs increased efficiency by 76, 128.9, and 154.3%, while the area increased by 12, 22.3, and 32.6% and the static power consumption increased by 5.1, 9.2, and 13.3%. The results showed a significant improvement in performance with only a slight increase in the area. Due to the limited area of chips, a two-thread microcontroller architecture was used for the IC design and tape-out. TSMC's 180nm process with a chip area of 1190 × 1190 μm at 133 MHz was used in this study.

Keywords: multithreading; microcontroller unit (MCU); thread-level parallelism (TLP); arithmetic logic unit (ALU)

1. Introduction

A microcontroller unit includes a core, memory, input/output interfaces, and external devices for multiple control and coordination capabilities. The instruction set architecture adopted in this study is the open-source RISC-V [1], originating from the University of California, Berkeley. It is designed based on the principle of reduced instruction set computer (RISC), which has a simplified instruction set, neat instruction format, and easy implementation of the pipeline architecture, all of which contribute to the concise code, fast computation speed, and small architecture volume. The instruction set of RISC-V includes the base integer instruction set and the standard extended instruction set. We implemented the 32-bit base integer instruction set and standard extension for integer multiplication and division instruction set (RV32IM) according to our needs, with 32 registers, 48 instructions, and 6 instruction formats. The instructions were divided into immediate addressing, register addressing, jump instructions, and load/store instructions [2,3].

To improve the speed and efficiency of the microcontroller unit (MCU), the design approach completes instructions in a single cycle to utilize pipelined instruction-level parallelism (ILP) in the same instruction set [4]. Although this improves the efficiency of execution significantly, it cannot meet the needs of current rapid developments. Therefore, hardware multithreading technology allows for breakthrough developments. A single thread is a production line that independently runs instructions, from fetch and decoding to execution. When the workload becomes heavy, a single production line cannot handle it, resulting in work fatigue, low efficiency, and other problems. Multiple threads, on the other hand, combine multiple identical single threads to disperse tasks into each thread, doubling the throughput at the same time. When applied to an MCU, multithreading utilizes the concept of thread-level parallelism (TLP) to compensate for the pause time or waiting for memory access that is caused by single-thread operation. This allows for the execution unit to remain operational without causing idle situations. Finally, the dispatch algorithm is used to determine which thread can send instructions to the execution unit for computation [5] for each thread to operate and determine the running speed.

A four-thread MCU with RISC-V is an instruction set containing one arithmetic logic unit (ALU) [6]. The efficient hint-based event (EHE) algorithm is combined with a new efficient event-based issue scheduling algorithm and is applied to the RISC-V multithreaded pipeline. Its concept is a combination of round robin and event-driven. In general, round-robin is used for dispatch, but when the next instruction that is executed is a jump instruction, it is changed to an event-driven mode for execution. This method is more efficient than simple round-robin and event-driven algorithms. Yuta and Nobuyuki proposed a real-time system-specific multithreaded RISC-V processor [7]. The proposed processor utilizes ILP and TLP, and if vector operations are added, data-level parallelism (DLP) is obtained, completing all forms of parallelization; Abdullah et al. proposed that the hardware conforms to the Pulpino processor and simultaneously supports internet of things (IoT) applications with interleaved multithreading [8]. Sylvain put forward a large-scale multithreaded RISC-V processor core called Simty and proposed an open platform with the same name for studying GPU microarchitectures and hybrid CPU-GPU microarchitectures [9].

We tested the architecture proposed in Ref. [6] using three benchmarks: Coremark, SHA, and Dijkstra. After converting them through the compiler into assembly and analyzing the types of instructions, 94% of the instructions were used in the ALU, which only accounted for 9.2% of the total area of the RISC-V architecture (Table 1). In a multithreaded processor, having only one ALU limits the efficiency of each thread's execution. Therefore, increasing the number of ALUs improves performance. Therefore, we created a RISC-V multithreaded pipeline architecture, which includes two to four ALUs, and individual benchmarks are used for area, power, and performance analysis. Based on the EHE algorithm for a single ALU, modifications were made to develop a dispatch algorithm for two to four ALUs. Multiple instructions were dispatched at the same time in parallel, thereby increasing efficiency. According to the test results, IPCs with two to four ALU architectures showed significant performance improvements, with slight increases in the microarea and static power consumption.

Table 1. Analysis of instruction types for benchmarks.

	Coremark	SHA	Dijkstra
ALU	94.3%	96.3%	95.7%
CSR	0.2%	0.5%	0.2%
MUL/DIV	1.7%	0%	0.5%
Others	3.8%	3.2%	3.6%

In this article, Section 2 elaborates on the multithreaded hardware architecture and pipelined RISC-V design; Section 3 introduces the patch algorithm; Section 4 presents the architecture implementation of this paper, which is validated using FPGA and implemented using very-large-scale integration (VLSI); Section 5 concludes this study and presents future research directions.

2. Architecture Design

Although the utilization rate of ALUs exceeds 90%, they only occupy less than 10% of the area; thus, it is speculated that increasing the number of ALUs can effectively improve the IPC. Therefore, we developed a four-thread RISC-V MCU with an *N* ALU architecture and dispatch algorithm. This architecture consists of three stages of pipelining, four hardware threads, and two to four ALUs. The RISC-V architecture including four hardware threads with one ALU is used as a control group to compare the performance, power consumption, and area. The four-thread RISC-V MCU with *N* ALUs is shown in Figure 1. The MCU design was used in different stages of three-stage pipelining.

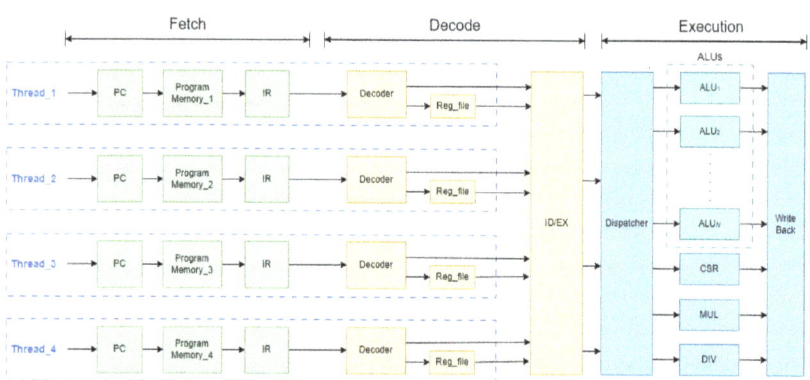

Figure 1. Architecture diagram of 4-thread RISC-V MCU with *N* ALUs, *N* = 1, 2, 3, or 4.

The instruction fetch stage is the first stage of the three-stage pipeline in the RISC-V MCU, which in most cases fetches instructions from the Program Memory based on the addresses stored in the program counter (PC). The instruction fetch stage is usually one of the key bottlenecks in processor performance, as it involves memory access and is relatively slow. Therefore, current processors use techniques such as branch prediction and instruction caching to enhance the speed of the instruction fetch stage, reduce pipeline stall, and improve processor throughput [7]. The PC in the instruction fetch stage of this architecture is the address of the instruction to be executed, which is fetched from the program memory and stored in the IR (instruction register), waiting for decoding in the next stage.

The instruction decoding stage is important in the RISC-V architecture. Due to its special streamlined design [10], there are more neat types of instruction formats, so there is no need to spend too many hardware resources in the decoding stage, and it executes quickly and concisely. At this stage, the processor decodes the instructions that are stored in the IR to distinguish the operation code (OPCode) and related information on the instructions, generate control signals, and read the value of the register file (Reg_file) [11] for subsequent instruction execution stages. Instruction decode/execution (ID/EX) stores disassembled instructions by category, such as the read register values, immediate values, etc., waiting to enter the instruction execution stage.

The instruction execution stage is the final stage in the processor, with various computing units for performing data operations on the data that were generated during the instruction decoding stage based on the corresponding OPCodes. The instruction includes addition, subtraction, multiplication, and division; data access; and other actions. The increase in the number of ALUs is achieved by modifications during the instruction execution stage, combined with the dispatch algorithm for multiple execution units. By combining round-robin and event-driven scheduling methods, appropriate instructions and the number of instructions are allocated to the corresponding execution units, enabling the MCU to execute normally and faster.

3. Dispatch Algorithm

The four-thread RISC-V MCU with one ALU adopts a configuration of multiple threads and a single execution unit. Each execution requires selecting one of the threads, and the algorithm used is the EHE algorithm. This is a combination of round-robin and event-driven triggering. In general, round-robin is used for dispatch, but when there are jump or branch instructions in the thread, it is executed in event-driven mode. This method was verified in Ref. [6] and was proven to be more efficient than using only round-robin algorithms or event-driven algorithms.

The architecture in this study is the four-thread RISC-V MCU with N ALUs, in which $N = 1, 2, 3$, or 4. A dispatch algorithm, that conforms to the new architecture was developed, and a maximum of N instructions was dispatched simultaneously. Firstly, the architecture fetches instructions from the Program Memory in advance to the ID/EX register, a process called prefetch, which requires two to four clock cycles before instruction dispatch. The dispatch algorithm uses the round-robin method to initialize the weights of four threads. The higher the weight value is, the higher the priority is.

3.1. Algorithm Steps

- The round-robin method is used to assign initial weights to four threads, with initial weight values including 4, 2, 1, and 0. The higher the weight value is, the higher the priority is. The initial weight values for the first round of the four threads are 4, 2, 1, and 0. The second-round values are 0, 4, 2, 1; the third-round values are 1, 0, 4, 2; and the fourth round consists of 2, 1, 0, and 4, and the weight values are iteratively initialized in this order, ensuring that each execution order receives a fair chance of execution.
- The DIV instruction for division is checked. The jump or branch instruction with control hazard is executed in the four threads. Therefore, the event-driven mode is switched to add 4 to the weight of the thread as the new weight to increase priority. If it is a DIV instruction, it increases the utilization rate of empty DIV units, and the completion time of the instruction is advanced. If it is a jump or branch instruction, it allows for more time to fetch the next instruction. It executes prefetch first and uses a countdown counter to record the number of clock cycles that are required for the thread to prefetch the next instruction.
- This dispatch algorithm limits the ability to dispatch up to N instructions at the same time. Among them, there is only one MUL, DIV, and CSR unit, so these three types of

instructions dispatch up to one each, and the rest must be ALU-type instructions. In addition, if the executing unit is a DIV unit, it takes three clock cycles to complete the execution. Therefore, the countdown counter is used to record the required number of clock cycles, while other executing units only need one clock cycle to complete the execution.

- At this point, the dispatch is completed. Then, step 1 is repeated to update the initial weights. However, if the prefetch countdown counter in step 2 does not reach 0, the instruction prefetching has not been completed. Therefore, the weight of this thread is set to 0, priority is given to threads that have been fetched but have not yet been executed, and steps 2, 3, and 4 are repeated to improve hardware resource utilization.

3.2. Pseudo Code of Dispatch Algorithm

The pseudo code in Algorithm 1 corresponds to the above algorithm steps and demonstrates how the dispatch algorithm operates.

Algorithm 1. Dispatcher for 4-thread RISC-V MCU with *N* ALUs

```
Input: N = 1, 2, 3, or 4
      prefetch = 2, 3, or 4 // clock number of prefetch
// Initializing weight
thread_weight [1] ← 4
thread_weight [2] ← 2
thread_weight [3] ← 1
thread_weight [4] ← 0
DIV_used_cnter ← 0
// Dispatching
while !four_threads_all_empty do
   for i = 1 to N do
      ALU_used[i] ← false
   end for
   MUL_used ← false
   CSR_used ← false
   if DIV_used_cnter = 0 then  // Executing division needs 3 clocks
      DIV_used ← false
   else
      DIV_used_cnter ← DIV_used_cnter − 1
   end if
   if thread_wait_cnter[num] ! = 0 then
      thread_wait_cnter[num] ← thread_wait_cnter[num] − 1
   end if
   for num = 1 to 4 do
      // Event-driven
      if thread_instruction[num] = Branch_instruction then
         thread_weight[num] ← thread_weight[num] + 4
         thread_wait_cnter[num] ← prefetch
      end if
   end for
   for i = 1 to N do
      num ← max(thread_weight [1..4])
```

Algorithm 1. *Cont.*

```
        thread_weight[num] ← 0
        if thread_instruction[num] = MUL_instruction and MUL_used = false then
            MUL ← thread_instruction[num]
            MUL_used ← true
        else if thread_instruction[num] = DIV_instruction and DIV_used = false then
            DIV ← thread_instruction[num]
            DIV_used ← true
            // Executing division needs 3 clocks
            DIV_used_cnter ← 3
        else if thread_instruction[num] = CSR_instruction and CSR_used = false then
            CSR ← thread_instruction[num]
            CSR_used ← true
        else then    // ALU_instruction
            for j = 1 to N do
                if ALU_used[j] = false then
                    ALU[j] ← thread_instruction[num]
                    ALU_used[j] = true
                end if
            end for
        end if
    end for
    // Round-robin
    thread_weight [1] ← thread_weight [4]
    thread_weight [2] ← thread_weight [1]
    thread_weight [3] ← thread_weight [2]
    thread_weight [4] ← thread_weight [3]
    // Checking whether prefetch is done
    for num = 1 to 4 do
        if thread_wait_cnter[num] ! = 0 then
            thread_weight[num] ← 0
        end if
    end for
end while
```

3.3. Illustrations

For the four-thread RISC-V MCU with two ALUS, a maximum of two instructions are dispatched simultaneously, where A represents general instructions, B represents jump and branch instructions, M represents multiplication instructions, D represents division instructions, and C represents CSR (control and status register) instructions.

As shown in Figure 2a, during the first dispatch, the initial weight of thread_1 is 4, thread_2 is 2, thread_3 is 1, and thread_4 is 0. No B instruction is detected, and the two highest priority instructions are both A instructions. Therefore, both thread_1 and thread_2 can be dispatched simultaneously.

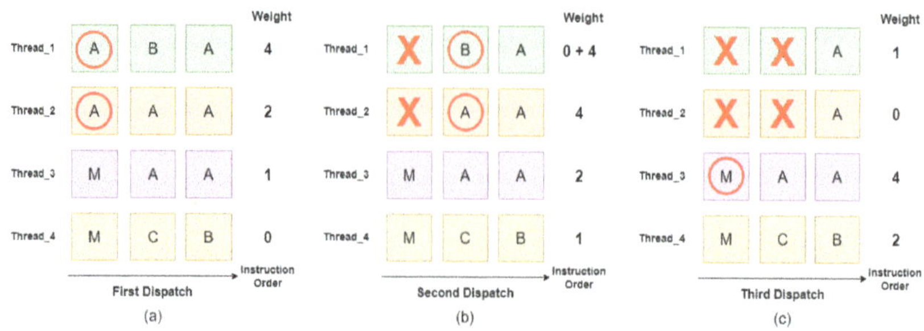

Figure 2. Dispatch algorithm examples, where (**a**) is the first dispatch, (**b**) is the second dispatch, and (**c**) is the third dispatch. The instruction marked with a cross has been executed, while the instruction enclosed in a circle is about to be executed.

As shown in Figure 2b, in the second dispatch, the initial weight of thread_1 is 0, thread_2 is 4, thread_3 is 2, and thread_4 is 1. However, thread_1 is a B instruction, and the weight of that thread is increased by 4 and does not include the M, D, and C instructions. Therefore, the highest weighted thread_1 and thread_2 are dispatched. In the third dispatch, the initial weight of thread_1 is 1, thread_2 is 0, thread_3 is 4, and thread_4 is 2 (Figure 2c). No B instruction is detected, but the highest weighted thread_3 and thread_4 are M instructions. Only the thread with the highest weight is selected, so thread_3 is dispatched.

4. Architecture Validation and Analysis

4.1. FPGA Verification

After the completion of the algorithm architecture design, it was implemented using the SystemVerilog hardware description language. The Segger Embedded Studio integrated development environment (IDE) V8.10a was to generate a RISC-V machine code for testing to verify whether the core architecture of the MCU meets the designed functions. After the register transfer-level (RTL) simulation using ModelSim 10.5b software, the architecture code was burned into Altera DE10 FPGA, and further validation was performed using the built-in SignalTap logic analysis in ModelSim. The verification was conducted as shown in Figures 3 and 4.

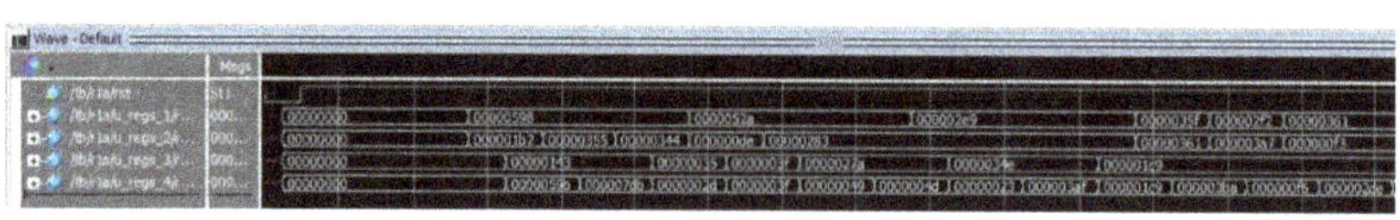

Figure 3. Coremark simulated waveform.

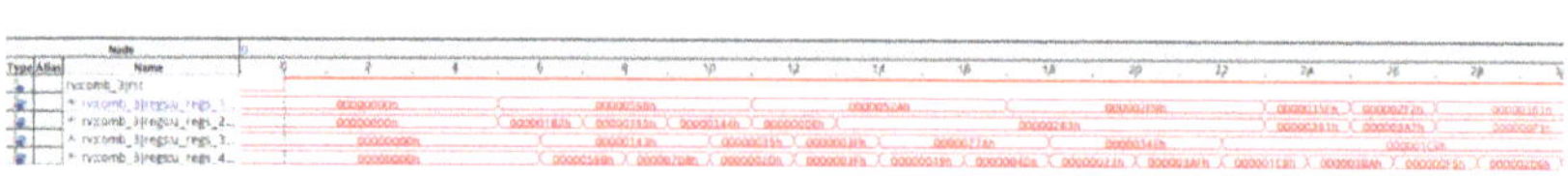

Figure 4. Coremark test waveform.

4.2. Architecture Analysis and Comparison

We implemented the four-thread RISC-V MCU with N ALUs using the 180 nm process of Taiwan Semiconductor Manufacturing Company (TSMC). Three key points of chip design were used as the benchmark for the analysis results, namely power consumption, performance, and area.

In the static power consumption analysis, the RISC-V architecture of one ALU was used as a benchmark to calculate the power consumption of the RISC-V architecture for other ALU quantities. As shown in Table 2, the static power consumption of the two-ALU RISC-V architecture was only 5.1% higher than that of the one-ALU RISC-V architecture, 9.2% higher than that of the three-ALU RISC-V architecture, and 13.3% higher than that of the four-ALU RISC-V architecture, with an increase of less than 31 mW.

Table 2. Static power consumption analysis of 4-thread RISC-V with *N* ALU architecture: *N* = 1, 2, 3, or 4.

	1 ALU [6]	2 ALUs	3 ALUs	4 ALUs
RISC-V(mW)	230.37	242.27	251.55	261.06
Power consumption comparison	100%	105.1%	109.2%	113.3%

In terms of the performance analysis, individual analyses were conducted with a required prefetch of two to four clock cycles, that is, the number of clock cycles that were required to prefetch instructions to the register. The analysis results are shown in Table 3. The architecture in Ref. [6] was set as the benchmark to calculate the performance multiples of different ALU architectures, as shown in Table 4. A best-case prefetch of two was used to illustrate that compared to the one-ALU RISC-V architecture, the two-ALU RISC-V architecture improved the IPC by 94.8%, the three-ALU RISC-V architecture improved the IPC by 162.3%, and the four-ALU RISC-V architecture improved it by 212.8%. In the worst-case prefetch, the two-ALU RISC-V architecture improved the IPC by 76%, the three-ALU RISC-V architecture improved it by 128.9%, and the four-ALU RISC-V architecture improved it by 154.3%. Increasing the number of ALUs effectively improved the efficiency.

Table 3. Performance analysis of 4-thread RISC-V with *N* ALU architecture: *N* = 1, 2, 3, or 4.

Prefetch	1 ALU [6]	2 ALUs	3 ALUs	4 ALUs
2	0.991	1.931	2.6	3.1
3	0.989	1.826	2.41	2.7
4	0.983	1.731	2.25	2.5

Table 4. Performance comparison of 4-thread RISC-V with *N* ALU architecture: *N* = 1, 2, 3, or 4.

Prefetch	1 ALU [6]	2 ALUs	3 ALUs	4 ALUs
2	1	1.948	2.623	3.128
3	1	1.846	2.437	2.73
4	1	1.76	2.289	2.543

Table 5 lists the cell areas of each architecture and also sets the architecture of [6] as the benchmark to compare the cell area sizes of each them. The two-ALU RISC-V architecture only increased by 12% in area, the three-ALU RISC-V architecture increased by 22.3%, and the four-ALU RISC-V architecture increased by 32.6%.

Table 5. Comparison of cell areas for 4-thread RISC-V with *N* ALU architecture: *N* = 1, 2, 3, or 4.

	1 ALU [6]	2 ALUs	3 ALUs	4 ALUs
Area (μm^2)	926,442	1,037,361	1,133,061	1,228,539

Finally, the prefetch was divided into two, three, and four to calculate the IPC increase per unit area (Table 6). The architecture proposed in this study effectively improved the IPC, and the area of improvement was not significant. From the perspective of efficiency and area, regardless of the prefetch, the architecture containing two ALUs was the best; from the perspective of high efficiency, the architecture that included four ALUs was the best.

Table 6. Improvements in IPCs and area ratio.

Prefetch	2 ALUs	3 ALUs	4 ALUs
2	8,474,653	7,787,280	6,981,201
3	7,546,047	6,877,392	5,663,743
4	6,743,660	6,132,059	5,021,566

4.3. Chip Implementation

We adopted the educational chip provided by the Taiwan Semiconductor Research Institute (TSRI) for the implementation of this architecture, which is produced using TSMC's T18 process. As it is an educational chip, the chip area is limited to 1190 × 1190 μm, which is not enough to achieve an MCU with four threads. Therefore, it was reduced to a version with two threads combined with two ALUs, and the cell-based design flow was performed using the EDA cloud provided by the TSRI. The chip layout is shown in Figure 5 and the chip specifications are shown in Table 7.

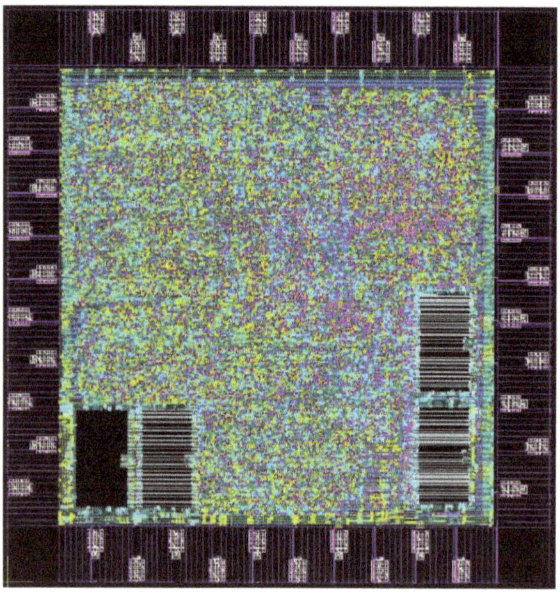

Figure 5. Layout diagram.

Table 7. Chip Specification.

Item	Specification
Process	TSMC 180 nm
Area	1190 μm × 1190 μm
Pin	40 pin
Core Power Pads	4 sets
Pad Power Pads	4 sets
Core voltage	1.8 V
Pad voltage	3.3 V
Power consumption	13.3 mW
Frequency	133 MHz

5. Conclusions

We developed an architecture to improve the performance of a four-thread RISC-V MCU. By increasing the number of ALUs and using the dispatch algorithm, the performance was improved in a limited area using the concepts of ILP and TLP. Compared to the original infrastructure, the IPCs with two to four ALU architectures increased by 76, 128.9, and 154.3%, with the area increasing by 12, 22.3, and 32.6%. The static power consumption increased by 5.1, 9.2, and 13.3%, achieving a significant improvement in performance with only a slight increase in area. In verification, the four-thread RISC-V MCU with *N* ALUs that was designed in this study operated correctly at a speed of 133 MHz through FPGA and cell-based design flow testing.

Algorithm modifications are necessary to find more appropriate dispatch methods for multiple threads. On the other hand, the RISC-V instruction set is currently continuously updating its functional modules to reserve space for self-defined instructions, such as adding a vector extension instruction set or customizing instructions and increasing the vector operation capabilities. By performing vector operations, convolution operations, and other operations with the same instructions and a large amount of data, the data-level parallelism (DLP) increased [7], while achieving ILP, TLP, and DLP. The MCUs had more comprehensive application upgrades and were optimized for applications in IoT and artificial intelligence.

Author Contributions: Conceptualization, M.-H.Y., Y.-H.L. and T.-F.L.; methodology, M.-H.Y., Y.-H.L. and T.-F.L.; software, T.-F.L.; validation, T.-F.L.; formal analysis, M.-H.Y., Y.-H.L. and T.-F.L.; resources, M.-H.Y. and Y.-H.L.; data curation, T.-F.L.; writing—original draft preparation, T.-F.L. and Y.-H.C.; writing—review and editing, M.-H.Y., Y.-H.L. and Y.-H.C.; visualization, Y.-H.C.; supervision, M.-H.Y., Y.-H.L., Y.-F.K. and C.-T.K.; project administration, T.-F.L.; funding acquisition, Y.-F.K. and C.-T.K. All authors have read and agreed to the published version of the manuscript.

Funding: This research was funded by the Bureau of Standards, Metrology, and Inspection (BSMI), Taiwan, under Grant Number 1D161121218-117.

Data Availability Statement: The data from this study can be obtained upon request from the corresponding author. They are not publicly accessible due to privacy concerns.

Conflicts of Interest: The authors declare no conflicts of interest.

References

1. Waterman, A.; Lee, Y.; Avizienis, R.; Patterson, D.A.; Asanovic, K. *The RISC-V Instruction Set Manual Volume II: Privileged Architecture Version 1.9*; University of California: Berkeley, CA, USA, 2016.
2. Lim, D.X.; Smitha, K.G. Pipelined MIPS Simulation: A plug-in to MARS simulator for supporting pipeline simulation and branch prediction. In Proceedings of the 2019 IEEE International Conference on Engineering, Technology and Education, Yogyakarta, Indonesia, 10–13 December 2019; pp. 1–7.
3. Curran, B.W.; Jacobi, C.; Bonanno, J.J.; Schroter, D.A.; Alexander, K.J.; Puranik, A.; Helms, M.M. The IBM z13 multithreaded microprocessor. *IBM J. Res. Dev.* **2015**, *59*, 1–13. [CrossRef]
4. Mendoza Escobar, J. *A Multithreading RISC-V Implementation for Lagarto Architecture*; Universitat Politècnica de Catalunya: Barcelona, Spain, 2020.
5. Das, A.; Jose, J.; Mishra, P. Data criticality in multithreaded applications: An insight for many-core systems. *IEEE Trans. Very Large Scale Integr. Syst.* **2021**, *29*, 1675–1679. [CrossRef]
6. Eni, Y.; Greenberg, S.; Ben-Shimol, Y. Efficient Hint-Based Event (EHE) Issue Scheduling for Hardware Multithreaded RISC-V Pipeline. *IEEE Trans. Circuits Syst. I Regul. Pap.* **2021**, *69*, 735–745. [CrossRef]
7. Nojiri, Y.; Yamasaki, N. A Design of Multithreaded RISC-V Processor for Real-Time System. In Proceedings of the 2023 Eleventh International Symposium on Computing and Networking Workshops (CANDARW), Matsue, Japan, 27–30 November 2023. [CrossRef]

8. Cheikh, A.; Cerutti, G.; Mastrandrea, A.; Menichelli, F.; Olivieri, M. The Microarchitecture of a Multi-threaded RISC-V Compliant Processing Core Family for IoT End-Nodes. In *ApplePies 2017. Lecture Notes in Electrical Engineering*; Springer: Berlin/Heidelberg, Germany, 2019; Volume 512.
9. Sylvain Collange. Simty: Generalized SIMT execution on RISC-V. In Proceedings of the CARRV 2017-1st Workshop on Computer Architecture Research with RISC-V, Boston, MA, USA, 14 October 2017.
10. Hennessy, J.L.; Patterson, D.A. *Computer Architecture: A Quantitative Approach*; Elsevier: Amsterdam, The Netherlands, 2011.
11. Lai, J.-Y.; Chen, C.-A.; Chen, S.-L.; Su, C.-Y. Implement 32-bit RISC-V Architecture Processor using Verilog HDL. In Proceedings of the 2021 International Symposium on Intelligent Signal Processing and Communication Systems, Hualien City, Taiwan, 16–19 November 2021.

engineering proceedings

MDPI

Proceeding Paper

The Use of Support Vector Machine to Classify Potential Customers for the Wealth Management of a Bank †

Chien-Hung Lai [1,*], Yi Lin [2], Ju-Wen Hsieh [3] and Yuh-Shyan Hwang [1]

1. Department of Electronic Engineering, National Taipei University of Technology, Taipei 106, Taiwan; yshwang@ntut.edu.tw
2. Department of Business Administration, Takming University of Science and Technology, Taipei 114, Taiwan; linyi@takming.edu.tw
3. Department of Wealth Management, Cathay United Bank, Taipei 110, Taiwan; nt19167@cathaybk.com.tw

* Correspondence: ee12245@ntut.edu.tw

† Presented at the 2024 IEEE 7th International Conference on Knowledge Innovation and Invention, Nagoya, Japan, 16–18 August 2024.

Abstract: We developed a method for the evaluation and selection of customer business analysis in two stages. First, using the bank's existing expert model, artificial rules of thumb were used to evaluate the value of each field of the data and establish screening rules. Secondly, the machine learning feature screening method was applied based on the customer's transaction data to find out whether the customer's contribution to the bank had a significant impact as a feature of the model. Based on the results, the best classification model was selected through data verification. The effectiveness of the proposed model was validated through actual case analysis, taking wealth management in banks as an example. The classification method, using support vector machines (SVMs), effectively assists banks in identifying potential customers efficiently and in planning to manage customers. This method helps to avoid the traditional blind spots, which emerge based on subjective judgment, and allows bank wealth managers to promote customer relationship management (CRM).

Keywords: SVM; wealth management; potential customer; CRM

1. Introduction

The stock markets in Taiwan and Japan have been booming. Banks' wealth management services aim to attract deposit holders to invest large amounts of money by trading trade stocks and bonds. Banks hope to create revenue by charging trading fees and deducting from customer surpluses. The key is to identify people who have money in accounts, are interested in investment, and have a mutual agreement with the bank. Such customers need to be identified as potential customers in order to promote wealth management services.

However, banks have many customers, and each person's financial management concept, income, and expenditure are different. It is difficult to select qualified people to be potential customers from many records. The application of machine learning in customer relation management (CRM) has been verified [1–6]. Previous methods solicited customers using marketing methods, such as calls, flyers, or questionnaires, but they were not effective. To improve effectiveness, banks target their customers. There are a variety of screening methods, and the conditions are different. Most of the screening conditions are based on the bank's financial relationship and the managers' decisions, which are subjective and lack mathematical logic [3]. The accuracy was only 56%. Therefore, it is not appropriate as a management strategy. The characteristics of the data must be engineered to achieve accuracy, as shown in Table 1.

Table 1. Performance after data engineering [3].

Algorithm	Accuracy	Recall	Precision
Logistic regression	98.49	99.56	97
Decision tree classifier	97.95	96.98	98.22
Support vector classifier	97.3	97.99	95.82
Random forest classifier	98.14	98.49	97.2
Multilayer perceptron classifier	99.41	98.93	99.68

Accuracy, recall, and precision are calculated using (1)–(3). The results are classified into true positives (TP), true negatives (TN), false positives (FP), and false negatives (FN). TP and TN describe the number of correct classifications. On the other hand, FP and FN represent the number of misclassifications.

$$\text{Accuracy} = \frac{\text{TP} + \text{TN}}{\text{TP} + \text{TN} + \text{FP} + \text{FN}}, \tag{1}$$

$$\text{Recall} = \frac{\text{TP}}{\text{FN} + \text{TP}}, \tag{2}$$

$$\text{Precision} = \frac{\text{TP}}{\text{FP} + \text{TP}}, \tag{3}$$

In this study, we cooperated with C Bank in Taiwan and used a machine learning algorithm to analyze the conditions for wealth management customer records. Features were extracted to build a model. From numerous customer records, potential customers can be identified to formulate business strategies.

2. Experiment Method

2.1. Analysis

Machine learning algorithms were proposed in Refs. [1–5] in order to study CRM and customer purchasing behavior. We used methods in Refs. [1,3]. Five algorithms were chosen, as listed in Table 1.

2.2. Algorithm

The multilayer perceptron classifier (MPC) is connected to multiple hidden layers [7]. Neurons in each layer have different weights, producing a function that expands the calculation of the output layer. As a result, an MPC cannot have transparency and interpretability [2]. Therefore, a decision tree classifier (DTC) and random forest classifier (RFC) were introduced [8]. RFC is regarded as a combination of many DTCs. Depending on entropy or Gini coefficient, DTC is used for classification to determine the maximum information gain, but it is not easy to determine the importance of features. RFC shows better results than DTC by establishing multiple DTCs and voting via multiple DTCs. Therefore, the more DTCs established, the better the RFC. However, as the amount of data increases, the depth and quantity of DTC also increase, and the execution efficiency of RFC is too reduced to process millions of data points.

Logistic regression [9] and support vector machine [10] are mostly used in non-linear classification methods. By finding out common conditions from the customer records of wealth management services, banks can use them as features to build models to predict potential customers. Since the model in this study needs to predict the results of binary classification, we consider the number of features and the size of the training sample with the following strategy.

- If n > m, either of the two is used.
- If n < m, SVM is used.

According to the observed situation, the number of customer records in the bank, m, must be larger than the number of fields in the customer information table, n. Therefore, based on the above strategy, we used SVM.

3. Implementation and Results

For feature selection, different screening methods are used according to supervised or unsupervised learning. Unsupervised learning removes features with low correlation to reduce interference from irrelevant features. The SVM algorithm belongs to supervised learning. It takes advantage of the overlapping characteristics of highly correlated features to retain highly correlated features and use them to find out effective fields for classifying potential customers. There are three types of feature screening methods used in supervised learning.

1. Filter Method

Each feature is scored according to divergence and correlation and sorted according to its importance. Then, the threshold is used to eliminate unimportant features in order to select required features [11,12]. Common filtering feature screening techniques include correlation and univariate features.

2. Wrapper Method

Based on the prediction results obtained using the algorithm, the quality of selected features is judged. Various feature combinations are tested to exclude unnecessary features until the best feature combination is selected [13]. The screening results are better than those of the filtering method but obtaining them is time-consuming [11,13]. Common packaging feature screening technologies include forward elimination, backward elimination, and recursive feature elimination (RFE).

3. Embedded Method

Machine learning algorithms and models are used for training. Through training, the contribution of each feature to the model is determined. A specific weight is used to represent its contribution. Features are sorted based on the weights. Although the screening effect of this is not better than that of the filtering method, it is more efficient than the packaging method [14]. Common embedding feature screening technologies include the assessment of feature importance and L1 regularization.

3.1. Implementation

We tracked the customers' recent transaction investments and the records of investment products on Bank C's official website to identify and label them. Labels include the customer's recent investment portfolio products, bonds, and time deposits, recently remitted foreign currency, browsed financial management, insurance, and a suspension for the invested funds. There are more than ten types of loss and profit suspension reminders for financial management. However, the primary consideration for financial managers is to identify potential customers with large investment intentions in bank products and with large pay subscription fees. Therefore, customers whose contribution in the past year (set as variable Y) was five times larger than the average contribution by all customers (set as variable X, that is, $Y \geqq 5X$) were defined as potential customers in this study.

We randomly selected more than 144,000 records from all customers with the following conditions: the asset balance was more than NT$3 million and Y ≧ 5X. We established a reference that can be used by various banks for wealth management services. We referred to segmentation, targeting, and positioning theory to determine factors [15], considering demographics, geographical location, beliefs and values, lifestyle, and consumer behavior. Based on the C Bank's company internal analysis, the following three aspects were defined:

- Basic customer information: In addition to age, gender, education, and occupation category, the customer needs to be qualified to invest in specific high-risk products and receive a "professional investor flag". This indicates whether the customer has a higher tolerance and participates in the bank's issuance of funds. Products with high investment returns contribute to the performance of banks' wealth management services. In addition, a "life cycle segmentation" method customized by Bank C is also added.
- Customer asset account: 9 fields are determined, including the total average asset balance, the foreign currency saving deposit balance, Taiwanese dollar deposits, foreign currency deposits, mutual fund assets under management, insurance, and structured notes. Based on the balance of each field, we identify customer investment behavior and preference characteristics. Factors such as the number of products held and whether someone is a check deposit holder are added to prove the customer's enthusiasm for investment and their risk-taking awareness.
- Customer management of bank transactions: The "bank-period-years" are used to determine the degree of customers' loyalty to the bank or represent customer stickiness. The "relationship manager's code" is also added as the experience level of the specialist who provides the service. The "customer's major district code" is used to determine the difference to marketing approaches from different geographic area.

Table 1 lists the 21 fields, as shown in Appendix A. The name, data type, and range of the data were used to train the model.

3.2. Results

More than 144,000 randomly selected items were confirmed by Bank C's wealth management department. Among them, 12,499 records (80%) were used as training data, and the other 3125 records (20%) were used as testing data, to train the SVM model developed in this study. Three methods were used according to different feature screening methods: a correlation analysis method, a feature importance assessment method, and L1 regularization. The selected fields were applied to the SVM model to test its accuracy and find the best feature screening method that is consistent with the current rules of the wealth management department. The accuracy and F1 score of the four classification models after training are listed in Table 2. Table 3 shows the retest results of the SVM models using a test set of 3125 records. The F1 score is calculated using (4). Figures 1–4 show the receiver operator characteristic curve (ROC) of the SVM model with four screening methods. Figures 5–8 the confusion matrixes.

$$\text{F1 Score} = \frac{2}{\left(\frac{1}{Precision} + \frac{1}{Recall}\right)} \tag{4}$$

Table 2. Comparison of models after training using SVM and classification methods.

Method	Chosen Fields	Accuracy	F1 Score
Rule of thumb	C2, C4, C5, C6, C7, C8, C9, C10, C11, C12, C13, C14, C15, C16, C17, C20, C21	0.83008	0.83

Table 2. *Cont.*

Method	Chosen Fields	Accuracy	F1 Score
Correlation	C2, C3, C4, C5, C6, C7, C8, C9, C10, C11, C14, C15, C16, C17, C18, C19, C20, C21	0.95424	0.95
Feature importance	C5, C8, C9, C10, C14, C16, C18	0.95296	0.95
L1 regularization	C8, C14, C16, C18	0.92064	0.92

Table 3. Comparison of results obtained by retesting four methods.

Method	AUC	Precision	Recall	F1 Score
Rules of thumb	0.7	0.35764	0.971	0.52274
Correlation	0.87	0.41748	0.936	0.57742
Feature importance	0.96	0.57538	0.977	0.72424
L1 regularization	0.82	0.33333	1	0.5

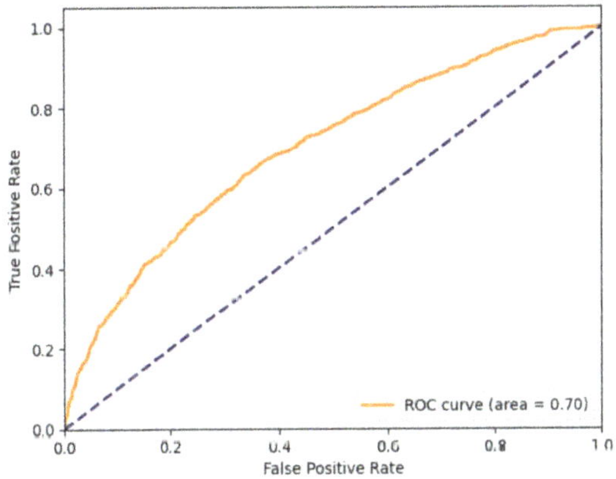

Figure 1. ROC of rules of thumb, dashed line represents the threshold.

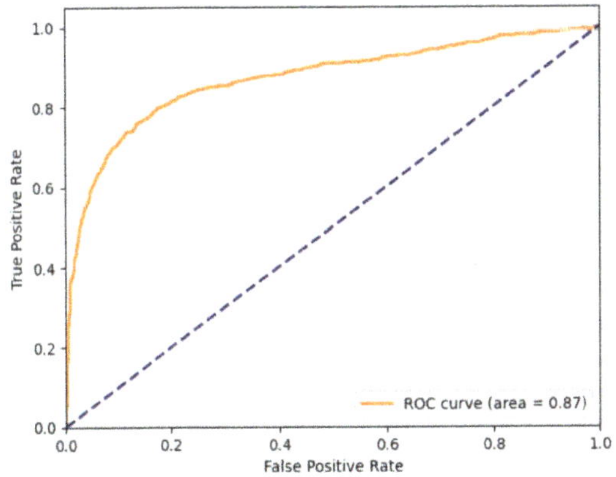

Figure 2. ROC of correlation, dashed line represents the threshold.

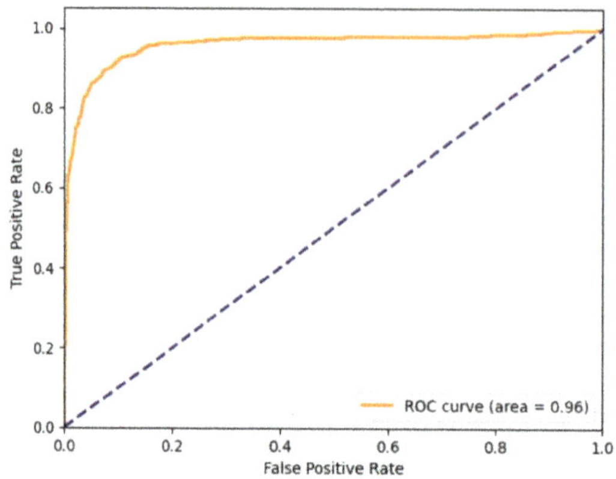

Figure 3. ROC of feature importance, dashed line represents the threshold.

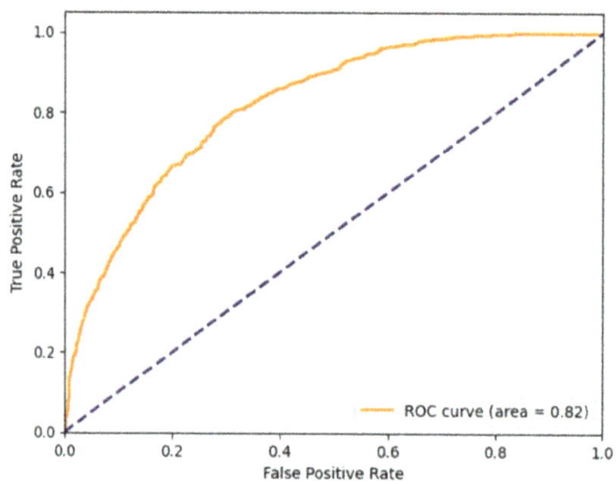

Figure 4. ROC of L1 regularization, dashed line represents the threshold.

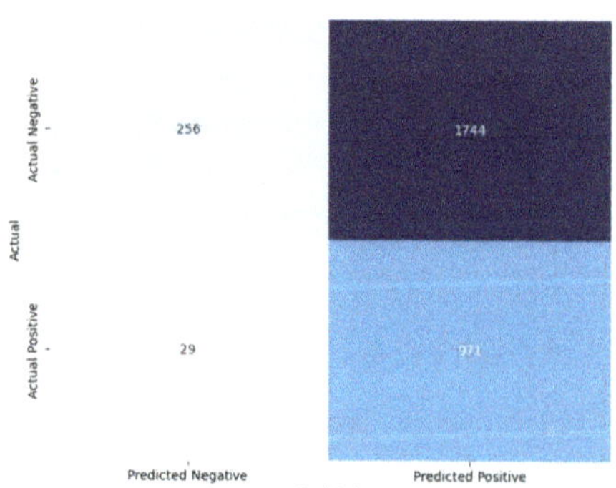

Figure 5. Confusion matrix of rules of thumb.

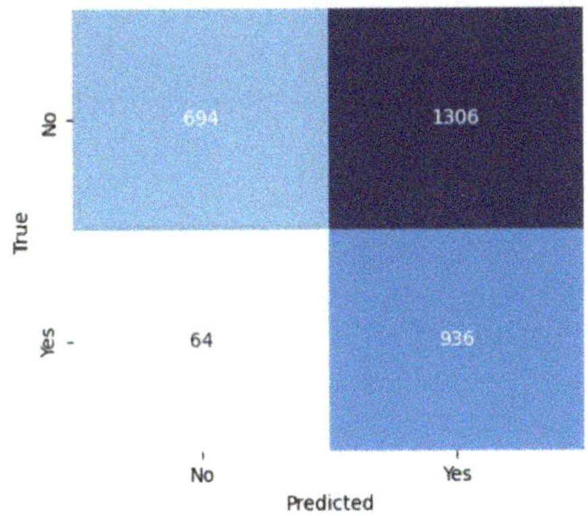

Figure 6. Confusion matrix of correlation.

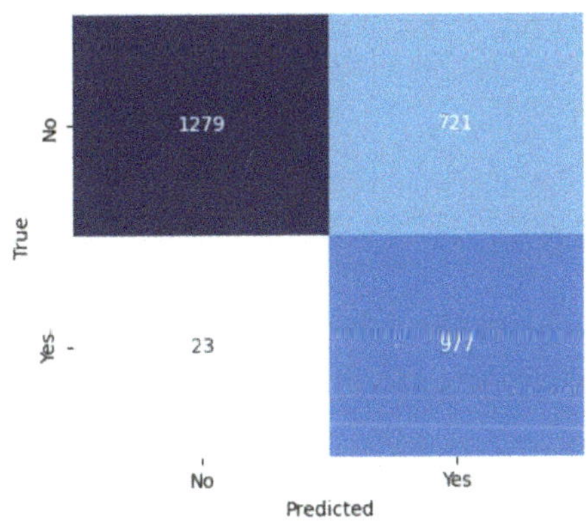

Figure 7. Confusion matrix of feature importance.

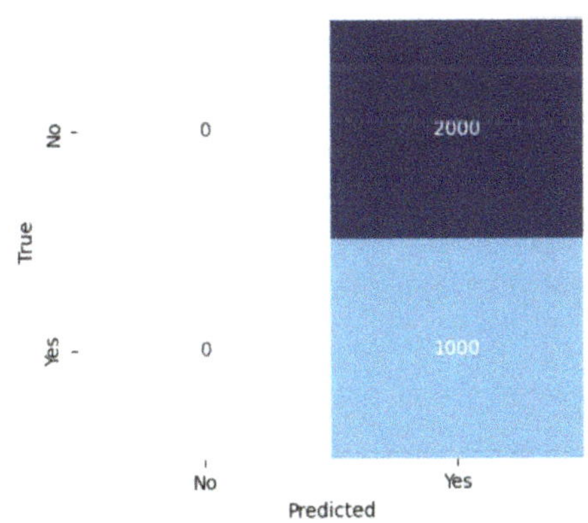

Figure 8. Confusion matrix of L1 regularization.

4. Conclusions

The SVM model of feature importance shows the best predictive ability, with an AUC of 0.977. It was the only model that judged TP and FP accurately with different thresholds. The AUC of the rule of thumb was 0.7, so the curvature of the ROC was less close to the boundary than seen in other models. The AUC of feature importance was 0.977, and the ROC was close to the boundary, which showed excellent discriminative power. The SVM model had an excellent ability to correctly judge the target category, but the best model misjudgment rate was about 60%. It is still challenging to accurately classify potential customers based on the results, but the model helps to identify non-potential customers, avoid ineffective customer management, and improve efficiency. The F1 score was used to judge the results, and the SVM model with feature importance showed the best prediction. The results are consistent with the judgment using the AUC.

We used the existing practices of Bank C as the experimental control group to observe whether the model enables efficient analysis. The accuracy of Bank C's current rules of thumb in the SVM model was 0.83. Although the accuracy did not reach 0, it still was satisfactory. After retesting, the AUC was 0.7, which showed that the current rules of thumb of Bank C were satisfactory. The accuracy of the SVM model was higher than 0.9, showing that the model improves the efficiency of avoiding subjective judgments. The number of important features was reduced compared with the 17 stipulated in Bank C's current rules.

The data contained invalid information due to subjective judgment. If an automated calculation mechanism is developed, the waste of computing resources can be reduced. The machine learning model can be incorporated into the future operating mechanisms of the bank to improve management efficiency.

Author Contributions: Conceptualization, C.-H.L. and J.-W.H.; methodology, C.-H.L.; software, C.-H.L. and J.-W.H.; validation, C.-H.L., J.-W.H. and Y.L.; formal analysis, Y.L.; investigation, J.-W.H. and Y.L.; resources, J.-W.H.; data curation, J.-W.H. and Y.L.; writing—original draft preparation, J.-W.H.; writing—review and editing, C.-H.L.; visualization, J.-W.H. and Y.L.; supervision, Y.-S.H.; project administration, Y.-S.H. All authors have read and agreed to the published version of the manuscript.

Funding: This research received no external funding.

Institutional Review Board Statement: This study did not require ethical approval.

Informed Consent Statement: Not applicable.

Data Availability Statement: The data cannot be obtained because the bank must protect customer privacy and has business secrets.

Acknowledgments: We sincerely thank the Wealth Management Department of Cathay Pacific Bank in Taiwan for providing all the assistance required during the writing of this article.

Conflicts of Interest: The authors declare no conflicts of interest.

Appendix A

Table A1. Current wealth management customer database fields of the Bank C.

Field Type	Field Name	Data Scope
Prediction (the answer)	C1, potential customer	No (0), yes (1)
Basic customer information	C2, age	20 to 70
	C3, gender	Male (1), female (2), enterprise (0)
	C4, education	1. Ph.D. 2. Master. 3. Bachelor. 4. Associate bachelor. 5. High school. 6. Others.
	C5, professional investor flag	No (0), yes (1)
	C6, occupation	1. Unstable. 2. Public agencies. 3. Teaching position/business. 4. Housekeeping. 5. Financial industry. 6. Others

Table A1. *Cont.*

Field Type	Field Name	Data Scope
	C7, life cycle segmentation	1. Top rich class. 2. Rich second generation. 3. Middle class. 4. Young. 5. Salary class. 6. Asset appreciation. 7. Retired.
Customer asset account	C8, total average asset balance	3 to 615 million
	C9, number of products held	0 to 8
	C10, check deposit holder	No (0), yes (1)
	C11, Taiwan foreign currency saving deposit balance	0~220 million
	C12, Taiwanese dollar saving deposit balance	0~337 million
	C13, Taiwanese dollar time deposit balance	0~208 million
	C14, foreign currency saving deposit balance	0~169 million
	C15, foreign currency time deposit balance	0~350 million
	C16, mutual fund net assets under management	0~569 million
	C17, insurance balance	0~647 million
	C18, structural notes balance	0~584 million
Customer management of bank transactions	C19, bank-period-years	0~104 years
	C20, relationship manager's code	1. Financial relationship management SRM1 2. Financial relationship management SRM2 3. Financial relationship management RM1 4. Financial relationship management RM2 5. PFB 6. YS 7. Comprehensive relationship management SRM 8. Comprehensive relationship management RM 9. Private banking RM
	C21, customer's major district code	1. North District 1 2. North District 2 3. North District 3 4. North District 4 5. Tao-zhu District 6. Middle District 1 7. Middle District 2 8. South District

References

1. Amnur, H. Customer Relationship Management and Machine Learning Technology for Identifying the Customer. *JOIV Int. J. Inform. Vis.* **2017**, *1*, 12–15. [CrossRef]
2. Ledro, C.; Nosella, A.; Pozza, I.D. Integration of AI in CRM: Challenges and guidelines. *J. Open Innov. Technol. Mark. Complex.* **2023**, *9*, 100151. [CrossRef]
3. Houdhury, A.M.; Nur, K. A Machine Learning Approach to Identify Potential Customer Based on Purchase Behavior. In Proceedings of the International Conference on Robotics, Electrical and Signal Processing Techniques (ICREST), Dhaka, Bangladesh, 10–12 January 2019.

4. Velu, A. Machine Learning Techniques for Customer Relationship Management. *Int. J. Creat. Res. Thoughts (IJCRT)* **2021**, *9*, 753–763.
5. Ledro, C.; Nosella, A.; Vinelli, A. Artificial intelligence in customer relationship management: Literature review and future research directions. *J. Bus. Ind. Mark.* **2022**, *37*, 48–63. [CrossRef]
6. Hsieh, J.-W.; Lai, C.-H.; Hwang, Y.-S. Application of Artificial Intelligence Technology in Bank Wealth Management Customer Operation Analysis-Utilizing Support Vector Machine to Predict and Classify Potential Customers. Master's Thesis, Executive Master of Business Administration (EMBA). National Taipei University of Technology (NTUT), Taipei, Taiwan, 19 January 2024.
7. Pal, S.K.; Mitra, S. Multilayer perceptron, fuzzy sets, classification. *IEEE Trans. Neural Netw.* **1992**, *3*, 683–697. [CrossRef] [PubMed]
8. Esmaily, H.; Tayefi, M.; Doosti, H.; Ghayour-Mobarhan, M.; Nezami, H.; Amirabadizadeh, A. A Comparison between Decision Tree and Random Forest in Determining the Risk Factors Associated with Type 2 Diabetes. *J. Res. Health Sci.* **2018**, *18*, 412.
9. Zou, X.; Hu, Y.; Tian, Z.; Shen, K. Logistic Regression Model Optimization and Case Analysis. In Proceedings of the IEEE 7th International Conference on Computer Science and Network Technology, Dalian, China, 19–20 October 2019.
10. MHearst, M.A.; Dumais, S.T.; Osuna, E.; Platt, J.; Scholkopf, B. Support vector machines. *IEEE Intell. Syst. Their Appl.* **1998**, *13*, 18–28. [CrossRef]
11. Zhang, R.; Nie, F.; Li, X.; Wei, X. Feature selection with multi-view data: A survey. *Inf. Fusion* **2019**, *50*, 158–167. [CrossRef]
12. Chandrashekar, G.; Sahin, F. A survey on feature selection methods. *Comput. Electr. Eng.* **2014**, *40*, 16–28. [CrossRef]
13. Guyon, I.; Elisseeff, A. An introduction to variable and feature selection. *J. Mach. Learn. Res. (JMLR)* **2003**, *3*, 1157–1182.
14. Salesi, S.; Cosma, G.; Mavrovouniotis, M. TAGA: Tabu Asexual Genetic Algorithm embedded in a filter/filter feature selection approach for high-dimensional data. *Inf. Sci.* **2021**, *565*, 105–127. [CrossRef]
15. Moutinho, L. Strategic management in tourism. In *Segmentation, Targeting, Positioning and Strategic Marketing*; Moutinho, L., Ed.; CABI: Wallingford, UK, 2000; pp. 121–166.

engineering proceedings

MDPI

Proceeding Paper

Enhancing Fabric Detection and Classification Using YOLOv5 Models †

Makara Mao [1], Jun Ma [1], Ahyoung Lee [2] and Min Hong [3,*]

1 Department of Software Convergence, Soonchunhyang University, Asan-si 31538, Republic of Korea; makaramao07@gmail.com (M.M.); majun@sch.ac.kr (J.M.)
2 Department of Computer Science, Kennesaw State University, Marietta, GA 30060, USA; alee146@kennesaw.edu
3 Department of Computer Software Engineering, Soonchunhyang University, Asan-si 31538, Republic of Korea
* Correspondence: mhong@sch.ac.kr
† Presented at the 2024 IEEE 7th International Conference on Knowledge Innovation and Invention, Nagoya, Japan, 16–18 August 2024.

Abstract: The YOLO series is widely recognized for its efficiency in the real-time detection of objects within images and videos. Accurately identifying and classifying fabric types in the textile industry is vital to ensuring quality, managing supply, and increasing customer satisfaction. We developed a method for fabric type classification and object detection using the YOLOv5 architecture. The model was trained on a diverse dataset containing images of different fabrics, including cotton, hanbok, dyed cotton yarn, and a plain cotton blend. We conducted a dataset preparation process, including data collection, annotation, and training procedures for data augmentation to improve model generalization. The model's performance was evaluated using precision, recall, and F1-score. The developed model detected and classified fabrics with an accuracy of 81.08%. YOLOv5s allowed a faster performance than other models. The model can be used for automated quality control, inventory tracking, and retail analytics. The deep learning-based object detection method with YOLOv5 addresses challenges related to fabric classification, improving the abilities and productivity of manufacturing and operations.

Keywords: YOLOv5; fabric classification; object detection; deep learning

1. Introduction

Deep learning has improved rapidly recently. Image recognition and video classification are widely used within industry to judge the ripeness of fruits and vegetables [1]. Object detection methods have attracted much attention, as exemplified by the YOLO algorithm. The advent of YOLO has notably enhanced detection speeds, expanding the operation of deep learning into various domains.

In online shopping applications, platforms such as Amazon, eBay, and Taobao have seamlessly integrated YOLO for diverse functionalities [2]. In agriculture, YOLO is extensively applied to environmental monitoring, crop health assessment, and pest detection. In online shopping applications, image recognition is used in clothing design, coordinating accessories, categorizing items, and conducting analyses [3]. With the rapid growth of e-commerce, online shopping has provided consumers with convenience and cheaper products. In the apparel industry, engaging customers depends on visual elements such as color, style, and other features, highlighting the need for accurate image representation [4].

For online stores, showcasing garments' various color and style variations as well as many try-on pictures and accessory options is paramount. This facilitates a straightforward selection aligned with customers' preferences and enhances the shopping experience [5]. The applications of AI technologies such as YOLO extend beyond online shopping to clothing classification and detection, underscoring their versatility and potential in the industry [6]. Leveraging YOLO's real-time processing capabilities and high accuracy in object detection, online retailers can streamline the browsing experience, offer accurate product recommendations, and elevate customer satisfaction. Moreover, this integration empowers online retailers to optimize inventory management to enhance the shopping experience.

2. Related Works

Object detection and image classification have been advanced thanks to deep learning techniques. In particular, the YOLO models have garnered considerable attention due to their ability to execute object detection. Video classification using deep learning techniques is used to simulate clothing for customers [7]. Song et al. [8] proposed evolved YOLOv5-based object detection to address inaccurate detection. The enhanced model achieved a better object detection performance using network pruning and parameter optimization [8]. The eye-in-hand measurement method showed increased precision, recall, mean average precision (mAP) value, and F1 score in contrast with the original YOLOv5 models [8]. Li et al. [9] enhanced object detection and addressed illegal immigration, industrial scenarios, natural disasters, and locating missing individuals or items utilizing photography with an F11 4K PRO drone and the VisDrone dataset.

Aduen et al. [10] enhanced the YOLOv5 object detection of smaller objects, particularly in autonomous racing, proposing a modified YOLO-Z model. The model's mAP was enhanced by 6.9% for fewer objects at an intersection over union of 50% in object detection and its speed increased by 3 ms compared with the original YOLOv5. Hang et al. [11] proposed YOLOv5 for object detection in satellite images by introducing fusion layers and predicting heads. They used swin transformer prediction heads (SPHs) to reduce calculation intricacy and integrate normalization-based attention modules (NAMs) to improve attention performance. Evaluated on the NWPU-VHR10 and DOTA datasets, the modified model SPH-YOLOv5 outperformed the original YOLOv5 with an increase of 0.071 in mAP. Additionally, most YOLO versions enhance object detection based on machine learning.

3. Methodology

YOLOv5 is an advanced deep-learning technique that consists of an input, backbone, neck, and head. Among its variants, YOLOv5s is notable for its compact size, making it ideal for edge artificial intelligence (AI) and deployment on microcontroller units (MCUs). Deep learning models rely on GPU-powered systems for efficient computations. We used an open-source Jupyter Notebook (7.2.2) for model training and testing. Figure 1 shows the steps used for setup, configuration, and running process applications to create a new YOLO layer. Users need to confirm the graphics processing unit (GPU) and compute the unified device architecture (CUDA) settings before installing YOLO algorithms. Finally, the selected YOLO algorithm was trained and tested by using pre-defined weights.

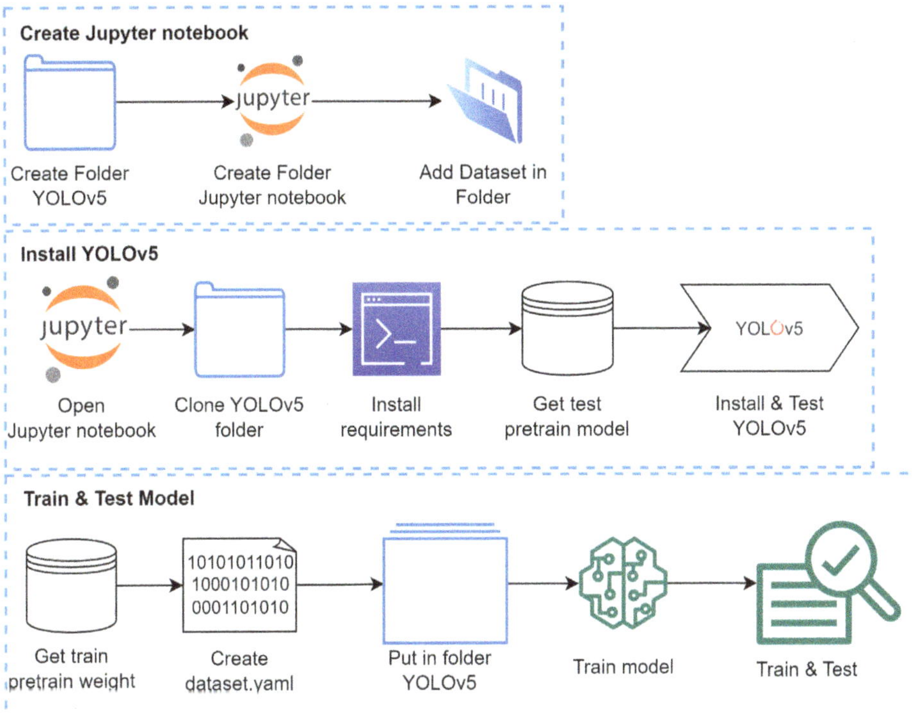

Figure 1. Implementation process for YOLOv5 model.

3.1. YOLO Algorithms

YOLO is a single-stage object detection algorithm renowned for its efficiency and effectiveness. Unlike traditional two-stage methods, YOLO formulates object detection as a regression problem. It predicts bounding boxes and class probabilities from entire images in a single evaluation, offering real-time inference capabilities. This approach enables YOLO to achieve a high accuracy while maintaining an impressive speed, making it well-suited for various applications such as autonomous vehicles, surveillance systems, and robotics. YOLO streamlines the object detection process by simultaneously handling localization and classification and provides a robust and efficient solution for computer vision tasks.

The YOLO series has evolved through Darknet powering versions 1 to 4, while YOLOv5 is built on PyTorch (2.5.1+cu121). YOLOv1, inspired by GoogLeNet, treated object detection as a regression task. YOLOv2 improved accuracy and speed by adopting the Darknet-19 framework, though challenges remained in detecting small objects. To address this, YOLOv3 introduced a feature fusion module and expanded its output to three dimensions, enhancing performance with smaller objects. Based on such advancements, YOLOv4 refined its detection accuracy. YOLOv5, the latest version, allows for more compact models.

The YOLO architecture is structured using the backbone, neck, and head. Figure 2 highlights the main differences among the various versions of YOLO. In YOLOv5, the FOCUS function enhances the processing speed by slicing and rearranging the input image. Spatial pyramid pooling (SPP) reduces image distortion in measurements, color, or culture. It accelerates the generation of candidate boxes while decreasing computational overheads.

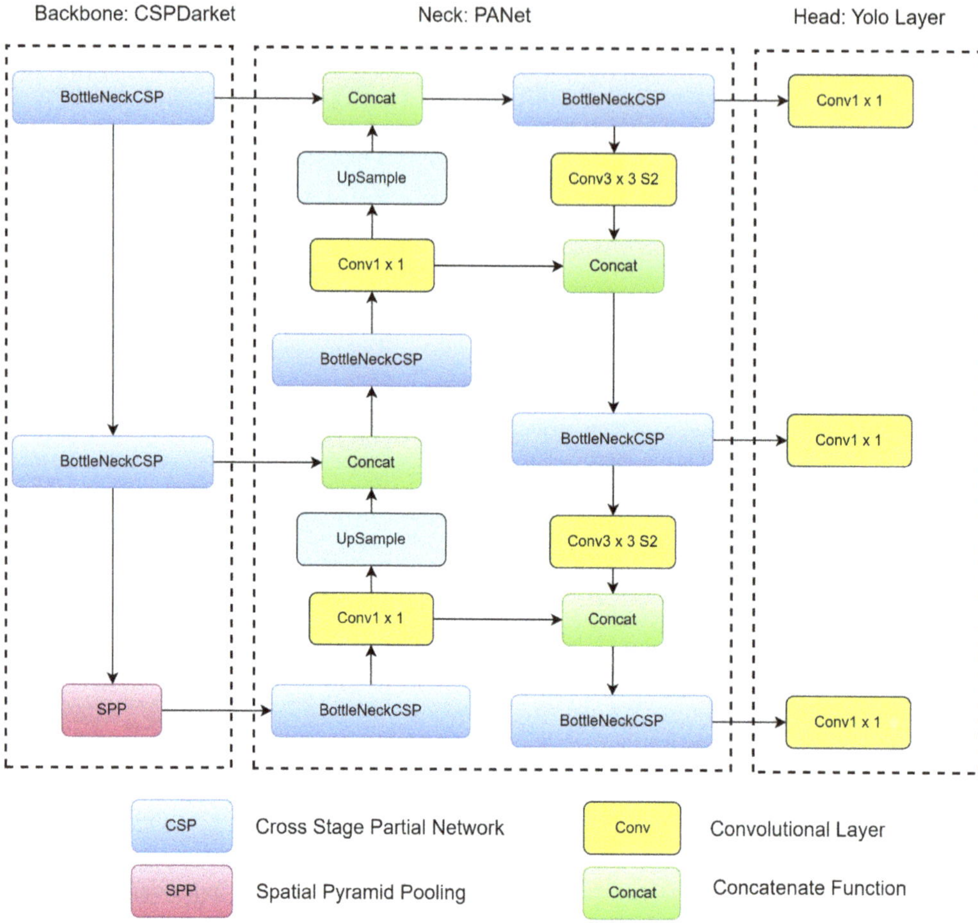

Figure 2. Architecture and processes of YOLOv5.

3.2. Dataset

Image datasets were collected using a camera, capturing data from the fabric category machine shown in Figure 3. The dataset comprised more than 5000 images and was categorized into five groups: plain cotton fabric, the hanbok fabric nobang, dyed cotton yarn, hanbok fabric, and a plain cotton blend fabric.

The dataset was split into training, validation, and testing sets. The training set was used to train the model, while the testing dataset was used for fine-tuning parameters and selecting the best model. Table 1 presents the distribution of samples in each category used in this study.

Table 1. Fabric dataset in this study.

Fabric Dataset	Training	Validation	Testing
Plain Cotton Fabric	756	216	108
Hanbok Fabric	945	270	135
Dyed Cotton Yarn	1029	294	147
Hanbok Fabric, Nobang	1197	342	171
Plain Cotton Blend Fabric	1470	402	201

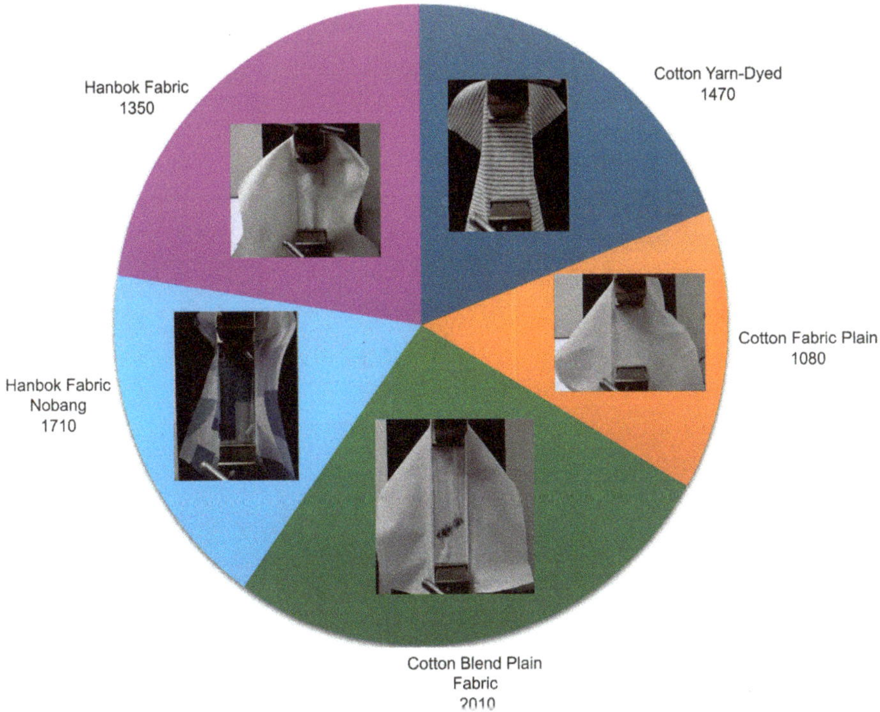

Figure 3. Images of samples of each category of fabric.

3.3. Image Labeling

LabelImg is an annotation tool developed in Python (3.12.4) that applies PyQt for its graphical interface annotations. It stores XML files in the PASCAL VOC format. Movere is also used in the YOLO and CreateML formats. The tool determines the object's area and class in labeling and selects an appropriate layout compatible with YOLO. This information includes the class, *X*, and *Y* coordinates and the distance and confines of objects Figure 4.

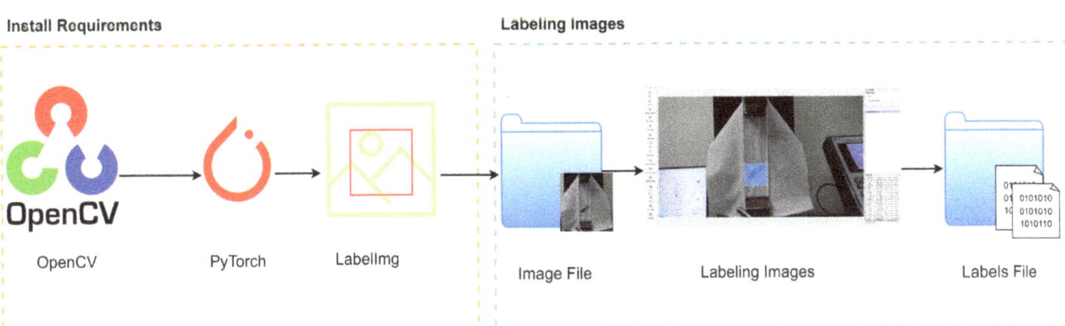

Figure 4. YOLOv5 labeling.

4. Result and Discussion

We used a Windows 10 system with an AMD Ryzen 9 5900X processor running at 3.70 GHz, 128 GB of RAM, and an NVIDIA GeForce RTX. PyTorch, TensorFlow, and other supporting libraries were used to develop the YOLOv5 with a backbone, neck, and head. In object detection, the spatial locations of objects are represented by encircling boxes. YOLO employs anchor boxes of different sizes and shapes to manage overlapping objects within images. Each anchor box is evaluated to determine which object's encircling box has the highest overlapping-to-non-overlapping ratio. In training, the expected encircling boxes are more continuous than the ground-truth labels, enhancing the detection performance where bounding boxes are color-coded by class, as shown in Table 2 and the results in Figure 5.

Table 2. Performance obtained for 5 and 100 epochs. Time spent by CPU and GPU.

Model Name	mAP@0.5 (%)	mAP@0.95 (%)	CPU Time (s)	GPU Time (s)
YOLOv5s	60.07	81.08	1805 s	665 s
YOLOv5n	45.07	81.02	3364 s	665 s
YOLOv5m	50.07	81.02	1.1169 s	784 s
YOLOv5l	50.08	81.02	1.8475 s	906 s
YOLOv5x	60.33	81.02	3.8888 s	1200 s

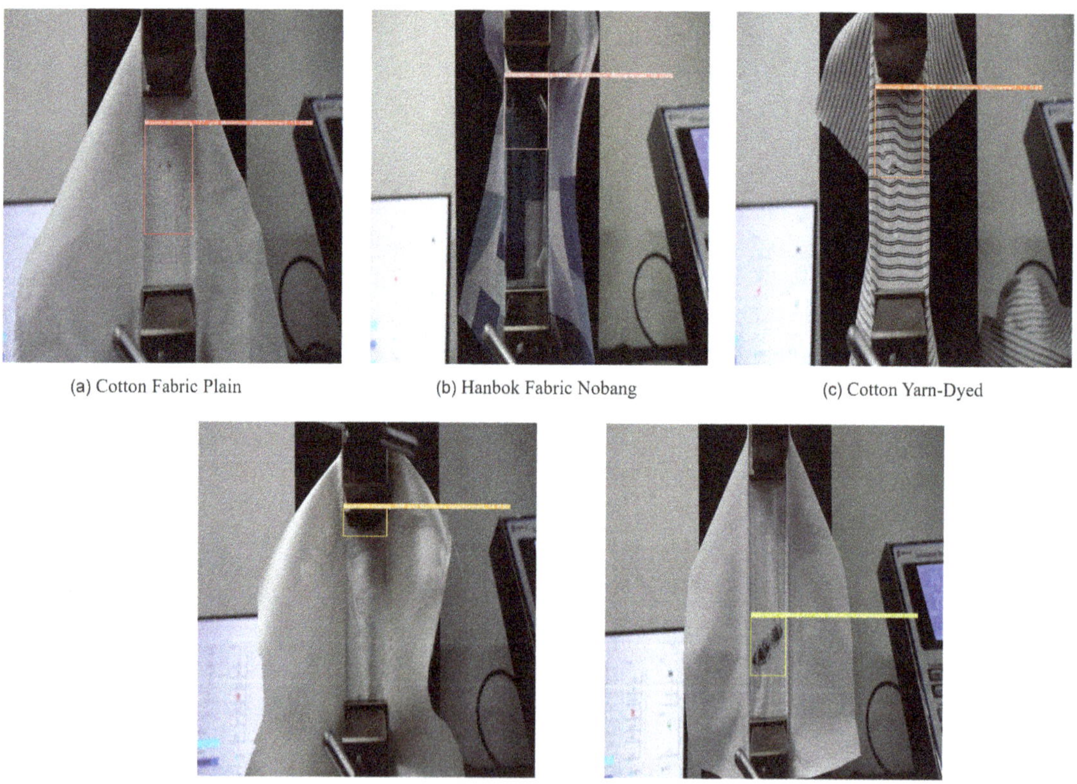

(a) Cotton Fabric Plain (b) Hanbok Fabric Nobang (c) Cotton Yarn-Dyed

(d) Hanbok Fabric (e) Cotton Blend Plain Fabric

Figure 5. Results from testing the types of fabrics.

5. Conclusions

We used the lightweight YOLOv5 algorithm for clothing recognition in various conditions and weight constraints. YOLOv5s, a single-stage object detection approach, enhances both the detection speed and size of the compact model. The model was developed and tested using the Jupyter Notebook in an open-source integrated development environment.

Image samples were taken using the LUMIX GH6 camera. Steps for building an efficient and cost-effective computing setup were developed. The experimental results highlighted the benefits of YOLOv5s regarding mAP, CPU and GPU operation times, and efficiency.

Author Contributions: M.H., conceptualization, project administration, writing—review and editing. J.M. and A.L., conceptualization, writing—review, and editing. M.M. format analysis, writing—original draft preparation, writing—review and editing. All authors have read and agreed to the published version of the manuscript.

Funding: This research was supported by the Basic Science Research Program through the National Research Foundation of Korea (NRF), funded by the Ministry of Education (NRF-2022R1I1A3069371), was funded by BK21 FOUR (Fostering Outstanding Universities for Research) (No.: 5199990914048), and was supported by the Soonchunhyang University Research Fund.

Institutional Review Board Statement: Not applicable.

Informed Consent Statement: Not applicable.

Data Availability Statement: Data available on request due to restrictions, e.g., privacy or ethical.

Conflicts of Interest: The authors declare no conflicts of interest.

References

1. Cheng, Z.Q.; Wu, X.; Liu, Y.; Hua, X.S. Video2shop: Exact matching clothes in videos to online shopping images. In Proceedings of the IEEE Conference on Computer Vision and Pattern Recognition, Honolulu, HI, USA, 21–26 July 2017; pp. 4048–4056.
2. Latha, R.S.; Sreekanth, G.R.; Rajadevi, R.; Nivetha, S.K.; Kumar, K.A.; Akash, V.; Bhuvanesh, S.; Anbarasu, P. Fruits and vegetables recognition using YOLO. In Proceedings of the International Conference on Computer Communication and Informatics (ICCCI), Hammamet, Tunisia, 25–27 January 2022; pp. 1–6.
3. Zhou, M.; Chen, G.H.; Ferreira, P.; Smith, M.D. Consumer behavior in the online classroom: Using video analytics and machine learning to understand the consumption of video courseware. *J. Mark. Res.* **2021**, *58*, 1079–1100. [CrossRef]
4. Yin, W.; Xu, B. Effect of online shopping experience on customer loyalty in apparel business-to-consumer ecommerce. *Text. Res. J.* **2021**, *91*, 2882–2895. [CrossRef]
5. Soegoto, E.S.; Marbun, M.A.S.; Dicky, F. Building the Design of E-Commerce. *IOP Conf. Ser. Mater. Sci. Eng.* **2018**, *407*, 012021. [CrossRef]
6. Boardman, R.; Chrimes, C. E-commerce is king: Creating effective fashion retail website designs. In *Pioneering New Perspectives in the Fashion Industry: Disruption, Diversity and Sustainable Innovation*; Emerald Publishing Limited: Bingley, UK, 2023; pp. 245–254.
7. Mao, M.; Va, H.; Lee, A.; Hong, M. Supervised Video Cloth Simulation: Exploring Softness and Stiffness Variations on Fabric Types Using Deep Learning. *Appl. Sci.* **2023**, *13*, 9505. [CrossRef]
8. Song, Q.; Li, S.; Bai, Q.; Yang, J.; Zhang, X.; Li, Z.; Duan, Z. Object detection method for grasping robot based on improved YOLOv5. *Micromachines* **2021**, *12*, 1273. [CrossRef] [PubMed]
9. Jung, H.K.; Choi, G.S. Improved yolov5: Efficient object detection using drone images under various conditions. *Appl. Sci.* **2022**, *12*, 7255. [CrossRef]
10. Benjumea, A.; Teeti, I.; Cuzzolin, F.; Bradley, A. YOLO-Z: Improving small object detection in YOLOv5 for autonomous vehicles. *arXiv* **2021**, arXiv:2112.11798.
11. Gong, H.; Mu, T.; Li, Q.; Dai, H.; Li, C.; He, Z.; Wang, W.; Han, F.; Tuniyazi, A.; Li, H.; et al. Swin-transformer-enabled YOLOv5 with attention mechanism for small object detection on satellite images. *Remote Sens.* **2022**, *14*, 2861. [CrossRef]

engineering proceedings

MDPI

Proceeding Paper

Preparation and Electrochemical Properties of Molybdenum Disulfide Nanomaterials †

Pin-Syuan Chen [1,*], Yi Hu [1], Si-Ying Li [1], Marta Mazurkiewicz-Pawlicka [2] and Artur Małolepszy [2]

1 Department of Mechanical and Materials Engineering, Tatung University, Taipei 104, Taiwan; huyi@gm.ttu.edu.tw (Y.H.); d10807002@ms.ttu.edu.tw (S.-Y.L.)
2 Faculty of Chemical and Process Engineering, Warsaw University of Technology, 02822 Warsaw, Poland; marta.pawlicka@pw.edu.pl (M.M.-P.); artur.malolepszy@pw.edu.pl (A.M.)
* Correspondence: d10707001@ms.ttu.edu.tw
† Presented at the 2024 IEEE 7th International Conference on Knowledge Innovation and Invention, Nagoya, Japan, 16–18 August 2024.

Abstract: As a transition metal chalcogenide, molybdenum disulfide is an important two-dimensional material. Due to its structural anisotropy, its different morphological structures impact performance. Therefore, improving existing preparation methods enhances its applications. Single-layer molybdenum disulfide is a direct bandgap semiconductor with excellent mechanical properties and chemical stability. We chose ammonium molybdate as the molybdenum source and L-cysteine as the sulfur source. By changing the pH and the reaction time in the environment, the hydrothermal method is used to synthesize the precursor and molybdenum disulfide with different morphologies to control its morphology. Electrochemical test results showed that the specific capacity of molybdenum disulfide synthesized at a current density of 0.6 A reaches 187.79 F/g at a reaction time of 24 h and a pH of 0.6. Its microstructure is in the shape of a flower ball, with a single piece size of about 50 nm and a thickness of about 5 nm. Its specific surface area reaches 36.88 m^2/g, which provides enough active sites.

Keywords: molybdenum disulfide; specific capacitance; nanomaterials

1. Introduction

Sustainable development has been emphasized. With economic progress, fossil fuels have been overexploited and excessively used. In the future, fossil fuels may become scarce, and the environmental problems could become worse. Therefore, new green alternative energy sources are necessary to avoid over-reliance on fossil energy, and energy storage is important.

Supercapacitors are promising energy storage devices due to their long cycle life [1], high power density [2], and fast charge and discharge [3,4]. They are widely used in hybrid vehicles and memory backup systems [5]. The morphology, porosity, and size of electrode materials influence the performance of supercapacitors. Two-dimensional metal sulfides are electrochemically active [6], making them suitable for use in supercapacitors due to their high conductivity and specific surface area. Among metal sulfides, molybdenum disulfide (MoS_2) has a graphene-like layered structure. Its particular form and unique properties attract research interest [7].

In this study, molybdenum disulfide was prepared using the hydrothermal method, and the relationship between its morphological structure and capacitance was studied by controlling the pH and reaction time.

2. Experiment

2.1. MoS_2 Nanomaterials

1 mmol amine molybdate is dissolved in 20 mL of deionized water and added 9 mL of ethylenediamine. Next, 3 N hydrochloric acid is added with magnetic stirring, and the pH of the solution is adjusted to 2. At the same time, 2.2726 g of cysteine is dissolved in 120 mL of deionized water. The solution is then mixed into two solutions. The prepared solution is transferred to a 200 mL polytetrafluoroethylene-lined reactor for reaction. The reaction temperature is set to 240 °C, and the reaction times are 6, 12, 18, and 24 h. The obtained product is washed thoroughly with deionized water and absolute ethanol, centrifuged, and dried at 60 °C for 24 h. Finally, the dried powder is heated at 700 °C for 2 h in nitrogen gas to obtain MoS_2 powder. The experimental methods corresponding to each sample number are shown in Table 1 below.

Table 1. Reaction time and pH for samples.

Sample	LMo-1	LMo-2	LMo-3	LMo-4	LMo-5	LMo-6
Time (hr)	6	12	18	24	24	24
pH	2	2	2	2	0.6	9.5

2.2. Characterization

The structure of the MoS_2 sample was characterized using an X-ray diffractometer and analyzed using a Raman spectrometer. The morphology of the MoS_2 samples was characterized using a scanning electron microscope, and the specific surface area and pore size distribution of the material were measured using a nitrogen adsorption and desorption-specific surface area analyzer. The electrochemical performance of the samples was tested under a conventional three-electrode system using CHI6081C. The prepared samples, conductive carbon black, and binder polyvinylidene fluoride (PVDF) were mixed evenly in the mass ratio of 80:10:10, drop-cast onto nickel foam, dried at 60 °C for 12 h, and used as the working electrode. The platinum sheet was used as the auxiliary electrode, Ag/AgCl was used as the reference electrode, and 3 M KOH was used as the electrolyte. The cyclic voltammetry curve was drawn at a potential range of 0–0.5 V with scan speeds of 10, 20, 50, 100, and 200 mV/s. Galvanostatic charge–discharge tests were conducted at a current density of 0.6 A/g to explore charge and discharge performance.

3. Results and Discussion

Figure 1 shows the XRD patterns of the products prepared at different reaction times when the hydrothermal reaction temperature is 240 °C, the pH is 2, and the molar ratio of molybdenum to sulfur is 1:2.4. The crystallinity of the MoS_2 crystal is poor. When the reaction time is increased to 24 h, the diffraction peak intensity of the sample LMo-4 becomes low.

Due to the growth of the nanosheet structure on the rod, its surface area increases, while defects in the surface structure cause the intensity of the diffraction peak to decrease [8,9]. After annealing at 700 °C for 2 h in the nitrogen atmosphere, the crystallinity increases. Figure 2 shows that the diffraction peaks at 14.1°, 33.2°, 39.7°, and 58.8°, correspond to the crystal planes of (002), (100), (103), and (110), which are consistent with the standard spectrum (JCPDS No.37- 1492). They show the characteristic peaks of hexagonal MoS_2 [10].

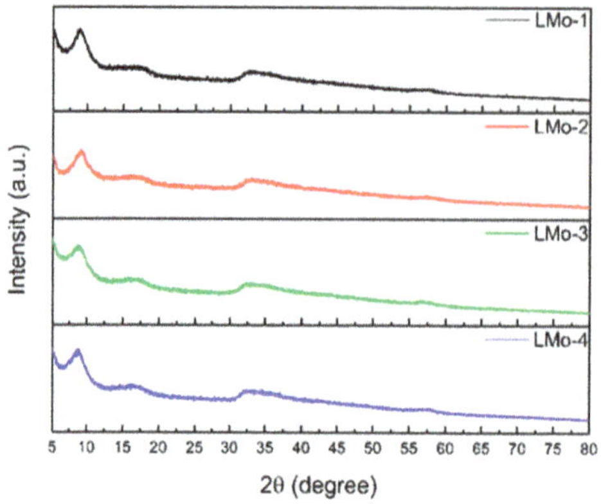

Figure 1. XRD patterns of LMo-1, LMo-2, LMo-3, LMo-4.

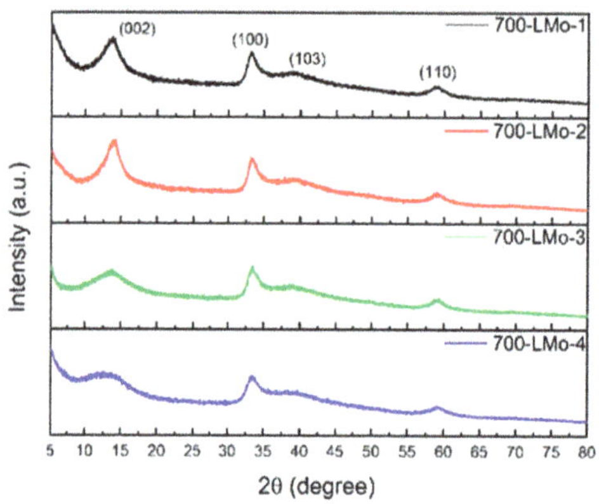

Figure 2. XRD patterns of LMo-1, LMo-2, LMo-3, LMo-4 after heating at 700 °C under N_2 atmosphere.

To understand the hydrothermal synthesis process and growth mechanism of MoS_2, the changes in morphology at different reaction heating times and pH values are studied. Figure 3a–d show MoS_2 prepared at pH value = 2. When the reaction time is 6 h, irregular structures are agglomerated, and the surface shows flakes attached. As the reaction time increases to 12 h, the original irregular structure gradually develops into a spherical shape, and the agglomeration phenomenon still exists. However, the surface flake boundaries become obvious (Figure 3b). When the reaction time reaches 18 h, spherical particles form in a spherical chain-like structure (Figure 3c). At the reaction time of 24 h, a rod-like structure appears (Figure 3d). The lamellar structure grows out of the rod-like crystal.

To explore the impact of the pH value on the morphology of the obtained product MoS_2, the microstructure of the precursor MoO_3 is observed. It is a rod-shaped structure with varying lengths and a diameter of 3−10 μm, a smooth surface, and no pattern structure (Figure 3g). Figure 3e shows MoS_2 prepared at pH = 0.6, where flower-like structures are formed by the self-assembly of many lamellae. The particle size is approximately 100 nm. Figure 3f shows no flaky structure on the particle surface. The surface is smooth compared to other samples. Figure 3h shows a commercial MoS_2 powder with a platelet structure of approximately 3 μm and a thickness of 100 nm. In the reaction process, the dissolution rate of MoO_3 increases as the pH value decreases. At pH = 2, the surrounding rod-like MoO_3 is observed due to less dissolved MoO_3. The concentration of molybdenum ions is high, so MoS_2 grows from the surface layer based on MoO_3. When the pH continues to decrease, the dissolution rate of MoO_3 accelerates, and the diffused molybdenum ions react more with the sulfur ions in the solution, forming a flower-ball-like structure composed of flakes of MoS_2. In acidic conditions, H+ accumulates on the surface and edge of the MoS_2 nanostructure. Due to the H+ charge shielding effect, many combinations occur between the edge and the surface and between the surface and the surface, preventing the crystal synthesis from becoming larger to form spherical MoS_2 [11].

Figure 3. SEM images of hydrothermal synthesis of all samples. (**a**) LMo-1, (**b**) LMo-2, (**c**) LMo-3, (**d**) LMo-4, (**e**) LMo-5, (**f**) LMo-6, (**g**) MoO_3, (**h**) commercial MoS_2 powder.

To confirm the composition of the material, Raman testing is performed on the synthesized material. Figure 4a shows the Raman spectrum excited at 532 nm wavelength. Figure 4b shows the vibration modes of two energy levels of MoS_2. MoS_2 prepared and the commercial MoS_2 show two characteristic peaks near 380 and 400 cm^{-1}. They correspond to the (E_{2g}^1) and (A_{1g}) energy levels of MoS_2, where E_{2g}^1 corresponds to the vibration mode of the two S atoms in the layer opposite to the Mo atom in the middle [12,13]. A_1 g reacts to the S atoms in a single out-of-plane vibration mode in the cell [14]. As the number of MoS_2 layers changes, the peaks corresponding to E_{2g}^1 and A_{1g} shift relatively. As the number of MoS_2 layers decreases, the A_{1g} vibration mode weakens. The peak position is red-shifted. Compared with the position of the A_{1g} characteristic peak of commercial MoS_2 (412 cm^{-1}), the A_{1g} characteristic peak appears at 405 cm^{-1} and is red-shifted.

(a)

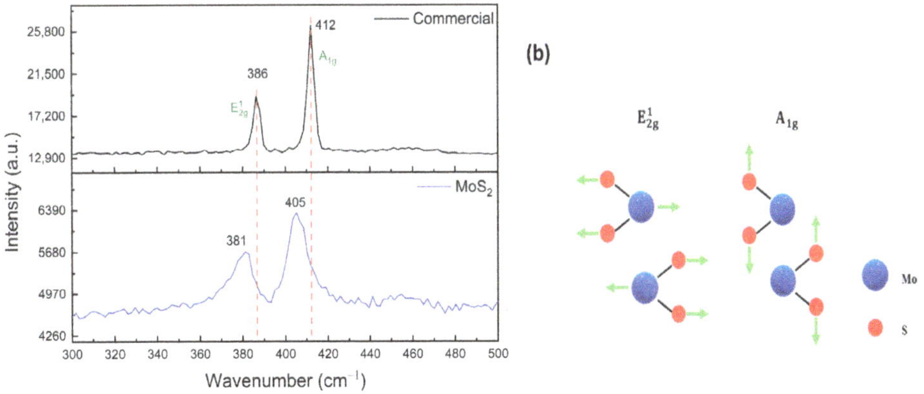

Figure 4. (**a**) Comparison of the Raman spectra of commercial MoS_2 and the MoS_2 prepared in this experiment. (**b**) Vibration modes of two energy levels of MoS_2.

Figure 5 shows the cyclic voltammogram curves of the samples in 1 mol/L KOH solution at a scan rate of 50 mV/s and a voltage of 0−0.5 V. Usually, transition metal oxides or hydroxides undergo redox reactions. A redox peak is observed in Figure 6, indicating a Faradaic capacitance behavior. Under the same scan rate conditions, the capacitance of the sample is proportional to the scan area of the cyclic voltammetry (CV) curve. The order of the scanning area is LMo-5 > LMo-4 > LMo-3 > LMo-6 > LMo-2 > LMo-1, which shows that the spherical sample LMo-5 has the largest specific capacitance due to the high specific surface area and excellent conductivity of the flower ball structure at the same scanning rate. To study the capacitive behavior of MoS_2 with different morphologies, a galvanostatic charge and discharge test (GCD) is performed (Figure 5). Figure 6 shows the galvanostatic charge and discharge curves of MoS_2 electrodes with different morphologies at a current density of 0.6 A/g. The discharge time of LMo-5 (144 s) is the longest. According to the specific capacitance C_sp = IΔt/(mΔV), the specific capacity of the MoS_2 electrode is calculated. The capacities of samples LMo-1 to LMo-6 are 28.19, 37.12, 73.17, 120.23, 187.79, and 42.56 F/g. There is an obvious charge–discharge plateau in the charge–discharge curve, indicating that it has a Faradaic capacitance behavior, which is consistent with the test results of the cyclic voltammetry curve.

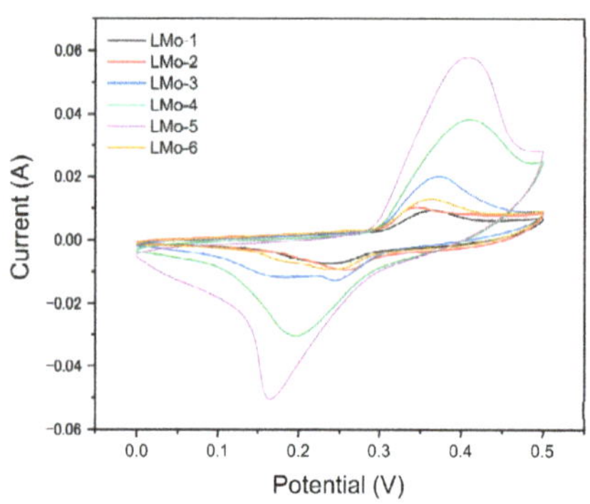

Figure 5. Cyclic voltammetry results for all samples.

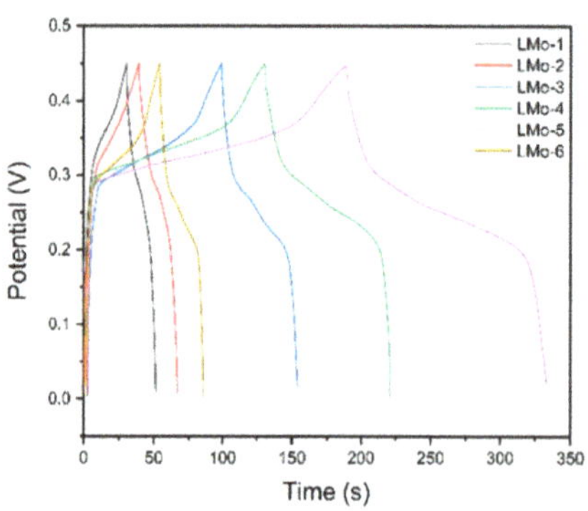

Figure 6. Galvanostatic charge and discharge test results of all samples.

4. Conclusions

To produce MoS_2, ammonium molybdate is used as the molybdenum source, and L-cysteine is used as the sulfur source. By hydrothermal heating at 240 °C and holding for 24 h, a self-assembly of many sheet-like structures is obtained. Spherical MoS_2 particle size is about 100 nm. The capacitance of MoS_2 with different morphologies is high enough to be used as supercapacitor electrode material. When the current density is 0.6 A/g, the discharge-specific capacity of the sample LMo-5 with a flower ball structure is 187.79 F/g. The structure with a highly specific surface area has a higher capacitance and conductivity as an excellent supercapacitor electrode material.

Author Contributions: P.-S.C.: wrote the paper, Y.H.: conceived and designed the analysis, S.-Y.L.: collected the data, M.M.-P.: contributed data or analysis tools A.M.: contributed data or analysis tools. All authors have read and agreed to the published version of the manuscript.

Funding: This research is supported by the National Science Council (NSC), Taiwan, under Grant Number: 111-2926-I-036-501-G.

Institutional Review Board Statement: Ethical review and approval were waived for this study, due to study not involving humans or animals.

Informed Consent Statement: This study not involving humans.

Data Availability Statement: Data is contained within the article.

Conflicts of Interest: The authors declare no conflict of interest.

References

1. Wang, Y.-X.; Seng, K.H.; Chou, S.-L.; Wang, J.-Z.; Guo, Z.; Wexler, D.; Liu, H.-K.; Dou, S.-X. Reversible sodium storage via conversion reaction of a MoS_2–C composite. *Chem. Commun.* **2014**, *50*, 10730. [CrossRef] [PubMed]
2. Zhang, Y.; Zeng, T.; Huang, D.; Yan, W.; Zhang, Y.; Wan, Q.; Yang, N. High-Energy-Density Supercapacitors from Dual Pseudocapacitive Nanoelectrodes, ACS Applied Energy Materials. *ACS Appl. Energy Mater.* **2020**, *3*, 10685. [CrossRef]
3. Wardani, V.; Rohmawati, L.; Setyarsih, W.; Alfarisi, D.; Subhan, A.J. Analysis of Charging/Discharging Supercapacitor Active Carbon/rGO Based on Natural Materials. *Phys. Conf. Ser.* **2020**, *1491*, 012044. [CrossRef]
4. Lu, X.; Zhai, T.; Zhang, X.; Shen, Y.; Yuan, L.; Hu, B.; Gong, L.; Chen, J.; Gao, Y.; Zhou, J.; et al. WO3–x@Au@MnO2 Core–Shell Nanowires on Carbon Fabric for High Performance Flexible Supercapacitors. *Adv. Mater.* **2012**, *24*, 938. [CrossRef] [PubMed]
5. Deng, Z.-Y.; Ferreira, J.M.F.; Sakka, Y.J. Hydrogen-Generation Materials for Portable Applications. *Am. Ceram. Soc.* **2008**, *91*, 3825. [CrossRef]
6. Simon, P.; Gogotsi, Y. Materials for electrochemical capacitors, Nature Materials. *Nat. Mater.* **2008**, *7*, 845. [CrossRef] [PubMed]
7. Whittingham, M.S. The hydrated intercalation complexes of the layered disulfides. *Mater. Res. Bull.* **1974**, *9*, 1681. [CrossRef]
8. Welkie, D.G.; Lagally, M.G. Analysis of surface structural defects by low energy electron diffraction. *Thin Solid Film.* **1982**, *93*, 219. [CrossRef]
9. Zhang, Y.; Wang, Q.; Li, C.; Piao, Y.; Hou, N.; Hu, K.J. Characterization of surface and subsurface defects induced by abrasive machining of optical crystals using grazing incidence X-ray diffraction and molecular dynamics. *Adv. Res.* **2022**, *36*, 51. [CrossRef] [PubMed]
10. Zhang, J.; Han, D.; Wang, Y.; Wang, L.; Chen, X.; Qiao, X.; Yu, X. Synergy between nanozymes and natural enzymes on the hybrid MoS_2 nanosheets/graphite microfiber for enhanced voltammetric determination of hydrogen peroxide. *Microchim. Acta* **2020**, *187*, 321. [CrossRef] [PubMed]
11. Ju, L.; Hong, D.; Jin, X.; Liu, H.; Yang, X.; Nie, L.; Liu, Q.; Gao, Z.; Zhu, W.; Wang, Y.; et al. Preparation and Study of Photocatalytic Properties of (M(M=Pt, Ag and Au)-TiO2)@MoS_2 Nanocomposites. *Inorganics* **2023**, *11*, 258. [CrossRef]
12. Pei, J.; Yang, J.; Xu, R.; Zeng, Y.-H.; Myint, Y.W.; Zhang, S.; Zheng, J.-C.; Qin, Q.; Wang, X.; Jiang, W.; et al. Exciton and Trion Dynamics in Bilayer MoS_2. *Small* **2015**, *11*, 6384. [CrossRef] [PubMed]
13. Golovynskyi, S.; Irfan, I.; Bosi, M.; Seravalli, L.; Datsenko, O.I.; Golovynska, I.; Li, B.; Lin, D.; Qu, J. Exciton and trion in few-layer MoS_2: Thickness- and temperature-dependent photoluminescence. *Appl. Surf. Sci.* **2020**, *515*, 146033. [CrossRef]
14. Ma, L.; Huang, G.; Chen, W.; Wang, Z.; Ye, J.; Li, H.; Chen, D.; Lee, J.Y.J. Cationic surfactant-assisted hydrothermal synthesis of few-layer molybdenum disulfide/graphene composites: Microstructure and electrochemical lithium storage. *Power Sources* **2014**, *264*, 262. [CrossRef]

engineering proceedings

MDPI

Proceeding Paper

Electrochromic Behavior of Manganese Oxides/Silver Thin Films from Electrochemical Deposition †

Yi Hu , Jiun-Shing Liu, Pin-Syuan Chen * and Si-Ying Li

Department of Mechanical and Materials Engineering, Tatung University, Taipei 10451, Taiwan; huyi@gm.ttu.edu.tw (Y.H.); scottliu@foiltec.net (J.-S.L.); g10507003@ms.ttu.edu.tw (S.-Y.L.)
* Correspondence: lovefafa0722@gmail.com

† Presented at the 2024 IEEE 7th International Conference on Knowledge Innovation and Invention, Nagoya, Japan, 16–18 August 2024.

Abstract: MnO_x thin films with silver additives were electrochemically deposited on an Indium Tin Oxide (ITO) substrate with silver acetate and potassium permanganate aqueous solution. The addition of Ag enhanced electrochromic behavior during cyclic voltammetry (CV). The morphology of the thin films was examined by using scanning electronic microscopy (SEM) and transmission electron microscopy (TEM). The chemical states of Mn and Ag ions on the surfaces of the thin films were examined using X-ray photoelectron spectroscopy (XPS). Spherical Ag_2O and Ag nanoparticles were homogeneously dispersed on the thin films. The electrochemistry of the thin films was examined by cyclic voltammetry in a conventional three-electrode system and an electrochemically tested system. The electrochromic behavior of the films was demonstrated through the cyclic voltammetry (CV) process in the KNO_3 electrolyte. The electrochromic behavior of the thin films depended on the redox reactions associated with the reaction between Ag and Ag_2O coupled with Mn^{4+} ions and Mn^{3+} ions in the KNO_3 electrolyte.

Keywords: thin film; electrochemical deposition; electrochromic

1. Introduction

The electrochemistry of Manganese oxide (MnO_x) electrodes has been extensively studied, starting from their application for active materials in catalysis, batteries, and supercapacitors [1–4]. Typically, MnO_x electrodes are prepared by galvanostatic electrolysis, and chemical vapor deposition, thermal decomposition, and electron beam evaporation have been proposed [5–7]. In our previous study, we applied the electrochemical deposition method to prepare MnO_x/Ag thin films [8]. According to the XPS analysis, thin films composed of Ag_2O and α-Mn_2O_3 nanoparticles had a lower Ag content, while those composed of Ag_2O_2 and β-MnO_2 had a higher Ag content. Ag nanoparticles in the thin film acted as a reduction agent and varied the valence of Mn from 4^+ to 3^+. The capacitance of the thin film increased with an increase in the ratio of Ag/Mn up to 8%. The impedance of the cell also increased with the Ag/Mn ratio. In subsequent research, we found that adding silver caused it to change its color in a cyclic voltammetry test due to its electrochromic behavior.

Electrochromic devices have attracted great attention recently due to the growing demand for smart electronics. Color electrochromic cells are operated in the visible light region and display their remaining capacity through instant color changes. Many materials have been explored with regard to this property, and MnO_x is one of the promising candidates. The electrochromic properties of MnO_x with various crystalline phases have been examined in aqueous [9–13] and nonaqueous [14–16] electrolytes. The electrochromic properties of materials depend on their crystalline structure and morphology. The electrochromic property of MnO_2-based materials is obtained through instantaneous extraction or insertion of electrons and electrolyte cations into or from the matrix. However, MnO is extensively used in supercapacitors owing to its high energy density, but it has a poor electrochromic performance.

In this study, nanocomposite thin films were obtained by electrochemical deposition on Indium Tin Oxide (ITO) glass substrate in $KMnO_4$ and $Ag(C_2H_5COO)$ aqueous solution. The electrochromic behavior of the thin film was demonstrated by cyclic voltammetry (CV) in the KNO_3 electrolyte.

2. Experiments

The raw materials used in the synthesis process-potassium permanganate ($KMnO_4$, ACS reagent and purity > 99%, Sigma Aldrich, St. Louis, MO, USA), silveracetate ($AgC_2H_3O_2$, purity > 99%, Sigma Aldrich) and Potassium nitrate (KNO_3, purity > 99%, Kingfex, Taichung, Taiwan). The nanocomposite thin films were obtained by electrodeposition with a potentiostatic method [8]. The thin films were examined by cyclic voltammetry with a three-electrode system using the electrochemically tested system (Electrochemical station 5000, Jiehan, Hsinchu, Taiwan). A platinum line and a standard calomel reference electrode (SCE) were used as the reference and the counter electrode, respectively. The cyclic voltammetry (CV) experiment was operated from 0 to 1.0 V in 0.1 M KNO_3 electrolyte at a scan rate of 20 mVs^{-1}. A field emission scanning electron microscope (FESEM, Hitachi S-4700) was used to investigate the morphology of the thin films. The state of the thin films was analyzed with transmission electron microscopy (TEM). The optical absorption spectra were recorded with a UV–visible 560 spectrophotometer (Jasco). The states of the Mn and Ag ions were examined by X-ray photoelectron spectroscopy (XPS, VG ESCA Scientific Theta Probe) with Al Kα (1486.6 eV), and the X-ray spot size was 15 μm.

3. Results and Discussion

Figure 1 shows a CV curve of the MnO_x electrode without the Ag additive in the KNO_3 electrolyte. The CV curve exhibits a quasi-rectangular shape with redox reactions at 0.2 and −0.23 V. The color of the film changed from brownish to dark brown following the first oxidation reaction at 0.2 V. As the voltage decreased, the film changed from dark brown to a brownish color under the reduction reaction at −0.23 V. The color change in the thin film is ascribed to the redox between Mn(3+) and Mn(4+) in the KNO_3 electrolyte [17,18].

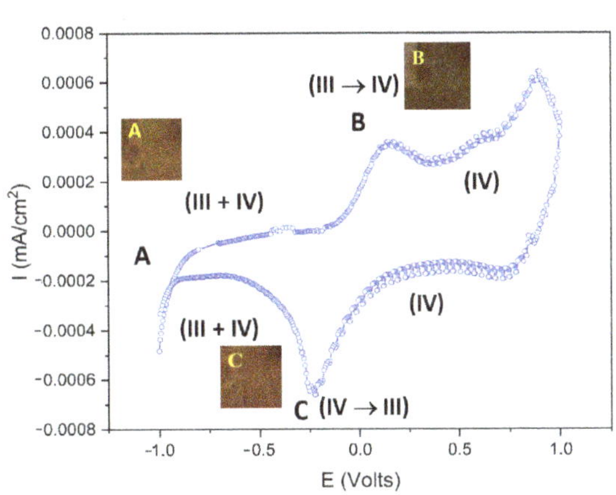

Figure 1. A CV curve of the MnO_x electrode at a scan rate of 0.5 mV/s in the KNO_3 electrolyte to analyze the valence state of Mn ions (inset: the color of the electrolyte near the electrode).

Figure 2 shows a CV curve of the MnO_x/Ag electrode and the valence state of the Mn and Ag ions. The CV curve of the MnO_x/Ag electrode is different from that of MnO_x in KNO_3. There are two oxidation reaction peaks located at 0.1 and 0.5 V, while the reduction peak is located at 0.25 V. During the oxidation process, the color of the electrode changed from a brownish color to black after peak I at a voltage of 0.1 V. The color of the electrode remained black after peak II at a voltage of 0.5 V and turned back to a brownish color after peak III at a voltage of −0.25 V.

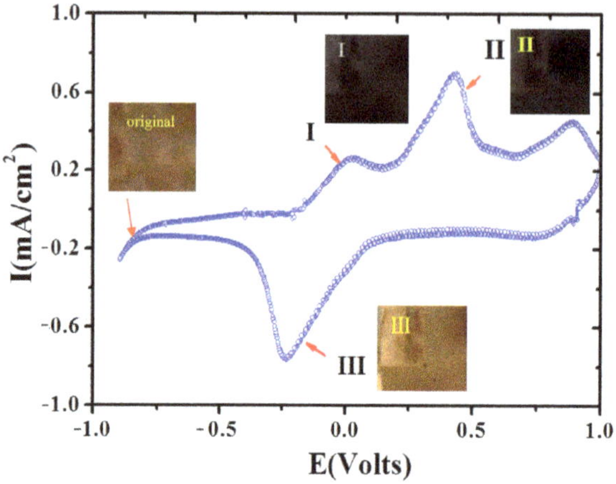

Figure 2. A CV curve of the MnO_x/Ag electrode at a scan rate of 0.5 mV/s in the KNO_3 electrolyte to analyze the valence state of Mn and Ag ions (inset: the color of the electrolyte near the electrode).

Figure 3 shows the changes in the absorbance of the thin films from the ultraviolet to visible wavelength at each step in the redox reaction. More light absorption of the thin films after the oxidation reaction was observed (peak I and peak II in the redox loop in Figure 2) at the 400 nm band. This absorption caused the film to turn black.

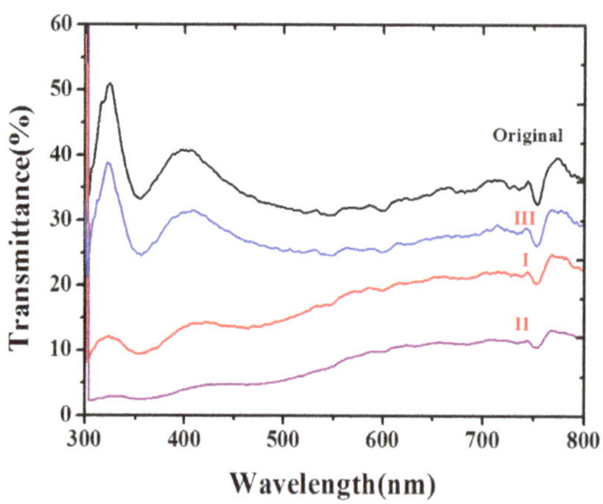

Figure 3. Changes in the absorbance of the thin films from the UV to visible wavelength at each step in the redox reaction. (I, II and III are related to the steps in Figure 2).

Figure 4 shows the XPS spectra of Mn $2p_{3/2}$ for the thin film at each redox step. The binding energy of Mn $2p_{3/2}$ was determined to be 640.9 eV for Mn^{2+}, 641.9 eV for Mn^{3+}, and 642.5 eV for Mn^{4+} which coincided with previous research [19]. The binding energy of Mn

$2p_{3/2}$ of the sample increased from 642.0 to 642.6 eV as the Ag content increased. The shift in the XPS peak to higher binding energy indicates that the oxidation state of manganese changed from Mn^{3+} to Mn^{4+} upon adding Ag. This was confirmed by the peak-fitting analysis of the XPS spectra, which indicated that the films primarily exhibited an oxidation state of Mn^{4+}, with a minor contribution from Mn^{3+} (Figure 4). The predominant phase of the thin film was amorphous MnO_2 (denoted as a-MnO_2). Therefore, the thin films were composed of Ag_2O nanoparticles of several nanoparticles dispersed homogeneously in the a-MnO_2 thin films.

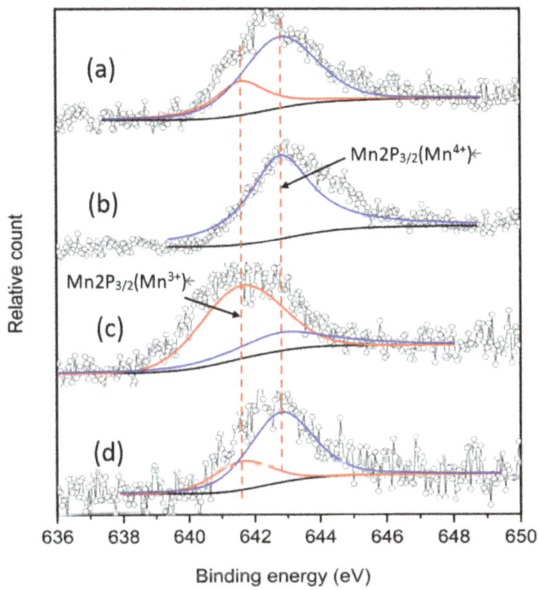

Figure 4. XPS spectra of Mn $2p_{3/2}$ for thin film at each redox step: (**a**) original, (**b**) peak I, (**c**) peak II, (**d**) peak III, as shown in Figure 2.

The XPS spectra of Ag $3d_{5/2}$ for the thin film with silver are shown in Figure 5. The peaks of the spectra were analyzed by peak-fitting, and the results suggested that only a single ionic state of Ag existed. The binding energy of Ag $3d_{5/2}$ in the matrix at 367.37 eV is shown in Figure 5a. This is attributed to the bonding between Ag and oxygen in Ag_2O [19]. In addition, a binding energy of 368.18 eV was shown for Ag $3d_{5/2}$ as the Ag nanoparticles precipitated on the surface of the thin film (Figure 4b), which was consistent with the value of pure Ag [20].

In this study, the predominant phase of the prepared thin film was MnO_x with a small amount of MnO_2. The valence state of the ions in the film was 93.8% for Mn^{3+}, 6.2% for Mn^{4+}, and 97.4% for Ag^+. The electrochromic process changed the valence state of Mn and Ag during cyclic voltammetry.

At peak I, the oxidation reaction at 0.1 V is as follows.

$$Mn^{3+} \rightarrow \tfrac{1}{4}Mn^{3+} + \tfrac{3}{4}Mn^{4+}, \text{ and } \tfrac{1}{2}Ag^{+} \rightarrow \tfrac{1}{2}Ag^{0} \quad (1)$$

At this time, Mn ions are oxidized at a lower electric potential. At the same time, half of the silver ions are reduced to silver.

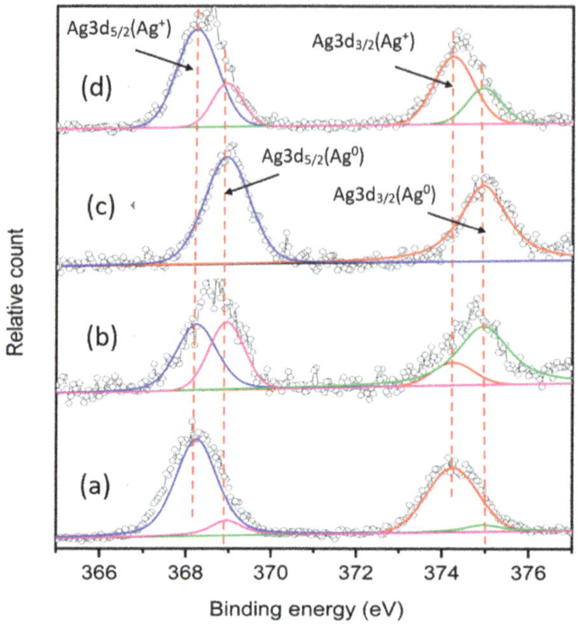

Figure 5. XPS spectra of Ag $3d_{5/2}$ for thin film with Ag content of 8% at (**a**) original, (**b**) peak I, (**c**) peak II, (**d**) peak III in the CV process, as shown in Figure 2.

At peak II, the redox reaction at 0.5 V is as follows.

$$¼Mn^{3+} \rightarrow ¼Mn^{4+}, \text{ and } ½Ag^{+} \rightarrow + ½Ag^{0} \quad (2)$$

Mn ions are completely oxidized at the original oxidation potential. At the same time, silver ions are completely reduced to silver.

At peak III, the redox reaction at −0.25 V is as follows. When the potential reached −0.25 V, a reduction peak appeared.

$$Mn^{4+} \rightarrow Mn^{3+}, \text{ and } Ag^{0} \rightarrow Ag^{+} \quad (3)$$

The above reactions were repeated 100 times in the CV process. The oxidation of silver ions accelerated the oxidation of manganese ions. Figure 6 shows SEM images of the thin films at each step of the redox reactions. The surface of the prepared thin film is smooth, as shown in Figure 6a. A lot of nanoparticles precipitated at the surface of the thin film after the oxidation reaction (peaks I and II in the redox loop in Figure 2) (Figure 6b,c). These nanoparticles are Ag with a size of about 20 nm. These nanoparticles disappeared after the oxidation reaction (peak III in the redox loop in Figure 2) (Figure 6d).

Figure 7 shows the TEM images and the electron diffraction patterns of the thin films and the dark particles before and after the redox reactions. The film is amorphous with the features in the diffraction pattern before oxidation, as shown in Figure 7a,b. Spherical nanoparticles are dispersed in the matrix of the thin film after the oxidation reaction (Figure 7c,d). The size of the nanoparticles (darker particles) in the thin film ranges from a few nanometers to 30 nanometers. The typical selected-area diffraction pattern in Figure 7d exhibits clear diffraction rings with interplanar spacing, which fits with the FCC structure of metallic silver.

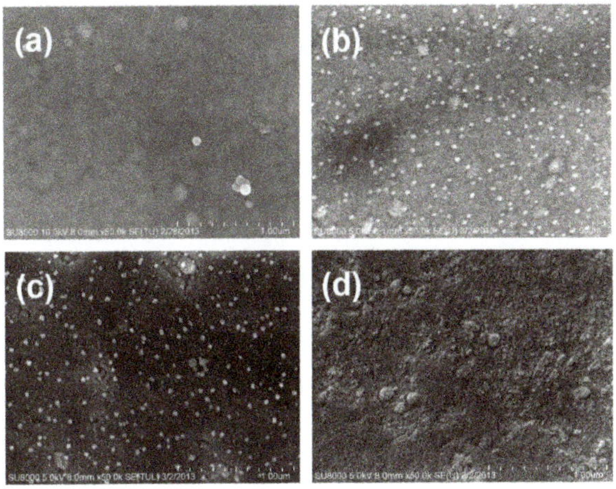

Figure 6. Images showing changes in the surface morphology of the thin films during the cyclic voltammetry process. (**a**) Original test piece; (**b**) I; (**c**) II; (**d**) III.

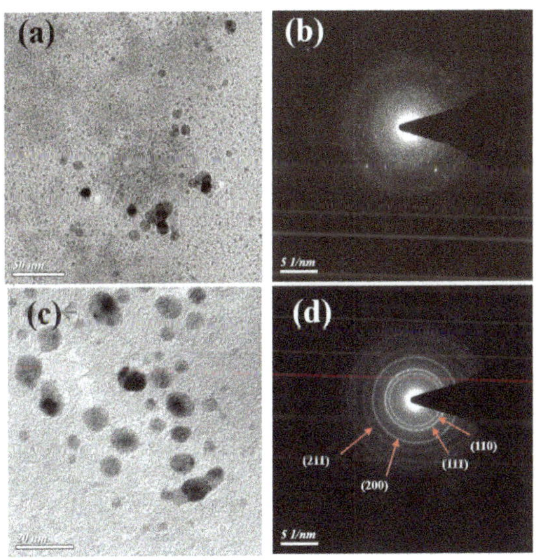

Figure 7. TEM images and electron diffraction patterns of the thin films (**a**,**b**) before and (**c**,**d**) after the redox reactions.

4. Conclusions

MnO_x/Ag nanocomposite films were obtained using silver acetate and potassium permanganate aqueous solution and electrodeposition methods. The electrochromic property of the thin films was observed in CV from 0 to 1.0 V with a scan rate of 20 mVs^{-1} in 0.1 M KNO_3 electrolyte. The electrochromic process was deduced from the changes in the ion valence of Mn and Ag and examined through XPS analysis using CV. Silver ions were reduced to silver metal nanoparticles when Mn was oxidized from +3 to +4. The size of the Ag nanoparticles was about 20 nm, and they disappeared after the oxidation reaction. The oxidation of silver ions accelerated the oxidation of manganese ions.

Author Contributions: Conceptualization, Y.H. and J.-S.L.; experiment J.-S.L., P.-S.C. and S.-Y.L.; investigation, Y.H. and J.-S.L.; analysis, P.-S.C.; writing—original draft, Y.H. and P.-S.C.; writing—review and editing, Y.H. All authors have read and agreed to the published version of the manuscript.

Funding: Grant from National Science and Technology Council (NSTC), Taiwan, under grant No. NSC 99-2221-E-036-013.

Institutional Review Board Statement: Ethical review and approval were waived for this study, due to study not involving humans or animals.

Informed Consent Statement: This study not involving humans.

Data Availability Statement: Data is contained within the article.

Acknowledgments: This research was supported by the National Science and Technology Council (NSTC).

Conflicts of Interest: The authors declare no conflict of interest.

References

1. Mayer, S.T.; Pekala, R.W.; Kaschmitter, J.L. The aerocapacitor: An electrochemical energy-storage device. *J. Electrochem. Soc.* **1993**, *140*, 446–451. [CrossRef]
2. Wei, J.; Nagarajan, N.; Zhitomirsky, I. Manganese oxide films for electrochemical supercapacitor. *J. Mater. Process. Technol.* **2007**, *186*, 356–361. [CrossRef]
3. Wu, M.; Snook, G.A.; Chen, G.Z.; Fray, D.J. Redox deposition of manganese oxide on graphite for supercapacitors. *Electrochem. Commun.* **2004**, *6*, 499–504. [CrossRef]
4. Jones, D.J.; Wortham, E.; Rozière, J.; Favier, F.; Pascal, J.L.; Monconduit, L. Manganese oxide nanocomposites: Preparation and some electrochemical properties. *J. Phys. Chem. Solids* **2004**, *65*, 235–239. [CrossRef]
5. Reddy, R.N.; Reddy, R.G. Sol-gel MnO_2 as an electrode material for electrochemical capacitors. *J. Power Source* **2003**, *124*, 330–337. [CrossRef]
6. Pang, S.C.; Anderson, M.A.; Chapman, T.W. Novel electrode materials for thin film supercapacitors: Comparison of electrochemical properties of sol-gel-derived and electrodeposited manganese dioxide. *J. Electrochem. Soc.* **2000**, *147*, 444–450. [CrossRef]
7. Jiang, J.H.; Kucernak, A. Electrochemical supercapacitor material based on manganese oxide: Preparation and characterization. *Electrochim. Acta* **2002**, *47*, 2381–2386. [CrossRef]
8. Hu, Y.; Liu, J.-S.; Huang, C.-L.; Liu, T.-C. Preparation of MnO_X Nano-supercapacitor with Silver Addition by Electrochemical Deposition. In Proceedings of the World Congress on Engineering, London, UK, 6–8 July 2011; Volume II.
9. Burke, L.D.; Murphy, O.J. Electrochromic behaviour of oxide films grown on cobalt and manganese in base. *J. Electroanal. Chem. Interfacial Electrochem.* **1980**, *109*, 373–377. [CrossRef]
10. Kanoh, H.; Hirotsu, T.; Ooi, K. Electrochromic Behavior of l-MnO2 Electrode Accompanying Li^+-Insertion in an Aqueous Phase. *J. Electrochem. Soc.* **1996**, *143*, 905–907. [CrossRef]
11. de Terresi, S.I.C.; Gorenstien, A. Electrochromic behaviour of manganese dioxide electrodes in slightly alkaline solutions. *Electrochim. Acta* **1991**, *37*, 2015–2019. [CrossRef]
12. Chigane, M.; Ishikawa, M.; Izaki, M. Preparation of Manganese Oxide Thin Films by Electrolysis/Chemical Deposition and Electrochromism. *J. Electrochem. Soc.* **2001**, *148*, D96. [CrossRef]
13. Seike, T.; Nagai, J. Electrochromism of 3d transition metal oxides. *Sol. Energy Mater.* **1991**, *22*, 107–117. [CrossRef]
14. Maruyama, T.; Osaki, Y. Electrochromic Properries of Manganese Oxide Thin Films Prepared by Chemical Vapor Deposition. *J. Electrochem. Soc.* **1995**, *142*, 3137. [CrossRef]
15. Long, J.W.; Qadir, L.R.; Stroud, R.M.; Rolison, D.R. Spectroelectrochemical investigations of cation-insertion reactions at sol-gel-derived nanostructured, mesoporous thin films of manganese oxide. *J. Phys. Chem. B* **2001**, *105*, 8712. [CrossRef]
16. Sakai, N.; Ebina, Y.; Takada, K.; Sasaki, T. Electrochromic Films Composed of MnO_2 Nanosheets with Controlled Optical Density and High Coloration Efficiency. *J. Electrochem. Soc.* **2005**, *152*, E384. [CrossRef]
17. Simon, P.; Gogotsi, Y. Materials for electrochemical capacitors. *Nat. Mater.* **2008**, *7*, 845–854. [CrossRef] [PubMed]
18. Li, D.; Meng, F.; Yan, X.; Yang, L.; Heng, H.; Zhu, Y. One-pot hydrothermal synthesis of Mn_3O_4 nanorods grown on Ni foam for high performance supercapacitor applications. *Nanoscale Res. Lett.* **2013**, *8*, 535–538. [CrossRef] [PubMed]

19. Foord, J.S.; Jackman, R.B.; Allen, G.C. An X-ray photoelectron spectroscopic investigation of the oxidation of manganese. *Philos. Mag.* **1984**, *A49*, 657–663. [CrossRef]
20. Schön, G. ESCA Studies of Ag, Ag_2O and AgO. *Acta Chem. Scand.* **1973**, *27*, 2623–2633. [CrossRef]

engineering proceedings

MDPI

Proceeding Paper

Enhanced Comparative Analysis of Pretrained and Custom Deep Convolutional Neural Networks for Galaxy Morphology Classification †

Tram Le *, Nickson Ibrahim, Thu Nguyen, Thanyaporn Noiplab, Jungyoon Kim and Deepshikha Bhati

Department of Computer Science, Kent State University, Kent, OH 44242, USA; nibrahim3@kent.edu (N.I.); tnguyen69@kent.edu (T.N.); tnoiplab@kent.edu (T.N.); jkim78@kent.edu (J.K.); dbhati@kent.edu (D.B.)
* Correspondence: hle12@kent.edu
† Presented at the 2024 IEEE 7th International Conference on Knowledge Innovation and Invention, Nagoya, Japan, 16–18 August 2024.

Abstract: Galaxy morphology classification is a crucial task in astronomy and astrophysics, providing information on galaxy formation and evolution. Traditionally, this classification has been a manual and labor-intensive process requiring significant astronomical expertise. However, advancements in artificial intelligence, particularly deep learning, offer more efficient and accurate solutions. We investigated the application of convolutional neural networks (CNNs) for galaxy morphology classification using the Galaxy10 DECals dataset. We developed and compared three models: a custom-built CNN using TensorFlow 2.18, a ResNet50 model initialized with random weights, and a pre-trained EfficientNetB5 model utilizing transfer learning, both implemented in PyTorch 2.6. Our results indicate that the custom model achieves an accuracy of 67%, while the ResNet50 and EfficientNetB5 models achieve 80 and 87% accuracy, respectively. The superior performance of the pre-trained Efficient-NetB5 model underscores the efficacy of transfer learning in astronomical image classification. These findings have significant implications for the application of deep learning techniques in astrophysical research.

Keywords: galaxy classification; image processing; convolutional neural networks; transfer learning

1. Introduction

Classifying galaxy morphology remains one of the most challenging tasks in astrophysics. This complexity arises not only from the inherent difficulty in categorizing galaxies based on their images but also from the noise and blurring in these images, which further complicates the classification process. Despite these challenges, recent studies using the Galaxy10 DECals dataset have achieved classification accuracies between 80 and 86%.

Driven by the potential to improve these results, this study aims to increase the accuracy through the application of convolutional neural networks (CNNs). Leveraging advanced machine learning techniques, we have developed a more robust model capable of better handling the intricacies of galaxy morphology. CNNs are particularly well-suited for image classification tasks because they can automatically detect important features without manual extraction, making them ideal for analyzing detailed and varied images of galaxies.

These networks use layers of convolutions that apply filters to input images, creating feature maps that capture essential patterns such as edges, shapes, and textures. For galaxy classification, CNNs can distinguish between the structural features of galaxies. Furthermore, machine learning techniques such as transfer learning can finetune pre-trained networks on new datasets, enhancing learning efficiency and accuracy.

Therefore, in this study, we refine a CNN architecture specifically tailored for classifying galaxies within the Galaxy10 DECals dataset. We leverage transfer learning to boost the performance of our CNN model, resulting in higher accuracy and efficiency. A detailed description of the dataset is provided for preprocessing techniques, model architecture, training process, and evaluation metrics. We conducted a thorough evaluation of our model's performance, including comparisons with existing benchmarks. The findings provide a reference to solve the current challenges in galaxy morphology classification and outline potential directions for future research.

2. Related Work

Recent advancements in galaxy morphology classification have improved accuracy and computational efficiency driven by deep learning architectures. Qian conducted a comparative analysis of VGG16, InceptionV3, and ResNet50, demonstrating that InceptionV3 excels across all performance metrics due to its robust inception modules adept at managing galaxy morphology complexities. However, the study's focus on a single dataset highlighted potential limitations in its generalizability across various astronomical imaging scenarios [1]. Wang highlighted the effectiveness of DenseNet201, which achieved an accuracy of 86% on the Galaxy10 DECals dataset, outperforming VGG16 and MobileNetV2. Despite its computational efficiency, the complexity of DenseNet201 resulted in longer training times and challenges in classifying underrepresented galaxy categories [2]. Premanand, Navneeth, et al. (2023) [3] evaluated ResNet, Parallel CNN, and VGG16 using the Star Galaxy Classification dataset and additional data from the Sloan Digital Sky Survey (SDSS). They found that Parallel CNN was the most effective, particularly in classifying galaxies and stars, achieving an accuracy of 90.08%. Each model's strengths were identified but a discussion was lacking on the broader applicability and implementation challenges [3].

Senel employed a hybrid approach using the Galaxy10 SDSS dataset, integrating CNNs with metaheuristic algorithms including grey wolf optimizer (GWO), particle swarm optimization (PSO), and artificial bee colony (ABC) to optimize neural network parameters. Although this approach led to a commendable 85% testing accuracy, it showed issues such as sample homogeneity and computational demands affecting diverse galaxy classifications [4]. Kalvankar, Pandit, and Parwate utilized Efficient NetB5 with the Galaxy Zoo 2 dataset and additional SDSS data, achieving impressive results with an accuracy of 93.7% and an F1 score of 0.8857. Despite its strengths, the need for deeper computational algorithm analysis and addressing dataset biases was noted [5].

Such studies deepen the understanding of galaxy morphology classification by highlighting the importance of sophisticated CNN architectures and dataset specificity in achieving high classification accuracy while models such as DenseNet201, InceptionV3, and EfficientNetB5 offer substantial advantages.

3. Dataset

The Galaxy10 DECals dataset [6] consists of 17,736 galaxy images from DECals (taken by the Dark Energy Camera) combined with their morphological labels from Galaxy Zoo DR2 (a volunteer citizen science project). Figure 1 shows sample galaxy images from the dataset. Each row represents a class. From top to bottom, their Galaxy10 DECals labels are class 6, 7, and 8.

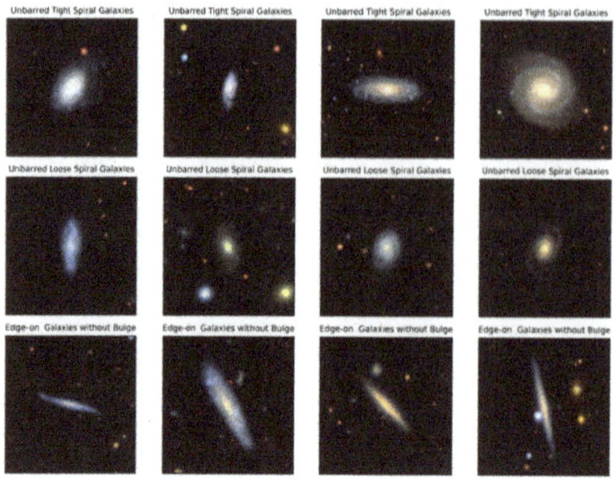

Figure 1. Plotting galaxy images from the dataset. Each row represents a class. From top to bottom, their Galaxy10 DECals labels are class 6, 7, and 8.

3.1. *Exploring*

The Galaxy10 DECals dataset has better resolution than the original SDSS dataset (Figure 2). Each image is 256 × 256 pixels on 3 spectral bands (g, r, and z) instead of the standard RGB. The images are categorized into 10 classes based on the shapes and features of the galaxies as follows:

- Class 0 (1081 images): Disturbed Galaxies;
- Class 1 (1853 images): Merging Galaxies;
- Class 2 (2645 images): Round Smooth Galaxies;
- Class 3 (2027 images): In-between Round Smooth;
- Class 4 (334 images): Cigar Shaped Smooth Galaxies;
- Class 5 (2043 images): Barred Spiral Galaxies;
- Class 6 (1829 images): Unbarred Tight Spiral Galaxies;
- Class 7 (2628 images): Unbarred Loose Spiral Galaxies;
- Class 8 (1423 images): Edgeon Galaxies without Bulge;
- Class 9 (1873 images): Edgeon Galaxies with Bulge.

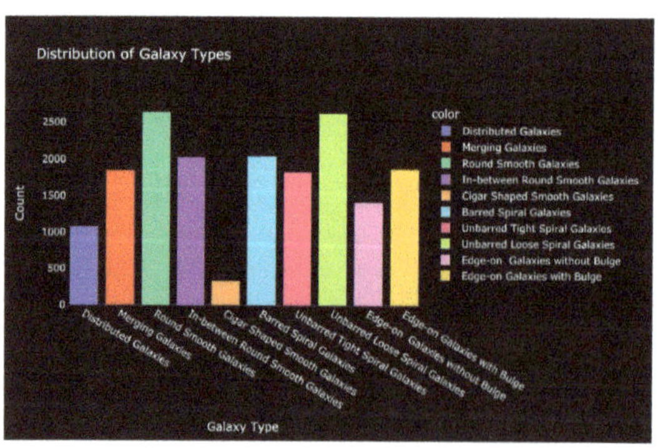

Figure 2. Original distribution of the classes in the Galaxy10 DECals dataset.

3.2. *Preprocessing*

3.2.1. Class Imbalance Mitigation

We employed data augmentation techniques to address the class imbalance issue, particularly focusing on the underrepresented class of "Cigar Shaped Smooth Galaxies".

By augmenting the existing dataset, we increased the sample size for this class while preserving the original features and characteristics of the images. The augmentation process significantly increased the number of images for the "Cigar Shaped Smooth Galaxies" class. By balancing the classes through augmentation, each class contained comparable samples. We observed a noticeable improvement in model accuracy, with an increase of approximately 24% across all classes.

3.2.2. Data Augmentation and Normalization

We standardized the image data by scaling pixel values from 0255 to 01. Then, we normalized each channel to a mean of 0 and a standard deviation of 1. This approach promotes faster convergence and improves accuracy. In addition, we experimented with different combinations of data augmentation methods on model performance, including rotation, resizing, flipping, shifting, zooming in and out, color jitter, and trivial augment wide. Specifically, we observed an increase in accuracy of approximately 10–12% when utilizing optimized augmentation strategies.

4. Methodology

In this study, we developed and compared three CNN-based models. The first model employs a foundational approach, utilizing convolutional layers to analyze image features. The second model is built with a more complex architecture without relying on pre-trained weights. The third model adopts transfer learning techniques to enhance performance.

4.1. Custom Model

We designed a custom CNN architecture for image processing, incorporating an input layer, four hidden layers, dropout, flatten, and a fully connected layer. The model, depicted in Figure 3, was constructed using the Sequential library, allowing for sequential layer stacking.

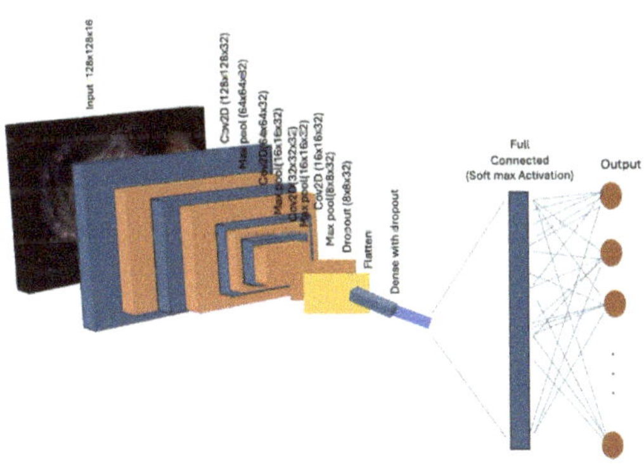

Figure 3. Custom-built CNN model architecture.

The input layer handles 128 × 128, 3-channel color images, initially resized from 256 × 256 images in the Galaxy10 DE Cals dataset. The first convolutional layer employs 16 filters, generating 16 feature maps (128 × 128). Node count increases in hidden layers to enable intricate feature detection. Each convolutional layer is followed by max pooling to downsample inputs while retaining critical features. A 2 × 2 kernel and a stride of 1 pixel in max pooling mitigate noise interference.

4.1.1. Activation Function:

Activation functions play an indispensable role in governing the activation of neurons within the neural network. In this instance, we opted for the Tanh activation function, a choice upheld across all layers except for the output layer. The final layer, responsible for delivering the probability distribution across output classes, leverages the softmax activation function to furnish the requisite probabilities.

4.1.2. Training, Testing, and Validation

Training of the model was orchestrated via the Adaptive Moment Estimation (ADAM) optimizer, a dynamic learning rate optimizer adept at steering the network toward minimal loss values. The training was executed in batches of 32, with each iteration guided by backward propagation and loss computation predicated on categorical cross-entropy. Early stopping was introduced to curtail training sessions upon the discernible absence of loss improvement, warranting a patience threshold of 5 epochs. Impressively, the custom model boasted 98,730 trainable parameters.

4.2. Tailored ResNet50 Based Model

Our model adapted the ResNet50 architecture for the Galaxy10 DECals dataset, leveraging its depth and residual connections for effective training [7]. ResNet50 comprises convolutional and pooling layers, followed by global average pooling and a fully connected layer.

Images were resized from $3 \times 256 \times 256$ to $3 \times 224 \times 224$ before input. We removed the final layer of ResNet50 to align better with our dataset. Following feature extraction, we added an adaptive average pooling layer to preserve spatial information. The flattened output was passed through fully connected layers with 512 and 128 neurons, each followed by ReLU activation to enhance the model's pattern learning ability. A dropout of 0.5 was applied after each layer to prevent overfitting. The architecture includes a fully connected layer for classifying into 10 classes, followed by softmax for a probability distribution. We initialized weights randomly to facilitate training from scratch, allowing the model to learn dataset-specific patterns.

4.3. Transfer Learning with EfficientNetB5 Model

Transfer learning uses knowledge from previously trained models to address similar problems to save computational costs and training time [8]. It uses the architecture and pre-trained weights of models trained on a large dataset, such as ImageNet [9].

In the method, we froze the early layers of the model so that their weights were not updated during the new training process. For finetuning, we selectively trained the latter layers to adapt the model to new tasks, thereby enhancing its performance on the datasets. This method allowed us to leverage the robust feature-extracting ability of the early layers while customizing the model for targeted applications through fine-tuning.

We experimented with various CNN architectures, including AlexNet [10], DenseNet121 [11], multiple ResNet models, and the EfficientNet [12] series from B1 to B4. EfficientNetB5 performs best in terms of accuracy and computational efficiency. Therefore, we chose EfficientNetB5 and used its pre-trained weights to optimize results without extensive training time. For architecture, we adapted the EfficientNetB5 backbone base model by removing the final layer to better tailor it to the dataset. We then incorporated an adaptive average pooling 2D layer. Then, the output from the adaptive pooling layer was flattened to convert the 2D feature maps into a 1D vector. The dropout rate was 0.5, and subsequent layers included a fully connected layer with 225 neurons and ReLU activation and a final output using softmax to categorize the classes.

5. Results

5.1. Effect of Data Augmentation on Accuracy

Data preprocessing was conducted including class imbalance mitigation and experimentation with data augmentation to increase the model accuracy. By addressing class imbalances and optimizing augmentation strategies, accuracy was enhanced by 15%. The distribution of images for each class before and after balancing is shown in Figure 4.

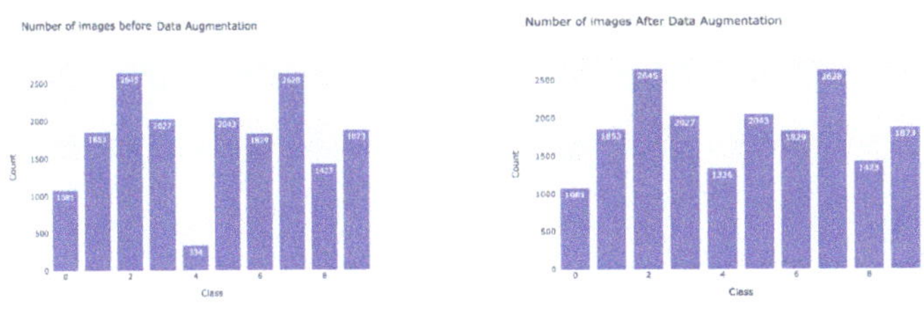

(a) (b)

Figure 4. Class distribution of the images before (**a**) and after (**b**) class imbalance mitigation.

5.2. Accuracy and Other Performance Metrics

Our custom model achieved an overall accuracy of 67% as shown in Figure 5.

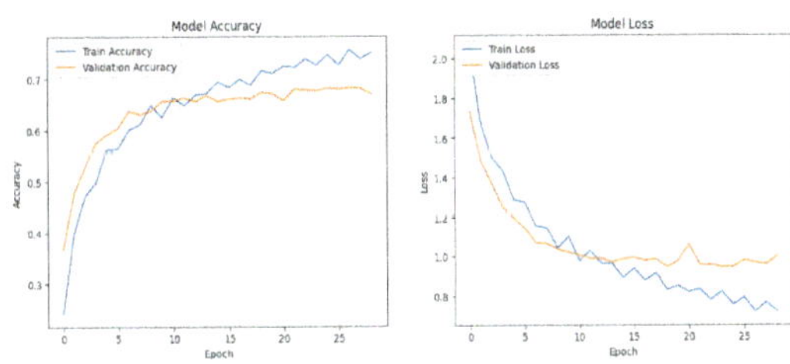

Figure 5. Loss and accuracy curves for training and validation of custom-built CNN model.

As shown in Table 1, the custom model showed an overall accuracy of 67% on 1874 samples distributed across 10 classes. Class 2 showed 83% precision and 84% recall, resulting in an 83% F1 score, while Class 3 presented with a 79% F1 score from 73% precision and 86% recall. In contrast, Class 0 underperformed with only a precision of 36% and a recall of 19%, yielding an F1-score of 25%. Classes 1, 5, 6, and 8 demonstrated moderate effectiveness with F1 scores between 54 and 74%. Figure 6 shows the loss/accuracy curve of the tailored ResNet50 model. The training and validation accuracy curves are closely aligned, with training slightly higher, suggesting minimal overfitting.

As shown in Table 2, the ResNet50 model shows an overall accuracy of 80% across 1774 samples in 10 classes. The model was trained on a dataset split into 90% for training and 10% for testing. The metrics show high precision and recall in most classes, particularly with Class 2 achieving the highest precision at 97%. Class 3 presents the highest recall at 94%. However, Class 0 had notably low recall 19%, indicating the model was difficult to identify this class, possibly due to its underrepresentation or more complex features compared to other classes. Model adjustments improve the detection accuracy for such underperforming classes.

Table 1. Results of performance metrics on multi-class—custom-built model.

Class	Precision	Recall	F1-Score	Support
0	0.36	0.19	0.25	118
1	0.57	0.56	0.56	181
2	0.83	0.84	0.83	268
3	0.73	0.86	0.79	182
4	0.9	0.82	0.86	128
5	0.78	0.56	0.65	231
6	0.66	0.55	0.6	183
7	0.45	0.67	0.54	260
8	0.72	0.76	0.74	139
9	0.74	0.74	0.74	184
Accuracy			0.67	1874
Macro avg	0.68	0.66	0.66	1874
Weighted avg	0.68	0.67	0.67	1874

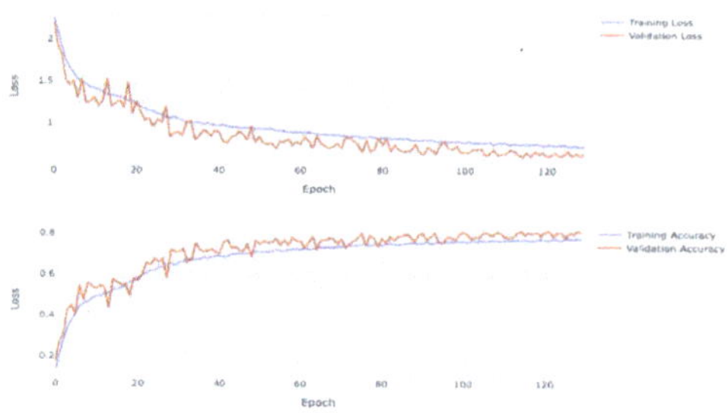

Figure 6. Loss and accuracy curves for train and validation of ResNet50 model.

Table 2. Results of performance metrics on multiclass ResNet50 model.

Class	Precision	Recall	F1-Score	Support
0	0.6	0.19	0.29	111
1	0.83	0.86	0.85	179
2	0.97	0.87	0.92	264
3	0.83	0.94	0.88	213
4	0.68	0.78	0.72	32
5	0.96	0.75	0.84	203
6	0.7	0.83	0.76	193
7	0.62	0.76	0.68	251
8	0.79	0.94	0.86	147
9	0.91	0.86	0.88	181
Accuracy			0.8	1774
Macro avg	0.79	0.78	0.77	1774
Weighted avg	0.81	0.8	0.8	1774

The transfer learning loss/accuracy decreased and stabilized. The training and validation accuracy curves remained close, with validation accuracy occasionally exceeding training accuracy in later epochs, suggesting effective learning of generalizable features without memorization. There was no sign of overfitting, and EfficientNetB5 performed well in both the training and testing phases (Figure 7).

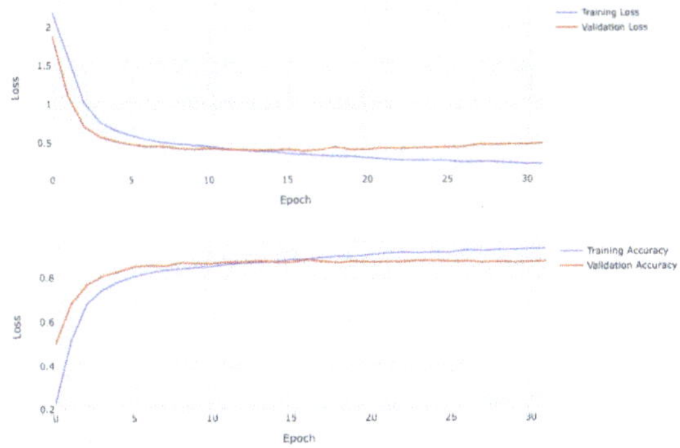

Figure 7. Loss and accuracy curves for train and validation of EfficientNetB5 model.

As shown in Table 3, the EfficientNetB5 model demonstrated strong performance with an 87% accuracy on a dataset of 3548 samples across 10 classes, split into 80% for training and 20% for testing. High precision and recall for Class 2 and 3 were as follows: F1-scores of 94%, a precision of 92%, and a recall of 96%. Class 4, despite having the fewest samples showed an F1 score of 77%. Class 7 presented the lowest F1 score at 78%, indicating variation in class performance. This showcases EfficientNetB5's effectiveness in managing multiclass classification with balanced results across various classes.

Table 3. Results of performance metrics on multiclass EfficientNetB5 model.

Class	Precision	Recall	F1-Score	Support
0	0.71	0.52	0.60	229
1	0.92	0.89	0.90	372
2	0.93	0.96	0.94	546
3	0.92	0.96	0.94	396
4	0.68	0.88	0.77	50
5	0.86	0.90	0.88	377
6	0.80	0.80	0.80	356
7	0.77	0.79	0.78	539
8	0.95	0.93	0.94	309
9	0.95	0.94	0.94	374
Accuracy			0.87	3548
Macro avg	0.85	0.86	0.85	3548
Weighted avg	0.87	0.87	0.87	3548

5.3. Model Performance

The confusion matrix for the first model serves as a discerning evaluation metric, elucidating the model's efficacy across distinct categories. The model's exemplary performance in discerning the "round smooth" category, achieved a commendable 73% precision, 86% recall, and an F1-score of 79%. Conversely, the model exhibited suboptimal performance in categorizing instances of the "Disturbed" category, manifesting a precision of 36%, recall of 19%, and an F1-score of 25%. The second model had diagonal dominance, where predicted labels matched the true labels for most classes, indicating good performance. Figure 8 shows that diagonal dominance was observed for the EfficientNetB5 model, indicating accurate predictions across various classes.

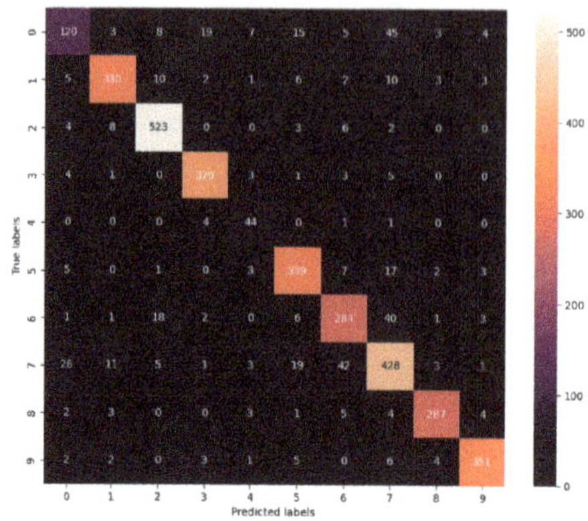

Figure 8. Confusion matrix displaying the classification results EfficientNetB5 model.

5.4. Hardware Specifications

To accelerate training, we utilized high-performance hardware configurations: a Tesla V100 GPU, equipped with 62 GB of RAM and 22.5 GB of GPU memory, with training durations exceeding 20 min. We use an NVIDIA A100 GPU, featuring 83.5 GB of RAM and 40 GB of GPU memory. Training times were approximately 1 h and 30 min for ResNet50 and 25 min for EfficientNetB5.

6. Discussion

Various CNN architectures including custom-built CNN, Tailored ResNet50, and Transfer Learning with EfficientNetB5 were experimented with to explore the potential and limitations of galaxy morphology classification. The custom-built CNN achieved a 67% overall accuracy. To address class imbalances, we employed targeted data augmentation strategies to bolster the representation of underrepresented classes. Augmenting training samples improved accuracy by 24%, notably benefiting previously lower-performing classes. This enhancement underscores the pivotal role of a balanced training set in optimizing deep learning models for diverse galaxy classification tasks. The tailored ResNet50 model showed improved overall accuracy (80%) and demonstrated strong performance across several classes. The consistently high scores in precision and recall for most classes indicate that the model's depth and residual connections effectively capture and leverage the complex features necessary for accurate classification.

The EfficientNetB5 model, utilizing transfer learning, showed the highest accuracy (87%) among the tested models. This superior performance is attributed to the model's ability to effectively generalize from pre-trained ImageNet weights, despite the domain differences between general objects and astronomical images. Notably, this model maintains high precision and recall across almost all classes, demonstrating robustness in handling diverse galaxy morphologies.

Table 4 shows the hyperparameters tuning for three different models: CustomBuilt CNN, Tailored ResNet50, and EfficientNetB5. Architecture and training strategies significantly impact the training duration and computational resources required. The strategic use of data augmentation and class rebalancing is essential in addressing inherent data biases, enhancing model training, and improving classification accuracy.

Table 4. Hyperparameters Tuning for Different Models.

Hyperparameters Tuning	Custom-Built CNN	Tailored ResNet50	EfficientNetB5
Batch Size	32	64	128
Weight	Initialized Randomly	Initialized Randomly	ImageNet
Learning Rate	1.00×10^{-3}	1.00×10^{-4}	1.00×10^{-5}
Optimizer	Adam	Adam	Adam
Loss Function	Cross Entropy	Cross Entropy	Cross Entropy
Momentum	beta_1 = 0.9	betas = (0.9, 0.999)	betas = (0.9, 0.999)
Weight Decay	N/A	1.00×10^{-2}	1.00×10^{-2}
Early Stop Trigger	patience = 5	patience = 15	patience = 10
Epoch	29	120	30
Parameters	98,730	24,624,074	28,804,069

7. Conclusions and Future Work

The significant role of CNNs in classifying galaxy morphology, a task where traditional human classification proves challenging due to the complexity and variability of galaxy images. CNNs, with their ability to autonomously extract and learn features, overcome these challenges, as demonstrated by the varied architectures that we built and analyzed. Addressing the noise in galaxy images is a priority for future work, as reducing noise could further improve model accuracy. Implementing advanced noise reduction techniques and possibly integrating denoising autoencoders in the pre-processing stage could prove beneficial. Furthermore, ongoing efforts to optimize the computational efficiency of our models are essential. Exploring more efficient network architectures or implementing more nuanced training strategies such as federated learning could reduce computational demands and make the training process more feasible on large datasets.

Author Contributions: Conceptualization, T.L., N.I., T.N. (Thu Nguyen), T.N. (Thanyaporn Noiplab), J.K. and D.B.; methodology, T.L., N.I., T.N. (Thu Nguyen), T.N. (Thanyaporn Noiplab), J.K. and D.B.; validation, T.L., N.I., T.N. (Thu Nguyen), T.N. (Thanyaporn Noiplab), J.K. and D.B.; formal analysis, T.L., N.I., T.N. (Thu Nguyen), T.N. (Thanyaporn Noiplab), J.K. and D.B.; investigation, T.L., N.I., T.N. (Thu Nguyen), T.N. (Thanyaporn Noiplab), J.K. and D.B.; resources, T.L., N.I., T.N. (Thu Nguyen), T.N. (Thanyaporn Noiplab), J.K. and D.B.; data curation, T.L., N.I., T.N. (Thu Nguyen), T.N. (Thanyaporn Noiplab), J.K. and D.B.; writing—original draft preparation, T.L., N.I., T.N. (Thu Nguyen), T.N. (Thanyaporn Noiplab), J.K. and D.B.; writing—review and editing, T.L., N.I., T.N. (Thu Nguyen), T.N. (Thu Nguyen) and D.B.; visualization, T.L., N.I., T.N. (Thu Nguyen), T.N. (Thanyaporn Noiplab) and D.B.; supervision, T.L., N.I., T.N. (Thu Nguyen), T.N. (Thanyaporn Noiplab), J.K. and D.B.; project administration, T.L., N.I., T.N. (Thu Nguyen), T.N. (Thanyaporn Noiplab), J.K. and D.B.; funding acquisition, T.L., N.I., T.N. (Thu Nguyen), T.N. (Thanyaporn Noiplab) and D.B. All authors have read and agreed to the published version of the manuscript.

Funding: This work was partially supported by funds from the Professional Activities Advisory Committee (PAAC) at Kent State University.

Institutional Review Board Statement: Not applicable.

Informed Consent Statement: Not applicable.

Data Availability Statement: Data are contained within the article.

Conflicts of Interest: The authors declare no conflict of interest.

References

1. Qian, Y. Performance comparison among vgg16, inceptionv3, and resnet on galaxy morphology classification. *J. Phys. Conf. Ser.* **2023**, *2580*, 012009. [CrossRef]
2. Wang, G. Galaxy morphology classification with densenet. *J. Phys. Conf. Ser.* **2023**, *2580*, 012064. [CrossRef]
3. Premanand, N.; Tarun, V.G.; Pawar, S.; Jawakar, D.N.; Deepa, S.; Jayapriya, J.; Vinay, M. Performance evaluation of convolutional neural networks for stellar image classification: A comparative study. In Proceedings of the 2023 International Conference on Data Science and Network Security (ICDSNS), Tiptur, India, 28–29 July 2023; IEEE: Piscataway, NJ, USA, 2023; pp. 1–6.
4. Şenel, F.A. A hyperparameter optimization for galaxy classification. *Comput. Mater. Contin.* **2023**, *74*, 4587.
5. Kalvankar, S.; Pandit, H.; Parwate, P. Galaxy morphology classification using efficientnet architectures. *arXiv* **2020**, arXiv:2008.13611.
6. Team, D.S.; Thakar, A.; Malof, J.; Wong, K.; Dey, A.; Schlegel, D. Galaxy10-DECaLS Dataset. 2021. Available online: https://astronn.readthedocs.io/en/latest/galaxy10.html#download-galaxy10-decals (accessed on 1 January 2025).
7. He, K.; Zhang, X.; Ren, S.; Sun, J. Deep residual learning for image recognition. In Proceedings of the IEEE Conference on Computer Vision and Pattern Recognition, Las Vegas, NV, USA, 27–30 June 2016; pp. 770–778.
8. Bengio, Y. Deep learning of representations for unsupervised and transfer learning. In Proceedings of the ICML Workshop on Unsupervised and Transfer Learning, Bellevue, WA, USA, 2 July 2011; JMLR Workshop and Conference Proceedings; pp. 17–36.
9. Krizhevsky, A.; Sutskever, I.; Hinton, G.E. Imagenet classification with deep convolutional neural networks. *Commun. ACM* **2017**, *60*, 84–90. [CrossRef]
10. Alom, M.Z.; Taha, T.M.; Yakopcic, C.; Westberg, S.; Sidike, P.; Nasrin, M.S.; Van Esesn, B.C.; Awwal, A.A.S.; Asari, V.K. The history began from alexnet: A comprehensive survey on deep learning approaches. *arXiv* **2018**, arXiv:1803.01164.
11. Huang, G.; Liu, Z.; Van Der Maaten, L.; Weinberger, K.Q. Densely connected convolutional networks. In Proceedings of the IEEE Conference on Computer Vision and Pattern Recognition, Honolulu, HI, USA, 21–26 July 2017; pp. 4700–4708.
12. Tan, M.; Le, Q. Efficientnet: Rethinking model scaling for con volutional neural networks. In Proceedings of the International Conference on Machine Learning, Long Beach, CA, USA, 9–15 June 2019; pp. 6105–6114.

engineering proceedings

MDPI

Proceeding Paper

Fast Collision Detection Method with Octree-Based Parallel Processing in Unity3D †

Kunthroza Hor [1], Taeheon Kim [1] and Min Hong [2,*]

[1] Department of Software Convergence, Soonchunhyang University, Asan-si 31538, Republic of Korea; zoroberto16@gmail.com (K.H.); taeheon1213@gmail.com (T.K.)
[2] Department of Computer Software Engineering, Soonchunhyang University, Asan-si 31538, Republic of Korea
* Correspondence: mhong@sch.ac.kr
† Presented at the 2024 IEEE 7th International Conference on Knowledge Innovation and Invention, Nagoya, Japan, 16–18 August 2024.

Abstract: Performing accurate and precise collision detection is a key to real-time applications in computer graphics, games, physical-based simulation, virtual reality, augmented reality, and research and development. Researchers have developed numerous methods to minimize computation time and enhance the accuracy of collision detection for pair-object collisions. Although the performance of the central processing unit (CPU) has significantly improved in recent years, it is still insufficient for many applications. In this study, we have developed an improved algorithm for geometric bounding volume hierarchy (BHV) in 3D spatial subdivisions using an Octree-based axis-aligned bounding box (AABB) structure. The AABB structure is used for collision detection and its computation by the central processing unit and graphic processing unit (GPU), which is implemented on the compute shader in Unity3D. AABB was defined as the maximum and minimum hexahedron within an object that is parallel to the coordinate axis. While GPU computing is essential for enhancing the object's performance. The proposed algorithm approaches Octree AABB-based GPU parallel processing to reduce the calculation or process of simulation for real-time collision detection and handles multiple computations. In the CPU environment, the algorithm spent 2.9 fps when simulating up to 20 objects of the Torus Model that contains 2.3 K vertices and 4.6 K triangles. In the GPU environment, it spent 635.62 fps with 20 objects, and the maximum number increased to 180 objects in real-time.

Keywords: collision detection; octree; GPU parallel processing; compute shader; unity3D

1. Introduction

As one of the most effective techniques for real-time applications, collision detection has been widely used, especially in computer graphics, games, computer animation, virtual reality, augmented reality, and 3D graphics [1]. To develop and detect a well-geometry structured 3D and an accurate real-time simulation, collision detection plays a critical role in intersecting and reaching objects adaptably and effortlessly. Collision detection is utilized to determine objects intersecting with each other in a simulated environment. The Octree axis-aligned bounding box (AABB) [2] represents a geometric bounding volume hierarchy (BVH) in 3D spatial subdivision for the simulation of the rigid objects that are characterized by a pair of objects intersection between the minimum point and maximum point [3]. If the minimum value of AABB exceeds the highest maximum value of the other along any axis, there is no collision along that particular axis. Conversely, if the maximum value of one AABB is less than the minimum value of the other along any axis, they cannot be colliding along that axis, as collision-free detection occurs at any pair between two objects. If the conditions do not meet the criteria for each of the three axes, the AABB intersections are considered to have collided.

Spatial subdivision has been widely utilized to optimize collision detection by reducing the number of collisions. The main approach is to divide the space into multiple cube regions and subsequently perform collision tests for primitives with the same cell using the Octree structure. Octree is a data structure that subdivides the space into eight equally sized cells or cube boundaries of the three-level hierarchical grid in a model space [4]. To develop and commercialize parallel processing, multicore processors are used as the graphic processing unit (GPU) for the computation of a massively parallel algorithm that allows for faster structure and quickly determines potential interactions between objects with each other at runtime [5]. GPU parallel processing can greatly enhance the processing speed for tasks that can be subdivided into smaller sub-tasks and computed on the GPU kernel program in the compute shader script for rendering.

2. Related Work

There are several studies on collision detection for rigid-body objects. Different algorithms are used to improve the efficiency of collision detection for rigid-body object simulation. Collision detection has been studied extensively in computer graphics, games, computer animation, and 3D graphics. Collision detection is the process of detecting when two or more objects and shapes are about to collide or have already collided. Aissa et al. [6] addressed the Octree-optimized microstructure of a discontinuous fiber for virtual sample generation to carry out precise prediction and simulation. The data of the tree structure are constructed recursively within a computational domain. The tree is constructed in refinement steps, dividing the computational domain into two to generate subdomains (children). This division process generates eight subdomains named Octree, with element allocation to the children managed using AABB. Serpa et al. [7] provided a complete testing technique that comprises several implementations in CPU and GPU environments. Related algorithms are categorized based on spatial and sorting methods. Spatial methods partition the problem using spatial data structures and employing a divide-and-conquer strategy. The data structure and implementation of the spatial structure are known as Grids, bounding-volume hierarchy (BHV), KD-Trees, Octrees, and BSP-Trees. The majority of the techniques reduce the search space by pruning collision pairs of object detections. The concept of GPUs is designed to execute large amounts of mathematical operations in parallel to efficiently operate the input data in a process called a graphics pipeline by utilizing scripts known as shaders [8]. Then, input data are employed to programmatically manipulate and expose it to the parallel, heavily multithreaded GPU hardware. GPU hardware and driver support are developed by introducing a new type of shader called the compute shader for general-purpose computations.

3. Methodology

The purpose of the implementation is to improve and represent collision detection between objects for geometric BHV in 3D spatial subdivision as a simple geometric primitive and vertices. The minimum and maximum bounding boxes are determined to include them in parallel hierarchical computation on the GPU. The algorithm comprises three methods.

3.1. Octree Structure

There are several spatial data structures to organize 2D and 3D data in computer graphics. Among these data structures, Octree is used widely and appears in many extensive application ranges, especially when positions must be accessed and manipulated.

Octree is a structure of increasingly dynamic and adaptable computations to decrease time complexity. Octree construction divides a three-dimensional space adaptively and the space of the children to the distribution of geometric primitives. It reduces the number of empty cells to save storage costs. With the advancement of programmable graphics hardware, Octrees can be constructed and traversed using GPU computation [9]. Octrees leverages the parallel computing feature of GPUs to enhance rendering and geometry processing. Its composition represents the full computational region of an object at the zero level of the octree hierarchy. Then, this region is subdivided into eight sub-regions of the children that present nodes in the first level. The eight sub-regions are defined by eight sub-region children that present nodes in the second level. The three levels of octree provide storage cost using reduced memory since the space partials without collision occurrence are not kept overlapped.

The Octree data structure hierarchically represent volume spaces. The space of the Octree needs to be traversed from the top to bottom to insert the object into the corresponding space nodes. Octants start from the root node in the zero level to the current eight nodes in the first level to the eight nodes in the second level of the spatial hierarchy. The root of the current nodes is defined by ordering the index number to fill the node regions inside the maximum of the Octree size as shown in Figure 1.

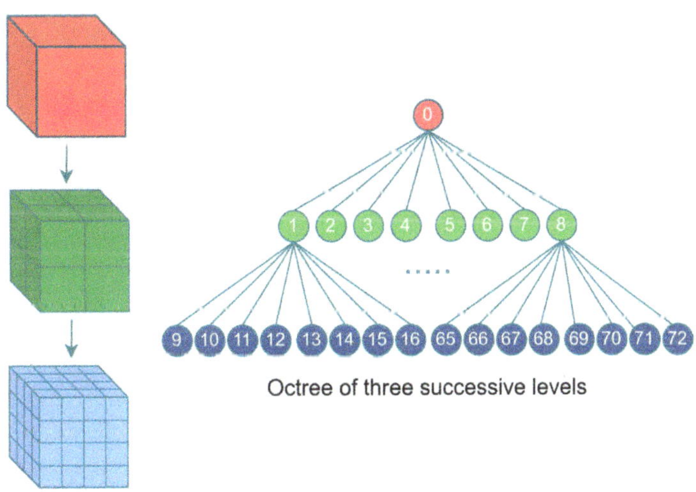

Figure 1. Structure of three successive levels in Octree.

3.2. AABB

AABB represents the basic structure. AABB is the most popular bounding box that has the maximum and minimum values of the vertex of the tetrahedron mesh on the x, y, and z coordinates. Its most outstanding feature is its ability to conduct rapid intersection, achieved by comparing coordinate values only. The method involves iterating through all vertices within the object's bounding box and then examining the projection range to detect overlaps between AABB intersections. This is achieved by comparing the two AABB bounding boxes along the three axes. The AABB is described in the x, y, and z directions to calculate the maximum and minimum along each axis by conducting an AABB intersection test with six comparison operations. The utilization of AABB and octree algorithms reduces test times and pair comparisons, thereby increasing overall efficiency at each level of the Octree structure. Each level has to be computed using six operations to find a collision index pair of objects in each level that are simultaneously intersecting and eliminated from further collision tests as presented in Figure 2.

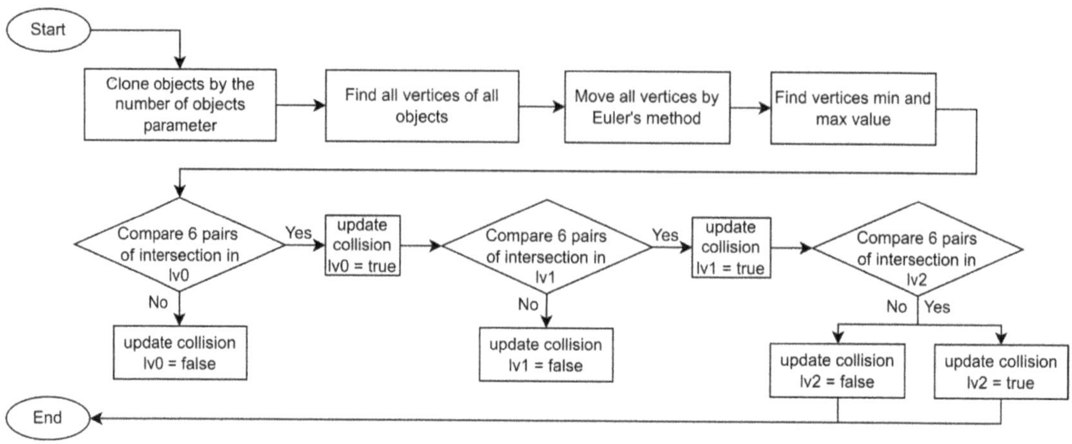

Figure 2. Six times of intersection comparison of minimum and maximum values.

After implementation, all combined vertices of all objects are moved by applying the explicit Euler physical-based method. The object and floor maximum and minimum bounding boxes are determined if there is an overlap between the AABB intersection. After collision detection occurs, the reverse velocity of particles colliding with the floor in the object is updated to make an object movement animation as shown in Figure 3.

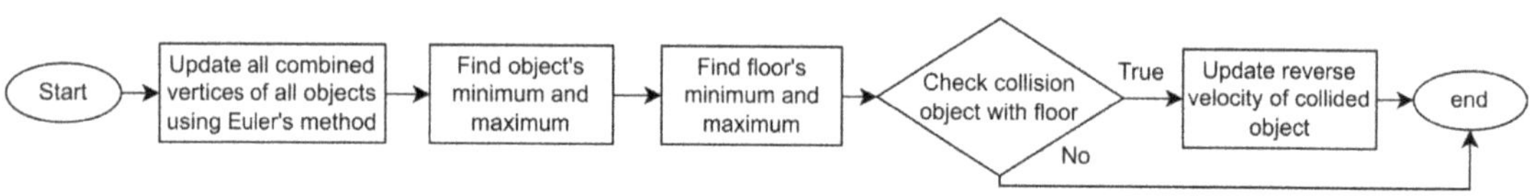

Figure 3. Checking collision detection of bounding box of objects with floor.

3.3. GPU Parallel Processing

Parallel hierarchical computation on the GPU speeds up performance, detection, and mathematical operations due to the nature of hardware platforms. The GPU-based algorithm utilizes parallelism to accelerate collision detection processes, such as general-purpose computing and simulation, by parallelizing the task massively and parallelly in hundreds of threads [10]. The Unity3D game engine is used widely in game development to render pipelines and conduct physical simulations using high-level shader language (HLSL). HLSL is essential for the GPU implementation and design of the programmable shader script operating on virtual works groups comprised of small execution units (cores). The C# programming language is used for scripting to perform operations sequentially. Its efficiency can be enhanced through multi-threaded computing. Shaders are programs tasked with the rendering of graphical data. Within Unity3D, the compute shader expands upon Microsoft Direct3D 11's technology to develop cross-platform applications. The Unity3D engine allows for the creation of a compute shader using a regular script. When invoking or running the application, it calls the dispatch method, which then activates the kernel program for execution within the C# script. The compute shader function is a kernel program specified by the number of workgroup blocks, with each workgroup block being generated by multiple threads. Moreover, the compute shader function as a kernel program is implemented by the number of threads with three parameters on the x, y, and z dimensions of the 3D vectors. The three parameters are represented as one-dimensional, two-dimensional, or three-dimensional computational tasks in parallel. The total number of threads is the product of the three parameters of the dispatch function. Therefore, the execution of all threads occurs in a non-sequential order, providing an advantage in performance, especially when handling extremely large data as demonstrated in Figure 4.

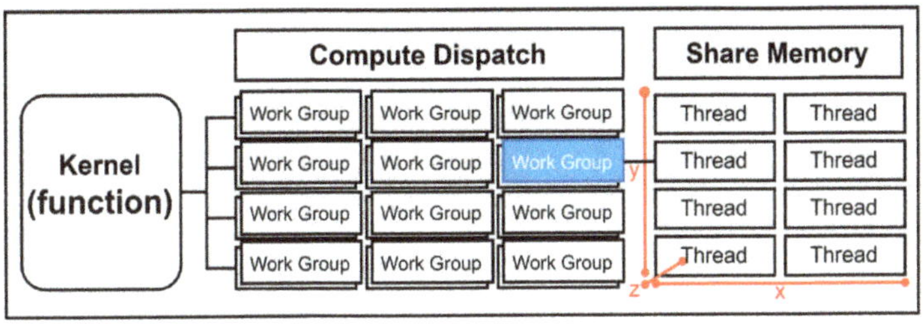

Figure 4. Compute shader elements of kernel in Unity.

4. Result and Discussion

The experiments were conducted using a Windows 10 Pro 64-bit system with an Intel CoreTM i7-7700 Processor with 3.60 GHz, 32 GB of RAM, and an NVIDIA GeForce RTX 3070Ti with 24 GB of V-RAM. C#, HLSL shader model 5.0 (Windows Graphics API), and Unity engine version 2020.3.8f1 were utilized to configure the GPU pipeline. CPU-based computations have limitations compared to GPU-based performance, as they are 26 times lower than the GPU, as provided in Table 1. The simulation is calculated 3 times for each object in the same condition, limited by 400 maximum frames in CPU, and limited by 2000 frames in GPU. In the case of FPS computation in the CPU, it is impossible to simulate insufficient performance. However, GPU-based implementation can handle the same situation more efficiently than CPU. In GPU computations for 10, 15, and 20 instances of each 3D model, minimal differences were observed due to the parallel processing capabilities of the GPU. When the number of objects increased to 180, there was a significant disparity in the results. Furthermore, the number of each model could not reach a larger number than 180 because the number of threads in the CPU count was over the allowed limit as shown in Figure 5.

Table 1. FPS comparison computation between CPU and GPU.

3D Model	Computation Comparison			
	Number of Objects	Average FPS in CPU	Average FPS in GPU	Times
Sphere 162 vertices 320 triangles	10	24.16	650.98	26
	15	12.34	645.25	52
	20	7.42	643.30	86
	180	–	532.71	–
Torus 2304 vertices 4608 triangles	10	6.83	638.34	93
	15	4.16	637.48	153
	20	2.9	635.62	219
	180	–	372.88	–
Bunny 2990 vertices 5976 triangles	10	5.51	635.25	115
	15	3.39	633.52	186
	20	2.39	629.19	263
	180	–	360.64	–
Armadillo 6362 vertices 12,720 triangles	10	2.91	623.26	214
	15	1.87	622.19	322
	20	–	619.23	–
	180	–	196.33	–

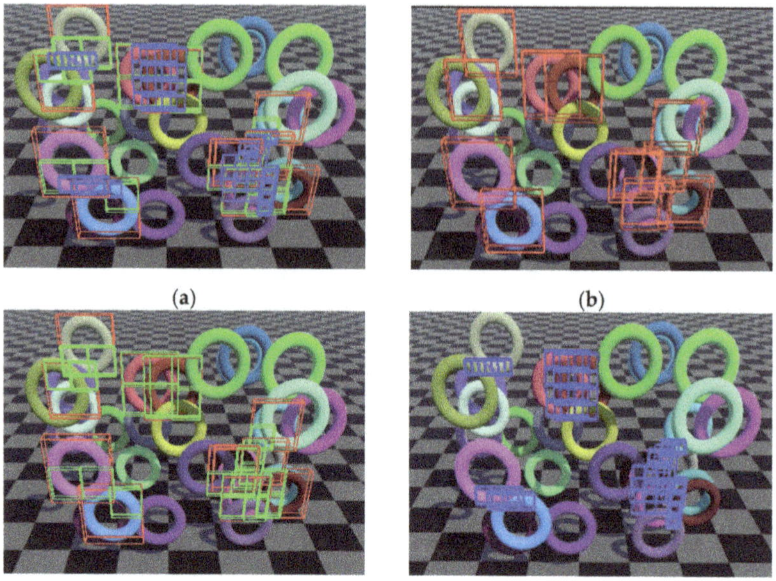

Figure 5. Experimental results from the Torus objects: (**a**) result of collision detection at all levels in Octree structure, (**b**) result of collision detection level zero, (**c**) result of collision detection level one, and (**d**) result of collision detection level two.

5. Conclusions

We developed the collision detection algorithm using the Octree-based AABB and compared the computation ability in CPU and GPU. The AABB algorithm calculated the bounding box for collision detection on objects by traversing all vertices and testing the intersection between the bounding boxes. The algorithm was implemented with an Octree-based structure for the all-pairs collision detection, the comparison of the two AABBs, and checking collision detection bounding boxes. Additionally, for parallel processing, a GPU compute shader was used. The computation speed of GPU-based was 26 times faster than CPU. In the FPS computation in CPU, the minimum cost of FPS was 1.87, and it was impossible to simulate insufficient performance. However, GPU-based implementation improved its performance in the same situation.

Author Contributions: M.H. provided conceptualization, project administration, and edited and reviewed the manuscript. T.K. analyzed and designed this study. K.H. implemented the simulation and wrote the original draft. All authors have read and agreed to the published version of the manuscript.

Funding: This research was supported by the Basic Science Research Program through the National Research Foundation of Korea (NRF) funded by the Ministry of Education under Grant NRF-2022R1I1A3069371, was funded by the BK21 FOUR (Fostering Outstanding Universities for Research) No. 5199990914048.

Institutional Review Board Statement: Not applicable.

Informed Consent Statement: Not applicable.

Data Availability Statement: No new data were created or analyzed in this study.

Conflicts of Interest: The authors declare no conflicts of interest.

References

1. Nie, Q.; Zhao, Y.; Xu, L.; Li, B. A survey of continuous collision detection. In Proceedings of the IEEE International Conference on Information Technology and Computer Application (ITCA), Guangzhou, China, 18–20 December 2020; pp. 252–257.
2. Ströter, D.; Mueller-Roemer, J.S.; Stork, A.; Fellner, D.W. OLBVH: Octree linear bounding volume hierarchy for volumetric meshes. *Vis. Comput.* **2020**, *36*, 2327–2340. [CrossRef]
3. Wang, M.; Chen, S.; Yang, Q. Design and application of bounding volume hierarchy collision detection algorithm based on virtual sphere. *J. Mech. Med. Biol.* **2019**, *19*, 1940044. [CrossRef]
4. Melero, F.J.; Aguilera, A.; Feito, F.R. Fast collision detection between high resolution polygonal models. *Comput. Graph.* **2019**, *83*, 97–106. [CrossRef]
5. Sung, M. Visibility-Based Fast Collision Detection of a large number of Moving Objects on GPU. *IEEE Access* **2023**, *11*, 49456–49463. [CrossRef]
6. Aissa, N.; Douteau, L.; Abisset-Chavanne, E.; Digonnet, H.; Laure, P.; Silva, L. Octree Optimized Micrometric Fibrous Microstructure Generation for Domain Reconstruction and Flow Simulation. *Entropy* **2021**, *23*, 1156. [CrossRef] [PubMed]
7. Serpa, Y.R.; Rodrigues, M.A.F. Broadmark: A Testing Framework for Broad-Phase Collision Detection Algorithms. *Comput. Graph. Forum* **2020**, *39*, 436–449. [CrossRef]
8. Mihai, C.C.; Lupu, C. Using graphics processing units and compute shader in real time multimodel adaptive robust control. *Electronic* **2021**, *10*, 2462. [CrossRef]
9. Yang, Y.; Chen, X.; Han, Y. Dadu-CD: Fast and efficient processing-in-memory accelerator for collision detection. In Proceedings of the ACM/IEEE Design Automation Conference (DAC), San Francisco, CA, USA, 20–24 July 2020; pp. 1–6.
10. Va, H.; Choi, M.H.; Hong, M. Real-time cloth simulation using compute shader in Unity3D for AR/VR contents. *Appl. Sci.* **2021**, *11*, 8255. [CrossRef]

engineering proceedings

MDPI

Proceeding Paper

Recommender System for Apparel Products Based on Image Recognition Using Convolutional Neural Networks †

Chin-Chih Chang [1,*], Chi-Hung Wei [2], Yen-Hsiang Wang [1], Chyuan-Huei Thomas Yang [3] and Sean Hsiao [4]

1 Department of Computer Science and Information Engineering, Chung Hua University, Hsinchu 30012, Taiwan; m11002012@chu.edu.tw
2 Ph.D. Program in Engineering Science, Chung Hua University, Hsinchu 30012, Taiwan; urderrick@gmail.com
3 School of Information Engineering, Shandong Vocational and Technical University of International Studies, Rizhao 276826, China; chyang@swut.edu.cn
4 Department of Computer Science and Information Engineering, Ming Chuan University, Taoyuan 48710, Taiwan; sean.hsiao@mail.mcu.edu.tw
* Correspondence: changc@chu.edu.tw

† Presented at the 2024 IEEE 7th International Conference on Knowledge Innovation and Invention, Nagoya, Japan, 16–18 August 2024.

Abstract: In e-commerce and fashion, personalized recommendations are used to enhance user experience and engagement. In this study, an apparel recognition and recommender system (ARRS) using convolutional neural networks (CNNs) was employed to analyze apparel images, extract features, and provide accurate recognition and recommendations. By learning patterns and features of clothes, the system enables robust recognition and personalized suggestions. The effectiveness of ARRS in recognizing apparel and generating relevant recommendations was validated. The system enhances user satisfaction and engagement on fashion e-commerce platforms.

Keywords: convolutional neural network; image recognition; recommender system

1. Introduction

Recent advancements in e-commerce technologies have revolutionized online shopping. As online retailers grow and apparel choices multiply, consumers can have personalized recommendations. To meet this demand, we improved the apparel recognition and recommendation processes of deep learning, particularly convolutional neural networks (CNNs) [1].

CNNs, inspired by visual perception, excel at extracting and learning intricate patterns from images, making them ideal for recognizing and categorizing apparel. We explored the design, implementation, and performance evaluation of an apparel recognition and recommender system (ARRS) using CNNs. By analyzing apparel images and extracting key features, the system generates personalized recommendations, aiming to enhance user satisfaction and engagement in fashion e-commerce [2,3]. The proposed system improves the efficiency and effectiveness of online fashion shopping.

This paper is structured as follows: Section 1 introduces the study, Section 2 presents related work, Section 3 outlines the research methodology, Section 4 discusses experiments and analysis, and Section 5 concludes the paper with future research directions.

2. Related Work

2.1. Deep Learning and CNN

Deep learning, a transformative force in artificial intelligence (AI), has revolutionized computer vision, natural language processing, and pattern recognition. Its success underpins innovations such as ChatGPT 3.5 and NVIDIA technologies [4,5]. At its core, deep learning uses neural networks, computational models inspired by the brain, consisting of interconnected layers that extract meaningful patterns from data [1].

CNNs, specialized for image analysis, apply learnable filters to capture patterns at different scales, such as edges and textures [2,3]. CNNs excel in autonomously learning hierarchical features, adjusting parameters through extensive training to accurately recognize complex objects. Key models, including ResNet50 [6,7], VGG19 [8], and YOLOv8 [9–11], exemplify CNNs' effectiveness, with their comparison shown in Table 1.

Table 1. Comparison of CNN models.

Model	ResNet50	VGG19	YOLOv8n
Architecture	50 layers	19 layers	225–230 layers
Convolutional filters	3 × 3	3 × 3	Varying sizes
Max pooling	2 × 2	2 × 2	Various sizes
Fully connected layers	Minimal (Often 1)	Large number	Minimal (might have none)
Top-5 accuracy on ImageNet	97.33%	97.07%	Not available
Data requirements	High (Tens of thousands)	High (Tens of thousands)	High (Tens of thousands)
Computational complexity	Medium	High	Lower
Interpretability	Moderate	Limited	Moderate

ResNet50, a deep CNN known for its depth and effectiveness in computer vision, uses residual connections to avoid vanishing gradients, excelling on datasets such as ImageNet [6,7]. VGG19, with its 19-layer architecture, leverages small filters and max-pooling layers to capture complex patterns, making it effective for image classification tasks [8]. You Only Look Once (YOLO) is used for real-time object detection, predicting bounding boxes and class probabilities in a single pass, with YOLOv8 showing strong performance in both detection and classification [9–11]. In apparel recognition and recommendation, CNNs are vital for precise image analysis and personalized recommendations. Section 3 details their application in system design.

2.2. Recommender Systems

Recommender systems predict user preferences or ratings, playing a key role in e-commerce, music, and video platforms [12–14]. These systems include collaborative filtering, content-based filtering, or hybrids to analyze data for ratings and browsing history. Collaborative filtering looks at user or item similarities, while content-based filtering matches user preferences with item attributes. Hybrid approaches combine these methods for better accuracy. Machine and deep learning enhance these techniques for complex recommendations. Platforms including Amazon, Netflix, and Spotify use them to improve user experience and boost sales. In fashion e-commerce, combining CNN-based image analysis with collaborative filtering provides tailored apparel suggestions.

2.3. Recommenders Based on Deep Learning

Deep learning has become a powerful tool in recommendation systems, enabling personalized content suggestions through its ability to learn complex patterns from raw data [15,16]. Current approaches use neural networks alone or combined with traditional methods. For a more in-depth exploration of deep learning-based recommendation mechanisms, refer to [16].

We enhanced recommender systems with content filtering, collaborative filtering, and deep learning, particularly in integrating content-based features from textual, visual, or multimodal data sources. CNNs process apparel images, while natural language processing handles product descriptions or reviews, improving accuracy. The proposed system integrates deep learning for image analysis and collaborative filtering for recommendations, offering personalized suggestions in fashion e-commerce.

3. Research Methodology

3.1. System Design

This system contains three main modules: data learning, image recognition, and recommendation modules, as shown in Figure 1.

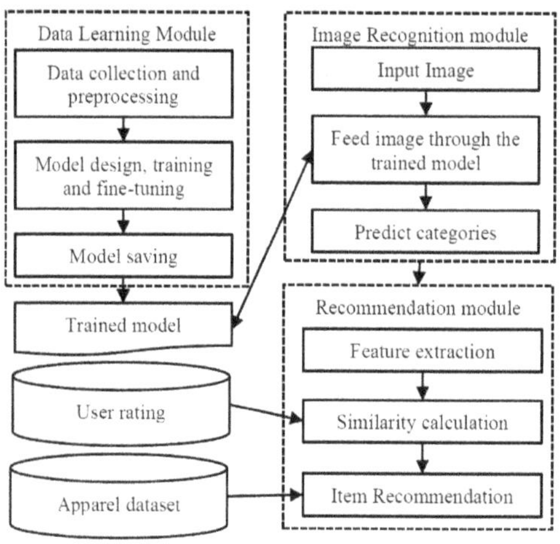

Figure 1. System architecture.

- The data learning module collects, preprocesses, and trains models, storing the trained model and weights for future use.
- The image recognition module processes image input and uses the trained model to predict categories.
- The image recommendation module extracts features, calculates similarity with user ratings, and provides personalized recommendations.

3.2. Data Collection and Preprocessing

The apparel dataset consists of 20 categories of apparel from the 108 categories provided by kaggle.com. The sources of the data are shown in Figure 2 and are organized as follows:

- The dataset comprises 20 categories of apparel specifically curated for this study.
- It is sourced from the "Fashion Product Images Dataset", specifically the 108-category dataset within it.
- The dataset is a collaboration with Param Aggarwal (Owner).
- The data originate from the Indian fashion e-commerce website: myntra.com.
- Dataset source link is https://www.kaggle.com/datasets/paramaggarwal/fashion-product-images-dataset (accessed on 20 March 2024).

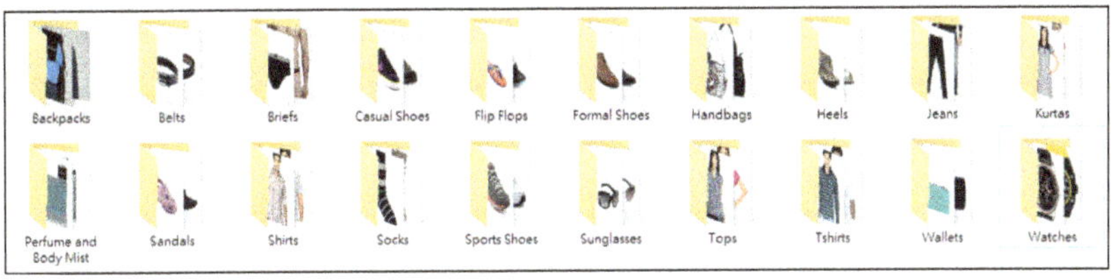

Figure 2. Apparel data source.

We collected preferences from 60 users who viewed a clothing photo dataset with 20 categories, each containing 600 photos, as shown in Figure 3. Users selected 100 favorites per category and put in "fashion_like_100", 10 favorites and put in "fashion_like_10", 5 and put in "fashion_like_5", and finally 1 favorite and put in "fashion_like_1". Preferences were stored in a CSV file with user IDs and liked item IDs, used as ratings in the recommender system to generate suggestions.

Figure 3. Apparel dataset.

In deep learning, data augmentation is a key to expanding the training dataset [17]. By exposing the model to a wider range of data, it enhances the model's ability to generalize, adapt to unseen data, and mitigate overfitting [18]. Data augmentation methods, such as scaling, shearing, and horizontal flipping, are commonly employed to achieve this. Scaling adjusts image sizes while preserving proportions, aiding the model in recognizing objects of varying sizes. Shearing alters image shapes to maintain the model's recognition capability for distorted parts. Horizontal flipping creates mirrored images, diversifying the data and enabling the model to recognize objects from different angles.

3.3. Model Design, Training, and Fine-Tuning

The proposed system adopts ResNet50, VGG19, YOLOv8n, and self-built CNN as deep-learning training modules. ResNet50 and VGG19 modules are chosen due to their advantages and nature along with the novelty of YOLOv8n. CNN's input image size is 224×224, comprising six layers. It initially splits into two branches: one with 64 3×3 convolutional kernels with a stride of 1, and another with 32 1×1 convolutional kernel. Following this, it integrates 64 3×3 convolutional kernels with a stride of 1, utilizing the ReLU activation function and Max pooling. Subsequently, it concatenates and processes with 128 3×3 convolutional kernels, ReLU activation function, and Max pooling. Finally, the 2D feature maps are transformed into 1D for output through fully connected layers. Figure 4 depicts this CNN architecture.

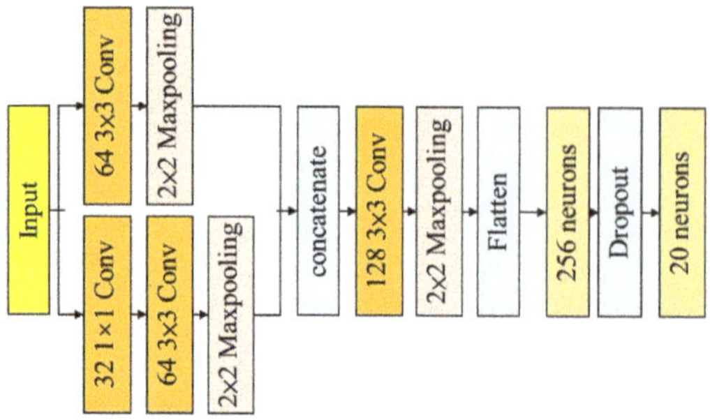

Figure 4. Self-built CNN model.

The implementation process of the learning procedure is as follows:

- Dataset construction: A dataset consisting of 20 apparel categories is established, divided into a 60:20:20 ratio for training, validation, and testing, respectively.
- ResNet50, VGG19, YOLOv8n, self-built CNN training: Each training module runs for 100 iterations, with 64 images processed per iteration, and validation is also conducted for 100 iterations with the same batch size.
- Testing: Testing involves processing 64 images at a time, with the final output being the accuracy rate.

3.4. Recommendation Engine

The recommendation module generates feature vectors using the selected model. In machine and deep learning, feature extraction is vital, transforming data into mathematical representations across multiple dimensions, each capturing distinct attributes or characteristics.

It calculates the similarity between the feature vectors of the input photo and photos in the database using cosine similarity is used to calculate similarity defined as

$$\text{Similarity} = cos(\theta) = \frac{A \cdot B}{||A||||B||} = \frac{\sum_{i=1}^{n} A_i \times B_i}{\sqrt{\sum_{i=1}^{n}(A_i)^2} \times \sqrt{\sum_{i=1}^{n}(B_i)^2}} \quad (1)$$

where A and B are feature vectors [19–21].

The average rating of photos in the database is multiplied by the similarity to obtain a score for similarity and rating. Because the system constantly provides recommendations, it can overlook certain items to be suggested. Therefore, mechanisms must be adopted to tackle this issue. In the system, we consider item popularity. To devise a strategy for improved performance, we developed the following strategies based on the selected top-k recommendations.

- For k = 1, items with high similarity scores but no ratings are prioritized at a 3% probability, with the remaining 97% of recommendations based on comprehensive scores.
- For k = 5, items with high similarity scores but no ratings are prioritized with a 20% probability, with the remaining 80% of recommendations based on scores.
- For k = 10, a 5% probability is allocated to include 1 item with a high similarity score but no rating in the recommendations, and a 90% probability to include 2 items with high similarity scores but no ratings. The remaining 5% of recommendations are based on comprehensive scores.

4. Experiments and Results

4.1. Experimental Device

The system was built using the following equipment, and the detailed development environment is presented in Table 2.

Table 2. System development environment.

Hardware	
Processor	Intel® Core™ i7-8086K
Memory	DDR4-32G
Graphics Card	NVIDIA GeForce GTX 1080Ti (11G)
Hardware	
Operating System	Windows 10
Development Tools	Anaconda3
Python Version	3.10

4.2. Model Training and Evaluation

In training deep learning networks, the accuracy rate varies as the number of iterations increases, ultimately reaching a saturation state of learning rate. Figures 5 and 6 depict the training graphs for ResNet50 and self-built CNN, respectively. The summary of training results is shown in Table 3. ResNet50, VGG19, and YOLOv8n show higher accuracy than self-built CNN.

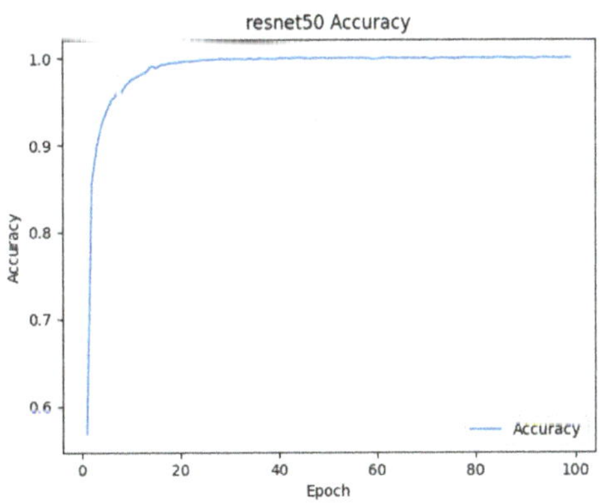

Figure 5. ResNet50 training graph.

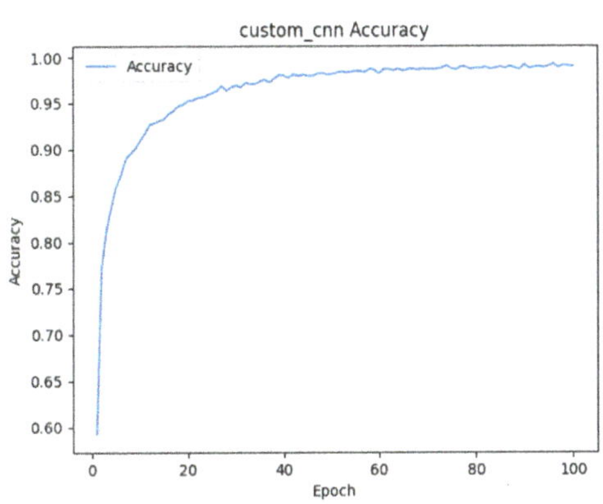

Figure 6. Self-built CNN training graph.

Table 3. Comparison of CNN models (Bold text highlights the highest value).

Training Model	Training Set Accuracy	Validation Set Accuracy	Test Set Accuracy
ResNet50	**0.9996**	0.921	**0.9288**
VGG19	0.9987	**0.9297**	0.9258
Self-built CNN	0.9897	0.8792	0.8842
YOLOv8n	0.9517	0.922	0.9233

4.3. Recommendation Evaluation

In evaluation [19,20], precision and recall are key metrics. Precision is the ratio of true positives to the sum of true and false positives, indicating the model's accuracy in predicting positive samples. Recall is the ratio of true positives to the sum of true positives and false negatives, reflecting the model's ability to capture positive samples. F1-score is the harmonic mean of precision and recall, balancing both metrics. We use precision at k, recall at k, and F1-score as evaluation metrics, focusing on the top k recommended items to calculate precision and recall.

- Precision at k is the value of successfully recommended items out of the total recommended items, where k is the total number of recommended items.

$$\text{Precision at k} = \frac{\text{\# of recommended items that has been rated}}{\text{\# of total recommended items}} \tag{2}$$

- Recall at k is the number of correctly predicted items among the recommended items.

$$\text{Recall at k} = \frac{\text{\# of items recommended and rated}}{\text{\# of all rated items}} \tag{3}$$

- F1-score is the harmonic mean of precision and recall.

$$\text{F1} - \text{score} = \frac{2 \times \text{Precision} \times \text{Recall}}{\text{Precision} + \text{Recall}} \tag{4}$$

The cosine similarity is used to compare four deep-learning models for collaborative filtering. Similarities are computed using a clothing dataset, and a comprehensive performance comparison is constructed using precision at k, recall at k, and F1-score. The summary of precision at k, recall at k, and F1-score is shown in Tables 4–6.

Table 4. Summary of precision At K (Bold text highlights the highest value).

Training Model	Top-1	Top-5	Top-10
ResNet50	0.96541	0.95925	0.81454
VGG19	0.97	0.96175	0.81567
Self-built CNN	0.96833	**0.96308**	**0.81625**
YOLOv8n	**0.97083**	0.96242	0.81333

Table 5. Summary of recall At K (Bold text highlights the highest value).

Training Model	Top-1	Top-5	Top-10
ResNet50	0.09943	0.18272	0.61035
VGG19	0.11136	0.21986	0.76828
Self-built CNN	0.12432	0.33533	**0.99172**
YOLOv8n	**0.14396**	**0.50067**	0.98776

Table 6. Summary of F1-score (Bold text highlights the highest value).

Training Model	Top-1	Top-5	Top-10
ResNet50	0.18029	0.307	0.69782
VGG19	0.19975	0.35791	0.79127
Self-built CNN	0.22035	0.49747	**0.89547**
YOLOv8n	**0.25074**	**0.65869**	0.8921

Top-10 recommended items present a more feasible solution. In the recommendations, the self-built CNN demonstrates superior precision and recall at K and F1-score. Excessively high precision in image accuracy may limit the variety of items for recommendation.

4.4. Computation Time

We apply four different training models individually to the recommendation system. Subsequently, we compare their operational speed. The summary of training time and recommendation time is shown in Table 7. YOLOv8n has the lowest training time, followed by the self-built CNN. YOLOv8n consistently has the lowest recommendation times across all top-K scenarios, followed by self-built CNN.

Table 7. Summary of training time (Bold text highlights the highest value).

Training Model	Training Set Accuracy	Validation Set Accuracy	Test Set Accuracy
ResNet50	**0.9996**	0.921	**0.9288**
VGG19	0.9987	**0.9297**	0.9258
Self-built CNN	0.9897	0.8792	0.8842
YOLOv8n	0.9517	0.922	0.9233

5. Conclusions and Future Work

We develop an efficient system that integrates apparel recognition with personalized recommendations using CNNs. While image feature extraction accuracy is important, overly precise accuracy may limit recommendation diversity. Among the four models tested, YOLOv8n shows the highest speed stability, followed by a custom-built CNN, with ResNet-50 and VGG19 trailing in training speed. Custom-built models are trained faster but with lower accuracy, highlighting the need to adjust model complexity based on specific requirements. The developed recommender system offers a foundation for exploring various deep learning frameworks and is adaptable to diverse image datasets.

It is still necessary to enhance recommendation precision by integrating contrastive language–image pretraining (CLIP), which can understand both text and images [22]. Fine-tuning CLIP on fashion descriptions and images can improve recommendation accuracy and enhance user satisfaction and engagement in fashion e-commerce.

Author Contributions: Conceptualization, C.-C.C.; Methodology, C.-C.C., C.-H.W., Y.-H.W. and C.-H.T.Y.; Writing—original draft preparation, C.-C.C.; writing—review and editing, S.H. All authors have read and agreed to the published version of the manuscript.

Funding: This research received no external funding.

Institutional Review Board Statement: Not applicable.

Informed Consent Statement: Not applicable.

Data Availability Statement: The data presented in this study are available upon request from the corresponding author.

Conflicts of Interest: The authors declare no conflicts of interest.

References

1. John, D. *Kelleher, Deep Learning*; The MIT Press: Cambridge, MA, USA, 2019.
2. LeCun, Y.; Bottou, L.; Bengio, Y.; Haffner, P. Gradient-based learning applied to document recognition. *Proc. IEEE* **1998**, *86*, 2278–2324. [CrossRef]
3. Huang, G.; Liu, Z.; van der Maaten, L.; Weinberger, K.Q. Densely Connected Convolutional Networks. In Proceedings of the 2017 IEEE Conference on Computer Vision and Pattern Recognition (CVPR), Honolulu, HI, USA, 21–26 July 2017; pp. 2261–2269.
4. Wu, T.; He, S.; Liu, J.; Sun, S.; Liu, K.; Han, Q.-L.; Tang, Y. A brief overview of ChatGPT: The history, status quo and potential future development. *IEEE/CAA J. Autom. Sin.* **2023**, *10*, 1122–1136. [CrossRef]
5. Mittal, S. A survey on optimized implementation of deep learning models on the NVIDIA Jetson platform. *J. Syst. Archit.* **2019**, *97*, 428–442. [CrossRef]
6. He, K.; Zhang, X.; Ren, S.; Sun, J. Deep Residual Learning for Image Recognition. In Proceedings of the IEEE Conference on Computer Vision and Pattern Recognition (CVPR), Las Vegas, NV, USA, 27–30 June 2016; pp. 770–778.
7. Koonce, B. ResNet 50. In *Convolutional Neural Networks with Swift for Tensorflow: Image Recognition and Dataset Categorization*; Apress: New York, NY, USA, 2021; pp. 63–72.
8. Zisserman, A.; Simonyan, K. Very deep convolutional networks for large-scale image recognition. *arXiv* **2014**, arXiv:1409.1556.
9. Redmon, J.; Divvala, S.; Girshick, R.; Farhadi, A. You Only Look Once: Unified, Real-Time Object Detection. In Proceedings of the IEEE Conference on Computer Vision and Pattern Recognition (CVPR), Las Vegas, NV, USA, 27–30 June 2016; pp. 779–788.
10. Jocher, J.G.; Stoken, A.; Borovec, J.; Changyu, L.; Hogan, A.; Diaconu, L.; Poznanski, J.; Yu, L.; Rai, P.; Ferriday, R.; et al. ultralytics/yolov5: V3. 0. *Zenodo* **2020**. [CrossRef]
11. Terven, J.R.; Cordova-Esparza, D.M. A comprehensive review of YOLO: From YOLOv1 to YOLOv8 and beyond. *arXiv* **2023**, arXiv:2304.00501.
12. Jannach, D.; Zanker, M.; Felfernig, A.; Friedrich, G. *Recommender Systems: An Introduction*; Cambridge University Press: Cambridge, UK, 2010.
13. Martin, F.J.; Donaldson, J.; Ashenfelter, A.; Torrens, M.; Hangartner, R. The big promise of recommender systems. *AI Mag.* **2011**, *32*, 19–27. [CrossRef]
14. Resnick, P.; Varian, H.R. Recommender Systems. *Commun. ACM* **1997**, *40*, 56–58. [CrossRef]
15. Cheng, H.-T.; Koc, L.; Harmsen, J.; Shaked, T.; Chandra, T.; Aradhye, H.; Anderson, G.; Corrado, G.; Chai, W.; Ispir, M.; et al. Wide & deep learning for recommender systems. In Proceedings of the 1st Workshop on Deep Learning for Recommender Systems, Boston, MA, USA, 15 September 2016; pp. 7–10.
16. Zhang, S.; Yao, L.; Sun, A.; Tay, Y. Deep Learning Based Recommender System: A Survey and New Perspectives. *ACM Comput. Surv.* **2020**, *52*, 1–38. [CrossRef]
17. Ko, B.; Ok, J. Time Matters in Using Data Augmentation for Vision-Based Deep Reinforcement Learning. *arXiv* **2021**, arXiv:2102.08581.
18. Yun, S.; Han, D.; Oh, S.J.; Chun, S.; Choe, J.; Yoo, Y. CutMix: Regularization Strategy to Train Strong Classifiers with Localizable Features. In Proceedings of the IEEE/CVF International Conference on Computer Vision (ICCV), Seoul, South Korea, 27 October–2 November 2019; pp. 6023–6032.
19. Zuva, K.; Zuva, T. Evaluation of Information Retrieval Systems. *Int. J. Comput. Sci. Inf. Technol. (IJCSIT)* **2012**, *4*, 35–43. [CrossRef]
20. Tan, P.-N.; Steinbach, M.; Kumar, V. *Introduction to Data Mining*, 2nd ed.; Pearson: London, UK, 4 January 2018.
21. Singhal, A. Modern Information Retrieval: A Brief Overview. *Bull. IEEE Comput. Soc. Tech. Comm. Data Eng.* **2001**, *24*, 35–43.
22. Li, Y.; Liang, F.; Zhao, L.; Cui, Y.; Ouyang, W.; Shao, J.; Yu, F.; Yan, J. Supervision exists everywhere: A data efficient contrastive language-image pre-training paradigm. *arXiv* **2021**, arXiv:2110.05208.

engineering proceedings

MDPI

Proceeding Paper

Comparison of Ultraviolet A/B and C Irradiation for Exosome Secretion Enhancement in HEK 293T Cell †

Ching-Chih Chan [1,*], Pohao Lin [2], Yi Xian [2], Ruey-Hwang Chou [3] and Yi-Jui Liu [4]

1 Electrical and Communication Engineering, Feng Chia University, Taichung 40724, Taiwan
2 Biomedical Informatics and Biomedical Engineering, Feng Chia University, Taichung 40724, Taiwan; s22551922@gmail.com (P.L.); max10533@gmail.com (Y.X.)
3 Institute of Cancer Biology, China Medical University, Taichung 40724, Taiwan; rhchou@mail.cmu.edu.tw
4 Department of Automatic Control Engineering, Feng Chia University, Taichung 40724, Taiwan; erliu@fcu.edu.tw
* Correspondence: yang5149@gmail.com
† Presented at the 2024 IEEE 7th International Conference on Knowledge Innovation and Invention, Nagoya, Japan, 16–18 August 2024.

Abstract: Exosomes, extracellular vesicles known for their stability, low immunogenicity, and excellent tissue penetration, are employed as delivery vehicles. These exosomes can traverse the tumor barrier and deliver therapeutic agents directly into pancreatic cancer cells. Targeted exosome vectors containing gene fragments to inhibit Kirsten rat sarcoma viral oncogene homolog (KRAS) activity are crucial for treating pancreatic tumors. Therefore, the content of the exosomes is critical. This study aims to compare the function of exosomes released by HEK-293T cells when exposed to ultraviolet A/B and ultraviolet C irradiation to determine its impact. HEK-293T cells were irradiated with ultraviolet A/B, and ultraviolet C for various indicated times, after which the cell count and exosome secretion were measured. Exosomes derived from HEK-293T cells were isolated through differential centrifugation and identified using four methods: cell counting, Bradford assay, nanoparticle tracking analysis (NTA), and Western blot analysis. Preliminary studies demonstrated that the cell count and Bradford assay expression were reduced in ultraviolet C compared to the control, with similar levels observed for ultraviolet A/B and the control. Exosome expression in Western blot analysis showed ultraviolet C, but a higher amount of ultraviolet A/B compared to the control. We introduce a comprehensive approach to ultraviolet irradiation, including ultraviolet A/B and ultraviolet C, which enhanced the secretion of exosomes by HEK-293T targeted vectors for KRAS inhibition in pancreatic cancer.

Keywords: HEK-293T; exosomes; ultraviolet A/B; ultraviolet C

1. Introduction

Pancreatic cancer exhibits high mortality and resistance to chemotherapy, primarily due to high metastasis, drug resistance, and the dense desmoplastic matrix [1], which is often driven by Kirsten rat sarcoma viral oncogene homolog (KRAS)-activating mutations. Exosomes are small membrane-bound vesicles that are secreted by various cells into the extracellular environment; extracellular vesicles known for their stability, low immunogenicity, and excellent tissue penetration, are employed as delivery vehicles. These exosomes can traverse the tumor barrier and deliver therapeutic agents directly into pancreatic cancer cells [2]. Targeted exosome vectors containing gene fragments to inhibit KRAS activity are crucial for treating pancreatic tumors. Therefore, the content of the exosomes is critical.

This study aims to compare the function of exosomes released by HEK-293T cells when exposed to ultraviolet A/B and ultraviolet C irradiation to determine its impact. HEK-293T cells were irradiated with ultraviolet A/B and ultraviolet C for various indicated times [3], after which the cell count and exosome secretion were measured. Exosomes derived from HEK-293T cells were isolated through differential centrifugation and identified using four methods: cell counting, Bradford assay, Nanoparticle tracking analysis (NTA), and Western blot analysis. Exosome expression in Western blot analysis showed extremely small amount in ultraviolet C, but a higher amount in ultraviolet A/B compared to the control [4]. A comprehensive approach is used for ultraviolet irradiation, including ultraviolet A/B and ultraviolet C, to enhance the secretion of exosomes by HEK-293T. The findings open a new direction for the mass production of exosome-targeted vectors for KRAS inhibition in pancreatic cancer [5].

2. Materials and Method

Exosomes derived from HEK-293T cells were isolated through differential centrifugation and identified using four methods (Figure 1): cell counting, Bradford assay, NTA, and Western blot analysis.

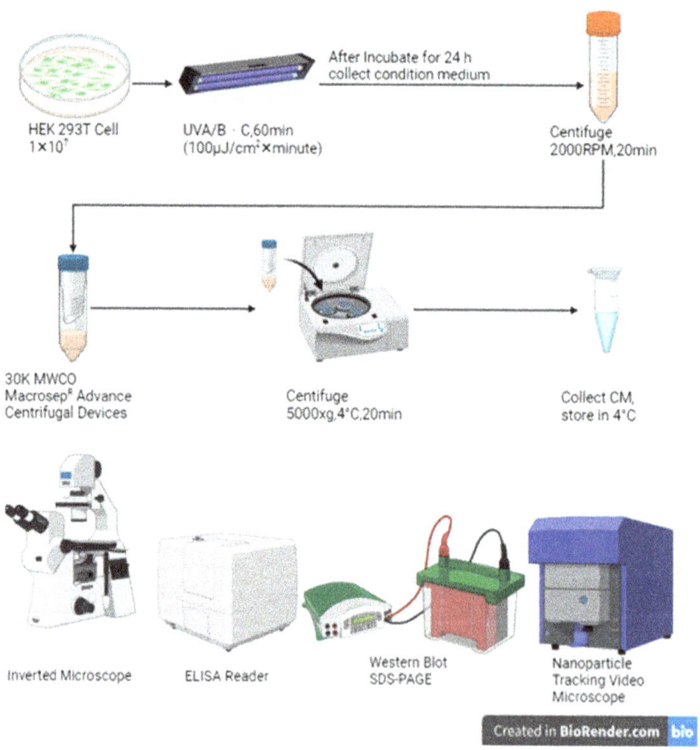

Figure 1. Materials and method of this study.

2.1. Cell Culture

The protocol for culturing the human embryonic kidney epithelial cell line HEK-293T was propagated in Dulbecco's modified Eagle's medium (DMEM, Thermo Fisher Scientific, Waltham, MA, USA) supplemented with 10% fetal bovine serum (FBS, Gibco, Miami, FL, USA) and 1% penicillin/streptomycin (P/S) medium and passaging to another new plate was performed when the cells reach 80% confluence.

- Day 1: Cells were planted at a density of 1×10^7 cells/mL in 10 cm dishes using DMEM medium supplemented with penicillin/streptomycin (P/S) and 10% fetal bovine serum (FBS); triple-plates cells were reserved for the control (Ct).
- Day 2: Irradiating the HEK-293T cells with ultraviolet A/B (UVP, CL-1000 Ultraviolet Crosslinker, Upland, CA, USA) and ultraviolet C (UVP, CL-1000 Ultraviolet Crosslinker, USA) light for 10/30/60 min; triple repeat. Then, the dishes were transferred to a cell culture incubator set at 37 °C with 5% CO_2 for 24 h.
- Day 3: The conditioned medium was collected from the dishes.

2.2. Cell Count and Cell Morphology

An electric inverted fluorescence microscope (Zeiss, Axio Vert.A1, Oberkochen, Germany) was used to evaluate cell morphology and calculate the remaining cell count and the viability of each experimental group using a hemocytometer for cell counting under different ultraviolet A/B and ultraviolet C irradiation doses.

2.3. Exosome Concentration

For all the different irradiation doses condition medium was centrifuged at 2000 RPM at 4 °C for 20 min to preliminarily separate cell debris. The debris free sample subsequently transfer to filtration was performed using a 0.45 μm syringe filter. The sample was concentrated 20-fold in a 30K concentrator tube at 5000× *g* at 4 °C for 20 min and stored in a refrigerator at 4 °C.

2.4. Bradford Assay

A series of protein standards with known concentrations (0.05, 0.1, 0.2, and 0.4 mg/mL) were prepared. The different irradiation ultraviolet A/B doses were irradiated for 0, 10, 30, and 60 min. The Bradford reagent of 20% Coomassie Brilliant Blue dye (Bio-Rad, Thermo Fisher Scientific, Waltham, MA, USA) was added to both standards and concentration exosomes. The mixtures were incubated were incubated for a brief period to allow dye–protein binding. Then, the absorbance of each sample was measured at a specific wavelength using a multifunctional full-spectrum analyzer (Bio Tek, Synergy 2, Winooski, VT, USA). A standard curve was plotted using the absorbance values of the standards to determine the protein concentration of the concentration exosomes.

2.5. Nanoparticle Tracking Analysis (NTA)

The nanoparticle tracking analyzer (Particle Metrix, ZetaView, Inning am Ammersee, Germany) was used to analyze the size distribution and concentration of exosomes to verify the purity of the material in the conditioned medium (CM).

2.6. Western Blot

Western blot analysis was performed to detect the expression of CD9, CD63, and CD81 in the samples extracted from HEK-293T cells, using a high-performance chemiluminescence analyzer (Chemi Doc Touch Imaging System, Bio-Rad, Hercules, CA, USA).

3. Results

The results were compared to the control with similar levels observed in previous studies, including ultraviolet A/B and ultraviolet C, which enhanced the secretion of exosomes by HEK-293T-targeted vectors for KRAS inhibition in pancreatic cancer. A comparison of ultraviolet A/B and ultraviolet C irradiation is shown in Figure 2. A multifunctional full-spectrum analyzer measures the protein concentrations recovered from lysates by Bradford assay(A). Cell numbers increased with increasing doses of ultraviolet A/B within a certain range, but cell numbers dropped rapidly with 1 mJ/cm^2 doses of ultraviolet C irradiation.

An inverted fluorescent microscope was used to see the cell morphology and alive/dead cell(B), that was exposed to varying doses of ultraviolet A/B and ultraviolet C irradiation, but the cells that were exposed to 1 mJ/cm^2 doses of ultraviolet C irradiation were all dead. Using a hemocytometer grid cell counter, the remaining cells and viability were measured. For varying doses of ultraviolet A/B, the viability was higher than 95% (C).

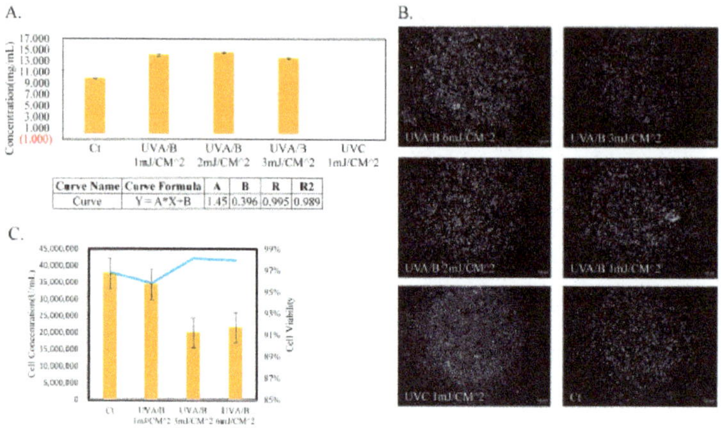

Figure 2. Comparison of UVA/B and UVC. (**A**) Using a series of protein standards with known concentrations to measure the protein concentration for different irradiation ultraviolet A/B and ultraviolet C of the concentration exosomes. (**B**) Using an electric inverted fluorescence microscope to evaluate cell morphology and alive/dead cell. (**C**) Using a hemocytometer for cell counting and under different ultraviolet A/B and ultraviolet C irradiation doses and calculate the remaining cell count and the viability of each experimental group.

The nanoparticle tracking analysis results are shown in Figure 3 in order to analyzing the size of concentration exosomes, the peak value are 113 nm and the full width at half maximum (FWHM) value are 111 nm, typically have a size distribution ranging from approximately 50 to 150 nm in diameter, the nanoparticle tracking analysis result, we can ensured the purityof the substances present in the conditioned medium (CM). The outer membrane of the exosomes secreted by HEK-293T cells expresses the proteins CD9, CD63, and CD81. The effect of varying doses of ultraviolet A/B irradiation on exosome production is shown in Figure 4. Using the Bradford assay to measure exosome concentration (A), the expression levels of CD9, CD63, and CD81 in Western blot analysis were analyzed (B), with the results aligning with the Bradford assay findings.

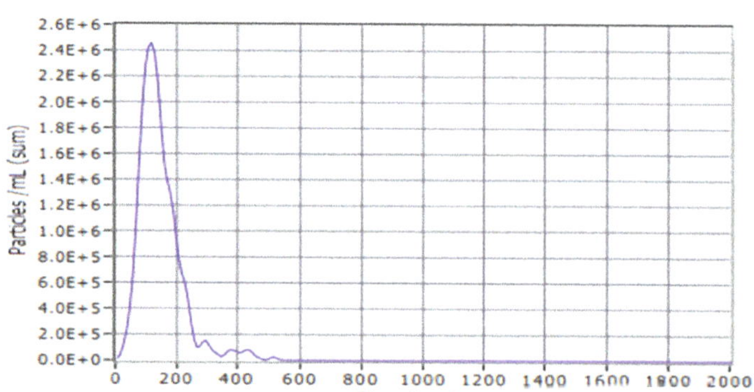

Figure 3. *Cont.*

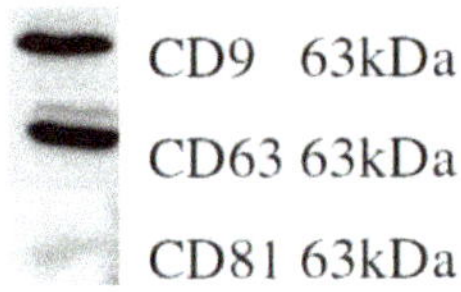

Diameter / nm	Particles/mL	FWHM / nm	Percentage
113	2.2×10^6	111	100

Figure 3. Characterization of exosomes.

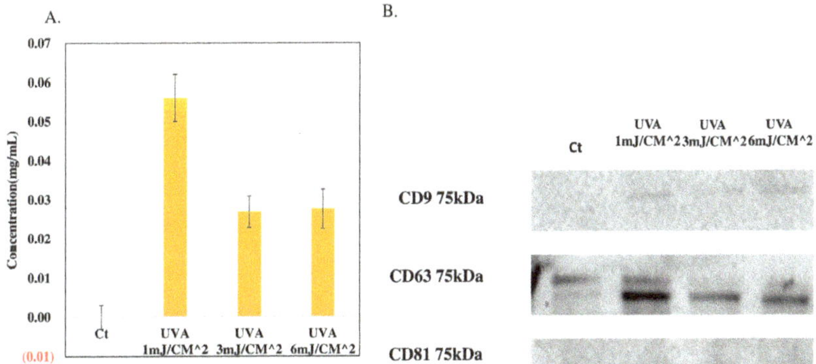

Curve Name	Curve Formula	A	B	R	R2
Curve	Y = A*X+B	1.34	0.348	0.993	0.986

Figure 4. Comparison of UVA/B and UVC. (**A**) Using a series of protein standards with known concentrations to measure the protein concentration for different irradiation ultraviolet A/B of the concentration exosomes. (**B**) The expression levels of CD9, CD63, and CD81 in Western blot analysis.

4. Conclusions

Preliminary studies demonstrated that cell count and Bradford assay expression were drop rapidly with 1 mJ/CM^2 doses of ultraviolet C irradiation, and the same result exposed to another doses. Using the Bradford Assay to measure the protein concentration of the concentration exosomes, and with the Western blot analysis to analyze the expression levels of CD9, CD63, and CD81. The results of comparing the two measurement methods both reach the maximum at the 1 mJ/CM^2 doses of ultraviolet A/B irradiation. This study introduces a comprehensive approach to ultraviolet A/B irradiation at the 1 mJ/CM^2 dose, which can maximum enhancement the secretion of exosomes by HEK-293T targeted vectors for KRAS inhibition in pancreatic cancer.

Author Contributions: Conceptualization, C.-C.C., P.L., R.-H.C., Y.X. and Y.-J.L.; methodology, C.-C.C., P.L., R.-H.C. and Y.-J.L.; software, C.-C.C., P.L. and R.-H.C.; validation, C.-C.C., P.L., R.-H.C. and Y.-J.L.; formal analysis, C.-C.C. and P.L.; investigation, C.-C.C. and P.L.; resources, C.-C.C., P.L., R.-H.C. and Y.-J.L.; data curation, C.-C.C. and P.L.; writing—original draft preparation, C.-C.C.; writing—review and editing, C.-C.C.; visualization, C.-C.C.; supervision, C.-C.C.; project administration, C.-C.C.; funding acquisition, C.-C.C. All authors have read and agreed to the published version of the manuscript.

Funding: This research received no external funding.

Institutional Review Board Statement: Not applicable.

Informed Consent Statement: Not applicable.

Data Availability Statement: Data are contained within the article.

Conflicts of Interest: The authors declare no conflict of interest.

References

1. Henry, B.M.; Skinningsrud, B.; Saganiak, K.; Pękala, P.A.; Walocha, J.A.; Tomaszewski, K.A. Development of the human pancreas and its vasculature—An integrated review covering anatomical, embryological, histological, and molecular aspects. *Ann. Anat.* **2019**, *221*, 115–124. [CrossRef] [PubMed]
2. Baert, A.L.; Delorme, G.; Van Hoe, L. (Eds.) The pancreas: Normal radiological anatomy and variants. In *Radiology of the Pancreas*; Springer: Berlin/Heidelberg, Germany, 1999; pp. 19–68.
3. Liu, J.; Ye, Z.; Xiang, M.; Chang, B.; Cui, J.; Ji, T.; Zhao, L.; Li, Q.; Deng, Y.; Xu, L.; et al. Functional extracellular vesicles engineered with lipid-grafted hyaluronic acid effectively reverse cancer drug resistance. *Biomaterials* **2019**, *223*, 119475. [CrossRef] [PubMed]
4. Shen, Z.; Sun, J.; Shao, J.; Xu, J. Ultraviolet B irradiation enhances the secretion of exosomes by human primary melanocytes and changes their exosomal miRNA profile. *PLoS ONE* **2020**, *15*, e0237023. [CrossRef] [PubMed]
5. Wang, C.; Li, N.; Li, Y.; Hou, S.; Zhang, W.; Meng, Z.; Wang, S.; Jia, Q.; Tan, J.; Wang, R.; et al. Engineering a HEK-293T exosome-based delivery platform for efficient tumor-targeting chemotherapy/internal irradiation combination therapy. *J. Nanobiotechnol.* **2022**, *20*, 247. [CrossRef] [PubMed]

engineering proceedings

MDPI

Proceeding Paper

RoBuCACO: ChatGPT-Based Educational Model for Creative Problem-Solving †

Jung-Suk Hyun [1] and Chan-Jung Park [2,*]

1 Department of Management Information Systems, Jeju National University, Jeju-si 63243, Jeju-do, Republic of Korea; jshyun@jejunu.ac.kr
2 Department of Computer Education, Jeju National University, Jeju-si 63243, Jeju-do, Republic of Korea
* Correspondence: cjpark@jejunu.ac.kr
† Presented at the 2024 IEEE 7th International Conference on Knowledge Innovation and Invention, Nagoya, Japan, 16–18 August 2024.

Abstract: Generative artificial intelligence (AI), including ChatGPT4o, is increasingly used across various sectors such as education. In this article, we introduce a new educational model, RoBuCACO, which combines the Butterfly Model for creative problem-solving in a ChatGPT-based framework. ChatGPT is trained on the Butterfly Model using Korean patent data to generate patent metadata. Users follow a structured learning process that includes the definition of roles for ChatGPT (Ro), learning the Butterfly Model (Bu), defining problems and contradictions (C), developing both abstract (A) and concrete solutions (C), and refining optimal (O) solutions. Korean patent metadata are used to obtain concrete solutions and collaborate with ChatGPT to iteratively refine optimal solutions.

Keywords: education model; contradiction; creative thinking; Butterfly Model; ChatGPT

1. Introduction

The history of science and technology is full of examples where contradictions were resolved to bring about innovation. For instance, James Watt solved the problem of needing a cylinder to be both hot and cold, Siemens addressed heating and cooling in the continuous process of flat glass production, Florey and Chain solved Fleming's problem of failing to extract penicillin by heating, Auguste Piccard was the first to ascend to the stratosphere in a closed gondola, Henry Bessemer developed a furnace that allowed input of iron ore while blocking external air, all resolving contradictory problems [1–4].

In art, literature, and logic, contradiction is also used as an important theme and methodology [5]. In the creative research of the natural and social sciences, the method of proof by contradiction is used to verify which theories are correct [6]. The paintings of René Magritte and the works of Gottfried Helnwein express contradictions between conflicting images and imaginations [7]. In literature, contradictions are used to reveal the conflicts and inner turmoil of characters, as presented in Shakespeare's *Hamlet* and Victor Hugo's *Les Misérables*, creating tension in pivotal moments when the protagonist faces a difficult problem. In the academic world, Georg Cantor, who proved the set size of the real numbers; Kurt Gödel with his incompleteness theorems; and Alan Turing, who developed the halting problem, all used proof by contradiction to validate their theories [8].

Despite the importance of contradictory problems, Western scholarship, which values rationality, has traditionally viewed contradictions as incorrect or impossible [9]. However, Genrich Altshuller and his colleagues in Russia, who were familiar with dialectics, readily accepted contradictions [10]. They realized that innovation occurs when contradictions are resolved, after analyzing creative patent specifications. They established TRIZ, an acronym for the Russian phrase meaning "theory of inventive problem solving", which looks at innovative problem-solving from the perspective of resolving contradictions. Owing to its unique perspective, TRIZ has developed its theories inductively, not deductively [11,12].

On the other hand, ChatGPT possesses exceptional abilities in translation and summarization [13]. It is particularly effective in translating and summarizing patent documents written in English, Chinese, Japanese, and Korean. In addition, it provides expert-level analysis similar to that of a professional patent attorney for basic technical questions and questions related to technology classification. Patent data are a treasure of human intellect and a form of big data, encapsulating the innovative journey of problem-solving that connects the past, present, and future. Previously, patent data were analyzed manually by professional patent attorneys, but with the advent of ChatGPT, it is possible to process and analyze patent big data [14]. Moreover, the analysis of patent big data using ChatGPT can derive concrete problem-solving solutions through human–computer interaction.

This study aimed to develop an educational model, RoBuCACO, that facilitates creative problem-solving by teaching students and ChatGPT the Butterfly Model for resolving contradictions. In the proposed model, we define a procedure where ChatGPT serves as a co-pilot to assist students in solving their problems. Introducing abstract logic symbols in the problem-solving process can succinctly represent the essence of the problem, allowing for easy resolution while avoiding trial and error. Defining symbols also prevents variations in interpretation among people, overcoming the ambiguity of everyday language. We define the components of contradiction problems and identify types of such problems based on the Butterfly Model utilizing propositional logic and symbolic logic to represent problems and derive generalized solution strategies. This enables students to specifically determine the appropriate solution strategies for the types of contradiction problems they need to address. By establishing the correct problem-solving strategies, it is possible to reduce the solution space and minimize trial and error in problem-solving [15]. The Butterfly Model offers a tool–object analysis that distinguishes types of contradiction problems and explores resources for problem-solving. By applying the results of a tool–object analysis along with the principles of resource utilization at the problem site, the scope of the problem space is reduced, enabling more efficient problem-solving [16–18].

In this research, the following research procedure was adopted: We trained ChatGPT on the contradiction-solving Butterfly Model and used South Korea's KIPRIS PLUS system to gather patent data related to a specific research topic. Additionally, we provided an educational model that defines the content students need to learn and the procedures for interacting with ChatGPT. Finally, we validated the procedure and described the significance of the educational model.

2. Butterfly Model for Contradiction Resolution

2.1. Butterfly Model

The contradiction-solving Butterfly Model, originally developed by the authors in [15,16], provides a structured approach designed to identify and resolve contradictions encountered in problem-solving. This model was built upon fundamental principles observed in various contradiction-resolution frameworks, which analyze core elements of a system: the *"desirable function (d)"*, the *"system state (s)"*, and the *"harmful function (h)"*. The desirable function (d) represents the primary objective or function of a system, such as improving energy efficiency or enhancing durability. The system state (s) refers to the current system conditions affecting performance, including physical parameters like temperature, pressure, and velocity. The harmful function (h) describes unintended side effects or negative consequences that may arise from changes in the system state [19–21].

The contradiction-solving Butterfly Model consists of a Butterfly Diagram for defining contradictions, a contradiction-problem-type matrix based on the Butterfly Diagram, and a Butterfly Algorithm derived from the matrix [17,18]. The Butterfly Diagram categorizes different problem types and suggests potential resolution strategies. The contradiction-problem-type matrix is shown in Table 1. The Butterfly Algorithm is structured as a tree and provides abstract solution strategies for given problems.

Table 1. Contradiction-problem-type matrix.

Title 1	$s \rightarrow h$	s ă$\leftarrow h$	$s \leftrightarrow h$
$d \rightarrow s$	$(d \rightarrow s) \wedge (s \rightarrow h)$	$(d \rightarrow s) \wedge (s \leftarrow h)$	$(d \rightarrow s) \wedge (s \leftrightarrow h)$
$d \leftarrow s$	$(d \leftarrow s) \wedge (s \rightarrow h)$	$(d \leftarrow s) \wedge (s \leftarrow h)$	$(d \leftarrow s) \wedge (s \leftrightarrow h)$
$d \leftrightarrow s$	$(d \leftrightarrow s) \wedge (s \rightarrow h)$	$(d \leftrightarrow s) \wedge (s \leftarrow h)$	$(d \leftrightarrow s) \wedge (s \leftrightarrow h)$

2.2. Butterfly Diagram

The Butterfly Diagram visually represents the relationships between the three elements (d, s, and h) and serves as a tool for contradiction analysis. This diagram illustrates how desired functions interact with system states and how these interactions can lead to undesired effects. This diagram likens the elements d, s, and h to the wings and body of a butterfly. The left wing shows the desirable function (d) and the changes in the system state (s) it causes, while the right wing displays the harmful functions (h) that may arise from the system state. The body of the butterfly illustrates how these elements interact, and strategies to adjust these interactions are sought to resolve the problem when necessary. According to the contraposition law in logic, $(d \rightarrow s) \equiv (\sim s \rightarrow \sim d)$, $(s \rightarrow h) \equiv (\sim h \rightarrow \sim s)$, and $(d \rightarrow h) \equiv (\sim h \rightarrow \sim d)$. This process is illustrated in Figure 1.

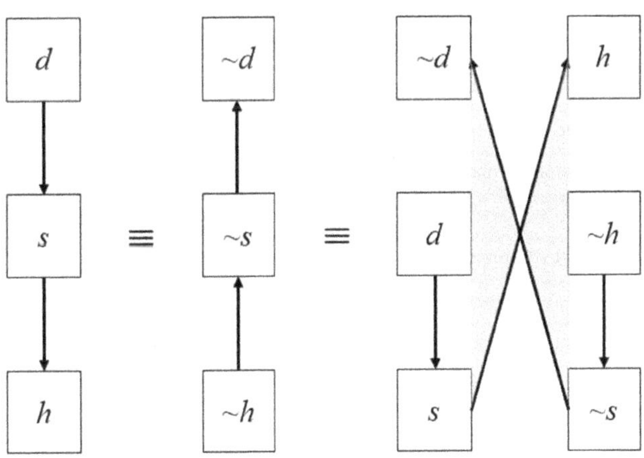

Figure 1. Butterfly Diagram.

Once a contradiction type is classified using the Butterfly Diagram, the problem-solving objective and abstract strategy can be determined, as summarized in Table 2. Then, the accuracy of the Butterfly Model is verified through empirical validation [15,16].

Table 2. Abstract problem-solving strategies.

Contradiction Problem Types	Problem-Solving Goal	Abstract Problem-Solving Strategy
$(d \rightarrow s) \wedge (s \rightarrow h)$	$d \oplus \sim h$	$s \oplus \sim s$
$(d \leftarrow s) \wedge (s \rightarrow h)$	$d \wedge \sim h$	$d \wedge \sim s$
$(d \leftrightarrow s) \wedge (s \rightarrow h)$	$d \oplus \sim h$	$s \oplus \sim s$
$(d \rightarrow s) \wedge (s \leftarrow h)$	$d \wedge \sim h$	$s \wedge \sim h$
$(d \leftarrow s) \wedge (s \leftarrow h)$	$d \wedge h$	$d \wedge s$
$(d \leftrightarrow s) \wedge (s \leftarrow h)$	$d \wedge \sim h$	$s \wedge \sim h$
$(d \rightarrow s) \wedge (s \leftrightarrow h)$	$d \oplus \sim h$	$s \oplus \sim s$
$(d \leftarrow s) \wedge (s \leftrightarrow h)$	$d \wedge \sim h$	$d \wedge \sim s$
$(d \leftrightarrow s) \wedge (s \leftrightarrow h)$	$d \oplus \sim h$	$s \oplus \sim s$

2.3. Butterfly Algorithm

The Butterfly Algorithm, based on the abstract problem-solving strategies derived from the Butterfly Diagram, applies the TRIZ separation principle and the tool–object analysis principle to propose the following specific problem-solving strategies (see also the following Algorithm 1) [19,22]. If the Butterfly algorithm is trained on ChatGPT, it is necessary because it helps generative AIs, like ChatGPT, achieve the objectives of this research more accurately.

Algorithm 1. Butterfly Algorithm

```
If (a conflict occurs at the same time) then
    if (the number of components causing the problem is single) then
        if (the component can be partially removed) then
            DO prior extraction of the component
        else DO alternative selection
    else
        if (the components serve a single purpose) then
            DO division and combination of components decomposing
a single system
        else
            if (there is homogeneity among the components) then
                DO division and combination of components for the whole/part
            else
                DO division and combination of components forming multiple
systems
else
    if (the number of components causing the problem is single) then
        if (the component can be partially removed) then
            DO posterior extraction of the component
        else DO time division and combination
    else
        if (the components serve a single purpose) then
            DO division and combination of components decomposing a single
            system
        else
            if (there is homogeneity among the components) then
                DO division and combination of components for the whole/part
            else DO division and combination of components forming multiple
systems
```

3. Educational Model: RoBuCACO

The educational model proposed in this study consists of the following steps: First, students define the role of ChatGPT in solving problems. Second, ChatGPT summarizes patent data to distinguish causes and effects, from which it identifies objectives and means while defining the structure of the metadata. Furthermore, it constructs a matrix of specific tools and targets derived from the defined objectives and means. Third, both students and ChatGPT are trained to understand the contradiction-solving Butterfly Model theory. Based on this, students define problems, and with the help of ChatGPT, they define contradictions. Fourth, based on the defined contradictions, students are encouraged to propose their problem-solving strategies. Fifth, the problem-solving strategies are presented to ChatGPT, which then uses the Butterfly Model for contradiction resolution, patent metadata, and the tool–object matrix to devise creative problem-solving alternatives. Sixth, students review the alternatives prepared by ChatGPT and generate improved problem-solving solutions based on the learned methods. Through iterative refinement, they present the best problem-solving solutions.

3.1. Role Definition (Ro)

To propose an educational model based on ChatGPT, students define the roles that ChatGPT must perform, such as patent specification translation and summarization, contradiction problem definition, and the suggestion of creative problem-solving strategies.

- ChatGPT is used to understand the context of the text and extract important information.
- ChatGPT defines the desirable function, the state resulting from the function, and the problems that the state causes in order to define contradictions.
- ChatGPT suggests various problem-solving strategies.

3.2. Learning About the Butterfly Model (Bu)

Students learn about the Butterfly Model to understand the given problem situation and creatively devise solutions from the perspective of contradiction resolution. In this process, they study types of contradictions, create Butterfly Diagrams, learn the separation principles, and study the 40 inventive principles [23].

3.3. Problem and Contradiction Definition (C)

In this research, to develop problem-solving strategies for creative problem-solving based on the contradiction-solving Butterfly Model, ChatGPT is trained on the theory of the Butterfly Model for contradiction resolution. When students who are not yet familiar with defining contradictions present a contradiction problem, ChatGPT assists in defining the contradiction. Additionally, it helps students create a Butterfly Diagram based on this definition. The students then develop problem-solving strategies based on the Butterfly Diagram [24].

3.4. Abstract Solution Strategy Definition (A)

Based on the contradictions and Butterfly Diagram defined by ChatGPT, the students use the Butterfly Model to identify abstract solution strategies on their own. By having the students understand the given problem from the perspective of contradictions on their own, they can gain a direction for problem-solving, which reduces trial and error [19].

3.5. Concrete Solution Strategy Finding (C)

Based on the learned patent data and the contradiction-solving Butterfly Model, ChatGPT assists students in developing problem-solving strategies and proposes creative solutions. This allows the students to evaluate their solutions independently and continuously improve new solutions. In this stage, students present the problem-solving directions they have identified, and ChatGPT, based on these directions, proposes specific solutions using the patent metadata it possesses.

3.6. Refinement for Finding the Optimal Solution (O)

Once ChatGPT proposes specific solutions, students review the alternatives and enhance the solutions through possible collaboration. This process is performed iteratively, allowing for interactions with ChatGPT to develop solutions from a creative perspective.

4. Generation and Creative Problem-Solving with RoBuCACO

KIPRIS Plus is a website that provides patent information utilization services in South Korea. It allows users to directly download data or use APIs to access all patent information made available by the patent office in real time (Figure 2). In this research, we used KIPRIS Plus to collect patent data and construct metadata. We used patents from KIPRIS Plus with an open API key for free, allowing up to 1000 inquiries per year.

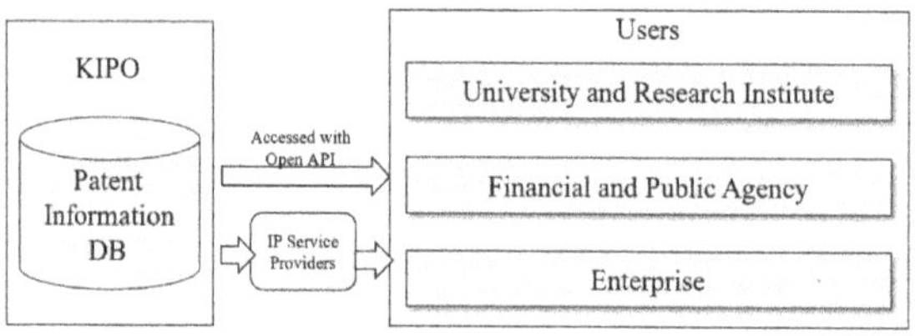

Figure 2. KIRPIS PLUS.

Next, to find problem-solving strategies for specific situations, patent data were summarized and analyzed using ChatGPT, and patent metadata were constructed. The process involved providing patent documents to ChatGPT to complete the analysis of cause and effect, objectives and means, as well as tools and targets. Metadata were then constructed with attributes for cause, effect, objective, means, tool, target, and keywords. After training ChatGPT on the Butterfly Model, a search for patents related to "electric scooter safety" was conducted. The analysis results of the patent documents showed contradictions appropriately, and the problem's cause, effect, objective, methods, and targets were identified (Figure 3).

The five alternatives were extracted from one of the uploaded patents as follows:

- Safety System: The patent discusses a safety system specifically designed for electric scooters.
- Detection System: Various detection systems, such as helmet-wearing and alcohol-level detection, are mentioned.
- Helmet Detection: A system for detecting whether a helmet is being worn by the rider.
- Alcohol Detection: The document details a system for detecting the alcohol level of the rider.
- Electric Scooter: The document focuses on technologies related to electric scooters, particularly their operation and safety features.

With the metadata constructed using ChatGPT, we defined contradictions to develop abstract solution strategies for new problems. A problem was solved to determine a support structure for young trees. While young trees need strong supports to prevent them from falling over during typhoons, such supports can hinder their growth. Therefore, we needed to create a support that would not inhibit growth while still preventing trees from falling over during strong winds. Since the learning of roles and the Butterfly Model was complete, the problem-solving process was initiated by defining the contradiction [17,18]. ChatGPT defined a contradiction as shown in Figure 4.

Public Patent 10-2023-0052417

Korean Intellectual Property Office(KR)
Public Patent Report (A)

(11 Publication No. 10-2023-0052417
(43 Publication Date 2023/04/20

(51) International Patent Classification
B62K 3/00 (2006.01) **B62J 45/41** (2020.01)
B62M 6/45 (2010.01)
(52) CPC Patent Classification
B62K 3/002 (2013.01)
B62J 45/41 (2022.08)
(21) Application Number **10-2021-0135465**
(22) Application Date 2021/10/13
Examination Request Date 2021/10/13

(71) **Applicant**
Hannam University Industry-Academic Cooperation Group
Yuseong-gu, Daejeon Metropolitan City
(72) **Inventor**
Chashin
Daejeon Metropolitan City, Daedeok-gu, Hannam-ro
Lee Yun-jae
Gyeonggi-do, Gwangju-si, Munhwa-ro
(74) **Agent**
Patent Law Firm Hyunmun

Total number of claims: 4
(54) **Title of Invention: Electric Scooter Rider Safety Protection System**

(a)

1. **Cause-Effect Analysis**:
   - **Cause**: Issues with non-compliance with safety regulations in road traffic, especially problems related to not wearing safety gear and drunk driving when using electric scooters.
   - **Effect**: Increased accident rates and resulting safety risks for users and pedestrians due to non-compliance with safety regulations.
2. **Objective-Means Analysis**:
   - **Objective**: To develop a system that enhances the safety of electric scooter riders and ensures compliance with traffic laws.
   - **Means**: Features such as helmet-wearing detection systems, alcohol level detection systems, and the detection of the rider's weight and position, which automatically check for compliance with safety regulations and take appropriate actions.
3. **Tool-Target Analysis**:
   - **Target**: Electric scooter riders and the electric scooter itself.
   - **Tool**: Sensors (for detecting helmet use, alcohol levels, and the weight and position of the rider), and computing units (software that decides on actions based on the detected information).

Butterfly Model's W, S, U:

- **Wanted Function (W)**: Ensuring safety and compliance with regulations while riding electric scooters.
- **State (S)**: The current operational state of the electric scooter and the behavior of the rider (whether helmets are worn, alcohol levels).
- **Unwanted Function (U)**: Occurrences of legal violations and safety hazards during the use of electric scooters.

(b)

Figure 3. Results generated from ChatGPT. (**a**) Patent upload. (**b**) Partial metadata for Korean patents.

Wanted Function (W): The support needs to be strong enough to prevent young trees from falling over during typhoons. This function focuses on providing sufficient stability to withstand extreme weather conditions.

State (S): The presence of a strong support structure for the young trees. This state represents the current condition where the trees are supported by structures intended to protect them from strong winds.

Unwanted Function (U): The strong support might inhibit the natural growth of the young trees. Although the supports are designed to offer stability, their rigidity or intrusive nature could restrict the tree's growth and development.

Figure 4. Contradictions defined by ChatGPT.

From the given W, S, and U, we drew a Butterfly Diagram for the problem [20]. Also, conflict identification and component analysis were defined as shown in Figure 5:

1. **Conflict Identification:**
   - **Conflict occurs at the same time**: The need for robust support to prevent young trees from falling during strong winds and the need to not hinder the tree's growth.
2. **Component Analysis:**
   - **Single component causing the problem**: The support structure itself.

Figure 5. Conflict identification and component analysis.

From the analysis, the abstract solution strategy for the problem was suggested as shown in Figure 6:

- **If the component can be partially removed**:
  - **Prior extraction of the component**: Design the support with adjustable or removable parts. For example, develop a support that can be adjusted in height and breadth as the tree grows. This could involve a telescopic design that can expand without being fully removed.

Figure 6. Abstract solution strategy.

Partially, ChatGPT described the method to implement the prior extraction of the component. Finally, a concrete solution was given for the problem. As a result, the following solutions were provided by ChatGPT.

- Adjustable Rings: Utilize adjustable support rings that can be expanded as the tree grows in girth. These rings should be attached to vertical stakes that can also be adjusted for height.
- Flexible Material: Construct the support rings and vertical stakes using a flexible material such as high-density polyethylene (HDPE) which has enough strength to withstand strong winds but enough flexibility to move slightly with the tree. This flexibility minimizes the risk of damaging the bark or restricting growth.
- Modular Setup: Design the support structure in segments or modules that can be easily added or removed. As the tree grows taller and thicker, additional modules can be attached to accommodate this growth without needing to replace the entire system.

The Butterfly Model was applied to solve a problem and invent a "one-way moving cable tie", for which we won a gold medal at an invention competition in Korea. The one-way moving cable tie acts as a strong support while allowing expansion only in the direction of the tree's growth. It does not stretch in the opposite direction. Considering the method, while ChatGPT may not have provided a specific method, it suggested the direction of problem-solving, potentially reducing the trial-and-error process that people might experience [6].

5. Conclusions

In this paper, we proposed RoBuCACO, an educational model for creative problem-solving based on ChatGPT. This model is built upon the Butterfly Model, which defines problems from the perspective of contradictions and provides directions for problem-solving while minimizing trial and error [15–19]. Although various creative problem-solving methods exist, the Butterfly Model serves as the foundation of the proposed educational model, as it uniquely guides users toward concrete solutions, distinguishing it from other approaches. Throughout this process, the model identifies similar patents related to the problem, constructs metadata for problem-solving, and utilizes ChatGPT to generate specific solutions. The problem-solver interacts iteratively with ChatGPT until the optimal solution is found [22,23]. This paper analyzed how ChatGPT contributes to the creative problem-solving process and confirmed its positive aspects. Since this research was conducted with students, it is essential to verify the educational effectiveness of the model after training [24]. The significance of this research lies in demonstrating the potential of generative AI, such as ChatGPT, to assist humans in creative problem-solving.

Author Contributions: Conceptualization, J.-S.H.; methodology and software, C.-J.P.; validation, J.-S.H.; writing-original draft preparation, C.-J.P.; writing-review and editing, J.-S.H. and C.-J.P. All authors have read and agreed to the published version of the manuscript.

Funding: This research received no external funding.

Institutional Review Board Statement: Not applicable.

Informed Consent Statement: Not applicable.

Data Availability Statement: The data used in the paper was dynamically and automatically generated by keyword search on the KIPRIS site. All users can access it through search on the website, https://plus.kipris.or.kr/eng/main.do, and the patent full-text can be downloaded as PDF and used as data.

Conflicts of Interest: The authors declare no conflict of interest.

References

1. Ikenson, B. *Patents: Ingenious Inventions How They Work and How They Came to Be*; Running Press: Philadelphia, PA, USA, 2012.
2. Watkins, P. *From Hell Island to Hay Fever*; Brown Dog Books: Bristol, UK, 2018.
3. Piccard, A. *Earth, Sky, and Sea*; Oxford University Press: Oxford, UK, 1956.
4. Bessemer, H. *Sir Henry Bessemer F R S: An Autobiography*; Offices of Engineering: Tokyo, Japan, 1905.
5. Pelkey, J. Greimas embodied: How kinesthetic opposition grounds the semiotic square. *Semiotica* **2017**, *214*, 277–305.
6. Polya, G. *How to Solve It: A New Aspect of Mathematical Method*; Princeton University Press: Princeton, NJ, USA, 1971.
7. Helnwein, G. Gespräch Mit Gottfried Helnwein Anlässlich der Eröffnung Seines Virtual Museums of Art in "Second Life". Available online: https://www.helnwein-museum.com/ (accessed on 5 February 2024).
8. Rucker, R. *Mind Tools: The Five Levels of Mathematical Reality*; Dover Publications, Incorporated: Mineola, NY, USA, 2013.
9. Peng, K.; Nisbett, R.E. Culture, dialectics, and reasoning about contradiction. *Am. Psychol.* **1999**, *54*, 741–754. [CrossRef]
10. Altshuller, G. *Creativity as an Exact Science: The Theory of the Solution of Inventive Problems*; CRC Press: Boca Raton, FL, USA, 1984.
11. Fey, V.; Rivin, E. *Innovation on Demand: New Product Development Using TRIZ*; Cambridge University Press: Cambridge, UK, 2005.
12. Zlotin, B.; Zusman, A.; Altshuller, G.; Philatov, V. *Tools of Classical TRIZ*; Ideation International Inc.: Southfield, MI, USA, 1999.
13. Zia-Ul-Haq, M.; Mujtaba, G. ChatGPT and patent eegime: Key legal concerns. *Trends Intellect. Prop. Res.* **2023**, *1*, 18–26.
14. Borger, J.G.; Ng, A.P.; Anderton, H.; Ashdown, G.W.; Auld, M.; Blewitt, M.E.; Brown, D.V.; Call, M.J.; Freytag, S.; Harrison, L.C.; et al. Artificial intelligence takes center stage: Exploring the capabilities and implications of ChatGPT and other AI-assisted technologies in scientific research and education. *Immunol. Cell Biol.* **2023**, *101*, 923–935. [CrossRef] [PubMed]
15. Hyun, J.S.; Park, C.J. Logical interpretation about problem types and solution strategies of the butterfly model for the automation of contradiction-based problem solving. In Proceedings of the 2014 IEEE International Conference on Teaching, Assessment and Learning for Engineering (TALE), Wellington, New Zealand, 8–10 December 2014; pp. 236–241.
16. Hyun, J.S.; Park, C.J. The butterfly algorithm: A contradiction solving algorithm based on propositional logic for TRIZ. *Int. J. Softw. Eng. Its Appl.* **2016**, *10*, 27–34.
17. Rajić, D. Mathematical-physical model of solving inventive problems. *FME Trans.* **2021**, *49*, 726–733. [CrossRef]
18. Rajić, D. Innovative synergism as a result of TRIZ and LT-system synthesis. In *Innovation as an Initiator of the Development "Innovations—Development Prospects"*; Faculty of Applied Management, Economy and Finance: Belgrade, Serbia, 2019; pp. 226–242.
19. Govindarajan, U.H.; Sheu, D.D.; Mann, D. Review of systematic software innovation using TRIZ. *Int. J. Syst. Innov.* **2019**, *5*, 72–90.
20. Jani, H.M. Teaching TRIZ problem-solving methodology in higher education: A review. *Int. J. Sci. Res.* **2013**, *2*, 98–103.

21. Wang, F.; Peng, Q.; Tan, R.; Wang, K.; Zhang, S.; Sun, J. An attribute-based method for identifying implicit contradictions in product development. *Comput.-Aided Des. Appl.* **2022**, *20*, 439–455.
22. Hyun, J.S.; Park, C.J. The Butterfly algorithm eliminating the brute-force mechanism in ARIZ for contradiction problem solving. In Proceedings of the 2015 8th International Conference on Security Technology (SecTech), Jeju, Republic of Korea, 25–28 November 2015; pp. 25–27.
23. Hyun, J.S.; Park, C.J. Innovative contradiction solving dimensions and algorithms. *J. Adv. Inf. Technol. Converg.* **2017**, *7*, 1–10.
24. Hyun, J.S.; Park, C.J. Effects of the Butterfly diagram education in primary and secondary schools on contradiction problem solving ability. In Proceedings of the 2015 IEEE Frontiers in Education Conference (FIE), El Paso, TX, USA, 21–24 October 2015; pp. 1–8.

engineering proceedings

MDPI

Proceeding Paper

The (*n*, *n*) Visual Multi-Secrets Sharing Scheme with Abilities of OR and XOR Decryption †

An-Hui Lo and Justie Su-Tzu Juan *

Department of Computer Science & Information Engineering, National Chi Nan University, Nantou 54561, Taiwan; s111321513@mail1.ncnu.edu.tw
* Correspondence: jsjuan@ncnu.edu.tw

† Presented at the 2024 IEEE 7th International Conference on Knowledge Innovation and Invention, Nagoya, Japan, 16–18 August 2024.

Abstract: Naor and Shamir introduced the fundamental concept of visual cryptography (VC) in 1994. In that model, the secret image is split into two meaningless shares, allowing the secret to be revealed and recognized by the human eye just by superimposing the two shares. Since then, many scholars have studied the VC problem. To improve the efficiency of transmitting secrets, multi-secret visual cryptography has been proposed to encrypt multiple secret images at the same time. On the other hand, with the reduction in hardware costs, research on XOR-based VC has become popular to address the issue of poor image quality in the recovered image of OR-based visual cryptography, though it requires computing equipment. Scholars also have developed VC schemes utilizing OR-based decryption (equivalent to traditional VC) and XOR-based decryption. These schemes recover secrets without additional hardware (using OR-based decryption) and provide higher-quality images when extra hardware is available (using XOR-based decryption). We propose a visual multi-secret sharing scheme (VMSSS) to encrypt multiple secret images into n (>1) shares. When all shares are collected, all original secret images are decrypted by using the OR or XOR operation.

Keywords: visual cryptography; secret sharing; OR operation; XOR operation; multi-secret

1. Introduction

Visual Cryptographic Scheme (VCS) is a method designed for securely sharing secret images. In 1994, Naor and Shamir introduced the widely recognized (t, n)-threshold VCS model, commonly referred to as (t, n) VCS [1]. This scheme encodes a binary secret image into n individual shares, each printed on a transparent sheet. The original image remains completely unrecoverable if fewer than t shares are combined. The decryption process is straightforward—when any t or more shares are overlaid using the 'OR' operation, the hidden image becomes visible to the human eye without requiring computational processing. Since no additional calculations or equipment are required for decryption, visual cryptography has many applications in life, such as lottery activities. Recently, many studies on merging with QR codes have been proposed.

Early implementations of VCS relied on pixel expansion, where each pixel in the secret image was mapped to multiple pixels in the generated shares. This approach inevitably led to distortion and an increase in image size upon reconstruction. This means that as more and more shares are formed, pixel expansion will make this approach practically infeasible (it could scale up a hundred or even a thousand times). To address this issue, Kafri and Keren proposed a secret-sharing technique based on the Random Grid (RG) method, which allows for black-and-white image encryption without pixel expansion [2]. Since their contribution, visual cryptography has continued to evolve, with researchers exploring new techniques to enhance efficiency and visual quality.

A visual multi-secret sharing scheme (VMSSS) differs from the previously mentioned VCS, as it simultaneously encrypts more than one secret into the same shares [3]. It reduces extra costs and enhances the efficiency of encryption. Many researchers have studied VMSSS in the past few years, focusing on limitations regarding the number and shape of secret images. At first, scholars designed a method to restore multiple images by rotating one of the shares, but this would limit the shape of the image (it must be a square) and the number of images (a maximum of four). Chang et al. [4] proposed a novel random grid-based VMSSS in 2018, which divides each secret image into several parts and then encrypts them into corresponding shares. By shifting one share by different pixels and superimposing it on the other share, the different secret image can be reconstructed separately.

In the context of visual shared secret schemes, contrast serves as the key metric for assessing the quality of the reconstructed image. It represents the distinction between the white and black regions of the secret image as observed in the restored version. Numerous definitions of contrast exist, with the most commonly used being presented in [5]. For any black and white image, the contrast value ranges from -1 to 1. In traditional VCS, the stacking operation is the OR operation, and the contrast of the reconstructed image reaches a maximum of 0.5. This low contrast limits many potential developments. To solve the above problems, XOR-based VCS was proposed [6]. The image reconstructed using the XOR operation can theoretically achieve a contrast value of 1, significantly enhancing its visual quality. However, it requires appropriate hardware equipment. Therefore, scholars have developed VC schemes using OR-based decryption and XOR-based decryption at the same time [7,8]. These schemes can recover secrets without additional hardware (using OR-based decryption) and provide higher-quality recovered images when light hardware is available (using XOR-based decryption).

VMSSS was previously designed as a method to encrypt multiple secret images into two shares, i.e., (2, 2) VCS, e.g., [4]. However, there is still a need for an (n, n) VCS that can encrypt multiple secret images simultaneously and recover the secret image by using OR or XOR operations without pixel expansion. This is useful in situations where more than two parties want to share multiple secrets. This allows for the recovery of secrets without extra hardware via OR-based decryption and offers higher-quality recovered images when additional hardware is used through XOR-based decryption. Huang et al. proposed an (n, n) XOR-based visual multi-secrets sharing scheme [9]. This method addressed the limitation in Chang et al.'s approach [4], which only encrypts the secret images into two shares, i.e., (2, 2) VMSSS, allowing for more than two participants. Although Huang et al.'s scheme secures the capability of XOR decryption, the OR operation can also be used in the decryption phase simultaneously, although the result is relatively poor. However, when $n > 3$, even the reconstruction result of the XOR operation is not satisfactory [9]. Therefore, we designed an improved method to solve this problem.

The remainder of this paper is organized as follows: Sections 2 and 3 cover the preliminaries and main results, respectively, while the conclusion is provided in Section 4.

2. Preliminaries

Huang et al. proposed an (n, n) XOR-based visual multi-secret scheme with meaningful shares [9]. Their approach incorporates three algorithms within the encryption process, whereas the decryption process consists of only a single algorithm. The encryption and decryption process are illustrated in Figures 1 and 2, respectively.

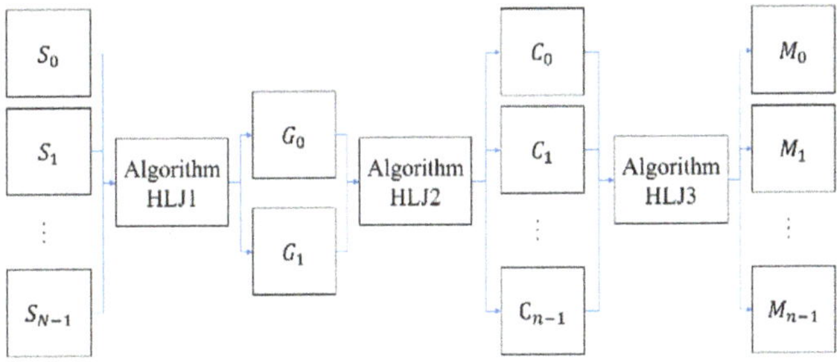

Figure 1. The encryption process model of Huang et al.'s scheme [9].

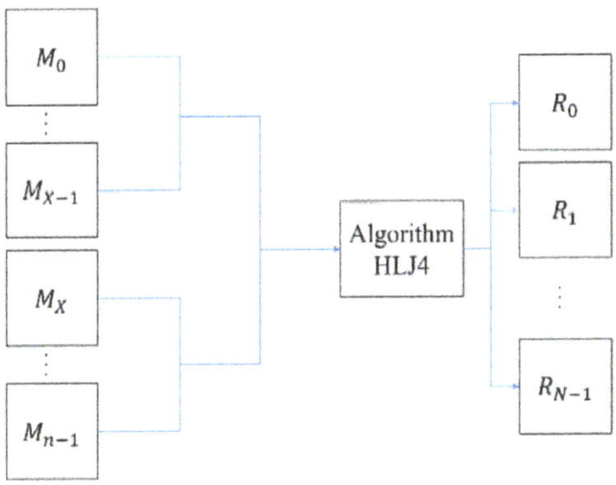

Figure 2. The decryption process model of Huang et al.'s scheme [9].

Reference [9] used Algorithm HLJ1 to enable all secret images to be restored using two shares, then expands these two shares into n shares (Algorithm HLJ2), and finally adds the camouflage algorithm (Algorithm HLJ3). When restoring with this scheme, n shares are stacked by using the XOR operation to restore—after shifting the last $N-X$ shares by $(i-1)w/p$ pixels (Figure 3). All the shares are superimposed to perform the XOR operation and obtain the restored image i. Where N represents the number of secret images, $X = \lceil n/2 \rceil$, w is the width of the image, and p is a parameter that can be selected by the user.

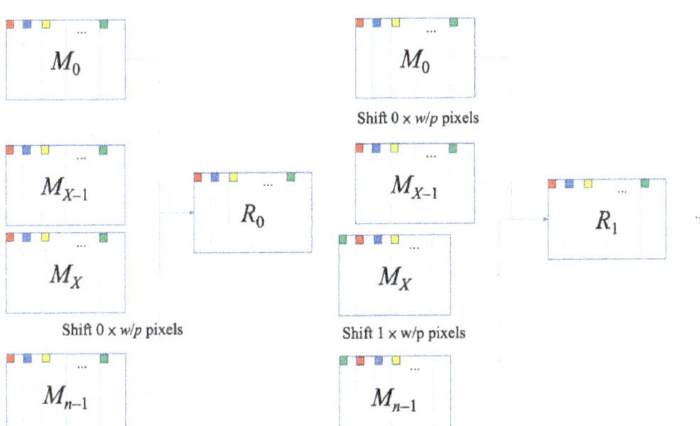

Figure 3. The illustrative diagram of Algorithm HLJ4 [9] (areas with small squares of the same color will be stacked in the same position).

Huang et al.'s scheme uses the parameter γ to control the proportion of camouflage in the shares. A higher γ value makes the shares more similar to the camouflage images. However, when other parameters remain constant, the restoration quality decreases as γ increases. If no camouflage is used ($\gamma = 0$), the restoration quality of the restored image becomes the highest. In addition, this scheme also can restore the secret image by using the OR operations, but the restoration quality becomes poor. Figure 4 presents the experimental results of Huang et al.'s scheme when $\gamma = 0$ with $N = 4$, $n = 4$, $p = 20$, $\gamma = 0$. Even though Huang et al.'s scheme performs well, its distortion increases, especially when N is larger. This is because it first uses the Chang et al.'s scheme [4] to encrypt all secret images into two shares, and then expand them into n shares. This process will cause great distortion, because only two shares can obtain information related to the original secrets, resulting in the fact that most of the pixels of those secret images cannot be accessed. That is, they are not encrypted into the share. Therefore, this paper develops a new scheme with similar goals to the scheme of Huang et al., but because it uses a direct encryption of all secret images into n shares, it can access most of the pixels of those secret images, which will greatly reduce the distortion and achieve better results.

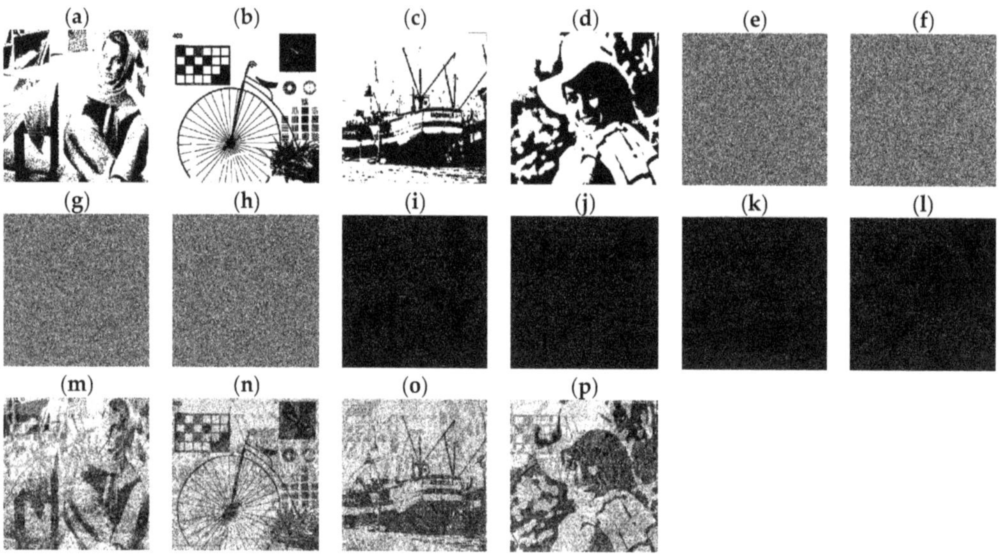

Figure 4. Experimental results by Huang et al.'s scheme with $N = 4$, $n = 4$, $p = 20$, $\gamma = 0$. (**a–d**) Secret images $S_0, \ldots,$ and S_3; (**e–h**) shares $M_0, \ldots,$ and M_3; (**i–l**) OR-restored images R_0, R_1, R_2, and R_3; and (**m–p**) XOR-restored images XR_0, XR_1, XR_2 and XR_3.

3. Main Results

The proposed VMSSS simultaneously encrypts N secret images and distributes n shares to n participants. After collecting the shares of all participants, these N secret images are decrypted using OR or XOR operations simultaneously. Figure 5 illustrates the model of the proposed encryption and decryption process. From the figure, we can clearly see that the proposed scheme directly uses Algorithm 1 to complete the encryption stage, which is relatively simple compared to the method in [9]. The decryption process is just to shift all the collected shares according to the rules separately and then stack them up.

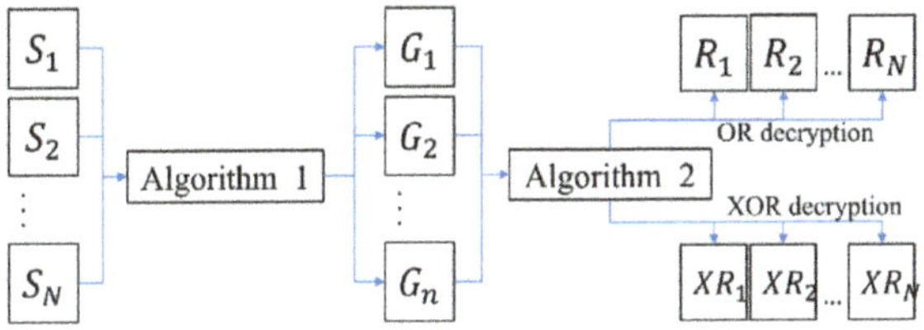

Figure 5. The model of the proposed scheme.

To simplify the expression, some notations and parameters are listed in Table 1. For the ease of comparison, we mostly use the same symbols and definitions as in [9]. This is also adopted by most of the literature.

Table 1. Some notations and parameters used in this work.

Notation	Description
0	A white pixel
1	A black pixel
S_i	The i-th secret image
G_i	The i-th meaningless image
R_i	The i-th reconstructed image by OR operation
XR_i	The i-th reconstructed image by XOR operation
$S(i,j)$	The value of the pixel at position (i,j) in image S
N	The quantity of secret images
n	The quantity of shares
p	The quantity of pieces into which an image will be divided
$\otimes$	OR operation
$\oplus$	XOR operation

3.1. Encryption Algorithm

Due to space limitations, this paper only discusses the case where $N = n$. Algorithm 1 encrypts N secret images (of the same size) on $N = n$ shares. In this way, any N participants can individually obtain their shares. During recovery, if any of them do not hand over the shares they have kept, then any secret image is not recovered. The main idea of Algorithm 1 is to learn from [4] and first design a method to move n shares when restoring the secret images. Then, for a group of pixels, a secret is randomly selected to be encrypted in order. That is, once a secret image and location are selected, the secret image and a whole set of pixels corresponding to the location will be encrypted into the shares at once.

To provide a clear understanding of Algorithm 1, the flowchart is shown in Figure 6. In Algorithm 1, Step 4 calculates the pixel selected at the beginning, which is the (i, j) position of the A-th share, and its result is the XOR result of the corresponding secret image pixel and all randomly generated pixels in Step 3. We divide each share into p equal parts, with the k-th part of the second share (G_1) to the k-th part of the N-th share (G_{N-1}) and the ($k + 1$)-th part of the first share (G_0) to form a group. Before processing the entire group, we complete the calculations for the redundant parts in Step 5 to fully process $p - 2$ groups in Step 6, which amounts to $(p - 2) \times N$ pixels. In Step 7, we calculate the remaining pixels (ranging from A to N, depending on the chosen A). The required calculations are finished once A, i, and j are selected. In other words, selecting A, i, and j generates $N - 1$ randomly produced pixels and $(p - 1) \times N + 1$ calculated pixels.

Algorithm 1. (N, N) Visual N Secret Sharing Scheme

Input: N secret images $S_0, S_1, \ldots, S_{N-1}$ of size $w \times h$ pixels, and a divisor of w: p (which must coprime to $N-1$).

Output: N shares $G_0, G_1, \ldots, G_{N-1}$ of size $w \times h$.

1. Randomly choose a number A, where $0 \leq A \leq N-1$.
2. Randomly choose a pixel (i,j), where $0 \leq i \leq w, 0 \leq j \leq h$.
3. If $A \neq 0$ do
   Randomly generated $N-1$ pixels
   $G_{(A+1)\%N}((i-w/p)\%w, j), G_{(A+2)\%N}((i-w/p)\%w, j), \ldots, G_{N-1}(i-w/p)\%w, j), G_0(i-w/p)\%w, j), G_1(i,j), G_2(i,j), \ldots, G_{A-1}(i,j)$ from $\{0,1\}$, and set
   $Result_{XOR} = G_{(A+1)\%N}((i-w/p)\%w, j) \oplus G_{(A+2)\%N}((i-w/p)\%w, j) \oplus \ldots \oplus G_{N-1}(i-w/p)\%w, j) \oplus G_0((i-w/p)\%w, j) \oplus G_1(i,j) \oplus G_2(i,j) \oplus \ldots \oplus G_{A-1}(i,j)$.
   Else do
   Randomly generated $N-1$ pixels
   $G_1(i,j), G_2(i,j), \ldots, G_{A-1}(i,j)$ from $\{0,1\}$, and set
   $Result_{XOR} = G_1(i,j) \oplus G_2(i,j) \oplus \ldots \oplus G_{A-1}(i,j)$.
4. $G_A(i,j) = S_A(i,j) \oplus Result_{XOR}$.
5. If $A \neq 0$ do
   For $l = 1$ to $N-A$ do
   $Result_{XOR} = Result_{XOR} \oplus G_{(A+l)\%N}((i-w/p)\%w, j)$.
   $G_{(A+l)\%N}(i,j) = S_{(A+l)\%N}(i,j) \oplus Result_{XOR}$.
   $Result_{XOR} = Result_{XOR} \oplus G_{(A+l)\%N}(i,j)$.
6. For $k = 1$ to $p-2$ do
   For $l = 1$ to N do
   $Result_{XOR} = Result_{XOR} \oplus G_{l\%N}((i+(k-1) \times w/p)\%w, j)$.
   $G_{l\%N}((i+k \times w/p)\%w, j) = S_{l\%N}((i+k \times w/p)\%w, j) \oplus Result_{XOR}$.
   $Result_{XOR} = Result_{XOR} \oplus G_{l\%N}((i+k \times w/p)\%w, j)$.
7. If $A = 0$ do
   For $l = 1$ to N do
   $Result_{XOR} = Result_{XOR} \oplus G_{l\%N}((i+(p-2) \times w/p)\%w, j)$.
   $G_{l\%N}((i+(p-1) \times w/p)\%w, j) = S_{l\%N}((i+(p-1) \times w/p)\%w, j) \oplus Result_{XOR}$.
   $Result_{XOR} = Result_{XOR} \oplus G_{l\%N}((i+(p-1) \times w/p)\%w, j)$.
   Else do
   For $l = 1$ to A do
   $Result_{XOR} = Result_{XOR} \oplus G_l((i+(p-2) \times w/p)\%w, j)$.
   $G_l((i+(p-1) \times w/p)\%w, j) = S_l((i+(p-1) \times w/p)\%w, j) \oplus Result_{XOR}$.
   $Result_{XOR} = Result_{XOR} \oplus G_l((i+(p-1) \times w/p)\%w, j)$.
8. Repeat Steps 1–7 until every pixel of shares $G_0, \ldots, G_{N-1}$ has been processed.

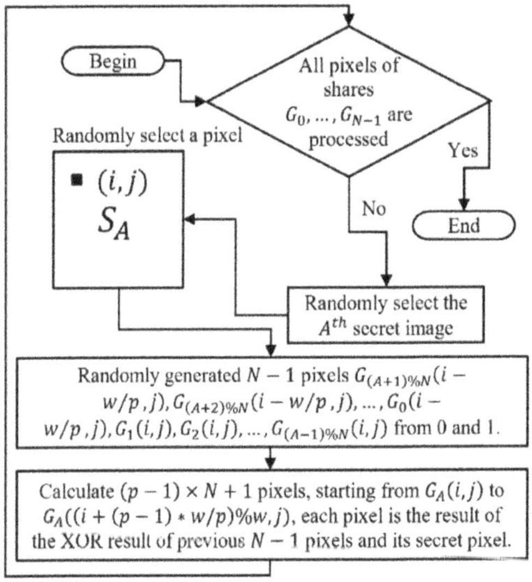

Figure 6. A flow chart of Algorithm 1.

3.2. Decryption Algorithm

Algorithm 2 decrypts to reconstruct the image. To restore any secret image, all shares must be collected and all secret images can be restored at the same time. In Algorithm 2, to reconstruct the initial secret image (S_0), all the shares are directly overlaid. To restore the second secret image (S_1), the second share (G_1) must be shifted cyclically w/p pixels to the left. This means that the leftmost w/p pixels of the second share must be moved and connected to the rightmost and stacked together with the other shares. Similarly, to reconstruct the third secret image (S_2), both the second and third shares (G_1, G_2) need to be shifted cyclically w/p pixels to the left. In general, to restore the t-th secret image, the second to the t-th shares must be shifted cyclically w/p pixels to the left ($1 < t \leq N-1$). The pixels moved to the left are added to the rightmost part of the share, where w is the width of the secret image (=the width of any share), and p is a parameter that is a divisor of the w chosen by the user. The schematic illustrating this process is shown in Figure 7.

Algorithm 2. Decryption Algorithm

Input: N shares $G_0, G_1, \ldots, G_{N-1}$ of size $w \times h$, a divisor of w: p.

Output: N reconstructed images $R_0, \ldots, R_{N-1}$ by OR operation and N reconstructed images $XR_0, \ldots, XR_{N-1}$ by XOR operation of size $w \times h$.

1. For each pixel in $\{(i,j) \mid 0 \leq i \leq w, 0 \leq j \leq h\}$, do Steps 2–3.
2. $R_0(i,j) = G_0(i,j) \otimes G_1(i,j) \otimes \ldots \otimes G_{N-1}(i,j)$.
   $XR_0(i,j) = G_0(i,j) \oplus G_1(i,j) \oplus \ldots \oplus G_{N-1}(i,j)$.
3. For $k = 1 \; to \; N-1$ do
   $R_k(i,j) = G_0(i,j) \otimes G_1((i + w/p)\%w, j) \otimes \ldots \otimes G_k((i + w/p)\%w, j) \otimes G_{k+1}(i,j) \otimes \ldots \otimes G_{N-1}(i,j)$.
   $XR_k(i,j) = G_0(i,j) \oplus G_1((i + w/p)\%w, j) \oplus \ldots \oplus G_k((i + w/p)\%w, j) \oplus G_{k+1}(i,j) \oplus \ldots \oplus G_{N-1}(i,j)$.
   End
4. Return $R_0, \ldots, R_{N-1}$ and $XR_0, \ldots, XR_{N-1}$.

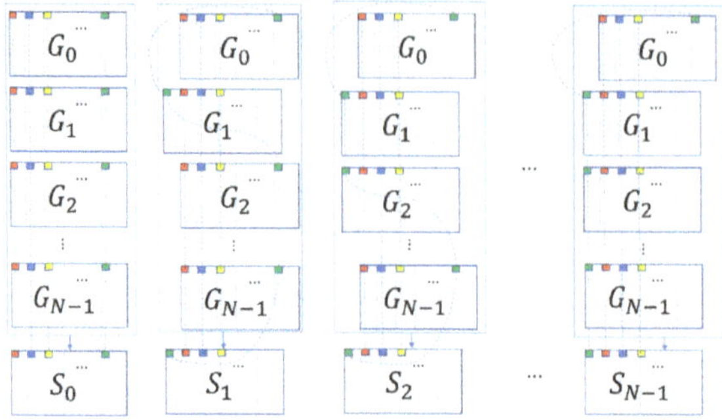

Figure 7. The schematic diagram of Algorithm 2 (areas with small squares of the same color will be stacked in the same position).

3.3. Experimental Results

This subsection provides two experiments for a subsequent comparison with the method proposed in [9]. In the initial experiment, we encrypted three secret images, whereas in the subsequent experiment, four secret images were encrypted. All images measure 500 × 500 pixels in size, with the parameter p consistently set to 20 across both experiments. The three secret images utilized in the initial experiment are depicted in Figure 8a–c, whereas the quartet of secret images employed in the subsequent experiment correspond to those illustrated in Figure 4a–d. Figure 9 shows the first experiment result by the proposed (n, n) VMSSS. Where the OR-reconstructed image refers to the image reconstructed through the OR operation, the XOR-reconstructed image refers to the image reconstructed through the XOR operation.

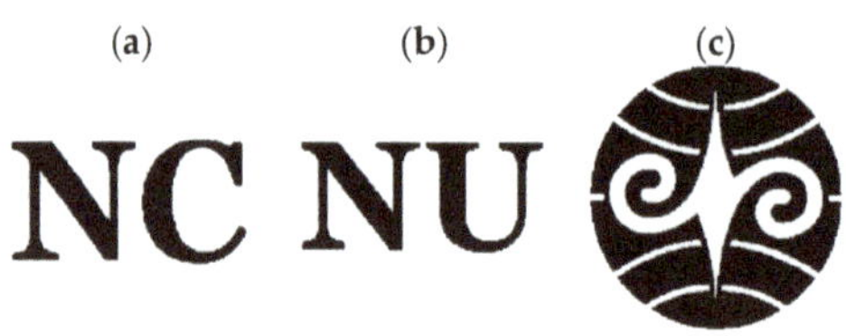

Figure 8. (**a**–**c**) Secret images S_0, S_1, and S_2 for the first experiment.

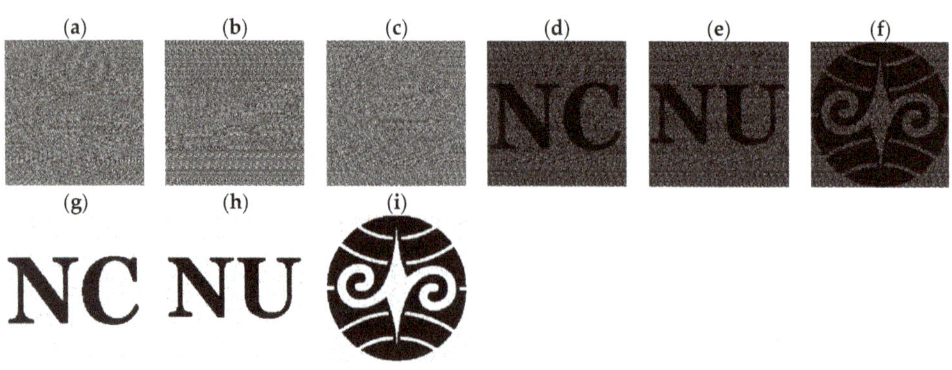

Figure 9. The first experiment results from the proposed (3, 3) VMSSS. (**a**–**c**) Shares G_0, G_1, and G_2; (**d**–**f**) OR-reconstructed images R_0, R_1, and R_2; and (**g**–**i**) XOR-reconstructed images XR_0, XR_1, and XR_2.

Figure 10 displays the results of the second experiment using the proposed (n, n) VMSSS. Since we used the same secret images for encryption, the experimental results can be directly compared with the results of [9] (Figure 4). The results demonstrate that the proposed scheme encrypts N secret images into N meaningless shares, ensuring that no visual information can be discerned from any individual share. However, when OR or XOR operations are applied, the original secret images can be clearly restored with high visibility. This confirms that the proposed method is an (n, n)-threshold visual multi-secret sharing scheme capable of concealing information completely during encryption while enabling a clear reconstruction during decryption.

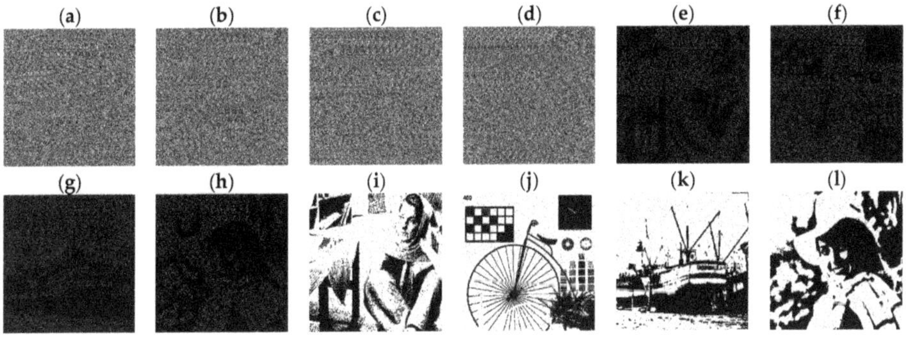

Figure 10. The second experiment results by the proposed (4, 4) VMSSS. **(a–d)** Shares G_0, G_1, G_2, and G_3; **(e–h)** OR-reconstructed images R_0, R_1, R_2, and R_3; and **(i–l)** XOR-reconstructed images XR_0, XR_1, XR_2, and XR_3.

3.4. Analysis

The effectiveness of a visual secret sharing scheme is often assessed by analyzing the quality of the reconstructed image. A key metric is often used to define the quality of reconstructed images. Contrast, often labeled α, with higher contrast values indicates a reconstructed image of a better quality [4]. The contrast value ranges between −1 and 1, with a contrast value of 1 indicating a perfect reconstruction, and 0 signifying that the reconstructed image is meaningless in relation to the original image.

Similar to [9], to articulate the concept of contrast, the average light transmittance $T(R)$ is employed, representing the ratio of white pixels to the total pixel count in image R. The notations $T[R[S(0)]]$ (and $(T[R[S(1)]]$, respectively) refers to the average transmittance of the area of image R that corresponds to the white (transparent) regions (and black (opaque) regions, respectively) of image S. This concept applies to both the secret image S and the reconstructed image R. Once the definition is understood, the average light transmittance for the proposed scheme is subsequently computed.

$$T_{OR}[R[S(1)]] = \frac{p \times N - N + 1}{p \times N} \times 0 + \frac{N-1}{p \times N} \times \frac{1}{2^N} = \frac{N-1}{2^N pN} \tag{1}$$

$$T_{OR}[R[S(0)]] = \frac{p \times N - N + 1}{p \times N} \times \frac{1}{2^{N-1}} + \frac{N-1}{p \times N} \times \frac{1}{2^N} = \frac{2pN - N + 1}{2^N pN} \tag{2}$$

$$T_{XOR}[XR[S(1)]] = \frac{p \times N - N + 1}{p \times N} \times 0 + \frac{N-1}{p \times N} \times \frac{1}{2} = \frac{N-1}{2pN} \tag{3}$$

$$T_{XOR}[XR[S(0)]] = \frac{p \times N - N + 1}{p \times N} \times 1 + \frac{N-1}{p \times N} \times \frac{1}{2} = \frac{2pN - N + 1}{2pN} \tag{4}$$

The average light transmittance formulas of this scheme based on the OR and XOR decryptions are (1) to (4). After calculating the average light transmittance, the theoretical average contrast value is determined according to the contrast equation $\alpha = \frac{T[R[S(0)]] - T[R[S(1)]]}{1 + T[R[S(1)]]}$ as in (5) and (6).

$$\alpha_{OR} = \frac{2(Np - N + 1)}{2^N Np + N - 1} \quad (5)$$

$$\alpha_{XOR} = \frac{2(Np - N + 1)}{2Np + N - 1} \quad (6)$$

Since both α_{OR} and α_{XOR} are greater than 0 (p is a positive integer), the correctness of the proposed scheme is also verified. In the following, we present a numerical analysis of the results from experiments one and two. The theoretical values and experimental data are listed in Table 2. The experimental data represent the average values from all reconstructed images. Compared with the previous result [9], the reconstructed image both by the OR and XOR operations of the proposed scheme has a higher contrast overall. The highest contrast of the restored image by the OR operation is $(4p—2)/(2^N Np + (N—2)p + 1)$, and that by the XOR operation of is $(4p—2)/(3Np—2p + 1)$ in [9]. When the above two formulas are compared with (5) and (6), the proposed scheme demonstrates a better performance than the one in Ref. [9] in both decryption methods. At the same N and p values, the proposed scheme performs satisfactorily, and the restored image using the XOR operation is almost completely restored as displayed in Table 2 (the value close to 1).

Table 2. The theoretical and average experimental contrasts of the restored images generated by experiment 1 and 2.

Experiment	Images	Contrasts		
		Theoretical	Experimental	[9]
1	$R_0, \ldots, R_2$	0.2407	0.2426	0.1557
	$XR_0, \ldots, XR_2$	0.9508	0.9506	0.5532
2	$R_0, \ldots, R_3$	0.1200	0.1252	0.0590
	$XR_0, \ldots, XR_3$	0.9448	0.9548	0.3881

4. Conclusions

In this paper, a novel (n, n) visual multi-secret sharing scheme with OR and XOR decryption capabilities is developed. The proposed scheme is to directly encrypt N secret images into n shares, so the recovered images are of a high quality. Especially when using XOR decryption, it is almost completely restored. In addition, the parameter p in the proposed VMSSS is determined by the user, enhancing the flexibility of the proposed scheme. Users can adjust the parameters according to their specific needs. The introduced VMSSS is both secure and efficient, boasting several key characteristics:

- An absence of pixel expansion;
- High visual quality in the reconstructed images;
- A flexible adjustability of the parameter p;
- The capability to encrypt multiple secrets at the same time;
- The secret image can be restored by employing either an OR operation or an XOR operation.

The comparison of the proposed VMSSS against prior research findings is shown in Table 3. Among them, [1,2,8] are schemes that can only encrypt a single secret image, while [4] supports decryption solely through the OR operation. Only [9] offers the same functionality as our proposed scheme; however, as discussed in the previous section, our reconstruction results are superior. Although this paper only shows the results when the quantity of secret images N is equal to the quantity of shares n ($N = n$), the other two (n, n) VMSSSs have also been designed when $N \neq n$ (for $N > n$ and $N < n$). This means that there can be no limit between the quantity of images and the quantity of participants, and the proposed (n, n) VMSSS can decrypt high-quality recovered images using the OR or

XOR operations. The proposed scheme has many development possibilities, including the transformation of nonsensical shares into meaningful ones, and using grayscale and color images for the encryption process. In addition, designing an MVSSS that can achieve a higher quality when using OR operation restoration is also an interesting topic that can be studied in the future.

Table 3. A comparative analysis of the proposed VMSSS against related works.

Schemes	Features		
	Absence of Pixel Expansion	Multiple Secrets	Decrypt by OR and XOR Operation
[1]	No	No	No
[2]	Yes	No	No
[4]	Yes	Yes	No
[8]	Yes	No	Yes
[9]	Yes	Yes	Yes
Ours	**Yes**	**Yes**	**Yes**

Author Contributions: Conceptualization, J.S.-T.J. and A.-H.L.; methodology, J.S.-T.J. and A.-H.L.; software, A.-H.L.; validation, J.S.-T.J.; formal analysis, J.S.-T.J. and A.-H.L.; investigation, A.-H.L.; data curation, A.-H.L.; writing—original draft preparation, A.-H.L.; writing—review and editing, J.S.-T.J.; visualization, A.-H.L.; supervision, J.S.-T.J.; project administration, J.S.-T.J.; funding acquisition, J.S.-T.J. All authors have read and agreed to the published version of the manuscript.

Funding: This work was partially supported by the National Science and Technology Council, R.O.C. under grant NSTC 112-2115-M-260-001-MY2.

Institutional Review Board Statement: Not applicable.

Informed Consent Statement: Not applicable.

Data Availability Statement: Data are contained within the article.

Conflicts of Interest: The authors declare no conflicts of interest.

References

1. Naor, M.; Shamir, A. Visual Cryptography. In *Advances in Cryptology—EUROCRYPT'94: Workshop on the Theory and Application of Cryptographic Techniques, Perugia, Italy, 9–12 May 1994*; Springer: Berlin/Heidelberg, Germany, 1995.
2. Kafri, O.; Keren, E. Encryption of pictures and shapes by random grids. *Opt. Lett.* **1987**, *12*, 377–379. [PubMed]
3. Lin, K.S.; Lin, C.H.; Chen, T.H. Distortionless visual multi-secret sharing based on random grid. *Inf. Sci.* **2014**, *288*, 330–346. [CrossRef]
4. Chang, J.J.-Y.; Huang, B.-Y.; Juan, J.S.-T. A New Visual Multi-Secrets Sharing Scheme by Random Grids. *Cryptography* **2018**, *2*, 24. [CrossRef]
5. Eisen, P.A.; Stinson, D.R. Threshold visual cryptography schemes with specified whiteness levels of reconstructed pixels. *Des. Codes Cryptogr.* **2002**, *25*, 15–61. [CrossRef]
6. Tuyls, P.; Hollmann, H.D.; Lint, J.V.; Tolhuizen, L.M.G.M. XOR-based visual cryptography schemes. *Des. Codes Cryptogr.* **2005**, *37*, 169–186.
7. Wu, X.; Sun, W. Random grid-based visual secret sharing with abilities of OR and XOR decryptions. *J. Vis. Commun. Image Represent.* **2013**, *24*, 48–62.
8. Lin, Y.-R.; Juan, J.S.-T. RG-Based (*k*, *n*)-Threshold Visual Cryptography with Abilities of OR and XOR Decryption. *Eng. Proc.* **2023**, *55*, 65. [CrossRef]
9. Huang, S.-Y.; Lo, A.-H.; Juan, J.S.-T. (*n*, *n*) XOR-based Visual Multi-secrets Sharing Scheme with Meaningful Shares. In Proceedings of the 2022 IEEE 5th International Conference on Knowledge Innovation and Invention (ICKII), Hualien, Taiwan; 2022; pp. 69–74. [CrossRef]

engineering proceedings

MDPI

Proceeding Paper

Review of Electronic Cooling and Thermal Management in Space and Aerospace Applications †

Kivilcim Ersoy

Aselsan A.S., Ankara 06200, Turkey; kivilcime@aselsan.com

† Presented at the 2024 IEEE 7th International Conference on Knowledge Innovation and Invention, Nagoya, Japan, 16–18 August 2024.

Abstract: The continuous miniaturization of electronics, high processing capacity, compact microelectronic devices, and high circuit density contribute to an increasing demand for the efficient cooling of electronics. For aerospace and space applications, where packaging and the optimal use of space, weight, and power are important, adequate and efficient cooling is a limiting factor due to the increased heat flux rates from compact-design electronic units. As a technology enabler, thermal management applications become important with the increasing demand for longer component operation times. This study aims to review the literature and the analysis results of thermal engineering applications on cooling of electronics and thermal management approaches in space and aerospace applications. Many advanced cooling applications with interdisciplinary advancements and their benefits are discussed.

Keywords: thermal management; electronic cooling

1. Introduction

Aerospace and space applications have advanced remarkably. More countries are involved in aerospace and space research. Aerospace and space systems necessitate high availability, low maintenance, and high performance in harsh environmental conditions. Space and aerospace applications demand packing mission equipment in small volumes to use limited space and power efficiently. For instance, unmanned aerial vehicles (UAVs) served for surveillance and reconnaissance missions initially. Nowadays, they are utilized for a wider range of missions, such as electronic support (ESM), self-protection, and electronic countermeasures (ECM), as well as many other kinds of suppression and destruction mission equipment with an efficient thermal management system of the packed equipment. Radar systems were used in World War II and scanned array radars and the detection possibilities of multiple targets have extended enormously. These improvements were possible due to the latest generation of multifunctional antennas, digitalization, advanced materials, and improvements in the electronics industry. In aerospace applications, the power range of radars and other similar mission equipment continues to increase, while their packaging and weight are becoming smaller. Therefore, advanced cooling techniques are used to improve heat transfer rates for more advanced aerial radar systems with higher power density. For satellite technologies, besides the decreasing cost of launches, the miniaturization of components plays an important role in the exponential increase in countries participating in space activities. Space applications are serving complex missions with the help of thermal engineering applications.

In this literature review, emerging cooling technologies are explained in cooling electronics, where chip-, component-, and board-level-cooling techniques are used. Platforms and system-level thermal management techniques are also addressed along with active and passive thermal management and their elements as an example of a satellite application. The applications of thermal engineering in aerospace applications are not limited to the examples given here.

2. Cooling Electronics

Moore's law states that the performance of processors per unit volume increases exponentially [1]. Both the International Renewable Energy Development Services and ASML Holding NV predict that process nodes reach 1 nm [2]. New device architectures increase functionality at historical rates. In addition, quantum computing accelerates this miniaturization [2]. Figure 1 from the International Roadmap for Devices and Systems shows the increasing effects of miniaturization and high-power density.

Year of Production	2020	2022	2025	2028	2031	2034
	G48M36	G45M24	G42M20	G40M16	G38M16T2	G38M16T4
Logic industry node range labeling	5	3	2.10	1.5	1.0 eq	0.7 eq
Logic device structure options	finFET	finFET LGAA	LGAA	LGAA	LGAA-3D	LGAA-3D
Mainstream device for logic	finFET	finFET	LGAA	LGAA	LGAA-3D	LGAA-3D
	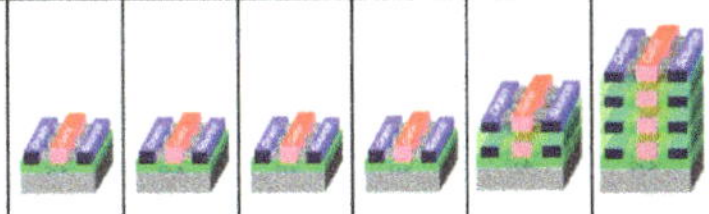					
Vdd (V)	0.7	0.7	0.65	0.65	0.6	0.6
Gate length (nm)	18	16	14	12	12	12
Number of stacked tiers	1	1	1	1	2	4
Number of stacked devices	1	1	3	3	4	4
Digital block area scaling - node to node		0.75	0.78	0.79	0.52	0.5
Cell height limitation - HD	M0	M0	M0	M0	M0	M0
SoC area scalling (stacked) - node to node		0.78	0.82	0.83	0.56	0.59
CPU frequency (GHz)	3.13	3.27	3.51	3.47	3.25	2.93
Frequency scaling - node to node		0.04	0.08	0.01	0.07	-0.09
CPU frequency at constant power density (GHz)	3.13	2.58	2.92	2.19	1.29	0.78
Power at iso frequency - node to node		-0.08	-0.26	-0.05	-0.07	-0.09
Power density - relative	1	1.27	1.2	1.58	2.52	3.73

Figure 1. Moore logic core device technology road map [2].

The downsizing of mission systems and miniaturization of new electronic components result in higher power density, higher heat generation density, and localized hotspots. This necessitates heat dissipation of high heat fluxes up to 200 W/cm^2 [3]. The temperature maxima and the heat fluxes are too high for GaN applications, which on one hand makes challenging operations possible, but it also requires advanced thermal management [4]. The ultra-high heat flux (greater than 1 kW/cm^2) and limited heat dissipation capacity of GaN-based devices result in significant performance degradation if the temperatures are not properly managed.

The performance and operation time of semiconductor devices drop with increasing temperatures. The percentage of temperature-related failures in electronics exceeds 55%, according to data supplied by the US Air Force. The operational reliability of semiconductors is directly related to temperature [5]. In fact, as depicted in Figure 2 the failure rate of electronic devices increases exponentially with increasing operating temperature according to MIL-HNBK2178 [6]. Operation time, reliability, and availability are important in aerospace and space applications, which makes solid thermal management even more important.

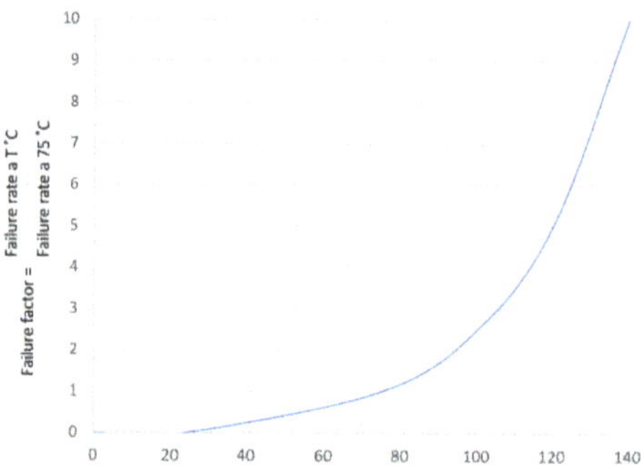

Figure 2. Failure rate vs. temperature graph of digital devices.

2.1. Cooling Techniques

Based on heat transfer effectiveness, cooling modes are classified into four general categories: radiation and free convection, forced-air cooling, forced liquid cooling, and liquid evaporation. The classifications are based on heat transfer mechanisms and coolants, as well as their cooling effectiveness. A comparison of the ranges of heat removal rates for the temperature difference between the heat transfer surfaces and the ambient temperature of 80 °C is illustrated in Figure 3 [7].

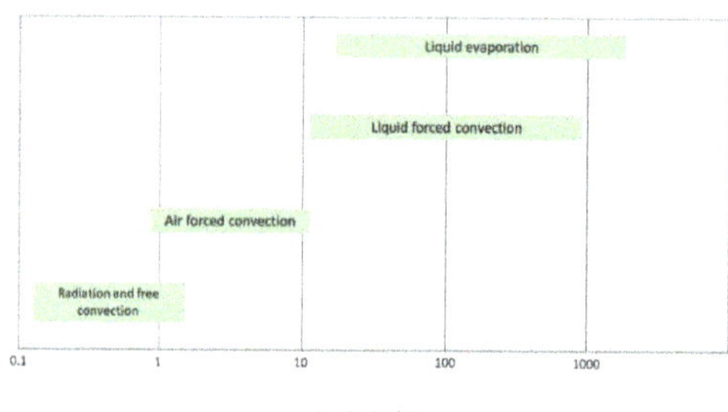

Figure 3. Heat transfer effectiveness of cooling methods [7].

In addition to the above-mentioned general categories, cooling methods are also classified into air cooling, liquid cooling, and refrigeration cooling based on the coolants used. The electronic devices or chips are cooled by transferring heat to the fluid. This heat dissipation can have phase transitions. This heat transfer is conducted by the forced convection of air and traditional heat fins and heat sinks. As conventional measures are not sufficient for removing heat from advanced devices, new approaches have been developed [8].

To supply sufficient cooling to high-heat-producing electronic equipment within small packaging, heat pipes, and conformal liquid blocks with mini-channels are utilized to enhance heat dissipation efficiently. Conventional cooling devices include external air-cooled heat sinks, which are insufficient to cool modern electronic devices and chips with high-power densities. Therefore, such devices need innovative mechanisms and techniques to enhance the heat removal rate, minimize their operating temperature, and maximize their longevity.

Emerging cooling methods are classified into various types such as heat pipes, heat pumps, microchannels, conformal cooling, spray cooling, phase change material (PCM)-based cooling, and thermoelectric cooling. Heat pipes and PCM-based cooling show great potential for the cooling of high-tech electronics and are used frequently.

2.1.1. Heat Pipes

The working principle of the heat pipe, as shown in Figure 4, is based on phase change in the working fluid. Heat pipes are promising for cooling electronic devices such as computers, laptops, telecommunication, and satellite modules [9]. They were introduced by General Motors in 1942 and developed and used by the Los Alamos National Lab and NASA in the 1960s and 1970s.

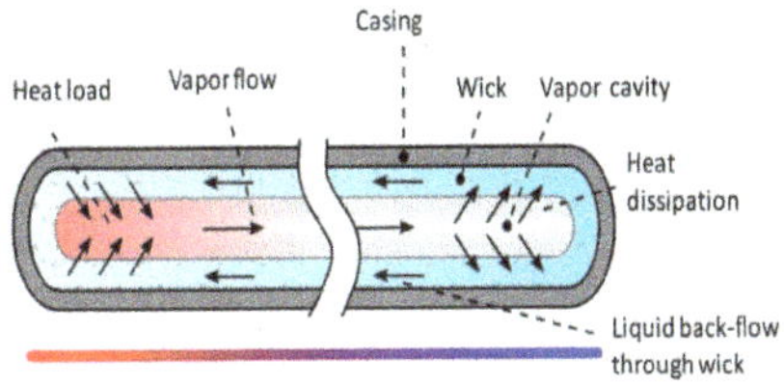

Figure 4. Working principle of heat pipes (Wikipedia).

Due to their extremely high effective thermal conductivity (up to several thousand times higher than even copper rods [10]) and very low effective thermal resistance (typically ranging from 0.05 to 0.4 °C/W [11]), heat pipes are one of the most feasible means of cooling high-heat-flux-generating electronic devices such as CPUs. Nowadays, numerous commercial cooling solutions for laptops, tablets, smartphones, as well as space applications involve heat pipes [12,13].

2.1.2. PCM

PCM absorbs energy at phase transitions to provide useful heat or cooling. PCMs have been studied for various applications including electronic cooling [6,14,15] as they have a high latent heat of fusion, high specific heat, controllable temperature stability, and small volume change during phase change. PCM is used in the efficient passive systems of many defense and aerospace applications, spacecraft and avionic thermal control, weapon systems, missile electronics, surveillance radars, and communication systems, as it can absorb a large quantity of heat without raising the temperature when the heat output occurs during the phase transition. Different designs of PCM for cooling microprocessors were studied by researchers [16–18]. The low thermal conductivity of PCM is one of the major limitations that hinder the effective heat transfer capability of the passive system. Several studies optimized thermal conductance with different-shaped (rectangular, circular, elliptical, triangular) pin-fin heat sinks using paraffin wax, n-Eicosane, aluminum, and rubber pad [17,19,20].

2.1.3. Microchannels

Microchannels are compact cooling elements that provide increased heat dissipation rates and reduce temperature gradients across electronic components. Tuckerman and Pease used microchannels in an electronic device to carry the energy from the chip to the environment, using water as a coolant. The heat transfer coefficient of heat dissipation increased as the flow channel diameter decreased [21,22].

Unlike traditional heat sinks that need a large surface area to increase heat dissipation rates, microchannels use small-diameter channels to increase the heat transfer coefficient by forcing the coolant to flow in close contact with the channel walls. Microchannel heat sinks have been optimized by replacing straight channels with different geometries [23], though the improvement in heat transfer performance was accompanied by a high pressure drop. Manifold layers have been introduced to compromise the high pressure drop and nonuniform temperature distributions. Harpole [24] proposed the concept of a manifold microchannel to cope with the flow maldistribution and large pressure drop in the microchannel. Microchannel heat sinks enhanced with manifold structures and jet impingement have great potential [25–27]. The microchannel heat sinks are efficiently used for GaN applications for embedded microchannel cooling.

Technologies such as additive manufacturing are used to fabricate monolithic microchannel heat sinks to reduce the pressure drop in the microchannel while enhancing heat transfer performance [28], whereas additive manufacturing enables the fabrication of complex geometries and enhances the permeability of structures. Additive manufacturing enables small fin wall sizes and makes it possible to fabricate more complex channel patterns [29].

2.1.4. Emergent Coolers

The removal of heat using microchannel heat sinks can be enhanced using smart fluids, i.e., nanofluids with effective thermophysical properties such as increased thermal conductivity. Mini-channel cooling systems with liquid coolants are efficient in cooling electronics [30]. The nanofluids are known as colloidal suspensions, where nanoparticles are dispersed in conventional base fluids. This is used to improve the heat transfer capacity of fluid by modifying their thermophysical suspension properties. Figure 5 depicts the thermal conductivity of conventional and innovative cooling fluids.

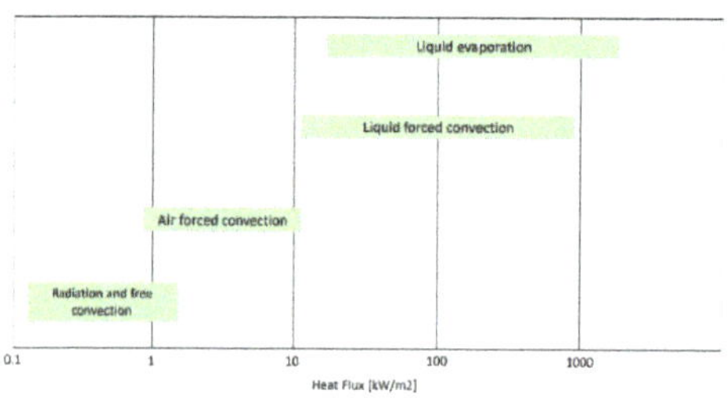

Figure 5. Thermal conductivity of conventional and nanofluids [31].

There are various studies conducted with nanofluids, which depicts that they have a potential in enhancing the cooling [32–34]

- Carbon-based nanoparticles: Carbon nanotubes and graphene are carbon-based nanoparticles employed for cooling in heat sinks. They significantly improve the thermal performance of liquid blocks due to their excellent thermal conductivities. However, they are not stable in aqueous media, because they are naturally hydrophobic and so cannot be dispersed in polar liquids. This hinders the efficient utilization of them. To overcome this challenge, they are functionalized through acid treatment and in this way, become hydrophilic. Nanofluids containing biologically produced graphene nanoplatelets are suggested as appropriate candidates for utilization in electronic cooling [35].
- Oxide nanoparticles: Alumina and titania nanoparticles are oxide nanoparticles applied in cooling electronics. Spherical oxide nanoparticles are employed to synthesize nanofluids in liquid blocks due to their stability, low cost, and thermal conductivity. Turgut et al. [36] investigated the performance of a commercial liquid cooling kit and found out that when using Al_2O_3 nanoparticles suspended in water, the temperature of the contact surface between the heater and the water block was 6.7% lower than that using water only. Abu-Nada et al. evaluated the heat transfer coefficient of Al_2O_3 nanofluids compared to water in a numerical investigation [37]. Bansal et al. evaluated the utilization of nanofluids for spray cooling processes, finding water-based alumina

solutions to be promising [38]. The positive results were confirmed by Wang and Xu, who enhanced the heat transfer coefficients and cooling [39].

- Magnetic nanoparticles: Magnetic nanofluids or ferrofluids are the suspensions of ferromagnetic nanoparticles (cobalt, iron, nickel, and their oxides) and non-magnetic base liquid. They have flowabilities of ordinary liquids and magnetic features similar to other magnetic materials simultaneously. This makes it possible to manage the flow, heat exchange, and motion of particles using external magnetic fields. Even if these suspensions can be used, few studies have been carried out on electronic cooling, as the electromagnetic EMI/EMC may be adversely affected.
- Hybrid nanofluids: Hybrid nanofluids are nanofluids that are produced with dispersion, unlike nanoparticles in a composite state or mixture form. The purpose of preparing hybrid nanofluids is to optimize the attributes of different nanoparticles while improving thermal conductivity.

It is important to control the form and size of the nanoparticles. When producing them, employing nanofluids in liquid blocks affects thermal performance considerably. The anisotropic behavior of nanofluids affects pumping power, and pressure loss and energy consumption need to be investigated for the use of nanofluids.

2.2. Design of Electronic Cooling

For electronic cooling, the thermal management design needs to keep the junction temperature less than 125 °C for SiC applications. For space, aerospace, and other military applications, GaN chips are used to tolerate much higher temperatures, up to 250 °C. Conservative design best practice needs a 10% safety zone, and the highest temperature needs to be considered for the worst operation ambient temperature.

The system can be modeled as depicted in Figure 6, including chip level, component level, board level, system level, and environment. Both top-down and bottom-up approaches can be used. A hybrid approach needs to be used, using the capabilities of each level and designing the system according to the application and its limits.

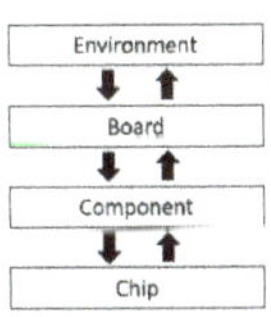

LEVEL	MODELING
Environment	Boundary Condition (Pressure and Temperature)
Cabinet	Interface to environment – balance energy
Board	Boundary Condition for component – P, V and T
Component	Interface to the board and coolant – energy balance
Chip	Junction temperature

Figure 6. Modeling thermal systems of electronics.

Thermal conditions on the chip and board levels are depicted in Figure 7. Knowing the chips' maximum junction temperatures, the dominant heat regimes (conduction), and the related variables (like thermal conductivity), the device can be designed.

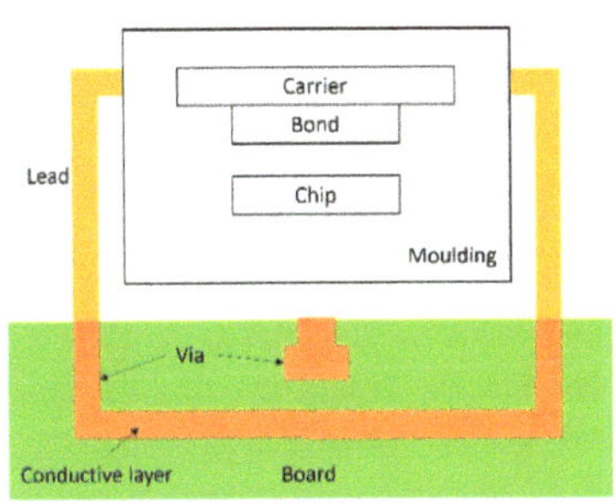

Figure 7. Thermal conditions on chip and board level.

The PCB architecture can be designed for air or liquid cooling. In both cases, the flow needs to be optimized by CFD computations to satisfy the cooling requirements and minimize the pressure drop. Both air and liquid applications are used in military aerospace applications [40]. Both liquid-cooled cold plate and air-cooled cold wall (using pin fins) applications are reported in Ref. [40]. The cooling technique between the board and the cabinet and/or environment varies according to applications.

According to the constraints, including the utilization area or platform (environment/packaging/power supply/weight and space constraints), the solution must be chosen utilizing conventional and innovative means as well as active and passive thermal management elements.

2.3. *Platform Thermal Management*

Aerospace and space platforms have extra power density. Satellites have unique challenges such as limited space and limited insulation. Small satellites often possess limited interior volume and valuable payload space. This spatial constraint mandates efficient and compact thermal solutions. Power availability is scarce and allocating power for thermal control systems must be balanced with other mission-critical functions. The heat created by mission electronics needs to be transferred in an efficient way to ease the dissipation. An innovative power-efficient solution becomes essential. Consequently, thermal management strategies are imperative to shield delicate instruments from thermal shock. The challenges of satellites are summarized in Table 1.

Table 1. Challenges of satellites [41].

Satellite Property
Low thermal mass
Limited external surface area
Limited volume
Limited power
High power density
Multi-layer insulation edge effects

Since the 2010s, smaller satellites have been manufactured at a minimum cost. Swarm concepts have been mentioned for cost-effective Earth-observing missions [42]. To reduce cost and development times, various methods have been used [43]. Satellites face extreme temperature fluctuations when exposed between the direct sunlight and the Earth's shadow. Thermal management serves to regulate temperature, minimize thermal gradients, efficiently dissipate the heat generated by electronics, keep thermal stresses as low as possible, and expand the lifespan [41]. Temperatures need to be regulated with passive and/or active thermal management technology and design methods or an appropriate combination of each. Satellites must operate in specified temperature ranges to maintain their accuracy and functionality. Maintaining uniform temperatures throughout the satellite prevents differential expansion and affects structural integrity and instrument alignment. Rapid temperature changes, such as those experienced during orbit transitions, cause thermal stress, which leads to mechanical failure or damage to sensitive components. Appropriate thermal management extends the operational life of satellites, ensuring missions for longer durations. Various thermal control designs [44–46] as well as thermal health control systems [47–49] have been introduced and developed in the last decade.

2.4. Thermal Management

The effective thermal management of satellites involves a combination of passive and active techniques, careful design considerations, and the utilization of specific materials and technologies tailored to the mission's requirements. These methods ensure that the satellite remains within its operational temperature limits, maximizes mission success, and delivers valuable scientific or technological data back to Earth.

2.4.1. Active Thermal Management

Active thermal management is used with active elements to regulate temperature and offers advantages including efficient heat dissipation, rapid response to dynamic thermal loads, and precise temperature control. Small satellites experience varying thermal loads due to changes in solar radiation, orbital conditions, orientation, and operational modes. Active thermal management systems can quickly respond to these dynamic loads and adjust cooling mechanisms accordingly. This rapid response prevents temperature fluctuation, thermal stress, and related potential failures. Moreover, if the heat needs to be continuously dissipated, active thermal management is indispensable. Active cooling methods, such as fans, pumps, and heat exchangers, enhance heat dissipation by actively moving air or coolants over heat-generating components. This efficient heat removal prevents the accumulation of excessive heat, reducing the risk of overheating and maintaining the satellite's operational efficiency.

Conventional elements such as fans, pumps, heat exchangers, and control units are utilized in satellites as well as thermoelectric coolers. Louvers or shutters are used on external surfaces to control the amount of heat radiated or absorbed. Louvers, consisting of blades and actuators, open or close based on temperature conditions.

Active Thermal Control Units (TCUs) are used to monitor and regulate the temperature of various components within a satellite. They actively control the operation of cooling systems, such as fans, pumps, or TECs, based on temperature sensors' feedback. TCUs ensure precise temperature regulation and adapt to changing thermal conditions. They play a vital role in actively managing the thermal environment of the satellite. Thermoelectric coolers (TECs) or Peltier devices transfer thermal energy from its low-temperature side to its high-temperature side by utilizing external electrical energy. TECs are often used in conjunction with heat sinks to dissipate collected heat. They have no moving parts and are high-reliability, lightweight, compact, and easy to operate and regulate. This makes TECs promising for the thermal management of spacecraft subsystems and components. Their successful utilization over a decade has been presented in various studies [50–52].

Active thermal management systems may have higher power consumption, weight, and volume demand compared to passive techniques. Additional components are added to the system, which creates new risks for reliability. However, the benefits of precise temperature control and efficient heat dissipation may outweigh these considerations in many satellite applications.

2.4.2. Passive Thermal Management

Passive thermal management techniques do not rely on complex moving parts like fans or pumps, which reduces the risk of mechanical failure and increases reliability. They require little to no power consumption since they do not rely on power input. Passive thermal elements possess reduced complexity and compact design to occupy minimal space, making them appropriate for small satellites or systems with limited physical dimensions. They provide long-term thermal control without the need for frequent maintenance or replacement. Overall, passive thermal management offers a reliable, low-power, compact, low-noise signature, and environmentally friendly solution for managing temperature in various applications, including satellites.

While passive thermal elements are simpler in design than active cooling methods, optimal performance requires careful design considerations. Factors such as material selection, surface area, geometry, and airflow patterns need multi-physics optimization to maximize heat dissipation. Specialized thermal coatings are applied to the satellite's external surfaces to control heat absorption and emission. These coatings maintain the desired thermal balance [53].

Multi-layer Insulation (MLI) blankets consist of multiple layers of reflective material separated by low-conductivity spacers. MLI is used to insulate sensitive components from extreme temperature fluctuations and reduce heat transfer through radiation. The reflective layers reflect solar radiation, while the spacers minimize conductive heat transfer. Heat pipes, as mentioned earlier, are passive devices that use phase-change principles and capillary action to transport heat. They are used for passive thermal management in small satellites [54,55]. Shukla showed various examples of heat pipes for aerospace and satellite applications [56]. Their utilization of nanofluids was presented by Mashei [57]. More advanced heat pipe techniques include nitrogen-charged cryogenic grooved heat pipes [58] and the utilization of different fluids of ammonia and methanol [13].

PCMs undergo a phase change (e.g., solid to liquid) while absorbing or releasing a significant amount of latent heat. They can be integrated into the satellite's structure or components to provide passive thermal buffering [59]. Their various applications have been investigated, developed, and used since then [60–62].

Passive techniques involve controlling the satellite's albedo (reflectivity) to influence the amount of solar radiation absorbed. This can be achieved through the choice of surface materials and coatings. Radiators are considered passive components when they are not deployable. These fixed radiators allow heat to be radiated away from the satellite's surface into space. Passive radiators are integrated into the satellite's structure on the outer surface. The satellite's structural components serve as passive thermal masses that stabilize temperature fluctuations by absorbing and releasing heat over time. Passive thermal management techniques are generally less adaptable to dynamic thermal loads or changing environmental conditions. They have slower heat dissipation rates than active cooling methods. They operate under specific parameters and may struggle to handle sudden increases in heat generation or varying ambient temperatures. Passive thermal management techniques may not effectively address localized temperature spikes. If heat-generating components are concentrated in specific areas, passive techniques struggle to distribute and dissipate the heat evenly. Still, due to the above-mentioned advantages, passive systems are preferable to be utilized as much as possible.

2.4.3. Best Practice for Small Satellite Thermal Management

Small satellites, named CubeSats, aim to pack more equipment into a smaller volume. This necessitates a mass-efficient design and increases thermal density. Overheated electronic boards and mission equipment cannot be replaced or serviced; therefore, the thermal solution must be robust and appropriate for the mission's lifespan. For CubeSats, due to limited power and budget, the passive techniques are preferred, as long as they are possible. It is crucial to optimize the passive thermal design concurrently with thermal, mechanical, and electrical input to deliver low-power, low-mass, small-volume, and reduced-power consumption versions of existing technologies.

In most space systems, all heat must be rejected via radiation along radiators on the outer surface. Therefore, the heat needs to be transferred to the radiators in an efficient way to ease dissipation. Both copper–water heat pipes and ammonia heat pipes are typically installed for board- and system-level heat transfer. A commonly used technology is PCM, which balances short-duration and duty cycle applications. PCMs' thermal storage provides a thermal buffer in duty cycle applications with high peak power. CubeSats experiencing transient power operation utilize PCM to meet thermal performance objectives while maintaining a low-complexity design that is optimized for mass and volume.

For duty cycle-operating components, storing thermal energy in small volumes is a desirable design option. The best practice is to utilize an intelligent combination of active and advanced passive elements such as oscillating heat pipes, miniature cryocoolers, micro louvers, and specific heat straps [63].

After the conceptual design phase and selection of appropriate thermal technology elements, detailed thermal analysis must be conducted to keep all components within the specific temperature range. Accurate yet cost-efficient numerical modeling of the system supported by tests is crucial to ensure reliability while avoiding over-design at the same time. Jurkowski et al. conducted sensitivity analysis (SA) and uncertainty quantification (UQ) to verify the thermal design of a nanosatellite data processing unit (DPU) [64].

Bonnici 2019 developed thermal finite element modeling for the thermal design of small satellites to calculate the transient heat load history on the pico-satellite's external surfaces due to environmental sources [65]. A thermal management system, consisting of a rollout deployable radiator, PCMs, and integrated heat pipes for high-power CubeSats was offered and proven to be feasible to be utilized in CubeSats [66].

3. Conclusions

A review of traditional and innovative cooling and thermal management techniques was conducted for cooling electronics in the aerospace and space industries. There are many methods to transfer and dissipate the created heat, each having advantages and drawbacks. Therefore, it is important to choose the right methods and elements for each case. The best practice needs to be chosen to provide thermal solutions.

Funding: This research received no external funding.

Informed Consent Statement: Not applicable.

Data Availability Statement: The original data presented in the study are openly available in the given references.

Conflicts of Interest: Kivilcim Ersoy was employed by the company (Aselsan A.S.). The author declares that the research was conducted in the absence of any commercial or financial relationships that could be construed as a potential conflict of interest.

References

1. Moore, G. Cramming More Components Onto Integrated Circuits. *Electronics* **1965**, *38*, 114. [CrossRef]
2. IRDS. *The International Roadmap for Devices and Systems*; IEEE: Piscataway, NJ, USA, 2020.
3. Kandlikar, S.G. Review and Projections of Integrated Cooling Systems for Three-Dimensional Integrated Circuits. *J. Electron. Packag.* **2014**, *136*, 024001. [CrossRef]
4. Al-Neama, A.F.; Kapur, N.; Summers, J.; Thompson, H.M. Thermal management of GaN HEMT devices using serpentine minichannel heat sinks. *Appl. Therm. Eng.* **2018**, *140*, 622–636. [CrossRef]
5. Garimella, S.V.; Persoons, T.; Weibel, J.A.; Gektin, V. Electronics Thermal Management in Information and Communications Technologies: Challenges and Future Directions. *IEEE Trans. Compon. Packag. Manuf. Technol.* **2016**, *7*, 1191–1205. [CrossRef]
6. *MILHDBK-2178B*; Reliability Prediction of Electronic Equipment. U.S. Department of Defense, NTIS: Springfield, VA, USA, 1974.
7. Scott, W.A. *Cooling of Electronic Equipment*; John Wiley and Sons: New York, NY, USA, 1974.
8. Tuckerman, D.; Pease, R. High-performance heat sinking for VLSI. *IEEE Electron. Device Lett.* **1981**, 2, 126–129. [CrossRef]
9. Faghri, A. Review and Advances in Heat Pipe Science and Technology. *J. Heat. Transf.* **2012**, *134*, 123001. [CrossRef]
10. El-Nasr, A.A.; El-Haggar, S.M. Effective thermal conductivity of heat pipes. *Heat. Mass. Transf.* **1996**, *32*, 97–101. [CrossRef]
11. Garner, S.D. Heat pipes for electronics cooling applications. *Electron. Cool.* **1996**, *2*, 18–23.

12. Chen, X.; Ye, H.; Fan, X.; Ren, T.; Zhang, G. A review of small heat pipes for electronics. *Appl. Therm. Eng.* **2016**, *96*, 1–17. [CrossRef]
13. Patel, V.K. An efficient optimization and comparative analysis of ammonia and methanol heat pipe for satellite appli-cation. *Energy Convers. Manag.* **2018**, *165*, 382–395. [CrossRef]
14. Tan, F.; Tso, C. Cooling of mobile electronic devices using phase change materials. *Appl. Therm. Eng.* **2004**, *24*, 159–169. [CrossRef]
15. Kandasamy, R.; Wang, X.-Q.; Mujumdar, A.S. Application of phase change materials in thermal management of electronics. *Appl. Therm. Eng.* **2007**, *27*, 2822–2832. [CrossRef]
16. Jaworski, M.; Domanski, R. A novel design for heat sink with PCM for electronic cooling. In Proceedings of the 10th International Conference on Thermal Energy Storage, Stockton, NJ, USA, 31 May–2 June 2006.
17. Ali, H.M.; Arshad, A.; Jabbal, M.; Verdin, P. Thermal management of electronics devices with PCMs filled pin-fin heat sinks: A comparison. *Int. J. Heat. Mass. Transf.* **2018**, *117*, 1199–1204. [CrossRef]
18. Righetti, G.; Zilio, C.; Doretti, L.; Longo, G.A.; Mancin, S. On the design of Phase Change Materials based thermal management systems for electronics cooling. *Appl. Therm. Eng.* **2021**, *196*, 117276. [CrossRef]
19. Wang, Y.-H.; Yang, Y.-T. Three-dimensional transient cooling simulations of a portable electronic device using PCM (phase change materials) in multi-fin heat sink. *Energy* **2011**, *36*, 5214–5224. [CrossRef]
20. Ashraf, M.J.; Ali, H.M.; Usman, H.; Arshad, A. Experimental passive electronics cooling: Parametric investigation of pin-fin geometries and efficient phase change materials. *Int. J. Heat. Mass. Transf.* **2017**, *115*, 251–263.
21. Garimella, S.V.; Sobhan, C.B. Transport in microchannels—A critical review. *Annu. Rev. Heat. Transf.* **2003**, *13*, 1–50. [CrossRef]
22. Khan, M.G.; Fartaj, A. A review on microchannel heat exchangers and potential applications. *Int. J. Energy Res.* **2011**, *35*, 553–582. [CrossRef]
23. Sui, Y.; Teo, C.; Lee, P.; Chew, Y.; Shu, C. Fluid flow and heat transfer in wavy microchannels. *Int. J. Heat Mass Transf.* **2010**, *53*, 2760–2772. [CrossRef]
24. Harpole, G.; Eninger, J. Micro-channel heat exchanger optimization. In Proceedings of the Seventh IEEE Semiconductor Thermal Measurement and Management Symposium, Phoenix, AZ, USA, 12–14 February 1991.
25. Plawsky, J.L.; Fedorov, A.G.; Garimella, S.V.; Ma, H.B.; Maroo, S.C.; Chen, L.; Nam, Y. Nano- and Microstructures for Thin-Film Evaporation—A Review. *Nanoscale Microscale Thermophys. Eng.* **2014**, *18*, 251–269. [CrossRef]
26. He, Z.; Yan, Y.; Zhang, Z. Thermal management and temperature uniformity enhancement of electronic devices by micro heat sinks: A review. *Energy* **2020**, *216*, 119223. [CrossRef]
27. Park, J.S.; Lee, D.K.; Yim, S.H. Cooling performance evaluation of impingement jet cooling passage on solid-state power amplifier in RF system. *Heliyon* **2022**, *8*, e12066. [CrossRef] [PubMed]
28. Kong, D.; Jung, E.; Kim, Y.; Manepalli, V.V.; Rah, K.J.; Kim, H.S.; Hong, Y.; Gil Choi, H.; Agonafer, D.; Lee, H. An additively manufactured manifold-microchannel heat sink for high-heat flux cooling. *Int. J. Mech. Sci.* **2023**, *248*. [CrossRef]
29. Panse, S.S.; Ekkad, S.V. Forced convection cooling of additively manufactured single and double layer enhanced microchannels. *Int. J. Heat. Mass. Transf.* **2021**, *168*, 120881. [CrossRef]
30. Arani, A.A.; Akbari, O.A.; Safaei, M.R.; Marzban, A.; Alrashed, A.A.; Ahmadi, G.R.; Nguyen, T.K. Heat transfer improvement of water/single-wall carbon nanotubes (SWCNT) nanofluid in a novel design of a truncated double-layered microchannel heat sink. *Int. J. Heat. Mass. Transf.* **2017**, *113*, 780–795. [CrossRef]
31. Souza, R.R.; Gonçalves, I.M.; Rodrigues, R.O.; Minas, G.; Miranda, J.; Moreira, A.L.; Lima, R.; Coutinho, G.; Pereira, J.; Moita, A.S. Recent advances on the thermal properties and applications of nanofluids: From nanomedicine to renewable energies. *Appl. Therm. Eng.* **2022**, *201*, 117725. [CrossRef]
32. Colangelo, G.; Favale, E.; Milanese, M.; de Risi, A.; Laforgia, D. Cooling of electronic devices: Nanofluids contribution. *Appl. Therm. Eng.* **2017**, *127*, 421–435. [CrossRef]
33. Ashrafmansouri, S.-S.; Esfahany, M.N. Mass transfer in nanofluids: A review. *Int. J. Therm. Sci.* **2014**, *82*, 84–99. [CrossRef]
34. Imani-Mofrad, P.; Heris, S.Z.; Shanbedi, M. Experimental investigation of the effect of different nanofluids on the thermal performance of a wet cooling tower using a new method for equalization of ambient conditions. *Energy Convers. Manag.* **2018**, *158*, 23–35. [CrossRef]
35. Sohel Murshed, S.M.; Nieto de Castro, C.A. A critical review of traditional and emerging techniques and fluids for electronics cooling. *Renew. Sustain. Energy Rev.* **2017**, *78*, 821–833. [CrossRef]
36. Turgut, A.; Elbasan, E. Nanofluids for electronics cooling. In Proceedings of the 2014 IEEE 20th International Symposium for Design and Technology in Electronic Packaging (SIITME), Bucharest, Romania, 23–26 October 2014; pp. 35–37.
37. Abu-Nada, E.; Oztop, H.F. Numerical Analysis of Al2O3/Water Nanofluids Natural Convection in a Wavy Walled Cavity, Numer. *Heat Transf. Part A Appl.* **2011**, *59*, 403–419. [CrossRef]
38. Aditya Bansal, I.I.I. Pyrtle Frank, Alumina Nanofluid for Spray Cooling Enhancement. In Proceedings of the ASME/JSME 2007 Thermal Engineering Heat Transfer Summer Conference collocated with the ASME 2007 InterPACK Conference, Vancouver, BC, Canada, 8–12 July 2007; pp. 797–803. [CrossRef]

39. Wang, X.; Xu, X.; Choi, S.U. Thermal Conductivity of Nanoparticle - Fluid Mixture. *J. Thermophys. Heat Transf.* **1999**, *13*, 474–480. [CrossRef]
40. Price, D.C. A review of selected thermal management solutions electronics for military system. *IEEE Trans. Compon. Pack. Technol.* **2003**, *26*, 26–39.
41. State-of-the-Art of Small Spacecraft Technology, Chapter 7 Thermal Control, NASA, 7.0 Thermal Control—NASA. Available online: https://www.nasa.gov/smallsat-institute/sst-soa/thermal-control/ (accessed on 1 February 2025).
42. Marinan, A.D. From Cubesats to Constellations: Systems Design and Performance Analysis. Ph.D. Thesis, Department Aeronautics Astronautics, Massachusetts Institute of Technology, Cambridge, MA, USA, 2013.
43. Zhang, X.; Wu, L.; Liu, H.; Feng, J.; Xu, M.; Rui, H. Research and Verification of Multi-Satellite Thermal Vacuum Test Method. In Proceedings of the 2020 International Conference on Sensing, Measurement & Data Analytics in the Ara of Artificial Intelligence (ICSMD), Xi'an, China, 15–17 October 2020.
44. Sun, R.; Wang, C.; Yin, M.; Yang, Z. Design method of thermal control for micro satellite with integrated structure. In Proceedings of the 2023 2nd International Symposium on Aerospace Engineering and Systems (ISAES), Nanjing, China, 19–21 May 2023; pp. 73–77.
45. Hu, G.J.; Chen, Y.J.; Wang, M.; Zhou, J.; Tao, J.S.; Li, Y.F.; Liu, B.L. Modular Thermal Control Technology and Its Application for Circular Low Earth Orbit Satellites. *J. Astronaut.* **2021**, *42*, 1396–1403.
46. Zhou, Y. A Balanced Satellite Thermal Control Optimization Method for Different Orbit and Attitude. *Chin. J. Space Sci.* **2016**, *36*, 380–385. [CrossRef]
47. Liu, L.; Zhang, X.; Feng, J. On-Orbit Health Management of Thermal Control Subsystem of a Micro Satellite. In Proceedings of the 2022 Global Reliability and Prognostics and Health Management (PHM-Yantai), Yantai, China, 13–16 October 2022; pp. 1–5.
48. Li, Y.; Huang, H. Current trends of spacecraft intelligent autonomous control. *Aerosp. Control Appl.* **2019**, *45*, 7–18.
49. Jianchao, F.; Xiaofeng, Z.; Xing, L. Spacecraft intelligent autonomous thermal control method based on space environment rediction. *Spacecr. Eng.* **2021**, *30*, 45–52.
50. Semena, N.P. The features of application of thermoelectric converters in spacecraft systems of temperature control. *Thermophys. Aeromechanics* **2013**, *20*, 211–222. [CrossRef]
51. Guo, D.; Sheng, Q.; Dou, X.; Wang, Z.; Xie, L.; Yang, B. Application of thermoelectric cooler in temperature control system of space science experiment. *Appl. Therm. Eng.* **2019**, *168*, 114888. [CrossRef]
52. Gilmore, D. *Spacecraft Thermal Control Handbook, Volume I: Fundamental Technologies*, Chapter 13: Thermoelectric Coolers; American Institute of Aeronautics and Astronautics (AIAA): Reston, VA, USA, 2002.
53. Liu, T.; Sun, Q.; Meng, J.; Pan, Z.; Tang, Y. Degradation modeling of satellite thermal control coatings in a low earth orbit environment. *Sol. Energy* **2016**, *139*, 467–474. [CrossRef]
54. Park, J.H. A study on thermal performance of heat pipe for optimal placement of satellite equipment. *ETRIJ* **1997**, *19*, 59–69. [CrossRef]
55. Hoa, C.; Demolder, B.; Alexandre, A. Roadmap for developing heat pipes for ALCATELSPACE's satellites. *Appl. Therm. Eng.* **2003**, *23*, 1099–1108. [CrossRef]
56. Shukla, K.N. Heat Pipe for Aerospace Applications—An Overview. *J. Electron. Cool. Therm. Control.* **2015**, *5*, 55065. [CrossRef]
57. Mashaei, P.R.; Shahryari, M. Effect of nanofluid on thermal performance of heat pipe with two evaporators; appli-cation to satellite equipment cooling. *Acta Astronaut.* **2015**, *111*, 345–355. [CrossRef]
58. Li, Y.; Li, N.; Jiang, Z.; Dong, D. Experimental study on the effect of the gas reservoir on the performance of a nitrogen-charged cryogenic grooved heat pipe. *Int. J. Refrig.* **2023**, *155*, 114–124. [CrossRef]
59. Abhat, A.; Groll, M. Investigation of phase change material /PCM/ devices for thermal control purposes in satellites. In Proceedings of the Thermophysics and Heat Transfer Conference, Boston, MA, USA, 15–17 July 1974. [CrossRef]
60. Fossett, A.J.; Maguire, M.T.; Kudirka, A.A.; Mills, F.E.; Brown, D.A. Brown, Avionics passive cooling with microencapsulated phase change materials. *J. Electron. Packag.* **1998**, *120*, 238–242. [CrossRef]
61. Kumar, R.; Misra, M.K.; Kumar, R.; Gupta, D.; Sharma, P.K.; Tak, B.B.; Meena, S.R. Phase change materials: Technology status and potential defense application. *Def. Sci. J.* **2011**, *61*, 576–582. [CrossRef]
62. Elshaer, A.M.; Soliman, A.M.; Kassab, M.; Mori, S.; Hawwash, A.A. Numerical study about thermal per-formance evaluation of PCM and PCM/fins composite-based thermal control module at microgravity conditions. Interna-tional Journal of Thermofluids. *Int. J. Thermofluids* **2023**, *20*, 100419.
63. Young, J. Advanced Concepts for Small Satellite Thermal Control. In Proceedings of the AIAA/USU Conference on Small Satellites, Technical Session 7: Advanced Technologies I, SSC21-VII-06, Logan, UT, USA, 7–12 August 2021.
64. Jurkowski, A.; Paluch, R.; Wójcik, M.; Klimanek, A. Thermal modelling of a small satellite data processing unit aided by sensitivity analysis and uncertainty quantification. *Int. J. Therm. Sci.* **2023**, *193*, 108514. [CrossRef]

65. Bonnici, M.; Mollicone, P.; Fenech, M.; Azzopardi, M.A. Azzopardi Analytical and numerical models for thermal related design of a new picosatellite. *Appl. Therm. Eng.* **2019**, *159*, 113908.
66. Yendler, B.; Meginnis, A.; Reif, A. Thermal Management for High Power Cubesats. In Proceedings of the AI-AA/USU Conference on Small Satellites, Pre-Conference Workshop Session VI: Advanced Concepts III, SSC20-WKVI-07, Logan, UT, USA, 1–6 August 2020.

engineering proceedings

MDPI

Proceeding Paper

Predicting Hit Songs Using Audio and Visual Features †

Cheng-Yuan Lee [1] and Yi-Ning Tu [2,*]

1 Kang Chiao International School, Taipei 231, Taiwan; leolee600627@gmail.com
2 Department of Statistics and Information Science, Fu Jen Catholic University, New Taipei City 242, Taiwan
* Correspondence: eniddu@gmail.com

† Presented at the 2024 IEEE 7th International Conference on Knowledge Innovation and Invention, Nagoya, Japan, 16–18 August 2024.

Abstract: Factors contributing to a song's popularity are explored in this study. Recent studies have mainly focused on using acoustic features to identify popular songs. However, we combined audio and visual data to make predictions on 1000 YouTube songs. In total, 1000 songs were grouped into two categories based on YouTube view counts: popular and non-popular. The visual features were extracted using OpenCV. These features were applied using machine learning algorithms, including random forest, support vector machines, decision trees, K-nearest neural networks, and logistic regression. Random forest performed the best, with an accuracy of 82%. Average accuracy increased by 9% in all models when using audio and visual features together. This indicates that visual elements are beneficial for identifying hit songs.

Keywords: hit song; machine learning; popularity; audio features; visual features

1. Introduction

In the traditional music industry, sales of record labels are the most direct route to financial success. Producers and singers aim to sell as many albums as possible. As the Internet is used widely, music streaming sites have become popular by offering songs for free. People can enjoy music without having to buy an album. Singers put their music online and websites pay them royalties for streams and downloads. Today, the main goal for singers and producers is to make songs that attract listeners. Figuring out what makes a song a hit is valuable in many areas. A song's popularity is determined by what listeners like, how catchy it is, and its commercial appeal.

Recent studies focus on extracting music audio with machine learning, such as decision trees, support vector machines (SVM)-based models, and logistic regression. In 2015, Anjana used Librosa to extract songs' audio waveforms to predict the song's popularity with the XGBoost classification algorithm. The second coefficient of Linear Predictive Coding has an accuracy of 71% [1]. In Mafas's study, audio features were combined with social media to predict a song's popularity. As a result, social media features are used to predict these factors with an accuracy increase of up to 10% [2]. In addition to audio features, visual features are an important element in predicting popularity. For instance, Elham used metadata, audio, and visual features to develop a movie recommender system [3].

In previous research, audio features of a song are used to train and predict a song's popularity with machine learning. However, no-one has studied the relationship between music videos and a song's popularity. Given that YouTube is an online platform, over half of people aged 13–36 listen to music on it. Therefore, a high-quality music video is the main factor for hit songs.

Using new methods, we predicted a song's popularity by extracting music video features from YouTube and combining them with an existing dataset that contains audio features. Datasets "Spotify and YouTube" from Kaggle were analyzed using OpenCV to extract the features of the music videos. Then, the dataset combined with audio and visual features was used to predict a song's popularity using machine learning techniques. Then, we explored which machine learning methods performed best for the proposed model and which visual elements enhanced the prediction of hit songs, and we provide suggestions to the music industry based on these results.

2. Literature Review

2.1. Music Popularity Prediction

Machine learning techniques have been used in various domains for prediction, especially in estimating the popularity of songs. Many researchers use different techniques to create a distinct prediction model. Ogihara highlighted the use of computational methods for genre classification. The introduction of Daubechies Wavelet Coefficient Histograms (DWCHs) is regarded as a significant development. Compared with existing techniques with algorithms such as SVMs and linear discriminant analysis, DWCHs have enhanced the accuracy of music genre classification [4].

Building on the computational approach, Dhanaraj and Logan predicted hit songs using automatic music analysis. They used a database containing 1700 songs with acoustic and lyric information to classify tracks as hits or non-hits using SVMs and boosting classifiers. They concluded that lyric-based features are more effective than acoustic features in identifying potential hits. However, the methods are not successful, with much room for improvement [5]. Borg and Hokkanen explored the relationship between acoustic features, genre, and YouTube view counts to predict hit songs. Their data were obtained from the Million Song Dataset with 10,000 songs and its pre-extracted features. By combining SVM and K-means clustering, an accuracy of 55% was achieved. The artist's popularity was found to be correlated with hit songs, regardless of the acoustic features [6]. Yang and Hu predicted the popularity of non-Western songs (Chinese songs in this case) using Western songs' mood classification models. They classified audio features for Western and Chinese music and compared the performance of mood classification. They considered psychoacoustic features to apply to Chinese songs with an accuracy of 74.7% [7].

As a modern technique, Yang et al. focused on using deep learning algorithms such as the convolutional neural network (CNN) model and the JYnet model to enhance the prediction of hit songs. The dataset was obtained from KKBOX with information about Taiwanese users' listening records from 2012 to 2013. The data were used to train shallow and deep learning models by framing it as a regression problem. Deep learning models performed better and highlighted the importance of how genres are proportioned in the learning algorithms [8].

In various studies, various data features with machine learning algorithms are used to predict hit songs. Rajyashree et al. used MIDI features [9], Yang et al. used high and low levels of audio features [10], Pham et al. combined songs with information such as artist names [11], Kim and Oh used granular acoustic features [12], and Nikas et al. used datasets of audio features, lyrics, and temporal features [13]. Mayerl et al. introduced ordinal classification to predict hit songs. They utilized two datasets containing 7736 and 73,432 songs to test classification, regression, and ordinal models. They suggested that ordinal classification has similar accuracy to classification and regression models and, in some cases, performed better [14].

These studies used machine learning to predict song popularity with various features such as acoustic properties, lyrics, and genre classifications. However, the primary focus was put on audio features to predict hits. Based on previous results, we incorporated visual features into audio features to enhance the predictive accuracy of hit songs.

2.2. Sound

Sound is characterized by several fundamental parameters essential for understanding its manipulation in various applications. The first one is frequency, which determines the pitch of a sound, and this is measured in Hertz (Hz). The second one is amplitude, indicating the volume or loudness of a sound, and this is measured in decibels (db). Waveform is measured by both frequency and amplitude, giving the shape of the sound wave [15]. Waveform represents sound in time and frequency domains. The time domain represents how the sound signal changes over time. Conversely, the frequency domain represents the breakdown of sound into its frequencies, each with its amplitude. Fourier Transform is used in the frequency domain by breaking down signals into the sum of sinusoidal functions. Sound can be captured using multiple digital formats, including WAV, MP3, AAC, and FLAC [16].

2.3. Visual Elements

Chang et al. introduced license plate recognition (LPR) with minimal environmental constraints. Using a dataset of 1088 images, they applied modules with color edge detectors to identify the image and number on the plate. The color detector's accuracy reached 93.7% [17]. Expanding the application of visual elements, Devi et al. used an image processing system to predict fruit yield based on its color and shape. Images of the fruit trees were taken and processed by OpenCV using an image processing algorithm. Machine learning algorithms achieved 94% accuracy [18]. Visual elements in agriculture are used in plant disease detection. Domingues used the online dataset PlantVillage with 18,160 images of tomatoes to extract features such as color distributions and texture patterns. SVM, Random Forest (RF), and Artificial Neural Networks (ANNs) were trained, achieving an accuracy above 90% in predicting tomato diseases [19]. Nageswaran et al. used image processing techniques such as noise reduction and feature extraction with the K-means technique to classify and predict lung cancer. They used a dataset of 83 CT scans from 70 patients. For this task, classification was performed using ANN, KNN, and RF. The ANN model showed the highest accuracy in predicting lung cancer.

Previous studies indicated that visual elements can be used for prediction with machine learning. They are not used to predict hit songs yet.

3. Materials and Methods

Figure 1 shows a flow chart of this research. Data were gathered from 'Spotify and YouTube' from 7 February 2023, including 20,717 songs on YouTube. The dataset was obtained from Kaggle. In total, 18 features were extracted, including 11 audio features and song information such as song title, artist, view count, etc. Data were collected from 13 April to 21 May 2023. After cleaning the dataset, 5 visual features were extracted using OpenCV and merged with audio features into a new dataset. The highest and lowest 500 songs in terms of YouTube views were labeled as popular and unpopular categories. Classification machine learning was used for audio and visual features.

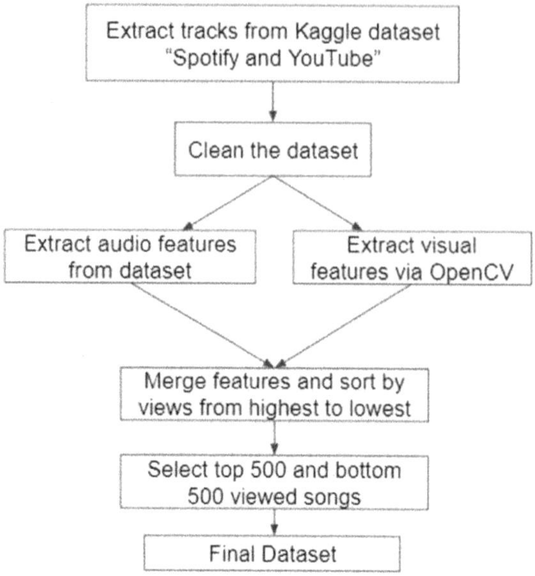

Figure 1. Research process of data preprocessing in this research.

Figure 2 shows the structure of this research. Two experiments were conducted to showcase the significance of visual features in prediction. The first dataset contained audio features. In total, 5 distinct machine learning models were used. The second dataset containing audio and visual features was processed with the same machine learning models.

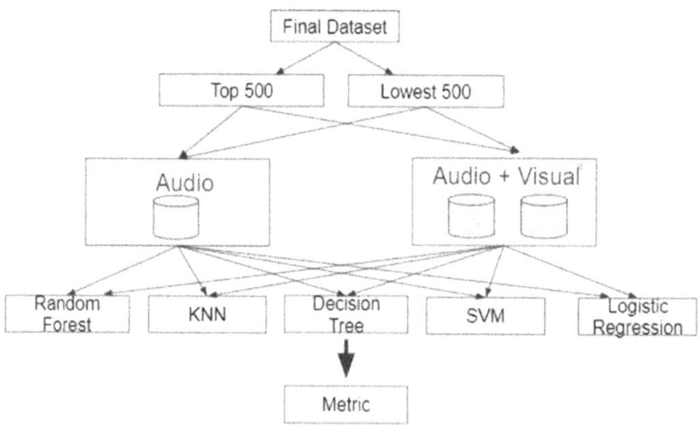

Figure 2. Experimental procedure flow chart.

3.1. Independent Variables

The 10 sets of audio features were determined from the dataset. A description of the features is shown in Table 1.

Table 1. Audio features.

Features	Description
Danceability	Describes how suitable a track is for dancing based on a combination of musical elements. A value of 0.0 is least danceable and 1.0 is the most danceable.
Energy	A measure from 0.0 to 1.0 which represents a perceptual measure of intensity and activity.
Key	The key the track is in. Integers map to pitches using standard Pitch Class notation.
Loudness	The overall loudness of a track in decibels (dB).
Speechiness	The presence of spoken words in a track. The more exclusively speech-like the recording (e.g., talk show, audio book, poetry), the closer to 1.0 the attribute value.
Acousticness	A confidence measure from 0.0 to 1.0 of whether the track is acoustic: 1.0 represents high confidence that the track is acoustic.
Liveness	The presence of an audience in the recording. Higher liveness values represent an increased probability that the track was performed live.
Valence	A measure from 0.0 to 1.0 describing the musical positiveness conveyed by a track. Tracks with high valence sound more positive (e.g., happy, cheerful, euphoric), while tracks with low valence sound more negative (e.g., sad, depressed, angry).
Tempo	The overall estimated tempo of a track in beats per minute (BPM).
Duration_ms	The duration of the track in milliseconds.

In total, 5 sets of visual features were acquired from OpenCV using the dataset YouTube playlist. The description of the features can be shown in Table 2.

Table 2. YouTube visual features.

Features	Description
Average brightness	The average luminance level across all frames of a video.
Average motion	The average magnitude of change between consecutive frames.
Average R value	The mean intensity of the red color across all pixels in a video.
Average G value	The mean intensity of the green color across all pixels in a video.
Average B value	The mean intensity of the blue color across all pixels in a video.

3.2. Dependent Variables

In total, 20,717 songs and 2079 artists were contained in the dataset. After eliminating invalid audio files and missing values, 18,774 songs remained. The top 500 highest-view-count videos were combined with the 500 lowest-view-count videos to create the final dataset. Cross-validation with 5 folds was performed on this dataset. Accuracy and F1-score were calculated for evaluating binary classifiers.

4. Results

Five different machine learning models were used to explore the effect of visual elements on hit song prediction, and the visual element correlation between popular and non-popular songs was analyzed.

4.1. Accuracy of Models

Table 3 summarizes the comparison results for the five models. The best-performing model is highlighted for each column. Out of the five models, RF exceeded the others by all four metrics. It outperformed in both audio features and audio combined with visual features, showing higher accuracy and F1-score. On the other hand, KNN had the poorest results across all metrics compared to the rest. There was an increase in accuracy after including visual features. Every model had an improvement ranging from 7 to 12%.

Table 3. Comparisons of average accuracy and F1-score of the five algorithms with audio and visual elements.

Algorithm	Audio Accuracy	Audio+Visual Accuracy	Audio F1-Score	Audio+Visual F1-Score
Support Vector Machine	72.57%	80.68%	71.98%	80.57%
Random Forest	73.87%	81.98%	73.57%	80.57%
Decision Tree	69.26%	76.47%	69.06%	76.40%
KNN	63.37%	74.77%	62.84%	74.29%
Logistic Regression	72.57%	80.78%	72.10%	80.69%
Average	70.33%	78.93%	69.91%	78.76%

4.2. Visual Features

Table 4 reveals significant differences in multiple characteristics. The t-test results indicate that all five features showed marked differences between the two groups from the low p-values. Table 4 also shows the difference in mean values for five visual features between popular and non-popular songs. Out of the five visual features, non-popular songs showed higher brightness and color values. However, popular songs had significantly higher motion values with an average of 7.78 compared to 3.18 in non-popular songs.

Table 4. Comparisons of average value of five visual elements with audio and visual elements.

Features	Non-Popular Songs			Popular Songs		
Average R Value	43 #2B0000	91 #5B0000	141 #8D0000	42 #2A0000	73 #490000	104 #680000
Average G Value	35 #002300	82 #005200	130 #008200	39 #002700	67 #004300	95 #005F00
Average B Value	32 #000020	80 #000050	128 #000080	36 #000024	65 #000041	94 #00005E
Average	(91,82,80)			(73,67,65)		

The following hypotheses were proposed and tested to determine significant differences in terms of brightness, motion, and RGB values. The first hypothesis (H1) suggested that the average brightness of non-popular songs is equal to the average brightness of popular songs.

Hypothesis 1:

- H_0: The average brightness of non-popular songs is equal to the average brightness of popular songs.
- $\left(average\ brightness\ \mu_{non\text{-}popular} = average\ brightness\ \mu_{popular}\right)$
- H_1: The average brightness of non-popular songs is not equal to the average brightness of popular songs.

- $\left(average\ brightness\ \mu_{non\text{-}popular} \neq average\ brightness\ \mu_{popular}\right)$

The average brightness for non-popular songs was 85.04, while for popular songs it was 68.93. The p-value was less than 0.05, which shows that the average brightness for popular songs was significantly lower than that of non-popular songs.

The second hypothesis (H2) suggested that the average motion of non-popular songs is equal to the average motion of popular songs.

Hypothesis 2:

- H_0: The average motion of non-popular songs is equal to the average motion of popular songs.
- $\left(average\ motion\ \mu_{non\text{-}popular} = average\ motion\ \mu_{popular}\right)$
- H_1: The average motion of non-popular songs is not equal to the average motion of popular songs.
- $\left(averager\ motion\ \mu_{non\text{-}popular} \neq average\ motion\ \mu_{popular}\right)$

The average motion for non-popular songs was 3.18, while for popular songs it was 7.78. The p-value of 8.10×10^{-64} was less than 0.05, which showed that the average motion for popular songs was significantly higher than that of non-popular songs.

The third hypothesis (H3) suggested that the average R value of non-popular songs was equal to that of popular songs.

Hypothesis 3:

- H_0: The average R value of non-popular songs is equal to the average R value of popular songs.
- $\left(average\ R\ value\ \mu_{non\text{-}popular} = average\ R\ value\ \mu_{popular}\right)$
- H_1: The average R value of non-popular songs is not equal to the average R value of popular songs.
- $\left(averager\ R\ value\ \mu_{non\text{-}popular} \neq average\ R\ value\ \mu_{popular}\right)$

The average R value for non-popular songs was 91.82, while for popular songs it was 73.81. The p-value was less than 0.05, which showed that the average R value for popular songs was significantly lower than that of non-popular songs.

The fourth hypothesis (H4) suggested that the average G value of non-popular songs was equal to the average G value of popular songs.

Hypothesis 4:

- H_0: The average G value of non-popular songs is equal to the average G value of popular songs.
- $\left(average\ G\ value\ \mu_{non\text{-}popular} = average\ G\ value\ \mu_{popular}\right)$
- H_1: The average G value of non-popular songs is not equal to the average G value of popular songs.
- $\left(averager\ G - value\ \mu_{non\text{-}popular} \neq average\ G\ value\ \mu_{popular}\right)$

The average G value for non-popular songs was 82.53, while for popular songs it was 67.08. The p-value was less than 0.05, which showed that the average G value for popular songs was significantly lower than that of non-popular songs.

The fifth hypothesis (H5) suggested that the average B value of non-popular songs is equal to the average B value of popular songs.

Hypothesis 5:

- H_0: The average B value of non-popular songs is equal to the average B value of popular songs.
- $\left(average\ B\ value\ \mu_{non\text{-}popular} = average\ B\ value\ \mu_{popular}\right)$
- H_1: The average B value of non-popular songs is not equal to the average B value of popular songs.
- $\left(averager\ B\ value\ \mu_{non\text{-}popular} \neq average\ B\ value\ \mu_{popular}\right)$

The average B value for non-popular songs was 80.37, while for popular songs it was 65.88. The p-value was less than 0.05, which showed that the average B value for popular songs was significantly lower than that of non-popular songs.

There were significant differences in RGB values between non-popular and popular songs. Interestingly, popular songs had lower RGB values than non-popular songs, considering the mean and standard deviation. For popular songs, the average R-value (73, #490000) was substantially lower than that for non-popular songs (91, #5B0000), similar to the upper bound (104, #680000 vs. 141, #8D0000) and lower bound (42, #2A0000 vs. 43, #2B0000), which refers to the mean R-value. Popular songs had higher intensity and saturation in their color elements. The mean and standard deviation values confirmed that these differences were statistically significant, which implied that brightness and color were influential factors in a song's popularity.

When inspecting the lower and upper values for popular and non-popular songs, popular songs showed a smaller difference between the bonds, while non-popular songs had a larger one. Popular songs showed colors within a narrower range, while non-popular songs had more diverse visual presentations.

5. Conclusions

Using acoustic and visual elements, we predicted the song's popularity. RF was the top-performing algorithm for predicting music popularity for all four metrics, with an accuracy of 81.98% and an F1-score of 81.85%. Visual elements in YouTube videos are beneficial for the overall prediction of hit songs, increasing accuracy by 81.85% and F1-score by 8.28%. The t-test results show that popular songs had higher motion compared to non-popular songs. Movement in a song's music video is crucial, indicating that dances and background changes are important in creating a hit song. Conversely, popular songs had lower RGB and brightness values than non-popular songs, indicating darker, dimmer colors. Additionally, popular songs showed a more consistent color in a narrower range, whereas non-popular songs showed diverse colors.

The dataset was uploaded in 2023. Thus, view counts may vary. In addition, songs are added or removed frequently to maintain an up-to-date dataset for further analysis. We used five visual features. Further research is necessary to extract more visual elements such as thumbnail quality, visual effects, or dance choreography. Additionally, the genres of the song needs to be added to predict its popularity.

Author Contributions: Conceptualization, C.-Y.L. and Y.-N.T.; methodology, C.-Y.L. and Y.-N.T.; software, C.-Y.L.; validation, C.-Y.L. and Y.-N.T.; formal analysis, C.-Y.L.; investigation, C.-Y.L.; resources, C.-Y.L.; data curation, C.-Y.L.; writing—original draft preparation, C.-Y.L.; writing—review and editing, Y.-N.T.; visualization, C.-Y.L.; supervision, Y.-N.T.; project administration, Y.-N.T. All authors have read and agreed to the published version of the manuscript.

Funding: This research received no external funding.

Institutional Review Board Statement: Not applicable.

Informed Consent Statement: Not applicable.

Data Availability Statement: Data sharing is not applicable to this article.

Conflicts of Interest: The authors declare no conflict of interest.

References

1. Anjana, S. A Robust Approach to Predict the Popularity of Songs by Identifying Appropriate Properties. Ph.D. Thesis, University of Colombo School of Computing, Colombo, Sri Lanka, 2021.
2. Yee, Y.K.; Raheem, M. Predicting music popularity using Spotify and YouTube features. *Indian J. Sci. Technol.* **2022**, *15*, 1786–1799. [CrossRef]
3. Motamedi, E.; Kholgh, D.K.; Saghari, S.; Elahi, M.; Barile, F.; Tkalcic, M. Predicting movies' eudaimonic and hedonic scores: A machine learning approach using metadata, audio and visual features. *Inf. Process. Manag.* **2024**, *61*, 103610. [CrossRef]
4. Dhanaraj, R.; Logan, B. Automatic Prediction of Hit Songs. In Proceedings of the 6th International Conference on Music Information Retrieval (ISMIR 2005), London, UK, 11–15 September 2005; pp. 488–491.
5. Li, T.; Ogihara, M.; Li, Q. A comparative study on content-based music genre classification. In Proceedings of the 26th Annual International ACM SIGIR Conference on Research and Development in Informaion Retrieval, Toronto, ON, Canada, 28 July–1 August 2003.
6. Borg, N.; Hokkanen, G. What Makes for a Hit Pop Song? What Makes for a Pop Song? Unpublished Thesis, Stanford University, CA, USA, 2011.
7. Yang, Y.H.; Hu, X. Cross-cultural Music Mood Classification: A Comparison on English and Chinese Songs. In Proceedings of the 13th International Society for Music Information Retrieval Conference (ISMIR 2012), Porto, Portugal, 8–12 October 2012; pp. 19–24. [CrossRef]
8. Yang, L.C.; Chou, S.Y.; Liu, J.Y.; Yang, Y.H.; Chen, Y.A. Revisiting the problem of audio-based hit song prediction using Convolutional Neural Networks. In Proceedings of the 2017 IEEE International Conference on Acoustics, Speech and Signal Processing (ICASSP), New Orleans, LA, USA, 5–9 March 2017.
9. Rajyashree, R.; Anand, A.; Soni, Y.; Mahajan, H. Predicting hit music using MIDI features and machine learning. In Proceedings of the 2018 3rd International Conference on Communication and Electronics Systems (ICCES), Coimbatore, India, 15–16 October 2018.
10. Zangerle, E.; Huber, R.; Vötter, M.; Yang, Y.H. Hit Song Prediction: Leveraging Low- and High-Level Audio Features. In Proceedings of the 20th International Society for Music Information Retrieval Conference (ISMIR 2019), Delft, The Netherlands, 4–8 November 2019.
11. Pham, B.D.; Tran, M.T.; Pham, H.L. Hit song prediction based on Gradient Boosting Decision tree. In Proceedings of the 2020 7th NAFOSTED Conference on Information and Computer Science (NICS), Ho Chi Minh City, Vietnam, 26–27 November 2020.
12. Kim, S.T.; Oh, J.H. Music intelligence: Granular data and prediction of top ten hit songs. *Decis. Support Syst.* **2021**, *145*, 113535.
13. Nikas, D.; Sotiropoulos, D.N. A machine learning approach for modeling time-varying hit song preferences. In Proceedings of the 2022 13th International Conference on Information, Intelligence, Systems & Applications (IISA), Corfu, Greece, 18–20 July 2022.
14. Vötter, M.; Mayerl, M.; Zangerle, E.; Specht, G. Song Popularity Prediction using Ordinal Classification. In Proceedings of the 20th Sound and Music Computing Conference (SMC 2023), Stockholm, Sweden, 12–17 June 2023.
15. Nagathil, A.; Martin, R. Evaluation of spectral transforms for Music Signal Analysis. In Proceedings of the 2013 IEEE Workshop on Applications of Signal Processing to Audio and Acoustics, New Paltz, NY, USA, 20–23 October 2013.
16. Weismer, G. Acoustic Phonetics. Available online: https://www.accessscience.com/content/article/a802380 (accessed on 26 February 2025).
17. Chang, T.; Chen, L.; Chung, Y.; Chen, S. Automatic license plate recognition. *IEEE Trans. Intell. Transp. Syst.* **2004**, *5*, 42–53. [CrossRef]
18. Devi, T.G.; Neelamegam, P.; Sudha, S. Image Processing System for automatic segmentation and yield prediction of fruits using open CV. In Proceedings of the 2017 International Conference on Current Trends in Computer, Electrical, Electronics and Communication (CTCEEC), Mysore, India, 8–9 September 2017.
19. Domingues, T.; Brandão, T.; Ferreira, J.C. Machine learning for detection and prediction of crop diseases and pests: A comprehensive survey. *Agriculture* **2022**, *12*, 1350. [CrossRef]

engineering proceedings

MDPI

Proceeding Paper

Development of Innovative 3D Spherical Retrieval System and Virtual Reality for Insomnia Prescriptions in Traditional Chinese Medicine †

Chia-Hui Shih [1,*], Geng-Hao Liu [2,3,*] and Ting-An Fan [3]

1 Institute of Information Management, Fu-Jen Catholic University, No. 510, Zhongzheng Rd., Xinzhuang Dist., New Taipei 242062, Taiwan
2 School of Traditional Chinese Medicine, Chang Gung University, No. 259, Wenhua 1st Rd., Guishan Dist., Taoyuan City 333323, Taiwan
3 Center for Traditional Chinese Medicine, Chang Gung Memorial Hospital, 8F., No. 123, Dinghu Rd., Guishan Dist., Taoyuan City 333008, Taiwan; 00037tim@gmail.com
* Correspondence: 159931@mail.fju.edu.tw (C.-H.S.); 8905033@cgmh.org.tw (G.-H.L.)
† Presented at the 2024 IEEE 7th International Conference on Knowledge Innovation and Invention, Nagoya, Japan, 16–18 August 2024.

Abstract: Insomnia is prevalent in modern society, and traditional Chinese medicine is gradually replacing Western medicine in its treatment. This study utilized insomnia symptoms and prescriptions from the "Dictionary of Chinese Medicine Prescriptions" to establish a 3D TCM spherical retrieval system, which intuitively displays the relationship between TCM prescriptions and insomnia symptoms. It enhances public and student interest in learning about TCM. Survey results indicate that the system effectively improves public knowledge of TCM and supports the United Nations Sustainable Development Goal 4: Quality Education.

Keywords: traditional Chinese medicine (TCM); spherical-type retrieval system (SRS); AI; MR; SDG 4; telemedicine

1. Introduction

Insomnia is a common symptom in modern individuals. In adults over the age of 35, the one-year incidence of acute insomnia is 27%, with nearly 20% of patients experiencing persistent sleep disturbances. Chronic insomnia significantly impacts one's quality of life, leading to impaired memory, lack of concentration, reduced work efficiency, emotional instability, and adverse effects on interpersonal relationships and social satisfaction. Chronic insomnia increases the risk of certain diseases and exacerbates the severity of existing conditions, including allergic diseases, asthma, cardiovascular diseases, metabolic syndrome, mental disorders, dementia, and sepsis [1]. Given the prevalence of insomnia symptoms and their serious impact on health, effective insomnia treatment has become an important medical concern.

Many patients choose alternative therapies, especially traditional Chinese medicine (TCM), known for its holistic and constitution-nourishing approach, to treat insomnia. According to statistics, insomnia was most diagnosed among patients seeking treatment at the TCM Department of Taipei Veterans General Hospital in 2002 [2]. The insurance database shows a gradual increase in the number of people seeking Chinese medicine treatment for insomnia from 2002 to 2004 [1]. This indicates widespread attention among Taiwanese patients to the application of TCM in the treatment of insomnia.

Considering the diversity in the composition and application of Chinese herbal formulas used in TCM for treating insomnia, numerous studies have been conducted to clarify their efficacy and scope of application. For instance, Taipei City Hospital published a study summarizing Chinese herbal formulas used for treating insomnia in Taiwan from 2003 to 2006, as shown in Table 1 [3].

Table 1. Medications used to treat sparse treatment in Taiwan (2003–2006) [3].

Chinese Herbal Formulae	Ingredients	n (%)
Jia-Wey-Shiau-Yau-San	Angelica sinensis, Atractylodes macrocephala, Paeonia lactiflora, Bupleurum chinense, Poria cocos, etc.	2485 (6.71)
Suan-Tsao-Jen-Tang	Ziziphus jujuba, Anemarrhena asphodeloides, Ligusticum wallichii, Poria cocos, Glycyrrhiza uralensis	2284 (6.17)
Tian-Wang-Bu-Xing-Dan	Rehmannia glutinosa, Panax ginseng, Poria cocos, Polygala tenuifolia, Acorus gramineus, etc.	1630 (4.40)
Qing-Xin-Lian-Zi-Yin	Nelumbo nucifera, Poria cocos, Astragalus membranaceus, Panax ginseng, Ophiopogon japonicus	1032 (2.79)

Table 1 provides statistical data on the most commonly used Chinese herbal medicines and formulas for insomnia, serving as a reference for clinical physicians. Although this data allows doctors to understand commonly used formulas, it does not delve into the relationship between different symptoms and the use of Chinese herbal formulas. Therefore, it does not provide individualized treatment approaches for different patients, nor does it comprehensively reflect the diversity of Chinese herbal formulas for treating insomnia. Additionally, there is a lack of an excellent retrieval system, and it does not visually present relevant information on TCM treatments for insomnia symptoms and medicines. To address the inefficiency of retrieval in Table 1, we established a 3D visualized spherical retrieval system (SRS) to improve the efficiency of retrieving TCM symptoms and medicines. Furthermore, by creating a virtual reality sphere (VRS), the public's learning interest in TCM can be increased in a novel and engaging way. This effort is related to the United Nations Sustainable Development Goal 4: Quality Education. Regardless of whether the users are TCM professionals or not, they can use the developed system to query information on TCM-related symptoms, corresponding Chinese medicines, and prescriptions.

2. Data Preprocessing

We selected 72 commonly used symptoms and 72 prescriptions from the insomnia prescription retrieval table in the Dictionary of TCM Formulas (Tables 2 and 3) and encoded them. Subsequently, we established basic data tables for these symptoms and medicines and calculated their similarities in preparation for data clustering.

Table 2. "Insomnia" formulary retrieval table in the TCM formulary dictionary (partial) [1].

Insomnia	01439	03290	03638	03652	03868
04345	04507	06334	07895	08214	09232
09762	09780	10424	10425	10428	10484
84176	87089	87502	87503	87504	87508
87510	88909	93382	93644	93849	93851
94347	95208	95815	96008		

Before calculating the similarities, it is important to compute the common medicines between pairs of symptoms. The more common medicines there are, the more similar the two symptoms are. Given a set of medicines $M = \{m_1, m_2, \ldots, m_{72}\}$ and a set of symptoms $S = \{s_1, s_2, \ldots, s_{72}\}$, the number of common medicines (i.e., co-occurrence medicines for a pair of symptoms) in the medicine set that can treat a pair of symptoms s and s' is calculated using (1) [4]. The computed results for common medicines are shown in Table 4. The data in Table 4 is then imported into the Orange software (Version 3.36.2) to cluster the 72 symptoms.

$$Co\text{-}medicine_{s,s'} = \sum_{m=1}^{72} \alpha_{m,s,s'}, \quad (1)$$

Table 3. Prescription page number table (partial) [1].

Number	Formula	Page
01439	Shi Wei Wen Dan Tang	168
03290	Ren Shen San	402
03638	Ren Shen Zhu Ye Tang	441
07895	Shang Xia Liang Ji Dan	1003
08214	Qian Li Liu Shui Tang	1041
09232	Xiao Suan Zao Tang	1176

Table 4. Co-medicine of symptom-to-symptom (partial) [1].

	s1	s2	s3	s4	s5
s1	0	0	0	0	0
s2	0	0	0	0	3
s3	0	0	0	1	0
s4	0	0	1	0	0
s5	0	3	0	0	0

3. Symptom Clustering

The k-means algorithm was used to cluster the symptoms. The k-means algorithm uses a greedy approach to improve clustering quality and approach a local optimum [5]. K-means is the most popular, simplest, and has low computational complexity among clustering algorithms [6]. Johannes et al. applied clustering methods to human grouping, while we clustered methods using data segmentation [7]. The purpose of clustering and solving in batches is to improve optimization efficiency and reduce the procurement and setup costs of computer hardware and software. We used the Orange software (https://orangedatamining.com/, accessed on 1 November 2023) for clustering. The process includes (1). data import, (2). setting attributes and target fields, (3). creating training and testing data, (4). building models, (5). evaluating models, and (6). examining predicted results (Figure 1) [8].

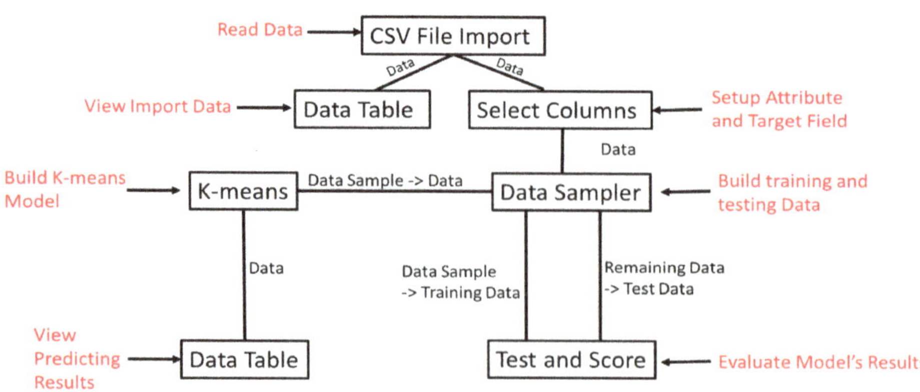

Figure 1. Procedure of k-mean clustering in Orange 3 [8].

We prepared a sphere with 144 grids. The 72 symptoms were divided into 3 clusters using the K-means algorithm and allocated to the upper hemisphere, while the lower hemisphere was allocated to the 72 medicines. Figure 2 shows the K-means clustering results for cluster C1. C1 contains 11 symptoms. Each symptom was assigned to one grid. To accommodate these symptoms, they were divided into 2 columns, with each column containing 6 grids, resulting in a total of 2 × 6 = 12 grids.

#	Symptom	Cluster	Silhouette
1	s31	C1	0.524702
2	s16	C1	0.565012
3	s21	C1	0.561783
4	s41	C1	0.475741
5	s44	C1	0.517913
6	s13	C1	0.466561
7	s14	C1	0.482402
8	s25	C1	0.462007
9	s43	C1	0.468212
10	s28	C1	0.560949
11	s40	C1	0.570792
12	s58	C1	0.610504

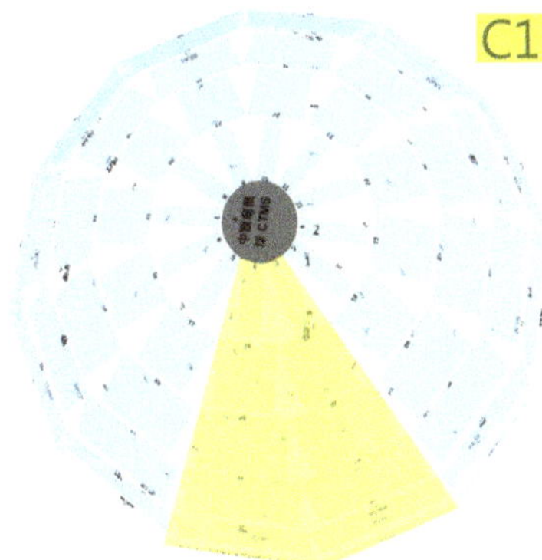

Figure 2. Distribution of cluster C1 on upper hemisphere. The Chinese text on the image is the name of this sphere: Traditional Chinese Medicine Syndrome and Medication Sphere.

The allocation results for the second cluster, C2, using K-means are shown in Figure 3. C2 contains 43 symptoms. To accommodate these symptoms, 8 vertical columns are used, each containing 6 grids, resulting in a total of 8 × 6 = 48 grids.

#	Symptom	Cluster	Silhouette
1	s58	C2	0.610504
2	s60	C2	0.601731
3	s59	C2	0.620185
4	s63	C2	0.642718
5	s71	C2	0.648413
6	s65	C2	0.643693
7	s67	C2	0.641931
…	…	…	…

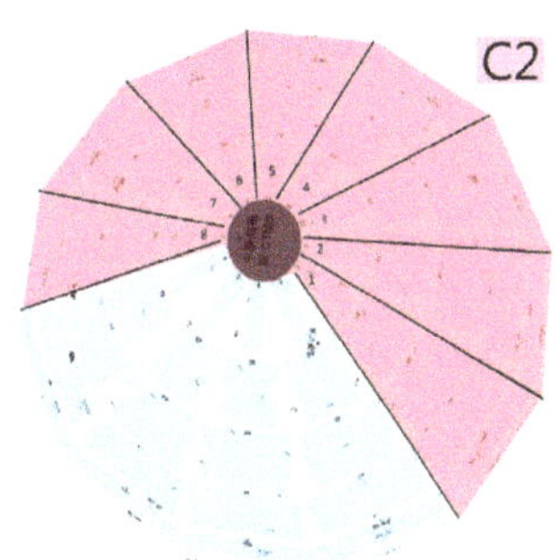

Figure 3. Distribution of cluster C2 on upper hemisphere. The Chinese text on the image is the name of this sphere: Traditional Chinese Medicine Syndrome and Medication Sphere.

The remaining symptoms that were not assigned to C1 and C2 were allocated to cluster C3. Thus, the allocation results of the symptoms on the upper hemisphere are shown in Figure 4. Grids of the same color on the sphere indicate the same cluster. Symptoms that share common medicines have higher similarity, and those with higher similarity are more likely to be grouped. Symptoms s40, s16, and s21 have the highest number of common medicines and are indeed grouped. This demonstrates the high accuracy of using k-means clustering.

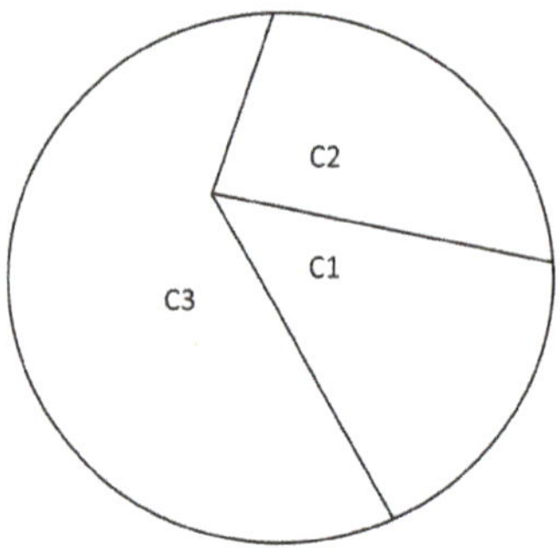

Figure 4. K-means clustering results in upper hemisphere.

Although the clustering results are highly accurate, dividing the symptoms into only three clusters does not adequately reflect the detailed characteristics of the symptoms and the specific functions of the medicines. Therefore, TCM practitioners need to categorize the symptoms into 12 categories, with each category assigned to the same vertical column. This approach makes it easier to search for and intuitively understand the main functions and directions of the medicines. Consequently, we reclassified the symptoms and medicines according to the recommendations of TCM experts. Finally, on the sphere, different colored grids represent different categories of symptoms and medicines.

4. TCM Database

After determining the positions of all symptoms and medicines in the sphere, we established a comprehensive TCM database containing all the data. Basic data tables for symptoms, medicines, and prescriptions were created with tables for grid numbers and coordinates on the sphere. The relationship between these tables was established. Figure 5 shows the snowflake schema diagram of the normalized relationships between the tables.

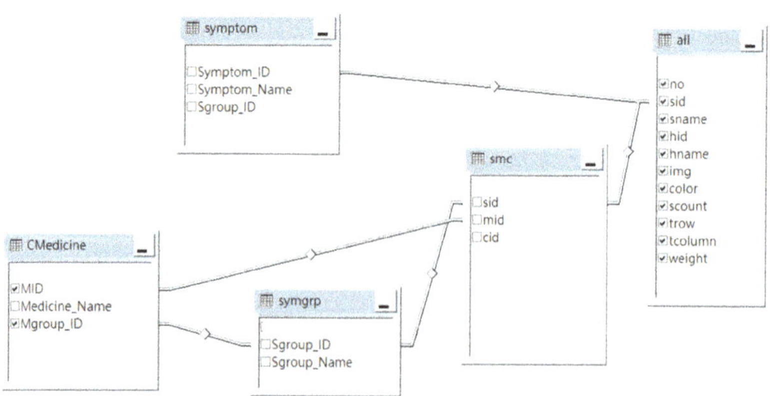

Figure 5. Entity-relationship diagram (ERD) of TCM database.

5. Retrieval System

After the TCM database was established, an online interactive SRS was created to help users query the corresponding medicines for TCM symptoms. This SRS uses technologies such as ASP.NET, JavaScript, Three.js, and Microsoft's SQL Server. It is built on a three-tier architecture. Users can access the system via a computer, smartphone, or tablet. The system's designed functions include the following: (1) query filtering based on symptoms or medicines; (2) zooming in and out and panning the sphere; (3) clicking on images on the sphere opens new pages with detailed information; (4) selecting a symptom from the dropdown menu to display the prescriptions for that symptom and rotating the sphere to position the symptom at the front center; (5) highlighting the relevant symptoms and corresponding medicines on the sphere in red when hovering the mouse over the menu, the hover function is triggered; and (6) visual analysis charts for prescriptions, such as word clouds and tree diagrams.

The URL for the completed TCM SRS is https://ctm.splab.fju.edu.tw/Ctm.aspx (Accessed on 22 February 2025). The website interface is shown in Figure 6. This URL automatically detects the access device, and if it is a mobile device (smartphone or tablet), the system automatically switches to the mobile version of the webpage. The mobile version interface is shown in Figure 7.

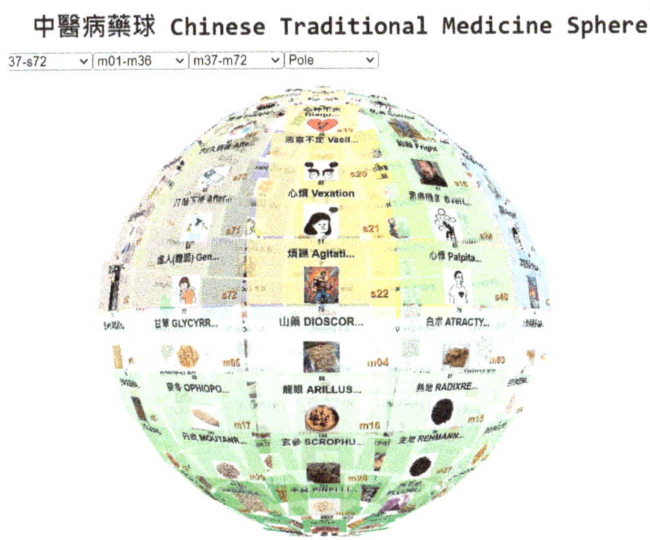

Figure 6. TCM SRS.

Additionally, for the desktop version, we developed an interactive highlight function (Hover). When users move the cursor over a symptom or medicine in the menu, the corresponding medicine names and their related symptom numbers on the sphere indicate the symptoms treatable by that medicine. Similarly, when the cursor hovers over a symptom, the background color of the symptom and its corresponding medicine numbers turn red on the sphere. This highlight function received high praise from participants during the survey (Figure 8).

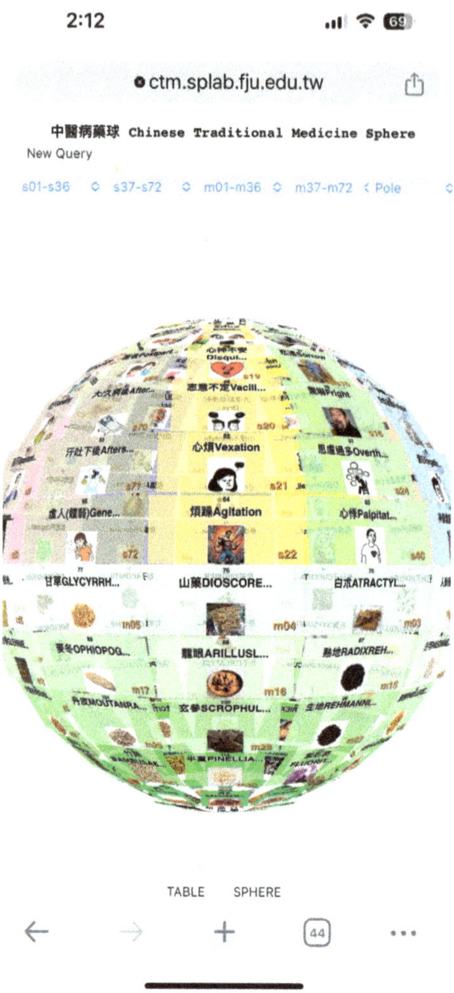

Figure 7. Mobile version of TCM SRS.

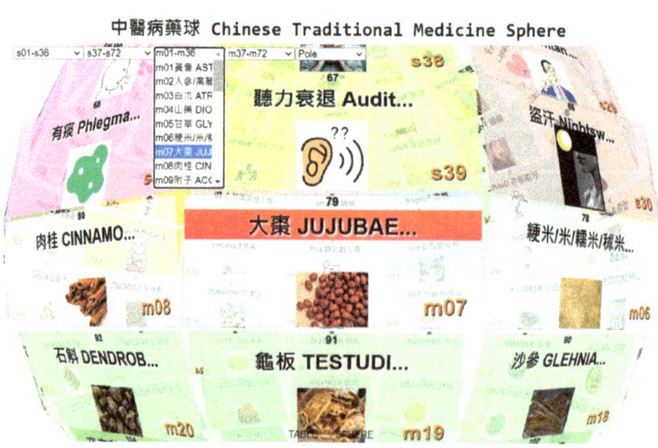

Figure 8. Highlight related objects (red) on the top menu.

The query interface design of this study allows users to find related medicines and prescriptions from symptoms or to find treatable symptoms from Chinese medicines. The query interface is shown in Figure 9.

Figure 9. Query user interface.

For consultations with patients in remote areas or those needing telemedicine, such as during the COVID-19 quarantine, we designed a telemedicine video consultation feature. Doctors need to click the "Video Consultation" button, and the system automatically activates the webcam. This allows TCM doctors to observe the patient's complexion, examine their tongue, and inquire about their symptoms, as shown in Figure 10.

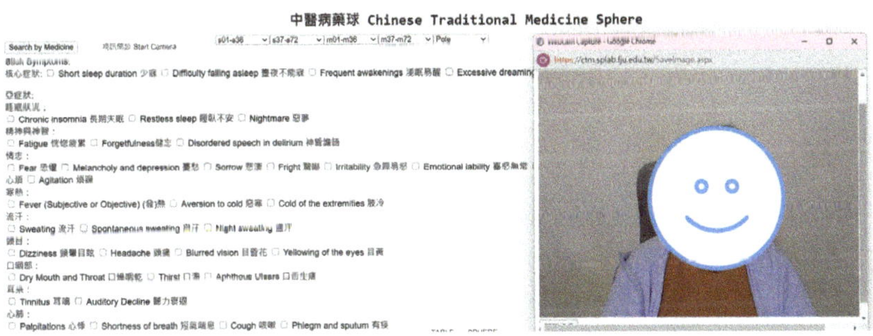

Figure 10. Telemedicine video consultation.

6. VRS

To enhance user interaction and increase interest in TCM, we developed a TCM VRS that users can immerse in using VR glasses or head-mounted devices. The system was developed using the Oculus Unity SDK and is compatible with Meta Quest 3 devices. The designed features of the VRS are as follows.

6.1. Spherical Information Display

In a 3D VR environment, a 3D spherical structure and 144 information panels about TCM are constructed.

6.2. Dynamic Double-Sided Browsing

Based on the user's position, whether inside or outside the sphere, the information panels dynamically adjust to display on the front or back, providing an optimal browsing experience.

6.3. *Smart Dragging*

Users can use the right-hand joystick and ray to drag information panels and rotate the sphere. To prevent the information panels from tilting or flipping during rotation, the sphere automatically aligns the selected information panel to an easy-to-read orientation, ensuring a smooth browsing experience.

6.4. *Position and Size Adjustment*

To allow users to browse and operate in their preferred way, the left-hand joystick can be used to adjust the sphere's position and size. The left-hand joystick's X button is used to set the sphere's center position on the ground at the user's feet, the Y button is used to set the sphere's center at the front of the left-hand controller, and the left-hand joystick is used to scroll up and down to adjust the sphere's size.

Figure 11 shows the development results and actual testing of the VR sphere. It includes a photo from one of the classes featuring the VRS experience. We involved a total of 4 classes where students engaged with the TCM VRS. Most students experienced VR for the first time and found the VRS extremely enjoyable. Students expressed interest in purchasing their own VR headsets, making it a perfect blend of education and entertainment.

Figure 11. TCM VRS.

7. Survey Result

We designed two types of questionnaires. The first questionnaire survey was carried out for professionals in TCM, students, and TCM doctors to gather their opinions on the TCM SRS and to conduct tests. A total of 38 questionnaire responses were received [1]. The second questionnaire survey was conducted for students majoring in information systems and interdisciplinary English programs, including three international students, office workers, and IT professionals. The 12 survey questions are shown in Table 5. We provided on-site demonstrations and operations of the SRS and VRS, allowing the participants to experience the systems and compare them with traditional 2D table queries. Finally, the participants filled out a 12-question feedback questionnaire. A total of 66 valid responses were received.

A total of 41 students accounted for 61% of the number of participants. A total of 11 (17%) were industrial and commercial staff. Five teachers accounted for 8%.

We conducted a student's *t*-test to compare the significance between two sets of data using TCM SRS and TCM Prescription Dictionary 2D Tables. The *t*-test results for Question 10, "TCM Sphere is easier than TCM Prescription Dictionary 2D Tables to find corresponding Chinese medicine and prescriptions (formulas) for insomnia symptoms", and Question 11, "I find TCM Sphere more interesting and enjoyable than TCM Prescription Dictionary 2D Tables", were a t-statistic of −3.110 and a p-value of 0.002. This indicates a significant difference in the evaluations between using TCM SRS and TCM Prescription Dictionary 2D Tables, as the p-value is less than 0.05. The null hypothesis was rejected, meaning the means between the two are different. These results suggested that participants perceived a significant difference in the ease of finding corresponding Chinese medicine and prescriptions (formulas) for insomnia symptoms and in the level of interest and enjoyment, favoring TCM SRS. The survey results for the 12 questions are shown in Table 6.

Table 5. A total of 12 survey questions.

No.	Question
1	The TCM Sphere left a deep impression on me.
2	The TCM VR Sphere allows users to immerse themselves in a virtual sphere or shrink it to browse like a globe. It is a novel and interesting interactive way to browse knowledge.
3	The TCM VR Sphere left a deep impression on me.
4	The TCM Sphere makes it easier to find TCM symptoms and related items compared to the TCM Prescriptions Dictionary in 2D table format.
5	The hover highlight function of the TCM Sphere on the computer version allows me to immediately know the correspondence between symptoms and Chinese medicines.
6	Searching visually with the TCM Sphere is more efficient than text-based search.
7	The TCM VR Sphere is interesting and fun.
8	The visual TCM Sphere can filter out uninteresting symptom items.
9	The TCM Sphere website and VR can increase my knowledge of TCM.
10	The TCM Sphere makes it easier to find insomnia-related Chinese medicines and prescriptions compared to the TCM Prescriptions Dictionary in 2D table format.
11	I find the TCM Sphere more interesting and fun than the TCM Prescriptions Dictionary in 2D table format.
12	The TCM Sphere is very valuable as a reference.

Table 6. Average scores of survey questions.

Question	Avg Score
Q11	4.47
Q2	4.44
Q7	4.41
Q12	4.29
Q3	4.23
Q5	4.21
Q8	4.2
Q9	4.17
Q1	4.11
Q10	4.06
Q4	4.02
Q6	3.91
AVG	4.21

In addition, interview results with TCM practitioners and professionals revealed that the SRS and VRS developed in this study serve as tools to assist TCM students in their studies, stimulating their interest in learning. However, due to differences between the medication sources and those currently in use, adjustments are needed for clinical application. Future system filtering using "AND" aligns with physicians' prescribing habits [1]. Additionally, TCM practitioners suggested that the VRS is interesting and recommended adding a feature for "synchronous consultation between physician and patient in a VR environment" to expand its applicability.

8. Conclusions

To address the challenge of retrieving information from 2D tables, we developed the TCM SRS. The TCM SRS is more user-friendly for finding symptoms corresponding to medications compared to traditional 2D tables. Additionally, to promote TCM knowledge, we introduced the VRS. VRS increases public and student interest and contributes to sustainable development.

Author Contributions: Conceptualization, C.-H.S. and G.-H.L.; methodology, C.-H.S.; software, C.-H.S. and T.-A.F.; validation, T.-A.F. and C.-H.S.; formal analysis, T.-A.F. and C.-H.S.; investigation, T.-A.F. and C.-H.S.; resources, G.-H.L. and C.-H.S.; data curation, T.-A.F.; writing—original draft preparation, C.-H.S. and T.-A.F.; writing—review and editing, C.-H.S., G.-H.L. and T.-A.F.; visualization, C.-H.S. and T.-A.F.; supervision, C.-H.S. and G.-H.L.; project administration, C.-H.S. and G.-H.L.; funding acquisition, C.-H.S. and G.-H.L. All authors have read and agreed to the published version of the manuscript.

Funding: This research was supported by the National Science and Technology Council, grant number NSTC 112-2410-H-030-099.

Institutional Review Board Statement: This study received approval from the Institutional Review Board (IRB) of the Chang Gung Medical Foundation (Approval No. 202400566B0).

Informed Consent Statement: Not applicable.

Data Availability Statement: No new data were created or analyzed in this study. Data sharing is not applicable to this article.

Acknowledgments: We thank Han-Lin Li and Ching-Ter Chang for their expert advice and guidance on the spherical search system.

Conflicts of Interest: The authors declare no conflicts of interest. The funders had no role in the design of the study; in the collection, analyses, or interpretation of data; in the writing of the manuscript; or in the decision to publish the results.

References

1. Fan, T.-A. Graduate Institute of Traditional Chinese Medicine. Master's Thesis, Chang Gung University, Taoyuan, Taiwan, 2024.
2. Chen, L.C.; Wang, B.R.; Chou, Y.C.; Tien, J.H. Drug utilization pattern of Chinese herbal medicines in a general hospital in Taiwan. *Pharmacoepidemiol. Drug Saf.* **2005**, *14*, 651–657. [PubMed]
3. Chen, L.C.; Chen, I.C.; Wang, B.R.; Shao, C.H. Drug-use pattern of Chinese herbal medicines in insomnia: A 4-year survey in Taiwan. *J. Clin. Pharm. Ther.* **2009**, *34*, 555–560. [PubMed]
4. Li, H.-L.; Shih, C.-H.; Hu, C.-C.; Yang, C. Forming a Therapeutic sphere to treat ailments using optimization techniques. *Int. Trans. Oper. Res.* **2022**, *30*, 3949–3978. [CrossRef]
5. Han, J.; Kamber, M.; Pei, J. *Data Mining: Concepts and Techniques, A Volume in The Morgan Kaufmann Series in Data Management Systems*, 3rd ed.; Elsevier eBook: Amsterdam, The Netherlands, 2012. [CrossRef]
6. Ikotun, A.-M.; Ezugwu, A.-E.; Abualigah, L.; Abuhaija, B.; Heming, J. K-means clustering algorithms: A comprehensive review, variants analysis, and advances in the era of big data. *Inf. Sci.* **2023**, *622*, 178–210. [CrossRef]
7. Johannes, K.; Jürgen, Z. A comparative study of item space visualizations for recommender systems. *Int. J. Hum.-Comput. Stud.* **2023**, *172*, 102987. [CrossRef]
8. Shih, C.H.; Chang, C.T. Constructing renewable energy sphere for efficient search. *J. Clean. Prod.* **2024**, *471*, 143412.

engineering proceedings

MDPI

Proceeding Paper

Digital Twin for Developing and Verifying Semiconductor Packaging License Models †

Lai-Chung Lee, Shou-Yen Zhao and Whei-Jane Wei *

Department of Multimedia and Game Development, Minghsin University of Science and Technology, Hsinchu 30401, Taiwan; lclee@must.edu.tw (L.-C.L.); shouyen@must.edu.tw (S.-Y.Z.)
* Correspondence: jwei0632@must.edu.tw

† Presented at the 2024 IEEE 7th International Conference on Knowledge Innovation and Invention, Nagoya, Japan, 16–18 August 2024.

Abstract: The traditional semiconductor packaging training process is time-consuming and carries the risk of damaging precision equipment due to improper operation. Additionally, the retirement of experienced trainers has led to loss of specialized training and testing expertise. To address these challenges, digital twin technology is applied to training packaging engineers. We conducted an empirical study at the packaging production line of the Minghsin University of Science and Technology to address talent training bottlenecks and imbalances between supply and demand. First, an integrated software and hardware system was designed by combining digital twin and mixed reality (MR). The development process of the digital twin system for the wafer-dicing machine includes on-site visits, machine operation instructions, certification content development, expert validity construction, small-scale testing and modifications. We compared the pre- and post-experiment scores of industry experts to evaluate the operation time of five participants and their feedback. Digital twin and MR for simulated training increased proficiency in operation. The digital twin training and certification model developed in this study improved students' pass rates in certification exams.

Keywords: digital twin; mixed reality semiconductor packaging; verification

1. Introduction

We analyzed the current situation of training for semiconductor packaging certification. The identified issues included the following: (1) limited skilled workforce to preserve and pass down knowledge and skills, (2) leveraging strong educators to teach logical thinking and exceptional key-point communication skills, and (3) time-consuming training and certification. By utilizing digital twin and artificial intelligence (AI) technologies, the experience of technicians can be preserved and transmitted to apprentices to teach valuable know-how.

Digital technology can be used to benefit students and colleagues by using virtual avatars on an integrated platform. In-depth interviews with scholars revealed that tasks such as bonding wire and changing ceramic nozzles are particularly challenging. Currently, candidate performance is evaluated based on scoring standards to save time and improve performance outcomes. Therefore, developing solutions using digital twin technology and enhancing the efficiency and effectiveness of training for certification in the semiconductor packaging industry is required.

2. Related Work

2.1. Digital Twin

The importance and impact of digital twin (DT) technology has become evident in product development, manufacturing, and service sectors. DT technology allows enterprises to create a digital replica that mirrors the physical object or system in a virtual environment. Digital twins can be updated in real-time to reflect the status and behavior of their physical counterparts. DT enhances the flexibility of product design and manufacturing efficiency and allows for more precise maintenance of products [1–3]. DT uses real-time simulation and predictive analysis to support every stage of the product lifecycle from design to production, and during product servicing. This technology is beneficial in identifying potential issues early as it reduces the need for physical testing and prototype development, thereby saving costs and accelerating time-to-market [4].

The application of DT in Industry 4.0 is significant. It integrates with technologies such as the Internet of Things (IoT), cloud computing, and AI using an efficient data and analytics network. This enables the manufacturing industry to monitor and optimize production processes in real time. DT enhances operational transparency that allows businesses to make faster, data-driven decisions [1]. Furthermore, DT also aids enterprises in transitioning towards sustainable production methods. For instance, in vertical farming, DT is used to simulate and optimize planting strategies to improve resource efficiency and crop yield [1].

In summary, the importance of DT technology lies in its ability to provide deep insights and predictions for accelerating innovation, improving efficiency, and achieving a higher degree of customer customization. As this technology continues to develop and be applied, its impact across various industries is expected to expand further.

2.2. Mixed Reality (MR)

MR combines virtual reality (VR) and augmented reality (AR), overlaying virtual information to provide an immersive experience. MR is transforming industrial training and operations. In semiconductor manufacturing, MR allows for operators and engineers to receive more intuitive training, familiarizing them with the operation of equipment such as wafer-slicing machines. MR is used to simulate work scenarios during training, enabling learners to master operational skills, enhance safety awareness, and reduce errors in actual operations. MR technology allows trainees to overlay virtual information in real-world environments. For example, operators can see the internal structure of machinery, operational processes, and real-time data through MR devices to more intuitively understand and master complex operational skills. Additionally, MR provides interactive educational content, such as step-by-step guidance and real-time feedback for trainees to practice their knowledge in real environments.

2.3. Training

DT and MR technologies, including the Meta Quest Pro, and semiconductor packaging equipment, such as wafer-slicing machines, show a tightly integrated and complementary relationship in creating an efficient and intelligent ecosystem in the industrial and technological fields. DT is used to create virtual models of physical devices. These models replicate the appearance of the physical devices and simulate their operational behaviors, making DT prominent in semiconductor manufacturing. The complex operations and high precision requirements of equipment such as wafer-slicing machines make DT an ideal tool for optimizing production processes, reducing downtime, and conducting preventative maintenance. By simulating real operational environments, engineers can test and adjust equipment in the virtual world to ensure optimal performance when operating in real-world settings.

The combination of DT technology with MR in training enhances training effectiveness. Virtual models of equipment generated by DT can be interactively manipulated within an MR environment, allowing users to practice in a virtual setting. For example, by using the Meta Quest Pro for simulating operations on a wafer-slicing machine, learners can perform cutting operations on a virtual model, observing and analyzing every detail of the cutting process. Virtual training helps learners better understand how the equipment operates and allows practice in a safe and low-cost environment. In the highly competitive and technology-intensive industry of semiconductor manufacturing, the integration of DT, MR, and advanced training equipment such as wafer-slicing machines is leading the Industry 4.0 revolution, driving the industry towards more intelligent and digital operations.

2.4. MR Training System

Astronaut training is complex and costly due to its specific infrastructure and equipment requirements. Hence, interactive user experiences must be simulated using immersive technologies to model the cockpit environment. Astronaut training is conducted to evaluate the integration of DT with immersive technologies for effective training. Although MR immersive technologies and DT offer advantages and applications, their synergistic integration can create a specialized training platform. Initially, by incorporating spatial elements, complex scenarios can be created to simulate and confirm the consistency of developed DT with actual systems, thus validating their practicality in astronaut training [5]. Using DT and MR for training in semiconductor packaging equipment such as wafer-slicing machines, die bonders, and wire bonders has advantages and disadvantages as follows.

2.4.1. Advantages

- Enhanced Training Efficiency and Safety: MR headsets provide an immersive training environment, allowing trainees to interact with 3D models of the equipment. This hands-on experience helps trainees to understand complex mechanical operations without risking damage to expensive equipment or accidents [6].
- Immediate Feedback: DT is used to simulate real-time scenarios and provide feedback, enhancing learning outcomes through visualized interfaces of operational results.
- Cost Reduction: Training on virtual models can reduce the risk of actual machine downtime, thus preventing production losses. This is particularly important in semiconductor manufacturing, where the costs of equipment downtime are very high [7].
- Improved Maintenance and Troubleshooting: DT is used to predict potential failures and maintenance needs based on real-time data. Training maintenance personnel with these predictive insights can prevent unexpected downtimes [8]. Trainees can learn how to diagnose and troubleshoot in a simulated environment, preparing for real scenarios without the fear of damaging equipment.

2.4.2. Disadvantages

- High Initial Investment: Implementing DT and MR requires substantial initial investments, including costs for software, hardware, and infrastructure. This may be a barrier for smaller companies [8].
- Maintenance and Updates: DT models and MR systems must be updated with data and software versions for with investment and technical expertise.
- Complexity and Data Demands: DT requires extensive data to accurately simulate real-world conditions. Collecting and processing this data is challenging in dynamic and complex manufacturing environments [7].

- Training and Adoption: MR headsets and DT interfaces have steep learning curves, potentially requiring additional training for employees [7].

In summary, although using DT and MR headsets in training for semiconductor packaging equipment offers advantages, efficiency, safety, and cost savings, there are challenges such as high initial costs, complex data management, and the need for specialized training and integration. Enterprises need to consider these factors when adopting these advanced training technologies.

2.5. Factory Valve Inspection

In factory settings, computer vision is used for status tracking, with Meta Quest technology informing users to what degree a valve or switch to be turned. Image recognition is used to transmit information to the MR system in a one-way interaction, mostly utilizing animations or step-by-step instructions. For example, the state of a rotating valve can be determined through image detection to decide the current attributes of pipelines or equipment. A virtual 3D model can be created to the right side of the actual electrical equipment at a 1:1 scale (Figure 1) which is a concept of DT. A comparative model needs to be built including physical devices to provide demonstrations or guidance in videos or on interfaces for one-on-one training or learning processes [9].

Figure 1. Concept of DT: The picture on the **left** is the digital twin of the wire cutting machine. The picture on the **right** is a digital twin of a six-axis machining center.

DT is efficient for the operation of certain equipment, whether for repeated training or assessment. For instance, DT allows for tracking and retrospective analysis due to prolonged operation or errors. Multiple machine MR connections allow for operating simultaneously, including inspecting valves, which indicates a chemical gas leak. This could happen due to natural disasters or human damage within the factory, leading to leaks. This training method determines valves requiring pressure reduction or pipelines to be isolated. It involves image recognition combined with MR training techniques. It also incorporates process diagrams or checkpoints to facilitate content input and verification (Figure 2).

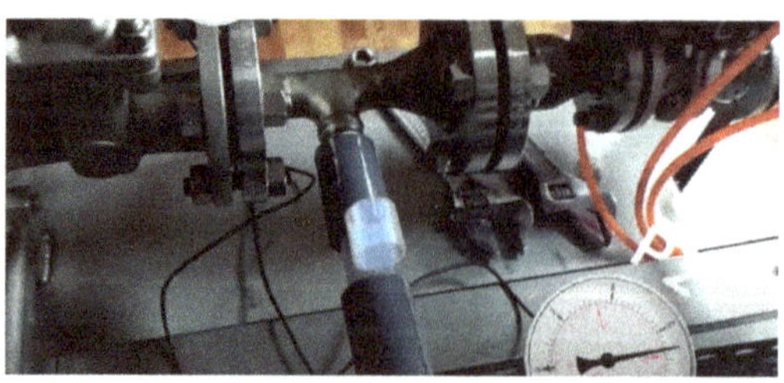

Figure 2. The virtual object in the picture is the transparent tube water flow effect test mode.

DT and MR have potential advantages in training for critical equipment in factories. DT technology creates an accurate digital replica of physical devices, enabling operators to perform operational simulations, troubleshoot, and optimize performance within a virtual environment. This provides a safe training setting, reducing the risk of damage to actual equipment from operational errors and records and analyzes operational data, offering training evaluations and recommendations for improvements [9]. DT and MR technologies enhance the quality and efficiency of packaging equipment training by providing a safe, efficient, and interactive environment, effectively addressing many issues inherent in traditional training methods.

3. Method and Implementation

3.1. Live Demonstration and Observation

To effectively master the operational details of the actual machines, on-site video recording and photography are required. The managers of the semiconductor institute-like production line are requested to operate all processes. The recordings are then transcribed into verbatim scripts and audiovisual files, which are finally reviewed by the managers for accuracy. In the operation, each item and test content must be verbally described to take notes and discuss scoring standards in real-time. In this study, a wafer-dicing machine was used for two examinations: the wafer-cutting process and the blade changing process of the dicing machine. Each major topic consists of five sub-questions, with each question scored out of 10 points, making a total of 100 points.

3.2. Development of DT and MR Content

We conducted a detailed analysis and study of the wafer-dicing machine to precise 3D modeling (Figure 3). Unity software was used for the development of the platform. It was used for its robust 3D rendering capabilities and the suite of tools and plugins that support various VR and MR devices. We used the Quest Pro, an advanced MR headset for high-resolution display and accurate spatial tracking capabilities for realizing highly interactive DT simulations.

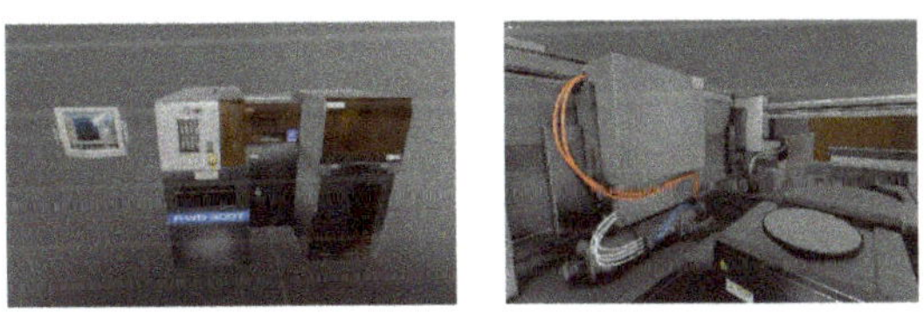

Figure 3. The picture on the **left** is a 3D digital simulation of a wafer-slicing machine. The picture on the **right** is a digital twin close-up of the process of replacing the cutting blade.

In a 3D modeling process, tracking technologies were used to ensure the accurate alignment of the model in physical space, enabling operators to interact with the model in the MR environment. This integration of technology allows for creating a digital twin of the wafer-dicing machine that is precise and interactive to enhance the training and operational understanding for users.

A real wafer-dicing machine was replicated 1:1 using DT. In collaboration with instructors, we collected detailed operations and safety precautions to design an interactive user interface that allows trainees to engage in immersive training using Quest Pro. Using the Unity3D development tool, we integrated the DT-built machine model with interactive operating steps and developed the MR application to realize interactive training scenarios.

In training, the spatial positioning of the MR device allows for moving the virtual training machine for synchronized operation. It also relocates the machine model in classrooms and dormitories, allowing trainees to conduct simulated training outside of a production-line-like environment. This flexibility enhances learning outcomes by providing practical experience in a variety of settings, ensuring that trainees can practice and refine their skills in diverse and controlled environments.

3.3. Validation

Once the system integration was completed, experts reviewed functionalities and steps of the UI interface displayed on screens. The invited industry experts from a leading domestic packaging equipment manufacturer possessed over 10 years of experience. They operated the MR simulation training system and provided feedback for necessary modifications. Managers from the production-line-like equipment site assisted instructors in training the participants. In using the MR headsets for operations, the following issues were identified.

- The decimal point appears too quickly after setting the dimensions of the wafer size.
- Clicking to start the wafer-feeding process requires an additional explanation in task box number five.
- Manual adjustment steps for horizontal alignment need a screen-operable option.
- After confirming F1 for horizontal alignment, there is a missing Display Change action.
- The sixth prompt box changed to "Click to Start".
- The screen displaying completion of blade changing (BBD value setup) appears too early.

These issues were addressed through iterative improvements to ensure the simulation training system meets the requirements and effectively prepares participants for certification.

3.4. Optimization and Testing of Interactive Interface

Following the instructions by industry instructors and specialists, we solved the issues such as changing the prompt window to manual horizontal alignment requiring checking both sides for horizontal alignment. The START button is used to begin wafer feeding (Figure 4). These iterative corrections involved multiple reviews and confirmations among the development team, industry instructors, and administrators to ensure the accuracy and rationality of the MR simulation system.

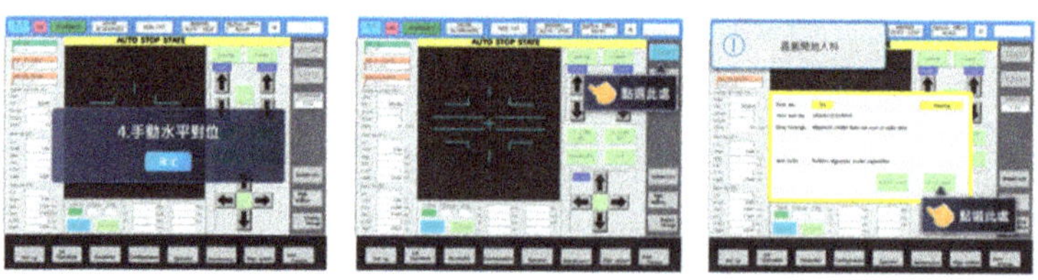

Figure 4. Corrections to the machine interface screen: Correction (**left**), Correction (**middle**), Correction (**right**).

3.5. Small-Scale Experiment

The participants were graduates from the Minghsin University of Science and Technology who majored in electrical engineering and were engaged in the experiments of this study. Before the experiment, each participant submitted an informed consent form and were compensated with a gift card. Before the participants began virtual operations, adjustments were made to the height of the machine's display screen due to variations in participant height. However, these adjustments were made after the training started, and the time taken to adjust the screen height was not counted in the overall time calculation.

The participants used the "think aloud" method or recited the steps aloud to reinforce learning. Before the experiment, an industry instructor asked the participants to operate the wafer-dicing machine to score in the pre-test. Each participant operated the actual machine and was trained using the MR simulation. Their performance was assessed again after the MR simulation training by the instructor.

After the experiment, each participant filled out the System Usability Scale (SUS) questionnaire. The survey was conducted to determine the user experience of the MR wafer-dicing simulation training system. The SUS uses a 5-point scale, where 1 represents "strongly disagree" and 5 represents "strongly agree". The participants selected multiple scores that best matched their experience.

4. Result and Analysis

The participants were being trained at the semiconductor academy to obtain certification in packaging equipment. Each participant completed five sessions of the MR simulation training system, with their times recorded over a week as shown in Table 1, which presents their proficiency and completion rates with the wafer-dicing machine.

Table 1. Overview of five students' MR simulation training times across five sessions.

Subjects	T1	T2	T3	T4	T5
S1	10 m 54 s	8 m 16 s	6 m 50 s	5 m 11 s	5 m 25 s
S2	10 m 23 s	8 m 11 s	4 m 37 s	5 m 31 s	4 m 46 s
S3	6 m 49 s	6 m 22 s	4 m 41 s	4 m 34 s	6 m 45 s
S4	6 m 57 s	6 m 26 s	5 m 2 s	5 m 13 s	5 m 13 s
S5	6 m 59 s	6 m 28 s	5 m 0 s	7 m 54 s	5 m 53 s
Average	8 m 24 s	7 m 9 s	5 m 14 s	5 m 41 s	5 m 36 s

The average time was 8 min and 24 s in the first session and decreased to 5 min and 36 s in the fifth session. This time reduction was due to their increasing familiarity with the system. They were initially less skilled but eventually adept at completing all tasks more efficiently. In each session, the participants employed the "think aloud" method or recited the operational steps to reinforce learning, further enhancing their mastery of the process.

Table 1 displays scores before and after five MR simulation training sessions under instructor observations. The scores were assigned by the industry instructor both before and after five sessions of MR simulation training. The "Pre-test Score" represents the initial score given by the instructor based on the student's performance in their first operation of the wafer-dicing machine simulation system. The "Post-test Score" shows the scores after completing five simulation training sessions, indicating significant overall improvement.

In initial scoring, the instructor pointed out common errors made by the participants, which affected their completion times and final scores and guided the participants' improvements throughout the training sessions.

Table 2 highlights the improvement in students' machine operation skills, with average scores rising from 76 before training to 93.6 after MR simulation sessions. This significant increase, confirmed by statistical tests, indicates the effectiveness of MR training in enhancing student proficiency. To assess the improvement of scores, a paired t-test was conducted. The p-value was 0.0136, indicating a statistically significant difference. This result implies that the participants significantly improved their skills after the training sessions. The Wilcoxon signed-rank test result showed a statistic of 0.0 and a p-value of 0.0625, suggesting a substantial improvement in their ability. The correlation coefficient between the average training times and scores was 0.133, suggesting that the reduction in practice time is not correlated with the improvement.

Table 2. Instructor-assigned scores for each student's real machine operation performance before and after training.

Subjects	Score 1	Score 2
S1	90	95
S2	60	91
S3	80	100
S4	70	86
S5	80	96
Average	76	93.6

The paired *t*-test and Wilcoxon sign rank test results indicate that the participants' scores were significantly improved after multiple training sessions. While there is correlation between the average training times and final scores, the relationship was not significant, indicating that reduced time is not necessarily correlated with scores.

Expert 1 said that the training system has an excellent foolproof design to prevent errors in interface operation and is user-friendly. Expert 3 noted that the system allows for practice anytime and anywhere, without location constraints. Expert 4 believed the system helps inexperienced learners understand machine operations. Expert 5 found the system convenient for learning, as it enables practice without a real machine.

In terms of learning effectiveness, the four experts felt that the training helped them understand machine operation and reinforced their knowledge of the human–machine interface operations. They had positive views on its application with excellent learning and revision resources for beginners or those unable to use the machines immediately. The system has an excellent user interface for simplifying use and enhancing understanding of procedures through on-screen explanations. However, deficiencies were observed, particularly in blade changing, which requires actual operation to fully grasp. A lack of tactile feedback due to the absence of physical interaction with machine parts was also observed.

The SUS scores for the MR version of the wafer-cutting machine simulation training system indicated high usability with high scores ranging from 62.5–95. The average score was 83. In the SUS, scores above 68 are regarded as good usability standards. Scores exceeding 83 indicate that the system's usability ranges from good to excellent, suggesting that most users find the system easy to use and well-suited to their needs.

The results of this study demonstrate a high user satisfaction with the system. The user interface is well-designed, aligning with the users' operational intuitions. This feedback is crucial for further refinements and expansions of the system's capabilities, meeting user expectations in practical training scenarios.

5. Conclusions

We developed a DT virtual training system to address insufficient hands-on training time on the actual machines. The process encompassed site visits to the pseudo-production line, observation and recording, simulation system planning and design, system development and debugging, and ultimately, practical verification. The participants underwent multiple training sessions to become familiar with the wafer-cutting machine's examination. A significant correlation between the pre- and post-test scores was observed; the more MR practice they had, the higher their operational scores. Expensive instrumentation and equipment lack an effective management system. If DT technology is effectively utilized, students can learn and practice anytime and anywhere, becoming proficient and ready for future opportunities.

Using this system, knowledge and skills through DT technology can be improved. Future research with more participants and extended duration is necessary to compare differences and correlation and enhance the impact of this simulation training system.

Author Contributions: Conceptualization, L.-C.L., S.-Y.Z. and W.-J.W.; methodology, L.-C.L., S.-Y.Z. and W.-J.W.; data curation, L.-C.L., S.-Y.Z. and W.-J.W.; formal analysis, L.-C.L., S.-Y.Z. and W.-J.W., investigation, L.-C.L., S.-Y.Z. and W.-J.W.; writing—original draft, L.-C.L., S.-Y.Z. and W.-J.W.; writing—review and editing, L.-C.L., S.-Y.Z. and W.-J.W. All authors have read and agreed to the published version of the manuscript.

Funding: This research was supported by the National Science and Technology Council for budget sponsorship of NSTC 112-2410-H-159-003.

Institutional Review Board Statement: The study was reviewed and approved by the ethics committee to ensure that it meets ethical guidelines for research involving human participants.

Informed Consent Statement: All participants provided informed consent, indicating that they agreed to participate with full understanding of the study's purpose, procedures, and any potential risks.

Data Availability Statement: Data supporting the study's findings is publicly available and can be provided upon request.

Acknowledgments: The authors not only appreciate grants from NSTC but also thank the iSynReal Company, including Tim Wang, Ruby Liu, and Ting-Yan Huang, for their full support of the research work.

Conflicts of Interest: The authors have no potential conflicts of interest, ensuring transparency in the study's results and interpretations.

References

1. Attaran, M.; Attaran, S.; Celik, B.G. The impact of digital twins on the evolution of intelligent manufacturing and Industry 4.0. *Adv. Comput. Intell.* **2023**, *3*, 11. [CrossRef] [PubMed]
2. Donoghue, I.; Hannola, L.; Papinniemi, J.; Mikkola, A. The Benefits and Impact of Digital Twins in Product Development Phase of PLM. In *Product Lifecycle Management to Support Industry 4.0. PLM 2018. IFIP Advances in Information and Communication Technology*; Chiabert, P., Bouras, A., Noël, F., Ríos, J., Eds.; Springer: Cham, Switzerland, 2018; Volume 540. [CrossRef]
3. Tao, F.; Cheng, J.; Qi, Q.; Zhang, M.; Zhang, H.; Sui, F. Digital twin-driven product design, manufacturing and service with big data. *Int. J. Adv. Manuf. Technol.* **2017**, *94*, 3563–3576. [CrossRef]
4. Alnowaiser, K.K.; Ahmed, M.A. Digital Twin: Current Research Trends and Future Directions. *Arab J. Sci. Eng.* **2023**, *48*, 1075–1095. [CrossRef]
5. Piñal, O.; Arguelles, A. Mixed reality and digital twins for astronaut training. *Acta Astronaut.* **2024**, *219*, 376–391. [CrossRef]
6. Haddod, F.; Dingli, A. *Intelligent Digital Twin System in the Semiconductors Manufacturing Industry, In AI in Industry 4.0. Studies in Computational Intelligence*; Dingli, A., Haddod, F., Klüver, C., Eds.; Springer: Cham, Switzerland, 2021; Volume 928. [CrossRef]
7. Jain, A. Future of Semiconductor Industry: Digital Twins Lead the Way. Available online: https://analyticsindiamag.com/it-services/the-future-of-the-semiconductor-industry-digital-twins-lead-the-way/ (accessed on 27 July 2024).
8. Land, I.; Niatas, J.; Serughetti, M. What Are Digital Twins? A Primer on Virtual Models. Synopsys, 2023. Available online: https://www.design-reuse.com/industryexpertblogs/54648/what-are-digital-twins-a-primer-on-virtual-models.html (accessed on 27 July 2024).
9. iSynReal, MR (Mixed Reality) Professional Tools and Course Development. Available online: https://www.isynreal.com/ (accessed on 17 July 2024).

engineering proceedings

MDPI

Proceeding Paper

Three-Dimensional Creation and Physical Movement in Art Therapy Using Virtual Reality Painting †

Chia-Chieh Lee and Min-Chai Hsieh *

Department of Multimedia and Animation, Tainan University of Technology, Tainan 710302, Taiwan; chiachiehba@gmail.com
* Correspondence: shiehminchai@gmail.com

† Presented at the 2024 IEEE 7th International Conference on Knowledge Innovation and Invention, Nagoya, Japan, 16–18 August 2024.

Abstract: Virtual Reality (VR) painting, an emerging form of artistic expression under 5G technology, showcases a broader range of expressive styles and dynamic visual effects compared to traditional computer graphics. The creative process in VR painting enhances spatial depth, exhibiting different spatial abilities and necessitating more physical movements, including hand controllers and foot movements in the virtual space. Furthermore, VR painting in art therapy encourages users to engage in physical activities, contributing to better emotional expression. This study involved digital-native users in VR painting, using Meta Quest 2 to operate Open Brush for their creations. Through observational methods, we examined user operational behaviors and conducted semi-structured interviews post-experiment to explore their painting performance and usage behaviors in the virtual environment. The results of this study indicate that VR painting enhances the sense of space and dynamic expression in creative work and improves users' emotional and physical engagement, providing new avenues for artistic expression. These findings contribute to improving the usability and application value of VR paintings.

Keywords: virtual reality; VR painting; usability; spatial cognition; physical movement

1. Introduction

Virtual Reality (VR) technology, as an emerging expressive digital art tool, offers unique advantages [1]. It provides a highly immersive environment, allowing individuals undergoing treatment to fully integrate into a virtual world, creating empathetic scenarios that reduce patient self-criticism. Additionally, VR technology simulates real-world situations, offering more personalized and targeted therapeutic approaches [2]. Moreover, the interactivity and real-time feedback within VR environments enable the dynamic and flexible therapeutic process and expand the possibilities for diverse and innovative psychological interventions.

The boundless nature of VR spaces increases physical engagement during VR operations compared to traditional tablet- or computer-based drawing. VR equipment demands the movement of upper and lower limbs when manipulating VR controllers. The immersive environment also influences users' spatial cognition and operational behaviors allowing users to experience a sense of liberation from real-world constraints, stressors, and tensions in an alternative or imaginary space [3]. Artwork created within VR generates distinct visual effects.

This study aims to investigate the creative behaviors and user experiences of multimedia animation students in virtual spaces. Students majoring in multimedia animation participated in virtual space creation activities. Through observation and in-depth interviews, their creative processes and behavioral manifestations were observed in virtual 3D spaces. We also observed users' operational behaviors using META QUEST 2 VR headsets, including their adaptation processes to virtual environments and interaction methods. We analyzed changes in students' physical movements and expressions when transitioning from 2D creation to 3D spaces to elucidate how VR technology influences their creative behaviors and bodily movements. The results provide a basis to enhance the future usability of VR drawings for creative purposes and physical stimulation.

2. Literature Review

2.1. VR Drawing

VR technology offers advantages in art and design: it enhances immersive experiences, enabling users to interact more realistically; facilitates personalized and multi-sensory design, providing visual, auditory, and tactile feedback; improves design efficiency by streamlining processes; and expands artistic expression through real-time interaction and liberation from physical constraints [4]. VR also accommodates diverse artistic textures, including special effects and animated visual elements [5]. Furthermore, users can maintain heightened focus during the creative process.

Once users become familiar with the controllers, they can completely control the device with uninterrupted engagement with the artistic flow. VR provides a novel private space, enabling users to enjoy freedom from real-world constraints, stressors, and tensions and facilitating entry into an alternative or imaginary space [3]. Users can better regulate spatial information and various factors through virtual environments, potentially alleviating anxiety and attention-related issues. VR virtual spaces allow for control over scene objects and their quantity, environmental information, and sky brightness (Figure 1).

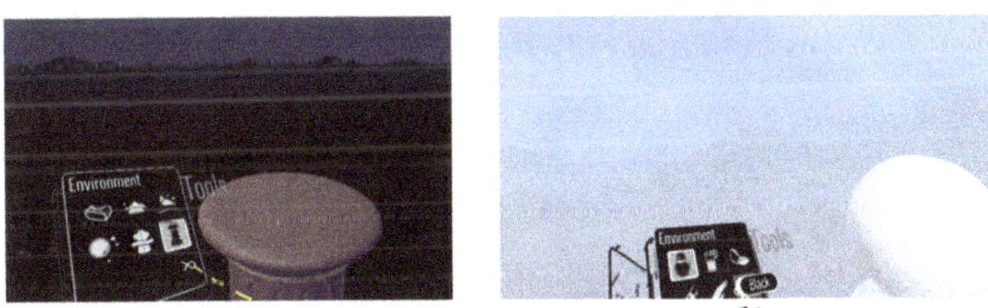

(a) (b)

Figure 1. Two levels of brightness in virtual environment in VR. (**a**) Low luminance and (**b**) high luminance.

2.2. Spatial Perception in Drawing

VR drawing offers a unique spatial experience for users to create freely within a three-dimensional environment. This immersive experience enhances spatial cognition and perceptual abilities. Through VR drawing, users can view and adjust their work from various angles and increase creative flexibility to understand spatial relationships. The sense of immersion and interactivity in VR environments significantly improves users' spatial navigation skills and creativity [1,2]. VR drawing also reduces the creative anxiety of users to focus more on exploration and creative expression [3]. These advantages highlight the extensive potential for VR technology applications in art and design. VR drawing incorporates spatial depth characteristics, enhancing creative flexibility compared to drawing on tablets and smartphones.

2.3. Physical Expression in VR Creation

In bodily movement, virtual environments offer expanded creative spaces, engaging participants in more extensive physical motions to create within surroundings. Hand movements in the air during VR creation are more physically demanding than traditional media, which may be challenging for stroke patients [6]. However, the full-body engagement required when operating VR equipment is highly enjoyable for users with intact motor abilities. The ability to erase portions of artwork in VR mitigates the sense of permanence associated with traditional artistic media. Consequently, participants can focus on exploration and creative expression without the anxiety of making irreversible mistakes [3]. This feature potentially enhances the therapeutic value of the creative process by reducing inhibitions and promoting more accessible artistic exploration.

3. Materials and Methods

We invited participants for the tree-house-person (THP) test [7]. The THP test is common in art therapy, featuring three distinct objects that allow for the exploration of spatial relationships, such as the distance between objects. In this study, participants completed the test using the Oculus Quest 2. We observed the participants' behavior during the test, including their spatial cognition and physical movements within the virtual space.

3.1. User Interviews

We conducted post-experiment interviews to understand the users' feelings about the VR space and gather usability feedback on the user experience with Openbrush v2.4. The user feedback and opinions on the VR drawing software allowed for further analysis and improvement of the operational process and user experience. The interview questions were categorized based on the study's focus on spatial cognition, operational processes, and user behavior.

- Spatial cognition and rotation behavior: How did you feel about spatial cognition and rotation behavior while operating the VR drawing software? Did you experience any dizziness or difficulty in orienting yourself?
- Drawing in virtual space: Could you easily find where to start drawing in the virtual space? Did you encounter any difficulties or challenges in drawing in a 3D space?
- Constructing 3D objects: Could you accurately construct 3D objects through your strokes while drawing in VR? Did you feel capable of transforming a 2D plane into a 3D space?
- Standing position and mobility: Did you find it difficult or uncomfortable to choose a standing position? Did you need to move your feet frequently?
- Physical comfort and movement: Were your physical movements comfortable during the drawing process? Did you use more physical movements? Did you feel any fatigue in your shoulders or arms?
- Favorite features in Openbrush: What features did you like the most while using Openbrush? These include brush types, colors, etc.
- Overall experience with VR drawing software: What are the strengths and weaknesses of your overall experience with the VR drawing software? Do you have any suggestions for the operational process and interface design?

3.2. Experiment

VR drawing software offers a unique 3D creative experience. Most users can adapt to and enjoy drawing with features such as dynamic brushes, special effects, and three-dimensional creation. However, several users face challenges with spatial cognition and initial operation. The interface layout for tool selection and function distribution has room for improvement: implementing shortcut keys and distinguishing between physical and virtual space boundaries. Additionally, prolonged use can lead to physical fatigue, particularly in the arms and shoulders. Table 1 summarizes the feedback from the participants.

Table 1. Interview results.

Aspect	Positive Feedback	Negative Feedback	Suggestions
Spatial Cognition and Orientation	Most users can orient themselves; the VR perspective is broader than a computer's	Some users have difficulty orienting; some experience dizziness	Provide more spatial orientation aids
3D Drawing Experience	Able to draw 3D objects; creates a sense of depth	Difficult to draw detailed 3D graphics; requires time to adapt	Improve fine drawing tools
Movement and Standing	Most users feel comfortable; the joystick can be used for movement	Concern about hitting physical objects; some need to move their feet	Enhance safety warning system
Physical Comfort	Overall comfortable	Arms and shoulders can get tired, which is inconvenient when selecting tools	Optimize the tool selection interface; consider adding drawing shortcuts
Feature Preference	Dynamic brushes and effects are prevalent; the rainbow feature is attractive	No significant negative feedback	Further enrich dynamic brush effects
Overall Experience	Well-developed features; helpful prompts	Initial operation is complex; functions are scattered and hard to find	Simplify novice operation process; integrate frequently used functions

4. Results

4.1. Adaptability

Several participants adapted quickly to the VR environment. For instance, Participant 2 stated, "I do not experience VR sickness, and the perspective is wider than on a computer. Orientation is easy and stable". However, Participant 1 reported, "I feel dizzy, and it is somewhat difficult to orient myself. I cannot quite grasp the sense of space". The responses indicate individual differences in adaptability to the VR space.

4.2. Positive 3D Drawing Experience

Most participants engaged in 3D drawing within the VR environment. Participant 1 mentioned, "I can use the brush to create objects in 3D space, easily drawing items". Participant 3 also said, "I can precisely draw 3D objects". The participants' responses demonstrate the VR drawing software's strengths in 3D creation.

4.3. Movement and Safety Issues Need Attention

While most participants had no issues with movement, safety concerns remain. Participant 3 noted, "I have to move around to draw the back of objects, and I am afraid of bumping into tables and chairs in the real space". Participant 6 also mentioned, "I am worried about hitting physical objects when moving". The participants' responses highlight the importance of ensuring a safe VR environment.

4.4. Prolonged Use and Physical Fatigue

Several participants reported fatigue after prolonged use. Participant 2 noted, "My arms feel sore because there are no shortcut keys, so I have to flip through tools manually.

My shoulders get tired easily". Participant 3 also said, "My arms feel sore, and since the starting position for drawing is too high, I have to keep lifting my arms to draw objects high up". The participants' responses suggest adding shortcut keys to the interface and an undoing/redoing movement.

4.5. Special Effect Brushes

Almost all participants expressed a liking for the special effect brushes. Participant 1 stated, "I like the bubble brush the most," Participant 2 liked the "bubble brush and dynamic effect brushes," and Participant 4 mentioned, "I like the moving brushes, such as stars and comets". The participants' responses show that special effect brushes are a significant highlight of the VR drawing experience.

4.6. Improvement in User Interface

Although the overall experience is good, the user interface has room for improvement. Participant 1 suggested, "Some functions could be displayed individually instead of hidden within the space. It might be hard for beginners to find them". Participant 2 also proposed, "Common drawing functions could be integrated". The participants' responses indicate that simplifying and optimizing the user interface can enhance the participant experience (Figure 2).

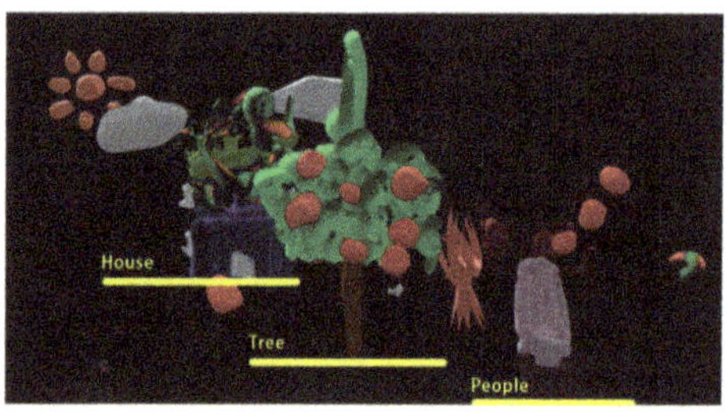

Figure 2. VR creation by Participant 3.

5. Conclusions

Interviews with users of VR drawing software showed that VR drawing software provided a new creative experience in 3D. However, when drawing 3D objects in virtual space, spatial orientation issues were determined. Participants encountered difficulty in positioning objects on the same horizontal plane, and the initial drawing point significantly influenced subsequent actions with the VR controller. To elevate the participant experience and broaden the horizons of VR drawing technology, the following need to be addressed:

- Developing more intuitive spatial orientation aids to introduce additional reference points for 3D space drawing to facilitate participant adaptation and utilization of the virtual space;
- Designing a safety warning system to mitigate the risk of participants colliding with their physical environment while immersed in the virtual space;
- Optimizing the tool selection interface to explore the integration of shortcuts or voice commands to minimize the physical fatigue associated with tool selection and switching;
- Enriching special effect brushes by expanding the variety of dynamic brushes while ensuring the ease of operation;
- Adding shortcut keys and reducing fatigue.

Participant feedback highlighted the importance of incorporating shortcut keys into the interface and introducing an undo/redo movement functionality.

In the long term, conducting a comparative study between virtual reality (VR) drawing and traditional drawing methods is crucial for evaluating their respective impacts on object creation and the enhancement of hand movements during the drawing process. The integration of spatial orientation within VR environments significantly influences physical interaction, presenting novel opportunities for user engagement. The outcomes of this study contribute meaningful insights into the usability of VR-based drawing systems and serve as a foundation for the future advancement of VR drawing technologies. The results of this research offer information on the usability of VR drawing creation and the future design of VR drawing.

Author Contributions: Conceptualization, C.-C.L.; methodology, C.-C.L.; software, C.-C.L. and M.-C.H.; validation C.-C.L.; formal analysis, C.-C.L.; investigation, C.-C.L.; resources, C.-C.L. and M.-C.H.; data curation, C.-C.L. and M.-C.H.; writing—original draft preparation, C.-C.L.; writing—review and editing, C.-C.L.; visualization, C.-C.L.; supervision, C.-C.L. and M.-C.H. All authors have read and agreed to the published version of the manuscript.

Funding: This research received no external funding.

Institutional Review Board Statement: Not applicable.

Informed Consent Statement: Not applicable.

Data Availability Statement: No new data were created or analyzed in this study. Data sharing is not applicable to this article.

Conflicts of Interest: The authors declare no conflict of interest.

References

1. Hacmun, I.; Regev, D.; Salomon, R. The principles of art therapy in VR. *Front. Psychol.* **2018**, *9*, 2082. [CrossRef] [PubMed]
2. Falconer, C.J.; Rovira, A.; King, J.A.; Gilbert, P.; Antley, A.; Fearon, P.; Ralph, N.; Slater, M.; Chris, R. Brewin Embodying self-compassion within VR and its effects on patients with depression. *BJPsych Open* **2016**, *2*, 74–80. [CrossRef] [PubMed]
3. Kaimal, G.; Carroll-Haskins, K.; Berberian, M.; Dougherty, A.; Carlton, N.; Ramakrishnan, A. VR in art therapy: A pilot qualitative study of the novel medium and implications for practice. *Art Therapy* **2020**, *37*, 16–24. [CrossRef]
4. Zhai, Q. Application of VR Technology in Art Design. In *Advances in Intelligent Systems and Computing*; Springer: Berlin/Heidelberg, Germany, 2021. [CrossRef]
5. Liu, H. Application and Development of VR Technology in Painting. *J. Phys. Conf. Ser.* **2021**, *1744*, 042225. [CrossRef]
6. Alex, M.; Wünsche, B.C.; Lottridge, D. VR art-making for stroke rehabilitation: Field study and technology probe. *Int. J. Hum. Comput. Stud.* **2021**, *145*, 102481. [CrossRef]
7. Piotrowski, C. Drawing techniques in assessment: A summary review of 60 survey-based studies of training and professional settings. *J. Indian Acad. Appl. Psychol.* **2016**, *42*, 220.

MDPI AG
Grosspeteranlage 5
4052 Basel
Switzerland
Tel.: +41 61 683 77 34

Engineering Proceedings Editorial Office
E mail: engproc@mdpi.com
www.mdpi.com/journal/engproc

www.ingramcontent.com/pod-product-compliance
Lightning Source LLC
Chambersburg PA
CBHW041937260926
51850CB00018B/236
* 9 7 8 3 7 2 5 8 5 7 7 3 9 *